STUDENT'S SOLUTIONS MANUAL

SALVATORE SCIANDRA

FINITE MATHEMATICS WITH APPLICATIONS IN THE MANAGEMENT, NATURAL, AND SOCIAL SCIENCES

Margaret L. Lial
American River College

Thomas Hungerford
St. Louis University

John Holcomb
Cleveland State University

Bernadette Mullins
Birmingham-Southern College

PEARSON

Boston Columbus Indianapolis New York San Francisco Upper Saddle River
Amsterdam Cape Town Dubai London Madrid Milan Munich Paris Montreal Toronto
Delhi Mexico City São Paulo Sydney Hong Kong Seoul Singapore Taipei Tokyo

The author and publisher of this book have used their best efforts in preparing this book. These efforts include the development, research, and testing of the theories and programs to determine their effectiveness. The author and publisher make no warranty of any kind, expressed or implied, with regard to these programs or the documentation contained in this book. The author and publisher shall not be liable in any event for incidental or consequential damages in connection with, or arising out of, the furnishing, performance, or use of these programs.

Reproduced by Pearson from electronic files supplied by the author.

Publishing as Pearson, 75 Arlington Street, Boston, MA 02116.

ISBN-13: 978-0-321-98632-0
ISBN-10: 0-321-98632-6

1 2 3 4 5 6 CRK 18 17 16 15 14

www.pearsonhighered.com

PEARSON

Table of Contents

Chapter 1 Algebra and Equations

Section 1.1 The Real Numbers

1. True. This statement is true, since every integer can be written as the ratio of the integer and 1.
For example, $5 = \frac{5}{1}$.

3. Answers vary with the calculator, but $\frac{2,508,429,787}{798,458,000}$ is the best.

5. $6(t+4) = 6t + 6 \cdot 4$
This illustrates the distributive property.

7. $-5 + 0 = -5$
This illustrates the identity property of addition.

9. $8 + (12 + 6) = (8 + 12) + 6$
This illustrates the associative property of addition.

11. Answers vary. One possible answer: The sum of a number and its additive inverse is the additive identity. The product of a number and its multiplicative inverse is the multiplicative identity.

For Exercises 13–15, let $p = -2$, $q = 3$ and $r = -5$.

13. $-3(p+5q) = -3[-2+5(3)] = -3[-2+15]$
$= -3(13) = -39$

15. $\frac{q+r}{q+p} = \frac{3+(-5)}{3+(-2)} = \frac{-2}{1} = -2$

17. Let $r = 3.8$.
$\text{APR} = 12r = 12(3.8) = 45.6\%$

19. Let $APR = 11$.
$APR = 12r$
$11 = 12r$
$\frac{11}{12} = r$
$r \approx .9167\%$

21. $3 - 4 \cdot 5 + 5 = 3 - 20 + 5 = -17 + 5 = -12$

23. $(4-5) \cdot 6 + 6 = -1 \cdot 6 + 6 = -6 + 6 = 0$

25. $8 - 4^2 - (-12)$
Take powers first.
$8 - 16 - (-12)$
Then add and subtract in order from left to right.
$8 - 16 + 12 = -8 + 12 = 4$

27. $\frac{2(-3) + \frac{3}{(-2)} - \frac{2}{(-\sqrt{16})}}{\sqrt{64} - 1}$
Work above and below fraction bar. Take roots.
$\frac{2(-3) + \frac{3}{(-2)} - \frac{2}{(-4)}}{8-1}$
Do multiplications and divisions.
$\frac{-6 - \frac{3}{2} + \frac{1}{2}}{8-1}$
Add and subtract.
$\frac{-\frac{12}{2} - \frac{3}{2} + \frac{1}{2}}{7} = \frac{-\frac{14}{2}}{7} = \frac{-7}{7} = -1$

29. $\frac{2040}{523}, \frac{189}{37}, \sqrt{27}, \frac{4587}{691}, 6.735, \sqrt{47}$

31. 12 is less than 18.5.
$12 < 18.5$

33. x is greater than or equal to 5.7.
$x \geq 5.7$

35. z is at most 7.5.
$z \leq 7.5$

37. $-6 < -2$

39. $3.14 < \pi$

41. a lies to the right of b or is equal to b.

43. $c < a < b$

45. $(-8, -1)$
This represents all real numbers between –8 and –1, not including –8 and –1. Draw parentheses at –8 and –1 and a heavy line segment between them. The parentheses at –8 and –1 show that neither of these points belongs to the graph.

–8 –1

47. $(-2,3]$

This represents all real numbers x such that $-2 < x \leq 3$. Draw a heavy line segment from –2 to 3. Use a parenthesis at –2 since it is not part of the graph. Use a bracket at 3 since it is part of the graph.

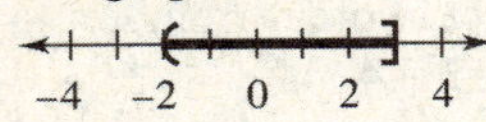

49. $(-2,\infty)$

This represents all real numbers x such that $x > -2$. Start at –2 and draw a heavy line segment to the right. Use a parenthesis at –2 since it is not part of the graph.

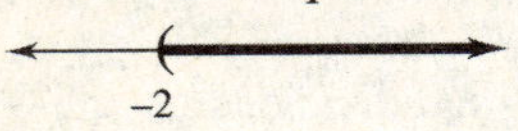

51. $|-9|-|-12| = 9-(12) = -3$

53. $-|-4|-|-1-14| = -(4)-|-15|$
$= -(4)-15 = -19$

55. $|5|\_\_|-5|$
$5\_\_5$
$5 = 5$

57. $|10-3|\_\_|3-10|$
$|7|\_\_|-7|$
$7\_\_7$
$7 = 7$

59. $|-2+8|\_\_|2-8|$
$|6|\_\_|-6|$
$6\_\_6$
$6 = 6$

61. $|3-5|\_\_|3|-|5|$
$|-2|\_\_3-5$
$2\_\_-2$
$2 > -2$

63. When $a < 7$, $a-7$ is negative.
So $|a-7| = -(a-7) = 7-a$.

Answers will vary for exercises 65–67. Sample answers are given.

65. No, it is not always true that $|a+b| = |a|+|b|$. For example, let $a = 1$ and $b = -1$. Then,
$|a+b| = |1+(-1)| = |0| = 0$, but
$|a|+|b| = |1|+|(-1)| = 1+1 = 2$.

67. $|2-b| = |2+b|$ only when $b = 0$. Then each side of the equation is equal to 2. If b is any other value, subtracting it from 2 and adding it to 2 will produce two different values.

69. 1; 2007

71. 9; 2003, 2004, 2005, 2006, 2008, 2009, 2010, 2011, 2012

73. 7; 2003, 2004, 2005, 2006, 2007, 2009, 2011

75. $|3.4-(-46.5)| = |49.9| = 49.9$

77. $|0.6-(4.4)| = |-3.8| = 3.8$

79. $|-10.8-(-46.5)| = |35.7| = 35.7$

81. 5; 2005, 2006, 2008, 2010, 2011

83. 5; 2007, 2008, 2009, 2010, 2011

Section 1.2 Polynomials

1. $11.2^6 \approx 1{,}973{,}822.685$

3. $\left(-\frac{18}{7}\right)^6 \approx 289.0991339$

5. -3^2 is negative, whereas $(-3)^2$ is positive. Both -3^3 and $(-3)^3$ are negative.

7. $4^2 \cdot 4^3 = 4^{2+3} = 4^5$

9. $(-6)^2 \cdot (-6)^5 = (-6)^{2+5} = (-6)^7$

11. $\left[(5u)^4\right]^7 = (5u)^{4\cdot 7} = (5u)^{28}$

13. degree 4; coefficients: 6.2, –5, 4, –3, 3.7; constant term 3.7.

15. Since the highest power of x is 3, the degree is 3.

17. $\left(3x^3+2x^2-5x\right)+\left(-4x^3-x^2-8x\right)$
$=\left(3x^3-4x^3\right)+\left(2x^2-x^2\right)+(-5x-8x)$
$=-x^3+x^2-13x$

19. $\left(-4y^2-3y+8\right)-\left(2y^2-6y+2\right)$
$=\left(-4y^2-3y+8\right)+\left(-2y^2+6y-2\right)$
$=-4y^2-3y+8-2y^2+6y-2$
$=\left(-4y^2-2y^2\right)+(-3y+6y)+(8-2)$
$=-6y^2+3y+6$

21. $\left(2x^3+2x^2+4x-3\right)-\left(2x^3+8x^2+1\right)$
$=\left(2x^3+2x^2+4x-3\right)+\left(-2x^3-8x^2-1\right)$
$=2x^3+2x^2+4x-3-2x^3-8x^2-1$
$=\left(2x^3-2x^3\right)+\left(2x^2-8x^2\right)+(4x)+(-3-1)$
$=-6x^2+4x-4$

23. $-9m\left(2m^2+6m-1\right)$
$=(-9m)\left(2m^2\right)+(-9m)(6m)+(-9m)(-1)$
$=-18m^3-54m^2+9m$

25. $(3z+5)\left(4z^2-2z+1\right)$
$=(3z)\left(4z^2-2z+1\right)+(5)\left(4z^2-2z+1\right)$
$=12z^3-6z^2+3z+20z^2-10z+5$
$=12z^3+14z^2-7z+5$

27. $(6k-1)(2k+3)$
$=(6k)(2k+3)+(-1)(2k+3)$
$=12k^2+18k-2k-3$
$=12k^2+16k-3$

29. $(3y+5)(2y+1)$
Use FOIL.
$=6y^2+3y+10y+5$
$=6y^2+13y+5$

31. $(9k+q)(2k-q)$
$=18k^2-9kq+2kq-q^2$
$=18k^2-7kq-q^2$

33. $(6.2m-3.4)(.7m+1.3)$
$=4.34m^2+8.06m-2.38m-4.42$
$=4.34m^2+5.68m-4.42$

35. $5k-[k+(-3+5k)]$
$=5k-[6k-3]$
$=5k-6k+3$
$=-k+3$

37. $R=5\,(1000x)=5000x$
$C=200{,}000+1800x$
$P=(5000x)-(200{,}000+1800x)$
$=3200x-200{,}000$

39. $R=9.75(1000x)=9750x$
$C=260{,}000+(-3x^2+3480x-325)$
$=-3x^2+3480x+259{,}675$
$P=(9750x)-(-3x^2+3480x+259{,}675)$
$=3x^2+6270x-259{,}675$

41. **a.** According to the bar graph, the net earnings in 2001 were \$265,000,000.

b. Let $x=3$.
$-1.48x^4+50.0x^3-576x^2+2731x-4027$
$=-1.48(3)^4+50.0(3)^3-576(3)^2+2731(3)-4027$
$=212.12$
According to the polynomial, the net earnings in 200 were approximately \$212,000,000.

43. **a.** According to the bar graph, the net earnings in 2010 were \$948,000,000.

b. Let $x=10$.
$-1.48x^4+50.0x^3-576x^2+2731x-4027$
$=-1.48(10)^4+50.0(10)^3-576(10)^2+2731(10)-4027$
$=883$
According to the polynomial, the net earnings in 2010 were approximately \$883,000,000.

45. Let $x = 13$.

$$-1.48x^4 + 50.0x^3 - 576x^2 + 2731x - 4027$$
$$= -1.48(13)^4 + 50.0(13)^3 - 576(13)^2 + 2731(13) - 4027$$
$$= 1711.72$$

According to the polynomial, the net earnings in 2013 will be approximately $1,711,720,000.

47. Let $x = 15$.

$$-1.48x^4 + 50.0x^3 - 576x^2 + 2731x - 4027$$
$$= -1.48(15)^4 + 50.0(15)^3 - 576(15)^2 + 2731(15) - 4027$$
$$= 1163$$

According to the polynomial, the net earnings in 2015 will be approximately $1,163,000,000.

For exercises 49–51, we use the polynomial $-.0057x^4 + .157x^3 - 1.43x^2 + 5.14x + 6.3$.

49. Let $x = 4$.

$$-.0057(4)^4 + .157(4)^3 - 1.43(4)^2 + 5.14(4) + 6.3$$
$$= 12.5688$$

Thus, there were approximately 12.6% below the poverty line in 2004. The statement is false.

51. Let $x = 3$.

$$-.0057(3)^4 + .157(3)^3 - 1.43(3)^2 + 5.14(3) + 6.3$$
$$= 12.6273$$

Let $x = 6$.

$$-.0057(6)^4 + .157(6)^3 - 1.43(6)^2 + 5.14(6) + 6.3$$
$$= 12.1848$$

Thus, there were 12.6% below the poverty line in 2003 and 12.2% below the poverty line in 2006. The statement is true.

For exercises 53–55, we use the polynomial $1 - .0058x - .00076x^2$.

53. Let $x = 10$.

$$1 - .0058x - .00076x^2$$
$$= 1 - .0058(10) - .00076(10)^2 = .866$$

55. Let $x = 22$.

$$1 - .0058x - .00076x^2$$
$$= 1 - .0058(22) - .00076(22)^2 = .505$$

For exercises 57, use $V = \frac{1}{3}h\left(a^2 + ab + b^2\right)$.

57. a. Calculate the volume of the Great Pyramid when $h = 200$ feet, $b = 756$ feet and $a = 314$ feet.

$$V = \frac{1}{3}(200)\left(314^2 + (314)(756) + 756^2\right)$$
$$\approx 60,501,067 \text{ cubic feet}$$

b. When $a = b$, the shape becomes a rectangular box with a square base, with volume b^2h.

c. If we let $a = b$, then $\frac{1}{3}h\left(a^2 + ab + b^2\right)$ becomes $\frac{1}{3}h\left(b^2 + b(b) + b^2\right)$ which simplifies to hb^2. Yes, the Egyptian formula gives the same result.

59. a. Some or all of the terms may drop out of the sum, so the degree of the sum could be 0, 1, 2, or 3 or no degree (if one polynomial is the negative of the other).

b. Some or all of the terms may drop out of the difference, so the degree of the difference could be 0, 1, 2, or 3 or no degree (if they are equal).

c. Multiplying a degree 3 polynomial by a degree 3 polynomial results in a degree 6 polynomial.

61. In order for the company to make a profit,

$$P = 7.2x^2 + 5005x - 230,000 > 0$$

Graph the function and locate a zero.

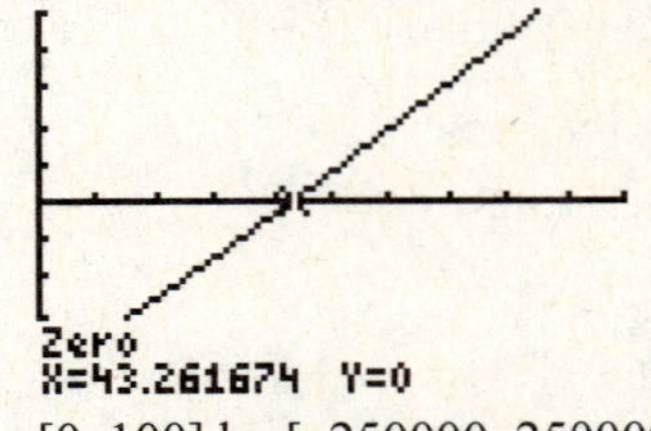

[0, 100] by [–250000, 250000]

The zero is at $x \approx 43.3$. Therefore, between 40,000 and 45,000 calculators must be sold for the company to make a profit.

Section 1.3 Factoring

1. $12x^2 - 24x = 12x \cdot x - 12x \cdot 2 = 12x(x-2)$

3. $r^3 - 5r^2 + r = r\left(r^2\right) - r(5r) + r(1)$
$= r\left(r^2 - 5r + 1\right)$

5. $6z^3 - 12z^2 + 18z$
$= 6z\left(z^2\right) - 6z(2z) + 6z(3)$
$= 6z\left(z^2 - 2z + 3\right)$

7. $3(2y-1)^2 + 7(2y-1)^3$
$= (2y-1)^2(3) + (2y-1)^2 \cdot 7(2y-1)$
$= (2y-1)^2[3 + 7(2y-1)]$
$= (2y-1)^2(3 + 14y - 7)$
$= (2y-1)^2(14y - 4)$
$= 2(2y-1)^2(7y-2)$

9. $3(x+5)^4 + (x+5)^6$
$= (x+5)^4 \cdot 3 + (x+5)^4(x+5)^2$
$= (x+5)^4\left[3 + (x+5)^2\right]$
$= (x+5)^4\left(3 + x^2 + 10x + 25\right)$
$= (x+5)^4\left(x^2 + 10x + 28\right)$

11. $x^2 + 5x + 4 = (x+1)(x+4)$

13. $x^2 + 7x + 12 = (x+3)(x+4)$

15. $x^2 + x - 6 = (x+3)(x-2)$

17. $x^2 + 2x - 3 = (x+3)(x-1)$

19. $x^2 - 3x - 4 = (x+1)(x-4)$

21. $z^2 - 9z + 14 = (z-2)(z-7)$

23. $z^2 + 10z + 24 = (z+4)(z+6)$

25. $2x^2 - 9x + 4 = (2x-1)(x-4)$

27. $15p^2 - 23p + 4 = (3p-4)(5p-1)$

29. $4z^2 - 16z + 15 = (2z-5)(2z-3)$

31. $6x^2 - 5x - 4 = (2x+1)(3x-4)$

33. $10y^2 + 21y - 10 = (5y-2)(2y+5)$

35. $6x^2 + 5x - 4 = (2x-1)(3x+4)$

37. $3a^2 + 2a - 5 = (3a+5)(a-1)$

39. $x^2 - 81 = x^2 - (9)^2 = (x+9)(x-9)$

41. $9p^2 - 12p + 4 = (3p)^2 - 2(3p)(2) + 2^2$
$= (3p-2)^2$

43. $r^2 + 3rt - 10t^2 = (r-2t)(r+5t)$.

45. $m^2 - 8mn + 16n^2 = (m)^2 - 2(m)(4n) + (4n)^2$
$= (m-4n)^2$

47. $4u^2 + 12u + 9 = (2u+3)^2$

49. $25p^2 - 10p + 4$
This polynomial cannot be factored

51. $4r^2 - 9v^2 = (2r+3v)(2r-3v)$

53. $x^2 + 4xy + 4y^2 = (x+2y)^2$

55. $3a^2 - 13a - 30 = (3a+5)(a-6)$.

57. $21m^2 + 13mn + 2n^2 = (7m+2n)(3m+n)$

59. $y^2 - 4yz - 21z^2 = (y-7z)(y+3z)$

61. $121x^2 - 64 = (11x+8)(11x-8)$

63. $a^3 - 64 = a^3 - (4)^3 = (a-4)\left(a^2 + 4a + 16\right)$

65. $8r^3 - 27s^3$
$= (2r)^3 - (3s)^3$
$= (2r-3s)\left[(2r)^2 + (2r)(3s) + (3s)^2\right]$
$= (2r-3s)\left(4r^2 + 6rs + 9s^2\right)$

67. $64m^3 + 125$
$= (4m)^3 + (5)^3$
$= (4m+5)\left[(4m)^2 - (4m)(5) + (5)^2\right]$
$= (4m+5)\left(16m^2 - 20m + 25\right)$

69. $1000y^3 - z^3$
$= (10y)^3 - (z)^3$
$= (10y - z)\left[(10y)^2 + (10y)(z) + (z)^2\right]$
$= (10y - z)\left(100y^2 + 10yz + z^2\right)$

71. $x^4 + 5x^2 + 6 = \left(x^2+2\right)\left(x^2+3\right)$

73. $b^4 - b^2 = b^2\left(b^2 - 1\right) = b^2(b+1)(b-1)$

75. $x^4 - x^2 - 12 = \left(x^2-4\right)\left(x^2+3\right)$
$= (x+2)(x-2)\left(x^2+3\right)$

77. $16a^4 - 81b^4 = \left(4a^2 - 9b^2\right)\left(4a^2 + 9b^2\right)$
$= (2a+3b)(2a-3b)\left(4a^2+9b^2\right)$

79. $x^8 + 8x^2 = x^2\left(x^6+8\right) = x^2\left(\left(x^2\right)^3 + 2^3\right)$
$= x^2\left(x^2+2\right)\left(x^4 - 2x^2 + 4\right)$

81. $6x^4 - 3x^2 - 3 = \left(2x^2+1\right)\left(3x^2-3\right)$ is not the correct complete factorization because $3x^2 - 3$ contains a common factor of 3. This common factor should be factored out as the first step. This will reveal a difference of two squares, which requires further factorization. The correct factorization is
$6x^4 - 3x^2 - 3 = 3\left(2x^4 - x^2 - 1\right)$
$= 3\left(2x^2+1\right)\left(x^2-1\right)$
$= 3\left(2x^2+1\right)(x+1)(x-1)$

83. $(x+2)^3 = (x+2)(x+2)^2$
$= (x+2)(x^2+4x+4)$
$= x^3 + 4x^2 + 2x^2 + 8x + 4x + 8$
$= x^3 + 6x^2 + 12x + 8,$
which is not equal to $x^3 + 8$. The correct factorization is $x^3 + 8 = (x+2)(x^2 - 2x + 4)$.

Section 1.4 Rational Expressions

1. $\dfrac{8x^2}{56x} = \dfrac{x \cdot 8x}{7 \cdot 8x} = \dfrac{x}{7}$

3. $\dfrac{25p^2}{35p^3} = \dfrac{5 \cdot 5p^2}{7p \cdot 5p^2} = \dfrac{5}{7p}$

5. $\dfrac{5m+15}{4m+12} = \dfrac{5(m+3)}{4(m+3)} = \dfrac{5}{4}$

7. $\dfrac{4(w-3)}{(w-3)(w+6)} = \dfrac{4}{w+6}$

9. $\dfrac{3y^2 - 12y}{9y^3} = \dfrac{3y(y-4)}{3y\left(3y^2\right)}$
$= \dfrac{y-4}{3y^2}$

11. $\dfrac{m^2 - 4m + 4}{m^2 + m - 6} = \dfrac{(m-2)(m-2)}{(m+3)(m-2)} = \dfrac{m-2}{m+3}$

13. $\dfrac{x^2 + 2x - 3}{x^2 - 1} = \dfrac{(x+3)(x-1)}{(x+1)(x-1)} = \dfrac{x+3}{x+1}$

15. $\dfrac{3a^2}{64} \cdot \dfrac{8}{2a^3} = \dfrac{3a^2 \cdot 8}{64 \cdot 2a^3} = \dfrac{3}{16a}$

17. $\dfrac{7x}{11} \div \dfrac{14x^3}{66y} = \dfrac{7x}{11} \cdot \dfrac{66y}{14x^3} = \dfrac{7x \cdot 66y}{11 \cdot 14x^3} = \dfrac{3y}{x^2}$

19. $\dfrac{2a+b}{3c} \cdot \dfrac{15}{4(2a+b)} = \dfrac{(2a+b) \cdot 15}{(2a+b) \cdot 12c} = \dfrac{15}{12c} = \dfrac{5}{4c}$

21. $$\frac{15p-3}{6}\div\frac{10p-2}{3}=\frac{15p-3}{6}\cdot\frac{3}{10p-2}=\frac{3(5p-1)\cdot 3}{3\cdot 2\cdot 2\cdot(5p-1)}=\frac{3(5p-1)\cdot 3}{3(5p-1)\cdot 2\cdot 2}=\frac{3}{4}$$

23. $$\frac{9y-18}{6y+12}\cdot\frac{3y+6}{15y-30}=\frac{9(y-2)}{6(y+2)}\cdot\frac{3(y+2)}{15(y-2)}=\frac{27(y-2)(y+2)}{90(y+2)(y-2)}=\frac{27}{90}=\frac{3}{10}$$

25. $$\frac{4a+12}{2a-10}\div\frac{a^2-9}{a^2-a-20}=\frac{4a+12}{2a-10}\cdot\frac{a^2-a-20}{a^2-9}=\frac{4(a+3)}{2(a-5)}\cdot\frac{(a-5)(a+4)}{(a+3)(a-3)}=\frac{4(a+3)(a-5)(a+4)}{2(a-5)(a+3)(a-3)}=\frac{2(a+4)}{a-3}$$

27. $$\frac{k^2-k-6}{k^2+k-12}\cdot\frac{k^2+3k-4}{k^2+2k-3}=\frac{(k-3)(k+2)}{(k+4)(k-3)}\cdot\frac{(k+4)(k-1)}{(k+3)(k-1)}=\frac{(k-3)(k+2)(k+4)(k-1)}{(k+4)(k-3)(k+3)(k-1)}=\frac{k+2}{k+3}$$

Answers will vary for exercises 29. Sample answer is given.

29. To find the least common denominator for two fractions, factor each denominator into prime factors, multiply all unique prime factors raising each factor to the highest frequency it occurred.

31. The common denominator is $35z$.
$$\frac{2}{7z}-\frac{1}{5z}=\frac{2\cdot 5}{7z\cdot 5}-\frac{1\cdot 7}{5z\cdot 7}=\frac{10}{35z}-\frac{7}{35z}=\frac{3}{35z}$$

33. $$\frac{r+2}{3}-\frac{r-2}{3}=\frac{(r+2)-(r-2)}{3}=\frac{r+2-r+2}{3}=\frac{4}{3}$$

35. The common denominator is $5x$.
$$\frac{4}{x}+\frac{1}{5}=\frac{4\cdot 5}{x\cdot 5}+\frac{1\cdot x}{5\cdot x}=\frac{20}{5x}+\frac{x}{5x}=\frac{20+x}{5x}$$

37. The common denominator is $m(m-1)$.
$$\frac{1}{m-1}+\frac{2}{m}=\frac{m\cdot 1}{m\cdot(m-1)}+\frac{(m-1)\cdot 2}{(m-1)\cdot m}=\frac{m}{m(m-1)}+\frac{2(m-1)}{m(m-1)}=\frac{m+2(m-1)}{m(m-1)}=\frac{m+2m-2}{m(m-1)}=\frac{3m-2}{m(m-1)}$$

39. The common denominator is $5(b+2)$.
$$\frac{7}{b+2}+\frac{2}{5(b+2)}=\frac{7\cdot 5}{(b+2)\cdot 5}+\frac{2}{5(b+2)}=\frac{35+2}{5(b+2)}=\frac{37}{5(b+2)}$$

41. The common denominator is $20(k-2)$.
$$\frac{2}{5(k-2)}+\frac{5}{4(k-2)}=\frac{8}{20(k-2)}+\frac{25}{20(k-2)}=\frac{8+25}{20(k-2)}=\frac{33}{20(k-2)}$$

43. First factor the denominators in order to find the common denominator.
$$x^2-4x+3=(x-3)(x-1)$$
$$x^2-x-6=(x-3)(x+2)$$
The common denominator is $(x-3)(x-1)(x+2)$.
$$\frac{2}{x^2-4x+3}+\frac{5}{x^2-x-6}=\frac{2}{(x-3)(x-1)}+\frac{5}{(x-3)(x+2)}=\frac{2(x+2)}{(x-3)(x-1)(x+2)}+\frac{5(x-1)}{(x-3)(x+2)(x-1)}$$
$$=\frac{2(x+2)+5(x-1)}{(x-3)(x+2)(x-1)}=\frac{2x+4+5x-5}{(x-3)(x-1)(x+2)}=\frac{7x-1}{(x-3)(x-1)(x+2)}$$

45. First factor the denominators in order to find the common denominator.

$y^2+7y+12=(y+3)(y+4)$

$y^2+5y+6=(y+3)(y+2)$

The common denominator is $(y+4)(y+3)(y+2)$.

$$\frac{2y}{y^2+7y+12}-\frac{y}{y^2+5y+6}$$
$$=\frac{2y}{(y+4)(y+3)}-\frac{y}{(y+3)(y+2)}$$
$$=\frac{2y(y+2)}{(y+4)(y+3)(y+2)}-\frac{y(y+4)}{(y+4)(y+3)(y+2)}$$
$$=\frac{2y(y+2)-y(y+4)}{(y+4)(y+3)(y+2)}=\frac{2y^2+4y-y^2-4y}{(y+4)(y+3)(y+2)}$$
$$=\frac{y^2}{(y+4)(y+3)(y+2)}$$

47. $\dfrac{1+\frac{1}{x}}{1-\frac{1}{x}}$

Multiply both numerator and denominator of this complex fraction by the common denominator, x.

$$\frac{1+\frac{1}{x}}{1-\frac{1}{x}}=\frac{x\left(1+\frac{1}{x}\right)}{x\left(1-\frac{1}{x}\right)}=\frac{x\cdot 1+x\left(\frac{1}{x}\right)}{x\cdot 1-x\left(\frac{1}{x}\right)}=\frac{x+1}{x-1}$$

49. $\dfrac{\frac{1}{x+h}-\frac{1}{x}}{h}$

The common denominator in the numerator is $x(x+h)$

$$\frac{\frac{1}{x+h}-\frac{1}{x}}{h}=\frac{\frac{x-(x+h)}{x(x+h)}}{h}=\frac{\frac{x-x-h}{x(x+h)}}{h}=\frac{\frac{-h}{x(x+h)}}{h}$$
$$=\frac{-h}{x(x+h)}\div h=\frac{-h}{x(x+h)}\cdot\frac{1}{h}$$
$$=\frac{-1}{x(x+h)}\text{ or }-\frac{1}{x(x+h)}$$

51. The length of each side of the dartboard is $2x$, so the area of the dartboard is $4x^2$. The area of the shaded region is πx^2.

a. The probability that a dart will land in the shaded region is $\dfrac{\pi x^2}{4x^2}$.

b. $\dfrac{\pi x^2}{4x^2}=\dfrac{\pi}{4}$

53. The length of each side of the dartboard is $5x$, so the area of the dartboard is $25x^2$. The area of the shaded region is x^2.

a. The probability that a dart will land in the shaded region is $\dfrac{x^2}{25x^2}$.

b. $\dfrac{x^2}{25x^2}=\dfrac{1}{25}$

55. Average cost = total cost C divided by the number of calculators produced.

$$\frac{-7.2x^2+6995x+230{,}000}{1000x}$$

57. Let $x = 10$. Then

$$\frac{.314(10)^2-1.399(10)+15.0}{10+1}\approx 2.95$$

The ad cost approximately $2.95 million in 2010

59. Let $x = 18$. Then

$$\frac{.314(18)^2-1.399(18)+15.0}{18+1}\approx 4.82$$

The cost of an ad will not reach $5 million in 2018.

61. Let $x = 11$. Then

$$\frac{.265(11)^2+1.47(11)+3.63}{11+2}\approx 3.99$$

The hourly insurance cost in 2011 was $3.99.

63. Let $x = 15$. Then

$$\frac{.265(15)^2+1.47(15)+3.63}{15+2}\approx 5.02$$

The hourly insurance cost in 2015 will be $5.02. The annual cost will be 5.02(2100) = $10,537.68

Section 1.5 Exponents and Radicals

1. $\dfrac{7^5}{7^3}=7^{5-3}=7^2=49$

3. $(4c)^2=4^2c^2=16c^2$

5. $\left(\dfrac{2}{x}\right)^5=\dfrac{2^5}{x^5}=\dfrac{32}{x^5}$

7. $\left(3u^2\right)^3\left(2u^3\right)^2=\left(27u^6\right)\left(4u^6\right)=108u^{12}$

9. $7^{-1}=\frac{1}{7^1}=\frac{1}{7}$

11. $-6^{-5}=-\frac{1}{6^5}=-\frac{1}{7776}$

13. $(-y)^{-3}=\frac{1}{(-y)^3}=-\frac{1}{y^3}$

15. $\left(\frac{4}{3}\right)^{-2}=\left(\frac{3}{4}\right)^2=\frac{9}{16}$

17. $\left(\frac{a}{b^3}\right)^{-1}=\left(\frac{b^3}{a}\right)^1=\frac{b^3}{a}$

19. $49^{1/2}=7$ because $7^2=49$.

21. $(5.71)^{1/4}=(5.71)^{.25}\approx 1.55$ Use a calculator.

23. $-64^{2/3}=-\left(64^{1/3}\right)^2=-(4)^2=-16$

25. $\left(\frac{8}{27}\right)^{-4/3}=\left(\frac{27^{1/3}}{8^{1/3}}\right)^4=\left(\frac{3}{2}\right)^4=\frac{3^4}{2^4}=\frac{81}{16}$

27. $\frac{5^{-3}}{4^{-2}}=\frac{4^2}{5^3}=\frac{16}{125}$

29. $4^{-3}\cdot 4^6=4^3=64$

31. $\frac{4^{10}\cdot 4^{-6}}{4^{-4}}=4^{10}\cdot 4^{-6}\cdot 4^4=4^8=65{,}536$

33. $\frac{z^6\cdot z^2}{z^5}=\frac{z^8}{z^5}=z^{8-5}=z^3$

35.
$$\frac{3^{-1}\left(p^{-2}\right)^3}{3p^{-7}}=\frac{3^{-1}p^{-6}}{3^1p^{-7}}=3^{-1-1}p^{-6-(-7)}$$
$$=3^{-2}p^1=\frac{1}{3^2}\cdot p=\frac{p}{9}$$

37. $\left(q^{-5}r^3\right)^{-1}=q^5r^{-3}=q^5\cdot\frac{1}{r^3}=\frac{q^5}{r^3}$

39.
$$\left(2p^{-1}\right)^3\cdot\left(5p^2\right)^{-2}=2^3\left(p^{-1}\right)^3(5)^{-2}\left(p^2\right)^{-2}$$
$$=2^3\left(p^{-3}\right)\left(\frac{1}{5^2}\right)\left(p^{-4}\right)$$
$$=2^3\left(\frac{1}{p^3}\right)\left(\frac{1}{5^2}\right)\left(\frac{1}{p^4}\right)$$
$$=\frac{8}{25p^7}$$

41.
$$(2p)^{1/2}\cdot\left(2p^3\right)^{1/3}=2^{1/2}p^{1/2}\cdot 2^{1/3}\cdot\left(p^3\right)^{1/3}$$
$$=2^{1/2}p^{1/2}\cdot 2^{1/3}\cdot p^1$$
$$=2^{5/6}p^{3/2}$$

43.
$$p^{2/3}\left(2p^{1/3}+5p\right)=p^{2/3}\left(2p^{1/3}\right)+p^{2/3}(5p)$$
$$=2p+5p^{5/3}$$

45.
$$\frac{\left(x^2\right)^{1/3}\left(y^2\right)^{2/3}}{3x^{2/3}y^2}=\frac{(x)^{2/3}(y)^{4/3}}{3x^{2/3}y^2}$$
$$=\frac{1}{3y^{2-4/3}}=\frac{1}{3y^{2/3}}$$

47.
$$\frac{(7a)^2(5b)^{3/2}}{(5a)^{3/2}(7b)^4}=\frac{7^2a^25^{3/2}b^{3/2}}{5^{3/2}a^{3/2}7^4b^4}=\frac{a^{2-\frac{3}{2}}}{7^2b^{4-\frac{3}{2}}}$$
$$=\frac{a^{1/2}}{49b^{5/2}}$$

49.
$$x^{1/2}\left(x^{2/3}-x^{4/3}\right)=x^{1/2}x^{2/3}-x^{1/2}x^{4/3}$$
$$=x^{7/6}-x^{11/6}$$

51.
$$\left(x^{1/2}+y^{1/2}\right)\left(x^{1/2}-y^{1/2}\right)=\left(x^{1/2}\right)^2-\left(y^{1/2}\right)^2$$
$$=x-y$$

53. $(-3x)^{1/3}=\sqrt[3]{-3x}$, (f)

55. $(-3x)^{-1/3}=\frac{1}{(-3x)^{1/3}}=\frac{1}{\sqrt[3]{-3x}}$, (h)

57. $(3x)^{1/3}=\sqrt[3]{3x}$, (g)

59. $(3x)^{-1/3}=\frac{1}{(3x)^{1/3}}=\frac{1}{\sqrt[3]{3x}}$, (c)

61. $\sqrt[3]{125}=125^{1/3}=5$

63. $\sqrt[4]{625} = 625^{1/4} = 5$

65. $\sqrt{63}\cdot\sqrt{7} = 3\sqrt{7}\cdot\sqrt{7} = 3\cdot 7 = 21$

67. $\sqrt{81-4} = \sqrt{77}$

69. $\sqrt{5}\sqrt{15} = \sqrt{75} = \sqrt{25\cdot 3} = \sqrt{25}\sqrt{3} = 5\sqrt{3}$

71. $\sqrt{50} - \sqrt{72} = 5\sqrt{2} - 6\sqrt{2} = -\sqrt{2}$

73. $5\sqrt{20} - \sqrt{45} + 2\sqrt{80}$
$= 5\cdot 2\sqrt{5} - 3\sqrt{5} + 2\cdot 4\sqrt{5}$
$= 10\sqrt{5} - 3\sqrt{5} + 8\sqrt{5} = 15\sqrt{5}$

75. $(\sqrt{5}+\sqrt{2})(\sqrt{5}-\sqrt{2}) = (\sqrt{5})^2 - (\sqrt{2})^2$
$= 5 - 2 = 3$

77. $\dfrac{3}{1-\sqrt{2}} = \dfrac{3}{1-\sqrt{2}}\cdot\dfrac{1+\sqrt{2}}{1+\sqrt{2}} = \dfrac{3(1+\sqrt{2})}{(1)^2-(\sqrt{2})^2}$
$= \dfrac{3(1+\sqrt{2})}{1-2} = \dfrac{3(1+\sqrt{2})}{-1}$
$= -3(1+\sqrt{2}) = -3 - 3\sqrt{2}$

79. $\dfrac{9-\sqrt{3}}{3-\sqrt{3}} = \dfrac{9-\sqrt{3}}{3-\sqrt{3}}\cdot\dfrac{3+\sqrt{3}}{3+\sqrt{3}} = \dfrac{27+9\sqrt{3}-3\sqrt{3}-3}{3^2-(\sqrt{3})^2}$
$= \dfrac{24+6\sqrt{3}}{9-3} = \dfrac{24+6\sqrt{3}}{6} = 4+\sqrt{3}$

81. $\dfrac{3-\sqrt{2}}{3+\sqrt{2}} = \dfrac{3-\sqrt{2}}{3+\sqrt{2}}\cdot\dfrac{3+\sqrt{2}}{3+\sqrt{2}}$
$= \dfrac{9-2}{9+6\sqrt{2}+2} = \dfrac{7}{11+6\sqrt{2}}$

83. $x = \sqrt{\dfrac{kM}{f}}$

Note that because x represents the number of units to order, the value of x should be rounded to the nearest integer.

a. $k = \$1, f = \$500, M = 100{,}000$
$x = \sqrt{\dfrac{1\cdot 100{,}000}{500}} = \sqrt{200} \approx 14.1$
The number of units to order is 14.

b. $k = \$3, f = \$7, M = 16{,}700$
$x = \sqrt{\dfrac{3\cdot 16{,}700}{7}} \approx 84.6$
The number of units to order is 85.

c. $k = \$1, f = \$5, M = 16{,}800$
$x = \sqrt{\dfrac{1\cdot 16{,}800}{5}} = \sqrt{3360} \approx 58.0$
The number of units to order is 58.

For exercises 85–87, we use the model
$\text{revenue} = 8.19x^{0.096}$, $x \geq 1$, $x = 1$ corresponds to 2001.

85. Let $x = 10$. Then $8.19(10)^{0.096} \approx 10.2$
The domestic revenue for 2010 were about $10,200,000,000.

87. Let $x = 15$. Then $8.19(15)^{0.096} \approx 10.6$
The domestic revenue for 2015 will be about $10,600,000,000.

For exercises 89–91, we use the model
$\text{death rate} = 262.5x^{-.156}$, $x \geq 1$, $x = 1$ corresponds to 2001.

89. Let $x = 11$. Then $262.5(11)^{-.156} \approx 180.6$
The death rate associated with heart disease in 2011 was approximately 180.6.

91. Let $x = 17$. Then $262.5(17)^{-.156} \approx 168.7$
The death rate associated with heart disease in 2017 will be approximately 168.7.

For exercises 93–95, we use the model
$\text{Pell Grant Aid} = 3.96x^{0.239}; x \geq 1, x = 1$ corresponds to 2001.

93. Let $x = 5$. Then $3.96(5)^{0.239} \approx 5.8$
According to the model, there were approximately 5,800,000 students receiving Pell Grants in 2005.

95. Let $x = 13$. Then $3.96(13)^{0.239} \approx 7.3$
According to the model, there will be approximately 7,300,000 students receiving Pell Grants in 2013.

For exercises 97–99, we use the model Annual CT scans $= 3.5x^{1.04}$, $x \geq 5$, $x = 5$ corresponds to 1995.

97. Let $x = 8$. Then $3.5(8)^{1.04} \approx 30.4$

The number of annual CT scans for 1998 were about 30,400,000.

99. Let $x = 22$. Then $3.5(22)^{1.04} \approx 87.1$

The number of annual CT scans for 2012 were about 87,100,000.

Section 1.6 First-Degree Equations

1.
$$\begin{aligned} 3x+8&=20 \\ 3x+8-8&=20-8 \\ 3x&=12 \\ \frac{1}{3}(3x)&=\frac{1}{3}(12) \\ x&=4 \end{aligned}$$

3.
$$\begin{aligned} .6k-.3&=.5k+.4 \\ .6k-.5k-.3&=.5k-.5k+.4 \\ .1k-.3&=.4 \\ .1k-.3+.3&=.4+.3 \\ .1k&=.7 \\ \frac{.1k}{.1}&=\frac{.7}{.1} \Rightarrow k=7 \end{aligned}$$

5.
$$\begin{aligned} 2a-1&=4(a+1)+7a+5 \\ 2a-1&=4a+4+7a+5 \\ 2a-1&=11a+9 \\ 2a-2a-1&=11a-2a+9 \\ -1&=9a+9 \\ -1-9&=9a+9-9 \\ -10&=9a \\ \frac{-10}{9}&=\frac{9a}{9} \Rightarrow -\frac{10}{9}=a \end{aligned}$$

7.
$$\begin{aligned} 2[x-(3+2x)+9]&=3x-8 \\ 2(x-3-2x+9)&=3x-8 \\ 2(-x+6)&=3x-8 \\ -2x+12&=3x-8 \\ 12&=5x-8 \\ 20&=5x \Rightarrow 4=x \end{aligned}$$

9. $\frac{3x}{5}-\frac{4}{5}(x+1)=2-\frac{3}{10}(3x-4)$

Multiply both sides by the common denominator, 10.

$$\begin{aligned} &10\left(\frac{3x}{5}\right)-10\left(\frac{4}{5}\right)(x+1) \\ &=(10)(2)-(10)\left(\frac{3}{10}\right)(3x-4) \end{aligned}$$

$$\begin{aligned} 2(3x)-8(x+1)&=20-3(3x-4) \\ 6x-8x-8&=20-9x+12 \\ -2x-8&=32-9x \\ -2x+9x&=32+8 \\ 7x&=40 \\ \frac{1}{7}(7x)&=\frac{1}{7}(40) \Rightarrow x=\frac{40}{7} \end{aligned}$$

11.
$$\begin{aligned} \frac{5y}{6}-8&=5-\frac{2y}{3} \\ 6\left(\frac{5y}{6}-8\right)&=6\left(5-\frac{2y}{3}\right) \\ 6\left(\frac{5y}{6}\right)-6(8)&=6(5)-6\left(\frac{2y}{3}\right) \\ 5y-48&=30-4y \\ 9y-48&=30 \\ 9y&=78 \\ y&=\frac{78}{9}=\frac{26}{3} \end{aligned}$$

13.
$$\begin{aligned} \frac{m}{2}-\frac{1}{m}&=\frac{6m+5}{12} \\ 12m\left(\frac{m}{2}-\frac{1}{m}\right)&=12m\left(\frac{6m+5}{12}\right) \\ (12m)\left(\frac{m}{2}\right)-(12m)\left(\frac{1}{m}\right)&=m(6m)+m(5) \\ 6m^2-12&=6m^2+5m \\ -12&=5m \\ \frac{1}{5}(-12)&=\frac{1}{5}(5m) \Rightarrow -\frac{12}{5}=m \end{aligned}$$

15. $\frac{4}{x-3} - \frac{8}{2x+5} + \frac{3}{x-3} = 0$

$\frac{4}{x-3} + \frac{3}{x-3} - \frac{8}{2x+5} = 0$

$\frac{7}{x-3} - \frac{8}{2x+5} = 0$

Multiply each side by the common denominator, $(x-3)(2x+5)$.

$$(x-3)(2x+5)\left(\frac{7}{x-3}\right) - (x-3)(2x+5)\left(\frac{8}{2x+5}\right) = (x-3)(2x+5)(0)$$

$$7(2x+5) - 8(x-3) = 0$$
$$14x + 35 - 8x + 24 = 0$$
$$6x + 59 = 0$$
$$6x = -59 \Rightarrow x = -\frac{59}{6}$$

17. $\frac{3}{2m+4} = \frac{1}{m+2} - 2$

$\frac{3}{2(m+2)} = \frac{1}{m+2} - 2$

$$2(m+2)\left(\frac{3}{2(m+2)}\right) = 2(m+2)\left(\frac{1}{m+2}\right) - 2(m+2)(2)$$

$$3 = 2 - 4(m+2)$$
$$3 = 2 - 4m - 8$$
$$3 = -6 - 4m \Rightarrow 9 = -4m \Rightarrow m = -\frac{9}{4}$$

19. $9.06x + 3.59(8x - 5) = 12.07x + .5612$

$$9.06x + 28.72x - 17.95 = 12.07x + .5612$$
$$9.06x + 28.72x - 12.07x = 17.95 + .5612$$
$$25.71x = 18.5112$$
$$x = \frac{18.5112}{25.71} = .72$$

21. $\frac{2.63r - 8.99}{1.25} - \frac{3.90r - 1.77}{2.45} = r$

Multiply by the common denominator (1.25)(2.45) to eliminate the fractions.

$$(2.45)(2.63r - 8.99) - (1.25)(3.90r - 1.77) = (2.45)(1.25)r$$

$$6.4435r - 22.0255 - 4.875r + 2.2125 = 3.0625r$$
$$1.5685r - 19.813 = 3.0625r$$
$$-19.813 = 1.494r$$
$$-\frac{19.813}{1.494} = \frac{1.494r}{1.494}$$
$$r \approx -13.26$$

23. $4(a + x) = b - a + 2x$

$$4a + 4x = b - a + 2x$$
$$4a = b - a - 2x$$
$$5a - b = -2x$$
$$\frac{5a-b}{-2} = \frac{-2x}{-2}$$
$$-\frac{5a-b}{2} = x \text{ or } x = \frac{b-5a}{2}$$

25. $5(b - x) = 2b + ax$

$$5b - 5x = 2b + ax$$
$$5b = 2b + ax + 5x$$
$$3b = ax + 5x$$
$$3b = (a+5)x$$
$$\frac{3b}{a+5} = \frac{(a+5)x}{a+5} \Rightarrow \frac{3b}{a+5} = x$$

27. $PV = k$ for V

$$\frac{1}{P}(PV) = \frac{1}{P}(k) \Rightarrow V = \frac{k}{P}$$

29. $V = V_0 + gt$ for g

$$V - V_0 = gt$$
$$\frac{V - V_0}{t} = \frac{gt}{t} \Rightarrow \frac{V - V_0}{t} = g$$

31. $A = \frac{1}{2}(B+b)h$ for B

$A = \frac{1}{2}Bh + \frac{1}{2}bh$

$2A = Bh + bh$ Multiply by 2.

$2A - bh = Bh$

$\frac{2A - bh}{h} = \frac{Bh}{h}$ Multiply by $\frac{1}{h}$.

$\frac{2A - bh}{h} = \frac{2A}{h} - b = B$

33. $|2h - 1| = 5$

$2h - 1 = 5$ or $2h - 1 = -5$

$2h = 6$ or $2h = -4$

$h = 3$ or $h = -2$

35. $|6 + 2p| = 10$

$6 + 2p = 10$ or $6 + 2p = -10$

$2p = 4$ or $2p = -16$

$p = 2$ or $p = -8$

37. $\left|\frac{5}{r-3}\right| = 10$

$\frac{5}{r-3} = 10$ or $\frac{5}{r-3} = -10$

$5 = 10(r-3)$ or $5 = -10(r-3)$

$5 = 10r - 30$ or $5 = -10r + 30$

$35 = 10r$ or $-25 = -10r$

$\frac{35}{10} = \frac{7}{2} = r$ or $\frac{-25}{-10} = \frac{5}{2} = r$

39. $1.250 = \frac{x}{8} \Rightarrow x = 10$

The stroke lasted 10 hours.

41. $-5 = \frac{5}{9}(F - 32)$

$-5\left(\frac{9}{5}\right) = \left(\frac{9}{5}\right)\left(\frac{5}{9}\right)(F - 32)$

$-9 = F - 32 \Rightarrow 23 = F$

The temperature –5°C = 23°F.

43. $22 = \frac{5}{9}(F - 32)$

$22\left(\frac{9}{5}\right) = \left(\frac{9}{5}\right)\left(\frac{5}{9}\right)(F - 32)$

$39.6 = F - 32 \Rightarrow 71.6 = F$

The temperature 22°C = 71.6°F.

45. $y = 1.16x + 1.76$

Substitute 13.36 for y.

$13.36 = 1.16x + 1.76$

$11.6 = 1.16x \Rightarrow 10 = x$

Therefore, the federal deficit will be $13.36 trillion in 2010.

47. $y = 1.16x + 1.76$

Substitute 19.16 for y.

$19.16 = 1.16x + 1.76$

$17.4 = 1.16x \Rightarrow 15 = x$

Therefore, the federal deficit will be $19.16 trillion in 2015.

49. $E = .118x + 1.45$

Substitute $2.63 in for E.

$2.63 = .118x + 1.45$

$1.18 = .118x \Rightarrow 10 = x$

The health care expenditures were $2.63 trillion in 2010.

51. $E = .118x + 1.45$

Substitute $3.338 in for E.

$3.338 = .118x + 1.45$

$1.888 = .118x \Rightarrow 16 = x$

The health care expenditures will be $3.338 trillion in 2016.

53. $.09(x - 2004) = 12y - 1.44$

Substitute .18 for y and solve for x.

$.09(x - 2004) = 12(.18) - 1.44$

$.09x - 180.36 = .72$

$.09x = 181.08$

$x = 2012$

18.0% of workers were covered in 2012.

55. $.09(x - 2004) = 12y - 1.44$

Substitute .21 for y and solve for x.

$.09(x - 2004) = 12(.21) - 1.44$

$.09x - 180.36 = 1.08$

$.09x = 181.44$

$x = 2016$

21% of workers will be covered in 2016.

57. $A = 4.35x - 12$

Substitute 20,777 for A and solve for x.

$20777 = 4.35x - 12$

$20789 = 4.35x \Rightarrow x \approx 4779$

California had approximately 4,779,000 tax returns filed in 2013.

59. $A = 4.35x - 12$
Substitute 13,360 for A and solve for x.
$13360 = 4.35x - 12$
$13372 = 4.35x \Rightarrow x \approx 3074$
Texas had approximately 3,074,000 tax returns filed in 2013.

61. $f = 800$, $n = 18$, $q = 36$

$$u = f \cdot \frac{n(n+1)}{q(q+1)} = 800 \cdot \frac{18(19)}{36(37)}$$
$$= 800 \cdot \frac{342}{1332} \approx 205.41$$

The amount of unearned interest is \$205.41.

63. Let x = the number invested at 5%.
Then $52{,}000 - x$ = the amount invested at 4%.
Since the total interest is \$2290, we have

$$.05x + .04(52{,}000 - x) = 2290$$
$$.05x + 2080 - .04x = 2290$$
$$.01x + 2080 = 2290$$
$$.01x + 2080 - 2080 = 2290 - 2080$$
$$.01x = 210$$
$$\frac{.01x}{.01x} = \frac{210}{.01}$$
$$x = 21{,}000$$

Joe invested \$21,000 at 5%.

65. Let x = price of first plot.
Then $120{,}000 - x$ = price of second plot.
$.15x$ = profit from first plot
$-.10(120{,}000 - x)$ = loss from second plot.

$$.15x - .10(120{,}000 - x) = 5500$$
$$.15x - 12{,}000 + .10x = 5500$$
$$.25x = 17{,}500$$
$$x = 70{,}000$$

Maria paid \$70,000 for the first plot and 120,000 – 70,000, or \$50,000 for the second plot.

67. Let x = average rate of growth of Tumblr.com.
Then $450{,}000 + x$ = average rate of growth of Pinterest.com.
$63x$ = visitors to Tumblr.com
$30(450{,}000 + x)$ = visitors to Pinterest.com.

$$63x = 30(450{,}000 + x)$$
$$63x = 13{,}500{,}000 + 30x$$
$$33x = 13{,}500{,}000$$
$$x = 409{,}091$$

Since x represents the average rate of growth of Tumblr.com, there was an average growth of 409,091 visitors.

69. The number of visitors to Tumblr.com was $63x = 63(409091) = 25{,}772{,}733$.

71. Let x = the number of liters of 94 octane gas;
200 = the number of liters of 99 octane gas;
$200 + x$ = the number of liters of 97 octane gas.

$$94x + 99(200) = 97(200 + x)$$
$$94x + 19{,}800 = 19{,}400 + 97x$$
$$400 = 3x$$
$$\frac{400}{3} = x$$

Thus, $\frac{400}{3}$ liters of 94 octane gas are needed.

73. Let x = number of miles driven

$$55 + .22x = 78$$
$$.22x = 23$$
$$x = 105$$

You must drive 105 miles in a day for the costs to be equal.

75. $y = 10(x - 75) + 100$
Substitute 180 in for y.

$$180 = 10(x - 75) + 100$$
$$80 = 10(x - 75)$$
$$8 = x - 75 \Rightarrow 83 = x$$

Paul was driving 83 mph.

77. Let x = the number of gallons of premium gas.
The number of gallons of regular gas = $15.5 - x$

$$3.80x + 3.10(15.5 - x) = 50$$
$$3.80x + 48.05 - 3.10x = 50$$
$$.7x = 1.95 \Rightarrow x = 2.8$$
$$15.5 - x = 15.5 - 2.8 = 12.7$$

Jack should get 2.8 gallons of premium gas and 12.7 gallons of regular gas.

Section 1.7 Quadratic Equations

1. $(x+4)(x-14)=0$

$x+4=0$ or $x-14=0$

$x=-4$ or $x=14$

The solutions are –4 and 14.

3. $x(x+6)=0$

$x=0$ or $x+6=0$

$x=-6$

The solutions are 0 and –6.

5. $2z^2=4z$

$2z^2-2z=0$

$2z(z-2)=0$

$2z=0$ or $z-2=0$

$z=0$ or $z=2$

The solutions are 0 and 2.

7. $y^2+15y+56=0$

$(y+7)(y+8)=0$

$y+7=0$ or $y+8=0$

$y=-7$ or $y=-8$

The solutions are –7 and –8.

9. $2x^2=7x-3$

$2x^2-7x+3=0$

$(2x-1)(x-3)=0$

$2x-1=0$ or $x-3=0$

$x=\frac{1}{2}$ or $x=3$

The solutions are $\frac{1}{2}$ and 3.

11. $6r^2+r=1$

$6r^2+r-1=0$

$(3r-1)(2r+1)=0$

$3r-1=0$ or $2r+1=0$

$r=\frac{1}{3}$ or $r=-\frac{1}{2}$

The solutions are $\frac{1}{3}$ and $-\frac{1}{2}$.

13. $2m^2+20=13m$

$2m^2-13m+20=0$

$(2m-5)(m-4)=0$

$2m-5=0$ or $m-4=0$

$m=\frac{5}{2}$ or $m=4$

The solutions are $\frac{5}{2}$ and 4.

15. $m(m+7)=-10$

$m^2+7m+10=0$

$(m+5)(m+2)=0$

$m+5=0$ or $m+2=0$

$m=-5$ or $m=-2$

The solutions are –5 and –2.

17. $9x^2-16=0$

$(3x+4)(3x-4)=0$

$3x+4=0$ or $3x-4=0$

$3x=-4$ $\quad$ $3x=4$

$x-\frac{4}{3}$ or $x=\frac{4}{3}$

The solutions are $-\frac{4}{3}$ and $\frac{4}{3}$.

19. $16x^2-16x=0$

$16x(x-1)=0$

$16x=0$ or $x-1=0$

$x=0$ or $x=1$

The solutions are 0 and 1.

21. $(r-2)^2=7$

$r-2=\sqrt{7}$ or $r-2=-\sqrt{7}$

$r=2+\sqrt{7}$ or $r=2-\sqrt{7}$

We abbreviate the solutions as $2\pm\sqrt{7}$.

23. $(4x-1)^2=20$

Use the square root property.

$4x-1=\sqrt{20}$ or $4x-1=-\sqrt{20}$

$4x-1=2\sqrt{5}$ or $4x-1=-2\sqrt{5}$

$4x=1+2\sqrt{5}$ or $4x=1-2\sqrt{5}$

The solutions are $\frac{1\pm2\sqrt{5}}{4}$.

25. $2x^2 + 7x + 1 = 0$
Use the quadratic formula with $a = 2$, $b = 7$, and $c = 1$.

$$x = \frac{-b \pm \sqrt{b^2 - 4ac}}{2a} = \frac{-7 \pm \sqrt{7^2 - 4(2)(1)}}{2(2)} = \frac{-7 \pm \sqrt{49 - 8}}{4} = \frac{-7 \pm \sqrt{41}}{4}$$

The solutions are $\frac{-7 + \sqrt{41}}{4}$ and $\frac{-7 - \sqrt{41}}{4}$, which are approximately –.1492 and –3.3508.

27. $4k^2 + 2k = 1$
Rewrite the equation in standard form.
$4k^2 + 2k - 1 = 0$
Use the quadratic formula with $a = 4$, $b = 2$, and $c = -1$.

$$k = \frac{-2 \pm \sqrt{2^2 - 4(4)(-1)}}{2(4)} = \frac{-2 \pm \sqrt{4 + 16}}{8} = \frac{-2 \pm \sqrt{20}}{8} = \frac{-2 \pm 2\sqrt{5}}{8} = \frac{2\left(-1 \pm \sqrt{5}\right)}{2 \cdot 4}$$

$$k = \frac{-1 \pm \sqrt{5}}{4}$$

The solutions are $\frac{-1 + \sqrt{5}}{4}$ and $\frac{-1 - \sqrt{5}}{4}$, which are approximately .309 and –.809.

29. $5y^2 + 5y = 2$
$5y^2 + 5y - 2 = 0$
$a = 5, b = 5, c = -2$

$$y = \frac{-5 \pm \sqrt{5^2 - 4(5)(-2)}}{2(5)} = \frac{-5 \pm \sqrt{25 + 40}}{10} = \frac{-5 \pm \sqrt{65}}{10} = \frac{-5 \pm \sqrt{65}}{10}$$

The solutions are $\frac{-5 + \sqrt{65}}{10}$ and $\frac{-5 - \sqrt{65}}{10}$, which are approximately .3062 and –1.3062.

31. $6x^2 + 6x + 4 = 0$
$a = 6, b = 6, c = 4$

$$x = \frac{-6 \pm \sqrt{6^2 - 4(6)(4)}}{2(6)} = \frac{-6 \pm \sqrt{36 - 96}}{12} = \frac{-6 \pm \sqrt{-60}}{12}$$

Because $\sqrt{-60}$ is not a real number, the given equation has no real number solutions.

33. $2r^2 + 3r - 5 = 0$
$a = 2, b = 3, c = -5$

$$r = \frac{-(3) \pm \sqrt{9 - 4(2)(-5)}}{2(2)} = \frac{-3 \pm \sqrt{9 + 40}}{4} = \frac{-3 \pm \sqrt{49}}{4} = \frac{-3 \pm 7}{4}$$

$$r = \frac{-3 + 7}{4} = \frac{4}{4} = 1 \text{ or } r = \frac{-3 - 7}{4} = \frac{-10}{4} = \frac{-5}{2}$$

The solutions are $-\frac{5}{2}$ and 1.

35. $2x^2 - 7x + 30 = 0$
$a = 2, b = -7, c = 30$

$$x = \frac{-(-7) \pm \sqrt{49 - 4(2)(30)}}{2(2)} = \frac{7 \pm \sqrt{-191}}{4}$$

Since $\sqrt{-191}$ is not a real number, there are no real solutions.

37. $1 + \frac{7}{2a} = \frac{15}{2a^2}$
To eliminate fractions, multiply both sides by the common denominator, $2a^2$.
$2a^2 + 7a = 15 \Rightarrow 2a^2 + 7a - 15 = 0$
$a = 2, b = 7, c = -15$

$$a = \frac{-7 \pm \sqrt{7^2 - 4(2)(-15)}}{2(2)} = \frac{-7 \pm \sqrt{49 + 120}}{4}$$

$$= \frac{-7 \pm \sqrt{169}}{4} \Rightarrow a = \frac{-7 + 13}{4} = \frac{6}{4} = \frac{3}{2} \text{ or}$$

$$a = \frac{-7 - 13}{4} = \frac{-20}{4} = -5$$

The solutions are $\frac{3}{2}$ and –5.

39. $25t^2 + 49 = 70t$

$25t^2 - 70t + 49 = 0$

$b^2 - 4ac = (-70)^2 - 4(25)(49)$
$= 4900 - 4900$
$= 0$

The discriminant is 0.
There is one real solution to the equation.

41. $13x^2 + 24x - 5 = 0$

$b^2 - 4ac = (24)^2 - 4(13)(-5)$
$= 576 + 260 = 836$

The discriminant is positive.
There are two real solutions to the equation.

For Exercises 43–45 use the quadratic formula:
$x = \dfrac{-b \pm \sqrt{b^2 - 4ac}}{2a}$.

43. $4.42x^2 - 10.14x + 3.79 = 0$

$$x = \frac{-(-10.14) \pm \sqrt{(-10.14)^2 - 4(4.42)(3.79)}}{2(4.42)}$$
$$\approx \frac{10.14 \pm 5.9843}{8.84} \approx .4701 \text{ or } 1.8240$$

45. $7.63x^2 + 2.79x = 5.32$

$7.63x^2 + 2.79x - 5.32 = 0$

$$x = \frac{-2.79 \pm \sqrt{(2.79)^2 - 4(7.63)(-5.32)}}{2(7.63)}$$
$$\approx \frac{-2.79 \pm 13.0442}{15.26} \approx -1.0376 \text{ or } .6720$$

47. **a.** Let $R = 450$ ft.
$450 = .5x^2 \Rightarrow 900 = x^2 \Rightarrow 30 = x$
The maximum taxiing speed is 30 mph.

b. Let $R = 615$ ft.
$615 = .5x^2 \Rightarrow 1230 = x^2 \Rightarrow 35 \approx x$
The maximum taxiing speed is about 35 mph.

c. Let $R = 970$ ft.
$970 = .5x^2 \Rightarrow 1940 = x^2 \Rightarrow 44 \approx x$
The maximum taxiing speed is about 44 mph.

49. **a.** Let $F = 12.6$.
$12.6 = -.079x^2 + .46x + 13.3$
$0 = -.079x^2 + .46x + .7$
Store $\sqrt{b^2 - 4ac} = \sqrt{(.46)^2 - 4(-.079)(.7)}$
$\approx .6579$ in your calculator.
By the quadratic formula, $x \approx 7.1$ or $x \approx$ -1.25. The negative solution is not applicable. There were 12,600 traffic fatalities in 2007.

b. Let $F = 11$.
$11 = -.079x^2 + .46x + 13.3$
$0 = -.079x^2 + .46x + 2.3$
Store $\sqrt{b^2 - 4ac} = \sqrt{(.46)^2 - 4(-.079)(2.3)}$
$\approx .9687$ in your calculator.
By the quadratic formula, $x \approx 9.04$ or $x \approx$ -3.22. The negative solution is not applicable. There were 11,000 traffic fatalities in 2009.

51. **a.** $A = .169x^2 - 2.85x + 19.6$
$A = .169(9)^2 - 2.85(9) + 19.6$
$A = 7.639$
The total assets in 2008 were 7,639,000,000,000.

b. $A = .169x^2 - 2.85x + 19.6$
$7.9 = .169x^2 - 2.85x + 19.6$
$0 = .169x^2 - 2.85x + 11.7$
$$x = \frac{2.85 \pm \sqrt{(-2.85)^2 - 4(.169)(11.7)}}{2(.169)}$$
≈ 7.07 or 9.8

The total assets were \$7.9 trillion in 2007 and 2009. Since we are looking before 2008, the answer is the yer 2007.

53. Triangle ABC represents the original position of the ladder, while triangle DEC represents the position of the ladder after it was moved.

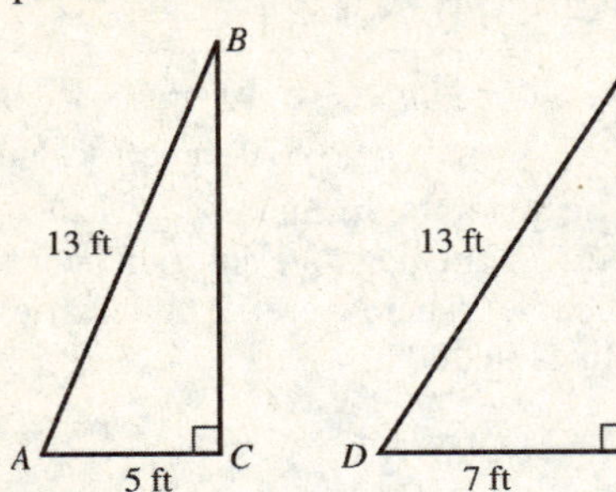

Use the Pythagorean theorem to find the distance from the top of the ladder to the ground.
In triangle ABC,

$13^2 = 5^2 + BC^2 \Rightarrow 169 - 25 = BC^2 \Rightarrow$
$144 = BC^2 \Rightarrow 12 = BC$

Thus, the top of the ladder was originally 12 feet from the ground.
In triangle DEC,

$13^2 = 7^2 + EC^2 \Rightarrow 169 - 49 = EC^2 \Rightarrow$
$120 = EC^2 \Rightarrow EC = \sqrt{120}$

The top of the ladder was $\sqrt{120}$ feet from the ground after the ladder was moved. Therefore, the ladder moved down $12 - \sqrt{120} \approx 1.046$ feet.

55. a. The eastbound train travels at a speed of $x + 20$.

b. The northbound train travels a distance of $5x$ in 5 hours.
The eastbound train travels a distance of $5(x + 20) = 5x + 100$ in 5 hours.

c.

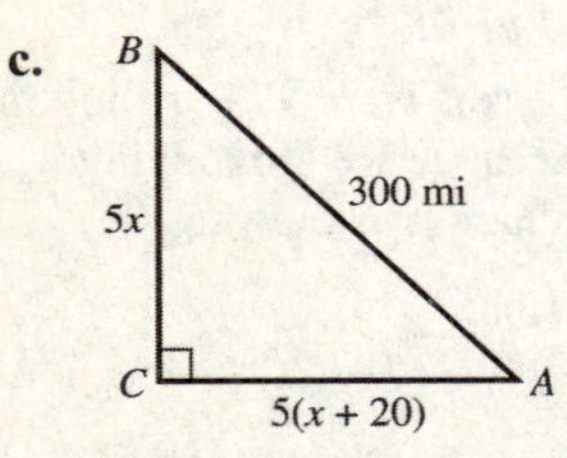

By the Pythagorean theorem,
$(5x)^2 + (5x + 100)^2 = 300^2$

d. Expand and combine like terms.

$$25x^2 + 25x^2 + 1000x + 10,000 = 90,000$$
$$50x^2 + 1000x - 80,000 = 0$$

Factor out the common factor, 50, and divide both sides by 50.

$$50(x^2 + 20x - 1600) = 0$$
$$x^2 + 20x - 1600 = 0$$

Now use the quadratic formula to solve for x.

$$x = \frac{-20 \pm \sqrt{20^2 - 4(1)(-1600)}}{2(1)}$$
$$\approx -51.23 \text{ or } 31.23$$

Since x cannot be negative, the speed of the northbound train is $x \approx 31.23$ mph, and the speed of the eastbound train is $x + 20 \approx$ 51.23 mph.

57. a. Let x represent the length. Then, $\frac{300 - 2x}{2}$ or $150 - x$ represents the width.

b. Use the formula for the area of a rectangle.
$LW = A \Rightarrow x(150 - x) = 5000$

c. $150x - x^2 = 5000$
Write this quadratic equation in standard form and solve by factoring.
$0 = x^2 - 150x + 5000$

$$x^2 - 150x + 5000 = 0$$
$$(x - 50)(x - 100) = 0$$
$$x - 50 = 0 \quad \text{or} \quad x - 100 = 0$$
$$x = 50 \quad \text{or} \quad x = 100$$

Choose $x = 100$ because the length is the larger dimension. The length is 100 m and the width is $150 - 100 = 50$ m.

59. Let x = the width of the uniform strip around the rug.

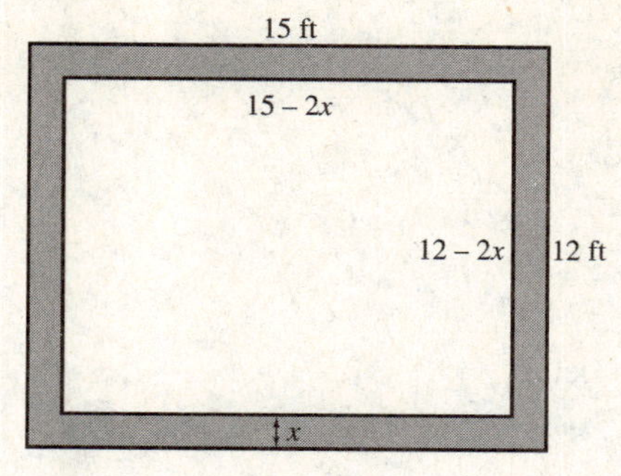

The dimensions of the rug are $15 - 2x$ and $12 - 2x$. The area, 108, is the length times the width.

Solve the equation.

$$(15-2x)(12-2x)=108$$
$$180-54x+4x^2=108$$
$$4x^2-54x+72=0$$
$$2x^2-27x+36=0$$
$$(x-12)(2x-3)=0$$
$$x-12=0 \quad \text{or} \quad 2x-3=0$$
$$x=12 \quad \text{or} \quad x=\frac{3}{2}$$

Discard $x = 12$ since both $12 - 2x$ and $15 - 2x$ would be negative.

If $x=\frac{3}{2}$, then

$$15-2x=15-2\left(\frac{3}{2}\right)=12$$

and $12-2x=12-2\left(\frac{3}{2}\right)=9$.

The dimensions of the rug should be 9 ft by 12 ft.

For exercises 61–65, use the formula $h=-16t^2+v_0t+h_0$, where h_0 is the height of the object when $t = 0$, and v_0 is the initial velocity at time $t = 0$.

61. $v_0=0,\ h_0=625,\ h=0$

$$0=-16t^2+(0)t+625 \Rightarrow 16t^2-625=0 \Rightarrow$$
$$(4t-25)(4t+25)=0 \Rightarrow t=\frac{25}{4}=6.25 \text{ or}$$
$$t=-\frac{25}{4}=-6.25$$

The negative solution is not applicable. It takes 6.25 seconds for the baseball to reach the ground.

63. a. $v_0=0,\ h_0=200,\ h=0$

$$0=-16t^2-(0)t+200 \Rightarrow 16t^2=200 \Rightarrow$$
$$t^2=\frac{200}{16} \Rightarrow t=\pm\frac{\sqrt{200}}{4} \approx \pm 3.54$$

The negative solution is not applicable. It will take about 3.54 seconds for the rock to reach the ground if it is dropped.

b. $v_0=-40,\ h_0=200,\ h=0$

$$0=-16t^2-40t+200$$

Using the quadratic formula, we have

$$t=\frac{-(-40)\pm\sqrt{(-40)^2-4(-16)(200)}}{2(-16)}$$
$$=-5 \text{ or } 2.5$$

The negative solution is not applicable. It will take about 2.5 seconds for the rock to reach the ground if it is thrown with an initial velocity of 40 ft/sec.

c. $v_0=-40,\ h_0=200,\ t=2$

$$h=-16(2)^2-40(2)+200=56$$

After 2 seconds, the rock is 56 feet above the ground. This means it has fallen $200 - 56 = 144$ feet.

65. a. $v_0=64,\ h_0=0,\ h=64$

$$64=-16t^2+64t+0 \Rightarrow$$
$$16t^2-64t+64=0 \Rightarrow$$
$$t^2-4t+4=0 \Rightarrow (t-2)^2=0 \Rightarrow t=2$$

The ball will reach 64 feet after 2 seconds.

b. $v_0=64,\ h_0=0,\ h=39$

$$39=-16t^2+64t+0 \Rightarrow$$
$$16t^2-64t+39=0 \Rightarrow (4t-13)(4t-3) \Rightarrow$$
$$t=\frac{13}{4}=3.25 \text{ or } t=\frac{3}{4}=.75$$

The ball will reach 39 feet after .75 seconds and after 3.25 seconds.

c. Two answers are possible because the ball reaches the given height twice, once on the way up and once on the way down.

In exercises 67–71, we discard negative roots since all variables represent positive real numbers.

67. $S = \frac{1}{2}gt^2$ for t

$2S = gt^2$

$\frac{2S}{g} = t^2$

$\sqrt{\frac{2S}{g}} \cdot \frac{\sqrt{g}}{\sqrt{g}} = t$

$\frac{\sqrt{2Sg}}{g} = t$

69. $L = \frac{d^4k}{h^2}$ for h

$Lh^2 = d^4k$

$h^2 = \frac{d^4k}{L}$

$h = \sqrt{\frac{d^4k}{L}} \cdot \frac{\sqrt{L}}{\sqrt{L}} = \frac{\sqrt{d^4kL}}{L}$

$h = \frac{d^2\sqrt{kL}}{L}$

71. $P = \frac{E^2R}{(r+R)^2}$ for R

$P(r+R)^2 = E^2R$

$P\left(r^2 + 2rR + R^2\right) = E^2R$

$Pr^2 + 2PrR + PR^2 = E^2R$

$PR^2 + \left(2Pr - E^2\right)R + Pr^2 = 0$

Solve for R by using the quadratic formula with $a = P$, $b = 2\text{Pr} - E^2$, and $c = \text{Pr}^2$.

$$R = \frac{-\left(2\text{Pr} - E^2\right) \pm \sqrt{\left(2\text{Pr} - E^2\right)^2 - 4P \cdot \text{Pr}^2}}{2P}$$

$$= \frac{-2\text{Pr} + E^2 \pm \sqrt{4P^2r^2 - 4\text{Pr}\,E^2 + E^4 - 4P^2r^2}}{2P}$$

$$= \frac{-2\text{Pr} + E^2 \pm \sqrt{E^4 - 4\text{Pr}\,E^2}}{2P}$$

$$= \frac{-2\text{Pr} + E^2 \pm \sqrt{E^2\left(E^2 - 4\text{Pr}\right)}}{2P}$$

$$R = \frac{-2\text{Pr} + E^2 \pm E\sqrt{E^2 - 4\text{Pr}}}{2P}$$

73. a. Let $x = z^2$.

$x^2 - 2x = 15$

b. $x^2 - 2x = 15$

$x^2 - 2x - 15 = 0$

$(x-5)(x+3) = 0$

$x = 5$ or $x = -3$

c. Let $z^2 = 5$.

$z = \pm\sqrt{5}$

75. $2q^4 + 3q^2 - 9 = 0$

Let $u = q^2$; then $u^2 = q^4$.

$2u^2 + 3u - 9 = 0 \Rightarrow (2u-3)(u+3) = 0$

$2u - 3 = 0$ or $u + 3 = 0$

$u = \frac{3}{2}$ or $u = -3$

Since $u = q^2$,

$q^2 = \frac{3}{2}$ or $q^2 = -3$

$q = \pm\sqrt{\frac{3}{2}}$ or $q = \pm\sqrt{-3}$ (not real)

$q = \pm\frac{\sqrt{3}}{\sqrt{2}} \cdot \frac{\sqrt{2}}{\sqrt{2}} = \pm\frac{\sqrt{6}}{2}$

The solutions are $\pm\frac{\sqrt{6}}{2}$.

77. $z^4 - 3z^2 - 1 = 0$

Let $x = z^2$; then $x^2 = z^4$.

$x^2 - 3x - 1 = 0$

By the quadratic formula, $x = \frac{3 \pm \sqrt{13}}{2}$.

$\frac{3+\sqrt{13}}{2} = z^2$

$\pm\sqrt{\frac{3+\sqrt{13}}{2}} = z$

Chapter 1 Review Exercises

1. 0 and 6 are whole numbers.

2. $-12, -6, -\sqrt{4}$, 0, and 6 are integers.

3. $-12, -6, -\frac{9}{10}, -\sqrt{4}, 0, \frac{1}{8}$, and 6 are rational numbers.

4. $-\sqrt{7}, \frac{\pi}{4}, \sqrt{11}$ are irrational numbers.

5. $9[(-3)4] = 9[4(-3)]$
Commutative property of multiplication

6. $7(4 + 5) = (4 + 5)7$
Commutative property of multiplication

7. $6(x + y - 3) = 6x + 6y + 6(-3)$
Distributive property

8. $11 + (5 + 3) = (11 + 5) + 3$
Associative property of addition

9. x is at least 9.
$x \geq 9$

10. x is negative.
$x < 0$

11. $|6-4| = 2,\ -|-2| = -2,\ |8+1| = |9| = 9,$
$-|3-(-2)| = -|3+2| = -|5| = -5$
Since $-5, -2, 2, 9$ are in order, then
$-|3-(-2)|, -|-2|, |6-4|, |8+1|$ are in order.

12. $-|\sqrt{16}| = -4,\ -\sqrt{8},\ \sqrt{7},\ |-\sqrt{12}| = \sqrt{12}$

13. $7 - |-8| = 7 - 8 = -1$

14. $|-3| - |-9+6| = 3 - |-3| = 3 - 3 = 0$

15. $x \geq -3$
Start at -3 and draw a ray to the right. Use a bracket at -3 to show that -3 is a part of the graph.

-3

16. $-4 < x \leq 6$
Put a parenthesis at -4 and a bracket at 6. Draw a line segment between these two endpoints.

-4 6

17. $\frac{-9+(-6)(-3)\div 9}{6-(-3)} = \frac{-9+18\div 9}{6+3} = \frac{-9+2}{9} = -\frac{7}{9}$

18. $\frac{20\div 4\cdot 2\div 5-1}{-9-(-3)-12\div 3} = \frac{5\cdot 2\div 5-1}{-9-(-3)-4}$
$= \frac{10\div 5-1}{-9+3-4} = \frac{2-1}{-6-4} = -\frac{1}{10}$

19. $\left(3x^4 - x^2 + 5x\right) - \left(-x^4 + 3x^2 - 6x\right)$
$= 3x^4 - x^2 + 5x + x^4 - 3x^2 + 6x$
$= \left(3x^4 + x^4\right) + \left(-x^2 - 3x^2\right) + (5x + 6x)$
$= 4x^4 - 4x^2 + 11x$

20. $\left(-8y^3 + 8y^2 - 5y\right) - \left(2y^3 + 4y^2 - 10\right)$
$= -8y^3 + 8y^2 - 5y - 2y^3 - 4y^2 + 10$
$= -8y^3 - 2y^3 + 8y^2 - 4y^2 - 5y + 10$
$= -10y^3 + 4y^2 - 5y + 10$

21. $(5k - 2h)(5k + 2h) = (5k)^2 - (2h)^2 = 25k^2 - 4h^2$

22. $(2r - 5y)(2r + 5y) = (2r)^2 - (5y)^2$
$= 4r^2 - 25y^2$

23. $(3x + 4y)^2 = (3x)^2 + 2(3x)(4y) + \left(4y^2\right)$
$= 9x^2 + 24xy + 16y^2$

24. $(2a - 5b)^2 = (2a)^2 - 2(2a)(5b) + (5b)^2$
$= 4a^2 - 20ab + 25b^2$

25. $2kh^2 - 4kh + 5k = k\left(2h^2 - 4h + 5\right)$

26. $2m^2n^2 + 6mn^2 + 16n^2 = 2n^2\left(m^2 + 3m + 8\right)$

27. $5a^4 + 12a^3 + 4a^2 = a^2\left(5a^2 + 12a + 4\right)$
$= a^2(5a + 2)(a + 2)$

28. $24x^3 + 4x^2 - 4x = 4x\left(6x^2 + x - 1\right)$
$= 4x(3x - 1)(2x + 1)$

29. $144p^2 - 169q^2 = (12p)^2 - (13q)^2$
$= (12p - 13q)(12p + 13q)$

30. $81z^2 - 25x^2 = (9z)^2 - (5x)^2$
$= (9z + 5x)(9z - 5x)$

31. $27y^3 - 1 = (3y)^3 - 1^3$
$= (3y - 1)\left[(3y)^2 + (3y)(1) + 1^2\right]$
$= (3y - 1)\left(9y^2 + 3y + 1\right)$

32. $125a^3 + 216 = (5a)^3 + (6)^3$
$= (5a + 6)\left[(5a)^2 - 5a(6) + 6^2\right]$
$= (5a + 6)\left(25a^2 - 30a + 36\right)$

33. $\frac{3x}{5} \cdot \frac{45x}{12} = \frac{3x \cdot 45x}{5 \cdot 12} = \frac{3 \cdot 5 \cdot 3 \cdot 3x^2}{4 \cdot 5 \cdot 3} = \frac{9x^2}{4}$

34. $\frac{5k^2}{24} - \frac{70k}{36} = \frac{5k^2 \cdot 3}{24 \cdot 3} - \frac{70k \cdot 2}{36 \cdot 2} = \frac{15k^2}{72} - \frac{140k}{72}$
$= \frac{15k^2 - 140k}{72} = \frac{5k(3k - 28)}{72}$

35. $\frac{c^2 - 3c + 2}{2c(c - 1)} \div \frac{c - 2}{8c} = \frac{(c - 1)(c - 2)}{2c(c - 1)} \cdot \frac{8c}{(c - 2)}$
$= \frac{8c(c - 1)(c - 2)}{2c(c - 1)(c - 2)} = \frac{8}{2} = 4$

36. $\frac{p^3 - 2p^2 - 8p}{3p\left(p^2 - 16\right)} \div \frac{p^2 + 4p + 4}{9p^2}$
$= \frac{p\left(p^2 - 2p - 8\right)}{3p(p + 4)(p - 4)} \cdot \frac{9p^2}{(p + 2)(p + 2)}$
$= \frac{p(p - 4)(p + 2) \cdot 9p^2}{3p(p + 4)(p - 4)(p + 2)(p + 2)}$
$= \frac{3p(p - 4)(p + 2) \cdot 3p^2}{3p(p - 4)(p + 2) \cdot (p + 4)(p + 2)}$
$= \frac{3p^2}{(p + 4)(p + 2)}$

37. $\frac{2m^2 - 4m + 2}{m^2 - 1} \div \frac{6m + 18}{m^2 + 2m - 3}$
$= \frac{2\left(m^2 - 2m + 1\right)}{(m + 1)(m - 1)} \cdot \frac{m^2 + 2m - 3}{6m + 18}$
$= \frac{2(m - 1)^2}{(m + 1)(m - 1)} \cdot \frac{(m + 3)(m - 1)}{6(m + 3)}$
$= \frac{2(m - 1)(m + 3) \cdot (m - 1)^2}{2(m - 1)(m + 3) \cdot 3(m + 1)} = \frac{(m - 1)^2}{3(m + 1)}$

38. $\frac{x^2 + 6x + 5}{4\left(x^2 + 1\right)} \cdot \frac{2x(x + 1)}{x^2 - 25}$
$= \frac{(x + 5)(x + 1) \cdot 2x(x + 1)}{4\left(x^2 + 1\right) \cdot (x + 5)(x - 5)}$
$= \frac{2(x + 5) \cdot x(x + 1)^2}{2(x + 5) \cdot 2\left(x^2 + 1\right)(x - 5)}$
$= \frac{x(x + 1)^2}{2\left(x^2 + 1\right)(x - 5)}$

39. $5^{-3} = \frac{1}{5^3}$ or $\frac{1}{125}$

40. $10^{-2} = \frac{1}{10^2}$ or $\frac{1}{100}$

41. $-8^0 = -\left(8^0\right) = -1$

42. $\left(-\frac{5}{6}\right)^{-2} = \left(-\frac{6}{5}\right)^2 = \frac{36}{25}$

43. $4^6 \cdot 4^{-3} = 4^{6+(-3)} = 4^3$

44. $7^{-5} \cdot 7^{-2} = 7^{-5+(-2)} = 7^{-7} = \frac{1}{7^7}$

45. $\frac{8^{-5}}{8^{-4}} = 8^{-5-(-4)} = 8^{-5+4} = 8^{-1} = \frac{1}{8}$

46. $\frac{6^{-3}}{6^4} = 6^{-3-4} = 6^{-7} = \frac{1}{6^7}$

47. $5^{-1} + 2^{-1} = \frac{1}{5} + \frac{1}{2} = \frac{7}{10}$

48. $5^{-2}+5^{-1}=\frac{1}{5^2}+\frac{1}{5}=\frac{1}{25}+\frac{1}{5}=\frac{6}{25}$

49. $\frac{5^{1/3}5^{1/2}}{5^{3/2}}=5^{1/3+1/2-3/2}=5^{-2/3}=\frac{1}{5^{2/3}}$

50. $\frac{2^{3/4}\cdot 2^{-1/2}}{2^{1/4}}=\frac{2^{1/4}}{2^{1/4}}=1$

51. $(3a^2)^{1/2}\cdot(3^2a)^{3/2}=3^{1/2}a\cdot 3^3a^{3/2}=3^{7/2}a^{5/2}$

52. $(4p)^{2/3}\cdot(2p^3)^{3/2}=4^{2/3}p^{2/3}\cdot 2^{3/2}\cdot p^{9/2}$

$=(2^2)^{2/3}p^{2/3}\cdot 2^{3/2}p^{9/2}$

$=2^{4/3}\cdot 2^{3/2}p^{2/3}p^{9/2}$

$=2^{17/6}p^{31/6}$

53. $\sqrt[3]{27}=3$

54. $\sqrt[6]{-64}$ is not a real number.

55. $\sqrt[3]{54p^3q^5}=\sqrt[3]{27\cdot 2p^3q^3q^2}$

$=\sqrt[3]{27p^3q^3}\cdot\sqrt[3]{2q^2}=3pq\sqrt[3]{2q^2}$

56. $\sqrt[4]{64a^5b^3}=\sqrt[4]{16a^4}\cdot\sqrt[4]{4ab^3}=2a\sqrt[4]{4ab^3}$

57. $3\sqrt{3}-12\sqrt{12}=3\sqrt{3}-12\sqrt{4\cdot 3}=3\sqrt{3}-12\cdot 2\sqrt{3}$

$=3\sqrt{3}-24\sqrt{3}=-21\sqrt{3}$

58. $8\sqrt{7}+2\sqrt{63}=8\sqrt{7}+2\sqrt{9\cdot 7}$

$=8\sqrt{7}+6\sqrt{7}=14\sqrt{7}$

59. $\frac{\sqrt{3}}{1+\sqrt{2}}=\frac{\sqrt{3}(1-\sqrt{2})}{(1+\sqrt{2})(1-\sqrt{2})}=\frac{\sqrt{3}-\sqrt{6}}{1-2}$

$=\frac{\sqrt{3}-\sqrt{6}}{-1}=\sqrt{6}-\sqrt{3}$

60. $\frac{4+\sqrt{2}}{4-\sqrt{5}}=\frac{(4+\sqrt{2})}{(4-\sqrt{5})}\cdot\frac{(4+\sqrt{5})}{(4+\sqrt{5})}$

$=\frac{16+4\sqrt{2}+4\sqrt{5}+\sqrt{10}}{16-(\sqrt{5})^2}$

$=\frac{16+4\sqrt{2}+4\sqrt{5}+\sqrt{10}}{11}$

61. $3x-4(x-2)=2x+9$

$3x-4x+8=2x+9$

$-x+8=2x+9$

$-1=3x\Rightarrow x=-\frac{1}{3}$

62. $4y+9=-3(1-2y)+5$

$4y+9=-3+6y+5$

$4y+9=2+6y$

$-2y=-7\Rightarrow y=\frac{7}{2}$

63. $\frac{2m}{m-3}=\frac{6}{m-3}+4$

$2m=6+4(m-3)$

$2m=6+4m-12$

$2m=4m-6$

$6=2m\Rightarrow 3=m$

Because $m=3$ would make the denominators of the fractions equal to 0, making the fractions undefined, the given equation has no solution.

64. $\frac{15}{k+5}=4-\frac{3k}{k+5}$

Multiply both sides of the equation by the common denominator $k+5$.

$15=4(k+5)-3k$

$15=4k+20-3k\Rightarrow k=-5$

If $k=-5$, the fractions would be undefined, so the given equation has no solution.

65. $8ax-3=2x$

$8ax-2x=3$

$x(8a-2)=3$

$x=\frac{3}{8a-2}$

66. $b^2x-2x=4b^2$

$(b^2-2)x=4b^2$

$x=\frac{4b^2}{b^2-2}$

67. $\left|\frac{2-y}{5}\right| = 8$

$\frac{2-y}{5} = 8$ or $\frac{2-y}{5} = -8$

$5\left(\frac{2-y}{5}\right) = 5(8)$ or $5\left(\frac{2-y}{5}\right) = -5(-8)$

$2 - y = 40$ or $2 - y = -40$

$-y = 38$ or $-y = -42$

$y = -38$ or $y = 42$

The solutions are –38 and 42.

68. $|4k + 1| = |6k - 3|$

$4k + 1 = 6k - 3$ or $4k + 1 = -(6k - 3)$

$4k + 1 = 6k - 3$ or $4k + 1 = -6k + 3$

$-2k = -4$ or $10k = 2$

$k = 2$ or $k = \frac{1}{5}$

The solutions are 2 and $\frac{1}{5}$.

69. $(b + 7)^2 = 5$

Use the square root property to solve this quadratic equation.

$b + 7 = \sqrt{5}$ or $b + 7 = -\sqrt{5}$

$b = -7 + \sqrt{5}$ or $b = -7 - \sqrt{5}$

The solutions are $-7 + \sqrt{5}$ and $-7 - \sqrt{5}$, which we abbreviate as $-7 \pm \sqrt{5}$.

70. $(2p + 1)^2 = 7$

Solve by the square root property.

$2p + 1 = \sqrt{7}$ or $2p + 1 = -\sqrt{7}$

$2p = -1 + \sqrt{7}$ or $2p = -1 - \sqrt{7}$

$p = \frac{-1 + \sqrt{7}}{2}$ or $p = \frac{-1 - \sqrt{7}}{2}$

The solutions are $\frac{-1 \pm \sqrt{7}}{2}$.

71. $2p^2 + 3p = 2$

Write the equation in standard form and solve by factoring.

$2p^2 + 3p - 2 = 0$

$(2p - 1)(p + 2) = 0$

$2p - 1 = 0$ or $p + 2 = 0$

$p = \frac{1}{2}$ or $p = -2$

The solutions are $\frac{1}{2}$ and –2.

72. $2y^2 = 15 + y$

Write the equation in standard form and solve by factoring.

$2y^2 - y - 15 = 0$

$(y - 3)(2y + 5) = 0$

$y = 3$ or $y = -\frac{5}{2}$

The solutions are 3 and $-\frac{5}{2}$.

73. $2q^2 - 11q = 21 \Rightarrow 2q^2 - 11q - 21 = 0 \Rightarrow$

$(2q + 3)(q - 7) = 0$

$2q + 3 = 0$ or $q - 7 = 0$

$q = -\frac{3}{2}$ or $q = 7$

The solutions are $-\frac{3}{2}$ and 7.

74. $3x^2 + 2x = 16 \Rightarrow 3x^2 + 2x - 16 = 0 \Rightarrow$

$(3x + 8)(x - 2) = 0$

$3x + 8 = 0$ or $x - 2 = 0$

$x = -\frac{8}{3}$ or $x = 2$

The solutions are $-\frac{8}{3}$ and 2.

75. $6k^4 + k^2 = 1 \Rightarrow 6k^4 + k^2 - 1 = 0$

Let $p = k^2$, so $p^2 = k^4$.

$6p^2 + p - 1 = 0$

$(3p - 1)(2p + 1) = 0$

$3p - 1 = 0$ or $2p + 1 = 0$

$p = \frac{1}{3}$ or $p = -\frac{1}{2}$

If $p = \frac{1}{3}$, $k^2 = \frac{1}{3} \Rightarrow k = \pm\sqrt{\frac{1}{3}} = \pm\frac{\sqrt{3}}{3}$

If $p = -\frac{1}{2}$, $k^2 = -\frac{1}{2}$ has no real number solution.

The solutions are $\pm\frac{\sqrt{3}}{3}$.

76. $21p^4 = 2 + p^2 \Rightarrow 21p^4 - p^2 - 2 = 0$

Let $u = p^2$; then $u^2 = p^4$.

$$21u^2 - u - 2 = 0$$
$$(3u-1)(7u+2) = 0$$
$$3u - 1 = 0 \quad \text{or} \quad 7u + 2 = 0$$
$$x = \frac{1}{3} \quad \text{or} \quad x = -\frac{2}{7}$$
$$p^2 = \frac{1}{3} \quad \text{or} \quad p^2 = -\frac{2}{7}$$

If $x = -\frac{2}{7}$, $p^2 = -\frac{2}{7}$ has no real number solution.

$$p = \pm\frac{1}{\sqrt{3}} = \pm\frac{\sqrt{3}}{3}$$

The solutions are $\pm\frac{\sqrt{3}}{3}$.

77. $p = \frac{E^2R}{(r+R)^2}$ for r.

$$p(r+R)^2 = E^2R$$
$$p\left(r^2 + 2rR + R^2\right) = E^2R$$
$$pr^2 + 2rpR + R^2p = E^2R$$
$$pr^2 + 2rpR + R^2p - E^2R = 0$$

Use the quadratic formula to solve for r.

$$r = \frac{-2pR \pm \sqrt{4p^2R^2 - 4p\left(R^2p - E^2R\right)}}{2p}$$
$$r = \frac{-2pR \pm \sqrt{4pE^2R}}{2p} = \frac{-pR \pm E\sqrt{pR}}{p}$$

78. $p = \frac{E^2R}{(r+R)^2}$ for E.

$$p(r+R)^2 = E^2R \Rightarrow E^2 = \frac{p(r+R)^2}{R} \Rightarrow$$
$$E = \pm\sqrt{\frac{p(r+R)^2}{R}} = \frac{\pm(r+R)\sqrt{pR}}{R}$$

79. $K = s(s-a)$ for s.

$$K = s^2 - as \Rightarrow s^2 - as - K = 0$$

Use the quadratic formula.

$$s = \frac{a \pm \sqrt{a^2 - 4(-K)}}{2} = \frac{a \pm \sqrt{a^2 + 4K}}{2}$$

80. $kz^2 - hz - t = 0$ for z.

Use the quadratic formula with $a = k$, $b = -h$, and $c = -t$.

$$z = \frac{-(-h) \pm \sqrt{(-h)^2 - 4(k)(-t)}}{2k}$$
$$= \frac{h \pm \sqrt{h^2 + 4kt}}{2k}$$

81. $|67 - (-44)| = |111| = 111\%$

82. $|-31 - (-45)| = |14| = 14\%$

83. Let x = the original price

$$x - .2x = 895$$
$$.8x = 895$$
$$x = 1118.75$$

The original price of the PC was \$1118.75.

84. Let x = the original price

$$x + .68x = 51$$
$$1.68x = 51$$
$$x = 30.36$$

The original price of the stock was \$30.36.

85. a. Let $O = 3.3$

$$3.3 = .2x + 1.5$$
$$1.8 = .2x$$
$$9 = x$$

The amount of outlays reached \$3.3 trillion in the year 2009.

b. Let $O = 3.9$

$$3.9 = .2x + 1.5$$
$$2.4 = .2x$$
$$12 = x$$

The amount of outlays reached \$3.9 trillion in the year 2012.

86. Let $R = 28.4$

$$28.4 = -3.6x + 64.4$$
$$-36 = -3.6x$$
$$10 = x$$

The amount of revenue from the general newspaper industry reached \$28.4 billion in the year 2010.

87. Let $x = 12$

$V = .89(12)^2 - 17.9(12) + 139.3$

$V = 52.66$

The total sales for the year 2012 was approximately 52, 660, 000.

88. **a.** Let $x = 10$

$A = -.023(10)^3 + .38(10)^2 + 3.29(10) + 107$

$A = 154.9$

The total sales for the year 2010 was approximately \$154, 900, 000, 000.

b. Through the use of graphing calculator, you can find that the function is greater than 140 when x is approximately 7, therefore it will pass \$140 billion in the year 2007.

89. **a.** Let $x = 9$

$$F = \frac{.004(9)^2 + 10.3(9) + 5.8}{9+1} = 9.8824$$

The total number of flights that departed in 2009 was approximately 9, 882, 400.

b. Through the use of graphing calculator, you can find that the function is greater than 10 when x is approximately 12, therefore it will pass 10 million flights in the year 2012.

90. **a.** Let $x = 9$

$$M = \frac{.122(9)^2 + 5.48(9) + 6.5}{9+2} \approx 5.97$$

The total amount of sales from manufacturing in 2009 was approximately \$5,970, 000, 000.

b. Through the use of graphing calculator, you can find that the function is greater than 6.0 when x is approximately 9, therefore it will pass \$6.0 trillion in the year 2009.

91. **a.** Let $x = 10$

$R = 4.33(10)^{0.66} \approx 19.79$

The total amount of research of development spending in 2000 was approximately \$19,790, 000.

b. Let $x = 20$

$R = 4.33(20)^{0.66} \approx 31.27$

The total amount of research of development spending in 2010 was approximately \$31,270, 000.

92. **a.** Let $x = 25$

$E = 91.5(25)^{0.07} \approx 114.6$

The total amount of employees in the service-providing industry in 2015 will be approximately 114,600, 000.

b. Through the use of graphing calculator, you can find that the function is greater than 115 when x is approximately 26, therefore it will pass 115 million employees in the year 2016.

93. Let x = the single interest rate

$$2000(.12) + 500(.07) = 2500x$$
$$275 = 2500x$$
$$.11 = x$$

Therefore, a single interest rate of 11% will yield the same results.

94. Let x = the amount of beef. Then $30 - x$ = the amount of pork.

$$2.8x + 3.25(30 - x) = 3.10(30)$$
$$2.8x + 97.5 - 3.25x = 93$$
$$-.45x = -4.5$$
$$x = 10$$

Therefore the butcher should use 10 pounds of beef and 30 – 10 = 20 pounds of pork.

95. **a.** Let $x = 19.47$

$P = 18.2(19.47)^2 - 26.0(19.47) + 789$

$P \approx 7{,}182$

The total number of patents for the state of New York was approximately 7,182.

b. Let $P = 3806$.

$$3806 = 18.2x^2 - 26.0x + 789$$
$$0 = 18.2x^2 - 26.0x - 3017$$

Using the quadratic formula, we have

$$x = \frac{26 \pm \sqrt{(-26)^2 - 4(18.2)(-3017)}}{2(18.2)}$$
$$= \frac{26 \pm \sqrt{220{,}313.6}}{2(18.2)} \approx -12.18 \text{ or } 13.61$$

The negative value is not applicable. Thus, the approximate population of Illinois was 13,610,000.

96. **a.** Let $x = 11$

$C = .22(11)^2 - 1.9(11) + 23.0$

$C \approx 28.72$

The total number of patents issued in California in 2011 was approximately 28,720.

b. Let $C = 24$.

$$24 = .22x^2 - 1.9x + 23.0$$
$$0 = .22x^2 - 1.9x - 1$$

Using the quadratic formula, we have

$$x = \frac{1.9 \pm \sqrt{(-1.9)^2 - 4(.22)(-1)}}{2(.22)}$$
$$= \frac{1.9 \pm \sqrt{4.49}}{2(.22)} \approx -0.5 \text{ or } 9.14$$

The negative value is not applicable. Thus, the most recent year that saw 24,000 patents in California was 2009.

97. Let x = the width of the walk.

The area $= (10+2x)(15+2x)-10(15)$
$= 150+20x+30x+4x^2-150$
$= 4x^2+50x$

To use all of the cement, solve

$200 = 4x^2+50x$

$0 = 4x^2+50x-200$

Using the quadratic formula, we have

$$x = \frac{-50 \pm \sqrt{(50)^2-4(4)(-200)}}{2(4)}$$
$$= \frac{-50 \pm \sqrt{5700}}{2(4)} \approx -15.6875 \text{ or } 3.1875$$

The width cannot be negative, so the solution is approximately 3.2 feet.

98. Let x = the length of the yard. Then $160 - x$ = the width of the yard. Area is equal to length times width.

The area $= lw$

$4000 = (x)(160-x)$

$0 = x^2-160x+4000$

Using the quadratic formula, we have

$$x = \frac{160 \pm \sqrt{(-160)^2-4(1)(4000)}}{2(1)}$$
$$= \frac{160 \pm \sqrt{9600}}{2(1)} \approx 31.01 \text{ or } 128.99$$

The length is longer than the width, so the length is approximately 129 feet and the width is approximately $160 - 129 = 31$ feet.

99. $v_0 = 150,\ h_0 = 0,\ h = 200$

$200 = -16t^2+150t+0 \Rightarrow$

$16t^2-150t+200 = 0 \Rightarrow$

$8t^2-75t+100 = 0 \Rightarrow t \approx 1.61 \text{ or } 7.77$

The ball will reach 200 feet on its downward trip after approximately 7.7 seconds.

100. $v_0 = 55,\ h_0 = 700,\ h = 0$

$0 = -16t^2-55t+700 \Rightarrow$

$16t^2+55t-700 = 0 \Rightarrow t \approx -8.55 \text{ or } 5.12$

The ball will reach the ground after approximately 5.12 seconds

Case 1 Consumers Often Defy Common Sense

1. The total cost to buy and run this electric hot water tank for x years is $E = 218 + 508x$.

2. The total cost to buy and run this gas hot water tank for x years is $G = 328 + 309x$.

3. Over 10 years, the electric hot water tank costs $218 + 508(10) = 5298$ or \$5298, and the gas hot water tank costs $328 + 309(10) = 3418$ or \$3418. The gas hot water tank costs \$1880 less over 10 years.

4. The total costs for the two hot water tanks will be equal when $218+508x = 328+309x \Rightarrow$ $199x = 110 \Rightarrow x \approx 0.55$.
The costs will be equal within the first year.

5. The total cost to buy and run this Maytag refrigerator for x years is $M = 1529.10 + 50x$.

6. The total cost to buy and run this LG refrigerator for x years is $L = 1618.20 + 44x$.

7. Over 10 years, the Maytag refrigerato costs $1529.10 + 50(10) = 2029.10$ or \$2029.10, and the LG refrigerator costs $1618.20 + 44(10) = 2058.20$ or \$2058.20. The LG refrigerator costs \$29.10 more over 10 years.

8. The total costs for the two refrigerators will be equal when $1529.10+50x = 1618.20+44x \Rightarrow$ $6x = 89.10 \Rightarrow x \approx 14.85$.
The costs will be equal in the 14th year.

Chapter 2 Graphs, Lines, and Inequalities

Section 2.1 Graphs, Lines, and Inequalities

1. (1, –2) lies in quadrant IV
(–2, 1)lies in quadrant II
(3, 4) lies in quadrant I
(–5, –6) lies in quadrant III

3. (1, –3) is a solution to $3x - y - 6 = 0$ because $3(1) - (-3) - 6 = 0$ is a true statement.

5. (3, 4) is not a solution to $(x-2)^2 + (y+2)^2 = 6$ because $(3-2)^2 + (4+2)^2 = 37$, not 6.

7. $4y + 3x = 12$
Find the y-intercept. If $x = 0$,
$4y = -3(0) + 12 \Rightarrow 4y = 12 \Rightarrow y = 3$
The y-intercept is 3.
Next find the x-intercept. If $y = 0$,
$4(0) + 3x = 12 \Rightarrow 3x = 12 \Rightarrow x = 4$
The x-intercept is 4.
Using these intercepts, graph the line.

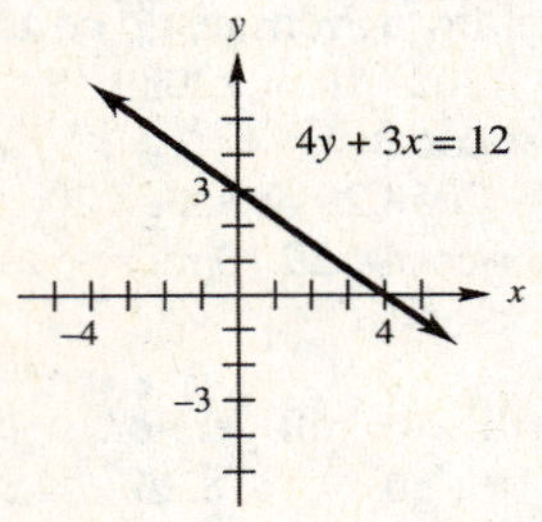

9. $8x + 3y = 12$
Find the y-intercept. If $x = 0$,
$3y = 12 \Rightarrow y = 4$
The y-intercept is 4.
Next, find the x-intercept. If $y = 0$,
$8x = 12 \Rightarrow x = \frac{12}{8} = \frac{3}{2}$
The x-intercept is $\frac{3}{2}$.
Using these intercepts, graph the line.

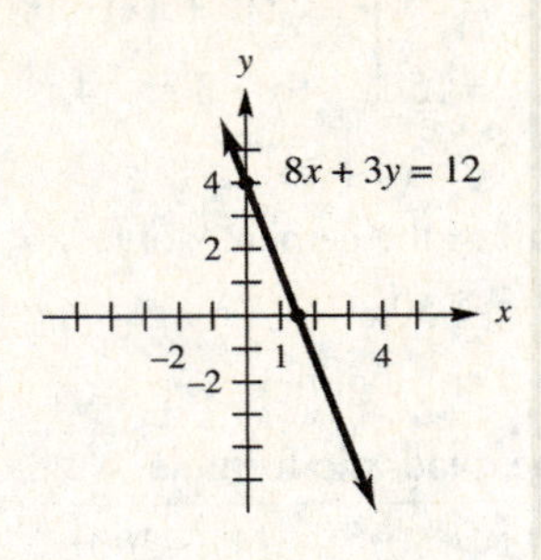

11. $x = 2y + 3$
Find the y-intercept. If $x = 0$,
$0 = 2y + 3 \Rightarrow 2y = -3 \Rightarrow y = -\frac{3}{2}$
The y-intercept is $-\frac{3}{2}$.
Next, find the x-intercept. If $y = 0$,
$x = 2(0) + 3 \Rightarrow x = 3$
The x-intercept is 3.
Using these intercepts, graph the line.

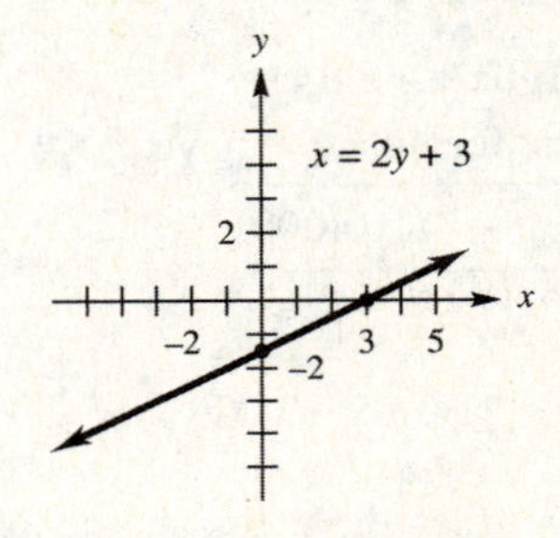

13. The x-intercepts are where the rays cross the x-axis, –2.5 and 3. The y-intercept is where the ray crosses the y-axis, 3.

15. The x-intercepts are –1 and 2. The y-intercept is –2.

17. $3x + 4y = 12$
To find the x-intercept, let $y = 0$:
$3x + 4(0) = 12 \Rightarrow 3x = 12 \Rightarrow x = 4$
The x-intercept is 4.
To find the y-intercept, let $x = 0$:
$3(0) + 4y = 12 \Rightarrow 4y = 12 \Rightarrow y = 3$
The y-intercept is 3.

19. $2x - 3y = 24$
To find the x-intercept, let $y = 0$:
$2x - 3(0) = 24 \Rightarrow 2x = 24 \Rightarrow x = 12$
The x-intercept is 12.
To find the y-intercept, let $x = 0$:
$2(0) - 3y = 24 \Rightarrow -3y = 24 \Rightarrow y = -8$
The y-intercept is –8.

21. $y = x^2 - 9$
To find the x-intercepts, let $y = 0$:
$0 = x^2 - 9 \Rightarrow x^2 = 9 \Rightarrow x = \pm\sqrt{9} = \pm 3$
The x-intercepts are 3 and –3.
To find the y-intercept, let $x = 0$:
$y = 0 - 9 = -9$
The y-intercept is –9.

23. $y = x^2 + x - 20$
To find the x-intercepts, let $y = 0$:
$0 = x^2 + x - 20 \Rightarrow 0 = (x+5)(x-4) \Rightarrow$
$x + 5 = 0 \Rightarrow x = -5$ or $x - 4 = 0 \Rightarrow x = 4$
The x-intercepts are –5 and 4.
To find the y-intercept, let $x = 0$:
$y = 0^2 + 0 - 20 = -20$
The y-intercept is –20.

25. $y = 2x^2 - 5x + 7$
To find the x-intercepts, let $y = 0$:
$0 = 2x^2 - 5x + 7$
This equation does not have real solutions, so there are no x-intercepts.
To find the y-intercept, let $x = 0$:
$y = 2(0)^2 - 5(0) + 7 = 7$
The y-intercept is 7.

27. $y = x^2$
x-intercept: $0 = x^2 \Rightarrow x = 0$
y-intercept: $y = 0$

x	y
–2	4
–1	1
0	0
1	1
2	4

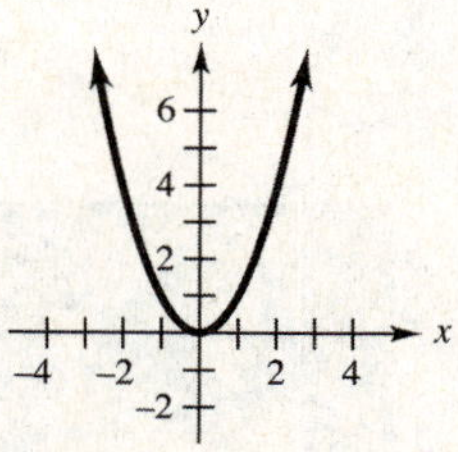

29. $y = x^2 - 3$
x-intercepts: $0 = x^2 - 3 \Rightarrow x^2 = 3 \Rightarrow x = \pm\sqrt{3}$
y-intercepts: $y = 0^2 - 3 = -3$

x	y
–3	6
–1	–2
0	–3
1	–2
3	6

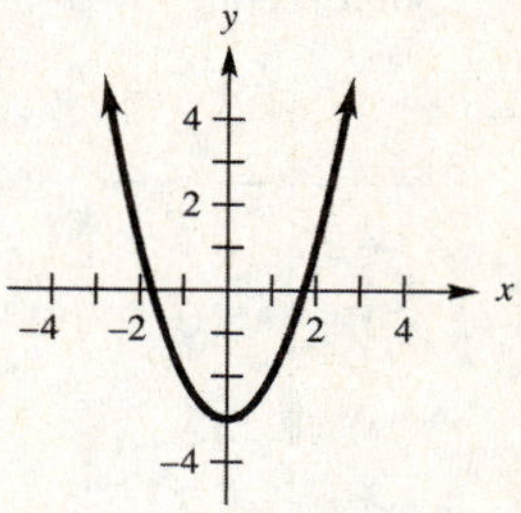

31. $y = x^2 - 6x + 5$
x-intercept:
$0 = x^2 - 6x + 5 \Rightarrow 0 = (x-1)(x-5) \Rightarrow$
$x - 1 = 0 \Rightarrow x = 1$ or $x - 5 = 0 \Rightarrow x = 5$
y-intercept: $y = (0)^2 - 6(0) + 5 = 5$

x	y
–2	21
–1	12
0	5
1	0
2	–3

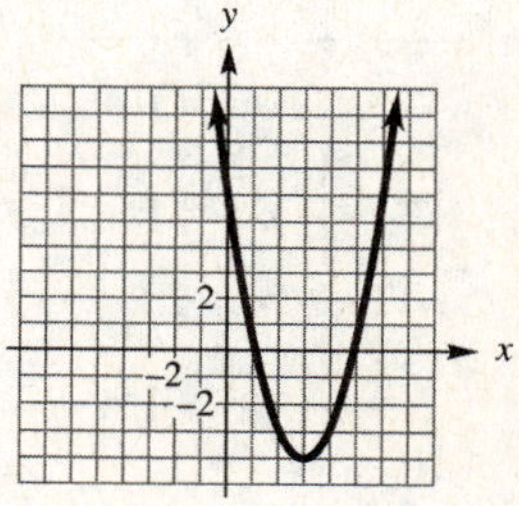

33. $y = x^3$
x-intercept: $0 = x^3 \Rightarrow x = 0$
y-intercept: $y = 0^3 \Rightarrow y = 0$

x	y
–2	–8
–1	–1
0	0
1	1
2	8

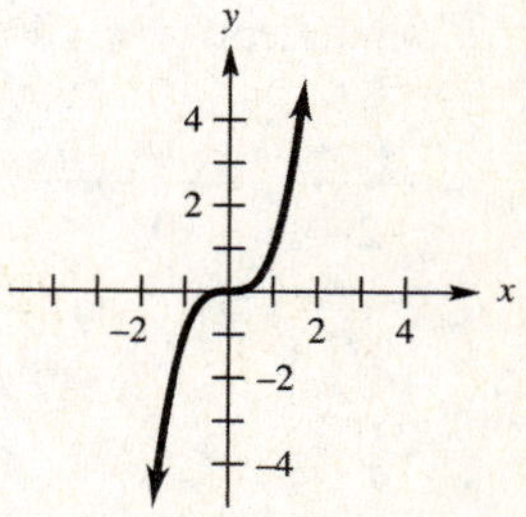

35. $y = x^3 + 1$

x-intercept:

$0 = x^3 + 1 \Rightarrow x^3 = -1 \Rightarrow x = \sqrt[3]{-1} = -1$

y-intercept: $y = 0^3 + 1 = 1$

x	y
–2	–7
–1	0
0	1
1	2
2	9

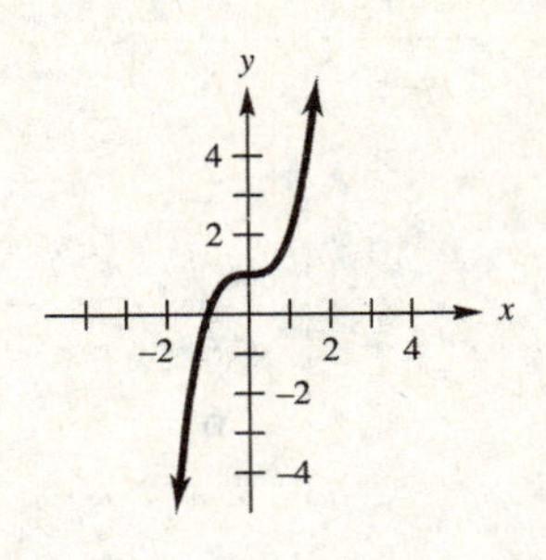

37. $y = \sqrt{x+4}$

x-intercept: $0 = \sqrt{x+4} \Rightarrow 0 = x + 4 \Rightarrow x = -4$

y-intercept: $y = \sqrt{0+4} = \sqrt{4} = 2$

x	y
–2	$\sqrt{2} \approx 1.4$
–1	$\sqrt{3} \approx 1.7$
0	2
2	$\sqrt{6} \approx 2.4$
5	3

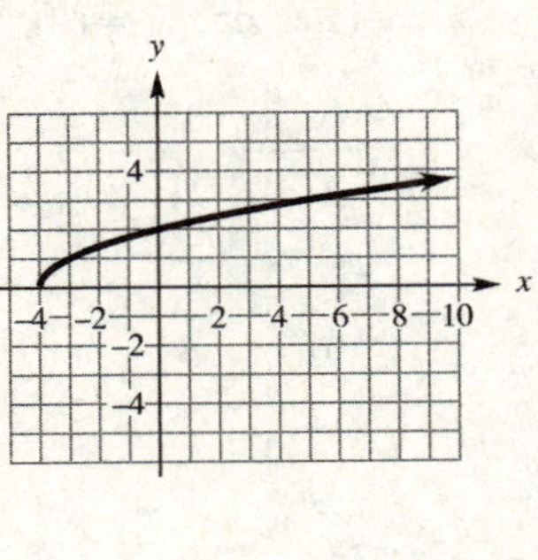

39. $y = \sqrt{4 - x^2}$

x-intercept:

$0 = \sqrt{4 - x^2} \Rightarrow 0 = 4 - x^2 \Rightarrow x^2 = 4 \Rightarrow$

$x = \pm\sqrt{4} = \pm 2$

y-intercept: $y = \sqrt{4} = 2$

x	y
–2	0
–1	$\sqrt{3} \approx 1.7$
0	2
2	$\sqrt{3} \approx 1.7$
5	0

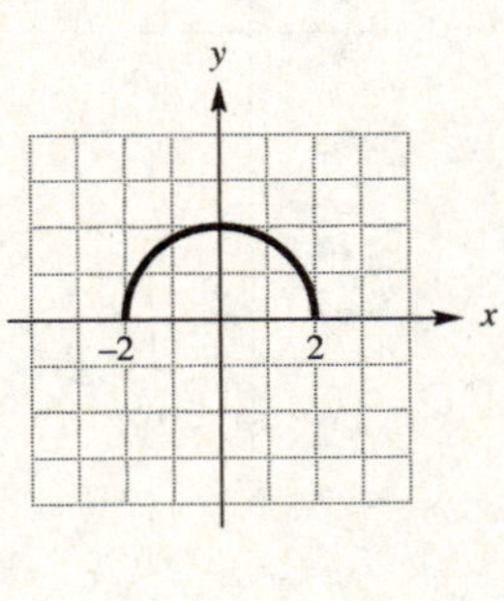

41. 2008; 20 million pounds

43. 2011

45. **(a)** about $1,250,000

(b) $1,750,000

(c) about $4,250,000

47. **(a)** about $500,000

(b) about $1,000,000.

(c) about $1,500,000.

49. beef, about 59 pounds; chicken, about 83 pounds; pork, about 47.5 pounds

51. 2001

53. about $512 billion

55. in 2008–2015

57. H–P, about $16.50; Intel, about $21

59. About $17.25 on Day 14

61. No

63. $y = x^2 + x + 1$

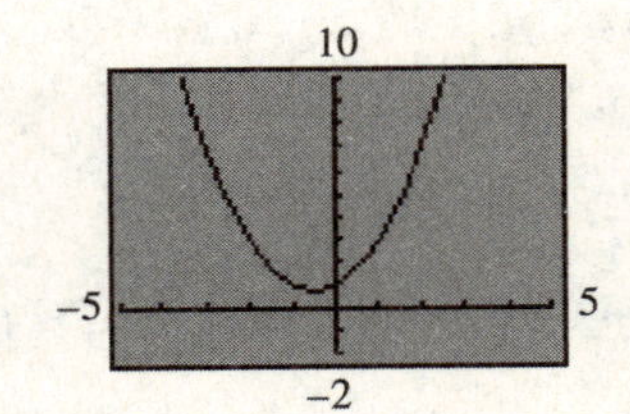

65. $y = (x-3)^3$

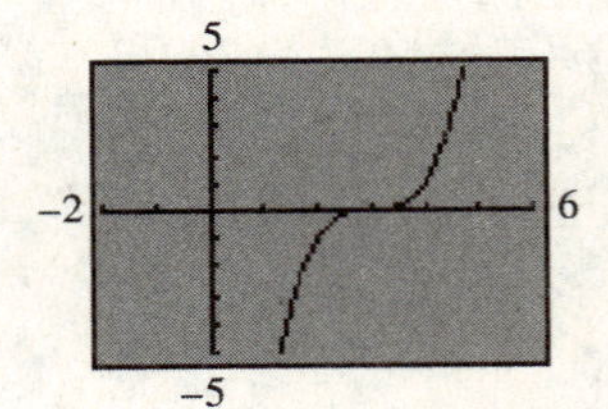

67. $y = x^3 - 3x^2 + x - 1$

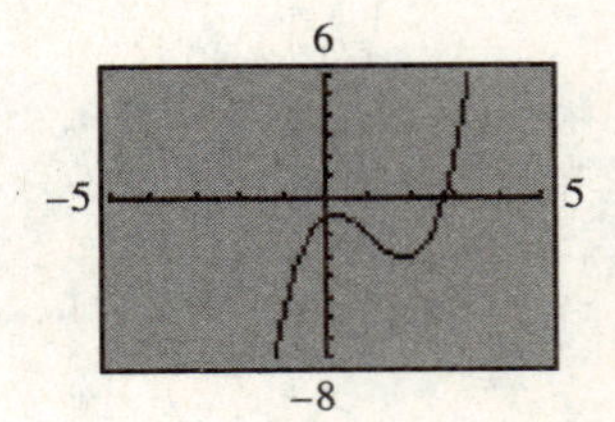

69. $y = x^4 - 2x^3 + 2x$

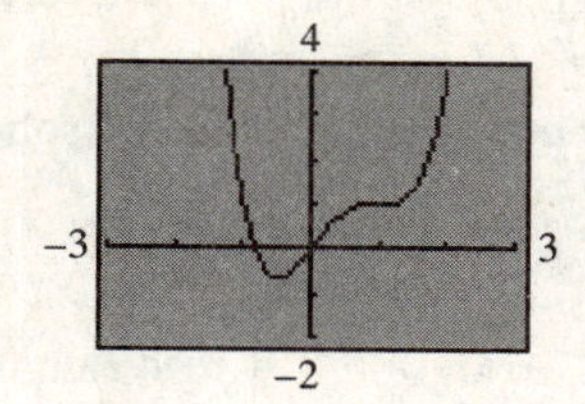

The "flat" part of the graph near $x = 1$ looks like a horizontal line segment, but it is not. The y values increase slightly as you trace along the segment from left to right.

71. $x \approx -1.1038$

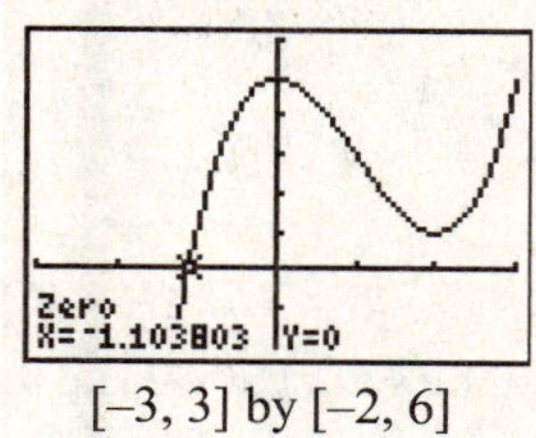

[−3, 3] by [−2, 6]

73. $x \approx 2.1017$

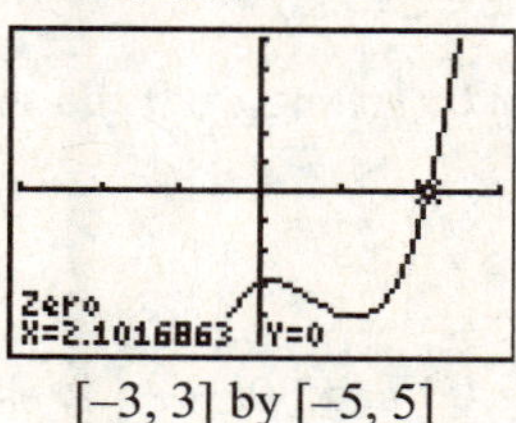

[−3, 3] by [−5, 5]

75. $x \approx -1.7521$

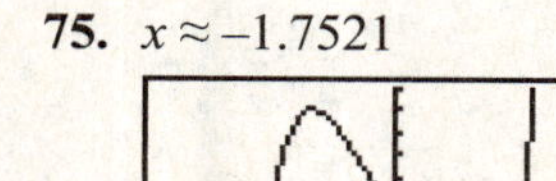

[−3, 3] by [−2, 12]

77. $y = .0556x^3 - 1.286x^2 + 9.76x - 17.4$

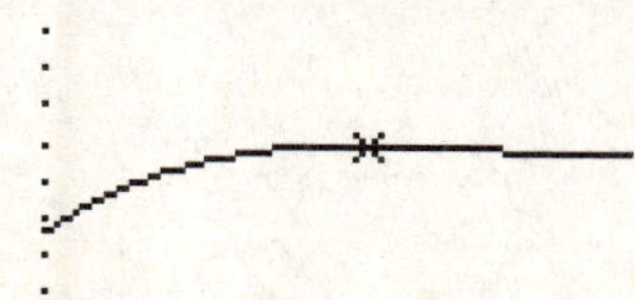

[4, 9] by [0, 10]

The maximum value of the total assets between 2005 and 2008 was approximately \$6.99 trillion.

79. Plot $y = .328x^3 - 7.75x^2 + 59.03x - 97.1$ on [6, 12] by [40, 50], then find the minimum of the curve.

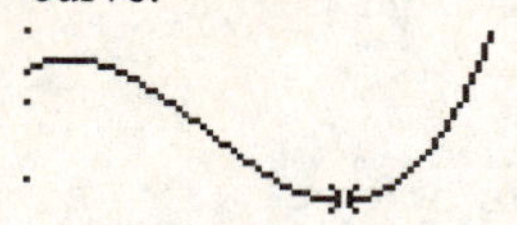

The minimum value of the household assets between 2007 and 2011 was approximately \$45.41 trillion.

Section 2.2 Equations of Lines

1. Through (2, 5) and (0, 8)

$$\text{slope } = \frac{\Delta y}{\Delta x} = \frac{8-5}{0-2} = \frac{3}{-2} = -\frac{3}{2}$$

3. Through (−4, 14) and (3, 0)

$$\text{slope } = \frac{14-0}{-4-3} = \frac{14}{-7} = -2$$

5. Through the origin and (−4, 10); the origin has coordinate (0, 0).

$$\text{slope } = \frac{10-0}{-4-0} = \frac{10}{-4} = -\frac{5}{2}$$

7. Through (−1, 4) and (−1, 6)

$$\text{slope } = \frac{6-4}{-1-(-1)} = \frac{2}{0}, \text{ not defined}$$

The slope is undefined.

9. $b = 5, m = 4$

$y = mx + b$

$y = 4x + 5$

11. $b = 1.5, m = -2.3$

$y = mx + b$

$y = -2.3x + 1.5$

13. $b = 4,\ m = -\frac{3}{4}$

$y = mx + b$

$y = -\frac{3}{4}x + 4$

15. $2x - y = 9$

Rewrite in slope-intercept form.

$-y = -2x + 9$

$y = 2x - 9$

$m = 2, b = -9.$

17. $6x = 2y + 4$
Rewrite in slope-intercept form.
$2y = 6x - 4 \Rightarrow y = 3x - 2$
$m = 3,\ b = -2.$

19. $6x - 9y = 16$
Write in slope-intercept form.
$$-9y = -6x + 16$$
$$9y = 6x - 16$$
$$y = \frac{2}{3}x - \frac{16}{9}$$
$$m = \frac{2}{3},\ b = -\frac{16}{9}.$$

21. $2x - 3y = 0$
Rewrite in slope-intercept form.
$3y = 2x \Rightarrow y = \frac{2}{3}x$
$m = \frac{2}{3},\ b = 0.$

23. $x = y - 5$
Rewrite in slope-intercept form.
$y = x + 5$
$m = 1,\ b = 5$

25. **(a)** Largest value of slope is at C.

(b) Smallest value of slope is at B.

(c) Largest absolute value is at B

(d) Closest to 0 is at D

27. $2x - y = -2$
Find the x-intercept by setting $y = 0$ and solving for x: $2x - 0 = -2 \Rightarrow 2x = -2 \Rightarrow x = -1$
Find the y-intercept by setting $x = 0$ and solving for y: $2(0) - y = -2 \Rightarrow -y = -2 \Rightarrow y = 2$
Use the points $(-1, 0)$ and $(0, 2)$ to sketch the graph:

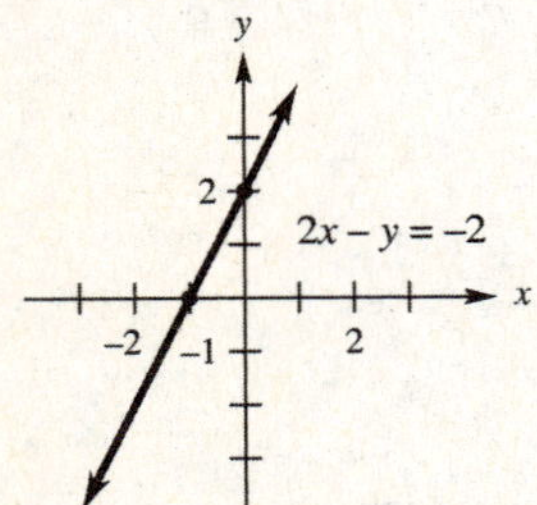

29. $2x + 3y = 4$
Find the x-intercept by setting $y = 0$ and solving for x: $2x + 3(0) = 4 \Rightarrow 2x = 4 \Rightarrow x = 2$
Find the y-intercept by setting $x = 0$ and solving for y: $2(0) + 3y = 4 \Rightarrow 3y = 4 \Rightarrow y = \frac{4}{3}$

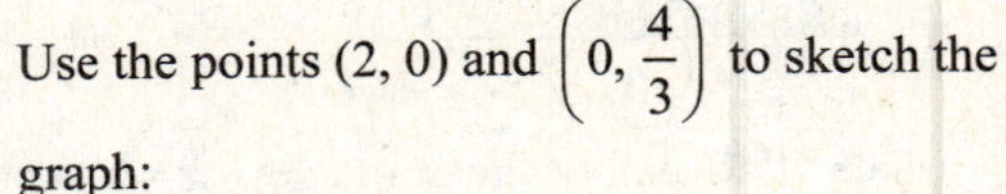
Use the points $(2, 0)$ and $\left(0, \frac{4}{3}\right)$ to sketch the graph:

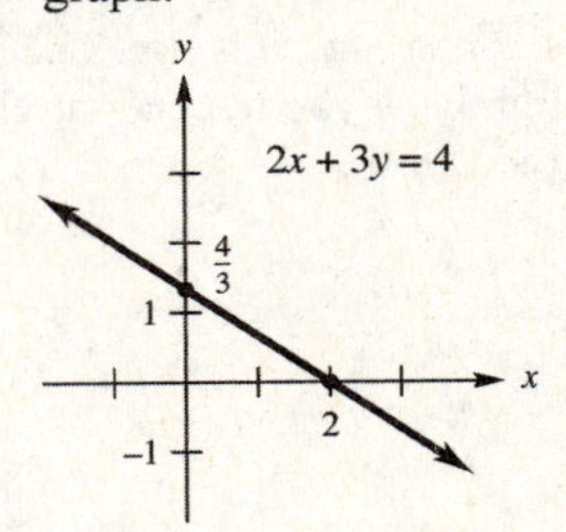

31. $4x - 5y = 2$
Find the x-intercept, by setting $y = 0$ and solving for x:
$4x - 5(0) = 2 \Rightarrow 4x = 2 \Rightarrow x = \frac{1}{2}$
Find the y-intercept by setting $x = 0$ and solving for y:
$4(0) - 5y = 2 \Rightarrow -5y = 2 \Rightarrow y = -\frac{2}{5}$
Use the points $\left(\frac{1}{2}, 0\right)$ and $\left(0, -\frac{2}{5}\right)$ to sketch the graph:

33. For $4x - 3y = 6$, solve for y.

$y = \frac{4}{3}x - 2$

For $3x + 4y = 8$, solve for y.

$y = -\frac{3}{4}x + 2$

The two slopes are $\frac{4}{3}$ and $-\frac{3}{4}$. Since

$\left(\frac{4}{3}\right)\left(-\frac{3}{4}\right) = -1$,

the lines are perpendicular.

35. For $3x + 2y = 8$, solve for y.

$y = -\frac{3}{2}x + 4$

For $6y = 5 - 9x$, solve for y.

$y = -\frac{3}{2}x + \frac{5}{6}$

Since the slopes are both $-\frac{3}{2}$, the lines are parallel.

37. For $4x = 2y + 3$, solve for y.

$y = 2x - \frac{3}{2}$

For $2y = 2x + 3$, solve for y.

$y = x + \frac{3}{2}$

Since the two slopes are 2 and 1, the lines are neither parallel nor perpendicular.

39. Triangle with vertices (9, 6), (–1, 2) and (1, –3).

a. Slope of side between vertices (9, 6) and (–1, 2):

$$m = \frac{6-2}{9-(-1)} = \frac{4}{10} = \frac{2}{5}$$

Slope of side between vertices (–1, 2) and (1, –3):

$$m = \frac{2-(-3)}{-1-1} = \frac{5}{-2} = -\frac{5}{2}$$

Slope of side between vertices (1, –3) and (9, 6):

$$m = \frac{-3-6}{1-9} = \frac{-9}{-8} = \frac{9}{8}$$

b. The sides with slopes $\frac{2}{5}$ and $-\frac{5}{2}$ are perpendicular, because $\frac{2}{5}\left(-\frac{5}{2}\right) = -1$. Thus, the triangle is a right triangle.

41. Use point-slope form with $(x_1, y_1) = (-3, 2)$, $m = -\frac{2}{3}$

$$y - y_1 = m(x - x_1)$$
$$y - 2 = -\frac{2}{3}(x - (-3))$$
$$y - 2 = -\frac{2}{3}(x + 3)$$
$$y - 2 = -\frac{2}{3}x - 2$$
$$y = -\frac{2}{3}x$$

43. $(x_1, y_1) = (2, 3)$, $m = 3$

$$y - y_1 = m(x - x_1)$$
$$y - 3 = 3(x - 2)$$
$$y - 3 = 3x - 6$$
$$y = 3x - 3$$

45. $(x_1, y_1) = (10, 1)$, $m = 0$

$$y - y_1 = m(x - x_1)$$
$$y - 1 = 0(x - 10)$$
$$y - 1 = 0 \Rightarrow y = 1$$

47. Since the slope is undefined, the equation is that of a vertical line through (–2, 12).

$x = -2$

49. Through (–1, 1) and (2, 7)

Find the slope.

$$m = \frac{7-1}{2-(-1)} = \frac{6}{3} = 2$$

Use the point-slope form with $(2, 7) = (x_1, y_1)$.

$$y - y_1 = m(x - x_1)$$
$$y - 7 = 2(x - 2)$$
$$y - 7 = 2x - 4$$
$$y = 2x + 3$$

51. Through (1, 2) and (3, 9)
Find the slope.

$$m = \frac{9-2}{3-1} = \frac{7}{2}$$

Use the point-slope form with $(1, 2) = (x_1, y_1)$.

$$y - y_1 = m(x - x_1)$$
$$y - 2 = \frac{7}{2}(x-1)$$
$$y - 2 = \frac{7}{2}x - \frac{7}{2}$$
$$y = \frac{7}{2}x - \frac{3}{2}$$
$$2y = 7x - 3$$

53. Through the origin with slope 5.

$(x_1, y_1) = (0, 0)$; $m = 5$

$$y - y_1 = m(x - x_1)$$
$$y - 0 = 5(x - 0) \Rightarrow y = 5x$$

55. Through (6, 8) and vertical.
A vertical line has undefined slope.

$(x_1, y_1) = (6, 8)$

$x = 6$

57. Through (3, 4) and parallel to $4x - 2y = 5$.
Find the slope of the given line because a line parallel to the line has the same slope.

$(x_1, y_1) = (3, 4)$

$$4x = 2y + 5$$
$$2y = 4x - 5$$
$$y = 2x - \frac{5}{2} \quad m = 2$$
$$y - y_1 = m(x - x_1)$$
$$y - 4 = 2(x - 3)$$
$$y - 4 = 2x - 6$$
$$y = 2x - 2$$

59. x-intercept 6; y-intercept –6
Through the points (6, 0) and (0, –6).

$$m = \frac{0-(-6)}{6-0} = \frac{6}{6} = 1$$

$(x_1, y_1) = (6, 0)$

$$y - y_1 = m(x - x_1)$$
$$y - 0 = 1(x - 6) \Rightarrow y = x - 6$$

61. Through (–1, 3) and perpendicular to the line through (0, 1) and (2, 3).
The slope of the given line is

$m_1 = \frac{1-3}{0-2} = \frac{-2}{-2} = 1$, so the slope of a line perpendicular to the line is $m_2 = \frac{-1}{1} = -1$.

$(x_1, y_1) = (-1, 3)$

$$y - y_1 = m(x - x_1)$$
$$y - 3 = -1(x - (-1))$$
$$y - 3 = -x - 1$$
$$y = -x + 2$$

63. Let cost x = 15,965 and life: 12 years. Find D.

$$D = \left(\frac{1}{n}\right)x = \frac{1}{12}(15{,}965) \approx 1330.42$$

The depreciation is \$1330.42 per year.

65. Let cost x = \$201,457; life: 30 years

$$D = \left(\frac{1}{n}\right)x = \frac{1}{30}(201{,}457) \approx 6715.23$$

The depreciation is \$6715.23 per year.

67. **a.** $x = 5$

$$y = 13.69(5) + 133.6 = 202.05$$

There were about \$202.05 billion or \$202,050,000,000 in sales from drug prescriptions in 2005.

b. $x = 10$.

$$y = 13.69(10) + 133.6 = 270.5$$

There were about \$270.5 billion or \$270,500,000,000 in sales from drug prescriptions in 2010.

c. $y = 340$

$$340 = 13.69x + 133.6$$
$$206.4 = 13.69x$$
$$15.1 \approx x$$

Sales from drug prescriptions will be about \$340 billion in the year 2015.

69. **a.** $x = 0$

$y = -1.8(0) + 384.6 = 384.6$

There were approximately 384.6 thousand or 384,600 employees working in the motion picture and sound industries in 2000.

b. $x = 10$

$y = -1.8(10) + 384.6 = 366.6$

There were approximately 366.6 thousand or 366,600 employees working in the motion picture and sound industries in 2010.

c. $y = 350$

$350 = -1.8x + 384.6$

$-34.6 = -1.8x \Rightarrow x \approx 19.2$

This corresponds to the year 2019. There will be approximately 350,000 employees working in the motion picture and sound industries in 2019.

71. **a.** The given data is represented by the points (5, 35.1) and (11, 29.7).

b. Find the slope.

$$m = \frac{29.7 - 35.1}{11 - 5} = \frac{-5.4}{6} = -0.9$$

$$y - y_1 = m(x - x_1)$$

$$y - 29.7 = -0.9(x - 11)$$

$$y - 29.7 = -0.9x + 9.9$$

$$y = -0.9x + 39.6$$

c. The year 2009 corresponds to $x = 9$.

$y = -0.9(9) + 39.6 = 31.5$

Total sales associated with lawn care were about $31.5 billion in 2009.

d. $y = 25$.

$25 = -0.9x + 39.6$

$-14.6 = -0.9x \Rightarrow x \approx 16.2$

This corresponds to the year 2016. Total sales associated with lawn care will reach $25 billion in 2016.

73. **a.** $(x_1, y_1) = (0, .4)$ and $(x_2, y_2) = (10, .75)$

Find the slope.

$$m = \frac{.75 - .4}{10 - 0} = \frac{.35}{10} = 0.035$$

The y-intercept is .4, so the equation of the line is $y = 0.035x + 0.4$.

b. The year 2014 corresponds to $x = 2014 - 2002 = 14$.

$y = 0.035(12) + 0.4 = 0.82$

In 2014, the price for chicken legs will be about $0.82 per pound.

75. **a.** $(x_1, y_1) = (0, 36845)$ and $(x_2, y_2) = (10, 27200)$

Find the slope.

$$m = \frac{27,200 - 36845}{10 - 0} = -964.5$$

The y-intercept is 36,845, so the equation is $y = -964.5x + 36,845$.

b. The year 2006 corresponds to $x = 2006 - 2000 = 6$.

$y = -964.5(6) + 36,845 = 31058$

In 2006, there were about 31,058 federal drug arrests.

77. **a.** The slope of $-.01723$ indicates that on average, the 5000-meter run is being run .01723 seconds faster every year. It is negative because the times are generally decreasing as time progresses.

b. $y = -.01723(2012) + 47.61 \approx 12.94$

The model predicts that the time for the 5000-m run will be about 12.94 minutes in the 2012 Olympics.

Section 2.3 Linear Models

1. **a.** Let (x_1, y_1) be (32, 0) and (x_2, y_2) be (68, 20).

Find the slope.

$$m = \frac{20 - 0}{68 - 32} = \frac{20}{36} = \frac{5}{9}$$

Use the point-slope form with (32, 0).

$$y - 0 = \frac{5}{9}(x - 32) \Rightarrow y = \frac{5}{9}(x - 32)$$

b. Let $x = 50$.

$$y = \frac{5}{9}(50 - 32) = \frac{5}{9}(18) = 10°\text{C}$$

Let $x = 75$.

$$y = \frac{5}{9}(75 - 32) = \frac{5}{9}(43) \approx 23.89°\text{C}$$

3. F = 867°

$$C = \frac{5}{9}(867 - 32) = \frac{5}{9}(835) \approx 463.89°C$$

5. Let $(x_1, y_1) = (6, 201.6)$ and $(x_2, y_2) = (11, 224.9)$. Find the slope.

$$m = \frac{224.9 - 201.6}{11 - 6} = 4.66$$

$$y - 201.6 = 4.66(x - 6)$$
$$y - 201.6 = 4.66x - 27.96$$
$$y = 4.66x + 173.64.$$

To estimate the CPI in 2008, let $x = 8$.

$$y = 4.66(8) + 173.64 \approx 210.92$$

To estimate the CPI in 2015, let $x = 15$.

$$y = 4.66(15) + 173.64 \approx 243.54$$

7. Let $(x_1, y_1) = (0, 6.0)$ and $(x_2, y_2) = (8, 6.5)$. Find the slope.

$$m = \frac{6.5 - 6.0}{8 - 0} = 0.0625$$

The *y*-intercept is 6.0, so the equation is $y = 0.0625x + 6.0$.

To estimate the number of employees working in the finance and insurance industries in 2010, let $x = 2010 - 2000 = 10$.

$$y = 0.0625(10) + 6.0 = 6.625$$

The number of employees was estimated to be 6.625 million in 2010.

9. Find the slope of the line.

$$(x_1, y_1) = (50, 320)$$
$$(x_2, y_2) = (80, 440)$$
$$m = \frac{y_2 - y_1}{x_2 - x_1} = \frac{440 - 320}{80 - 50} = \frac{120}{30} = 4$$

Each mile per hour increase in the speed of the bat will make the ball travel 4 more feet.

11. a. $y = -143.6x + 6019$

Data Point (x, y)	Model Point $(x, \hat{y})$	Residual $y - \hat{y}$	Squared Residual $(y - \hat{y})^2$
(7, 5036)	(7,5013.8)	22.2	492.84
(8, 4847)	(8, 4870.2)	–23.2	538.24
(9, 4714)	(9, 4726.6)	–12.6	158.76
(10,4589)	(10, 4583)	6	36
(11,4447)	(11,4439.4)	7.6	57.76

$y = -170.2x + 6250$

Data Point (x, y)	Model Point $(x, \hat{y})$	Residual $y - \hat{y}$	Squared Residual $(y - \hat{y})^2$
(7, 5036)	(7,5058.6)	–22.6	510.76
(8, 4847)	(8, 4888.4)	–41.4	1713.96
(9, 4714)	(9, 4718.2)	–4.2	17.64
(10,4589)	(10, 4548)	41	1681
(11,4447)	(11,4377.8)	69.2	4788.64

Sum of the residuals for model 1 = 0
Sum of the residuals for model 2 = 42

b. Sum of the squares of the residuals for model 1 = 1283.6
Sum of the squares of the residuals for model 2 = 8712

c. Model 1 is the better fit.

13. Plot the points.

L1	L2	L3
6	177.3	------
7	192	
8	194.7	
9	203	
10	203.6	
11	210	

L2(7) =

[5, 12] by [150, 220]

Visually, a straight line looks to be a good model for the data.

```
LinReg
 y=ax+b
 a=5.902857143
 b=146.592381
 r²=.914720892
 r=.9564104202
```

The coefficient of correlation is $r \approx .956$, which indicates that the regression line is a good fit for the data.

15. **a.** Using a graphing calculator, the regression-line model is $y = 5.90x + 146.59$.

b. The year 2015 corresponds to $x = 15$. Using the regression-line model generated by a graphing calculator, we have $y = 5.90(15) + 146.59 = 235.09$, or about \$235 billion in sales.

17. **a.** Using a graphing calculator, the regression-line model is $y = -3.96x + 73.98$.

b. The year 2016 corresponds to $x = 16$. Using the regression-line model generated by a graphing calculator, we have $y = -3.96(16) + 73.98 = 10.62$, or about \$10.62 billion in revenue.

19. **a.** Using a graphing calculator, the regression line model for estimated operating revenue (in billions of dollars) from internet publishing and broadcasting is given by $y = 2.37x - 2.02$.

b. Let $x = 12$ (2012).
$y = 2.37(12) - 2.02 = 26.42$ billion
Let $x = 14$ (2014).
$y = 2.37(14) - 2.02 = 31.16$ billion
The operating revenue was about \$26.42 billion in 2012 and will be about \$31.16 billion in 2014.

21. **a.** Using a graphing calculator, the regression-line model is $y = -2.318x + 55.88$.

b. $y = -2.318(6) + 55.88 \approx 41.972$. There were about 41,972 traffic fatalities in 2006.

c. Let $y = 28$.

$$28 = -2.318x + 55.88$$
$$-27.88 = -2.318x$$
$$x \approx 12.03$$

There were 28,000 traffic fatalities in the year 2012.

d. Using a graphing calculator, the coefficient of correlation is about –.972.

Section 2.4 Linear Inequalities

1. Use brackets if you want to include the endpoint, and parentheses if you want to exclude it.

3. $-8k \le 32$

Multiply both sides of the inequality by $-\frac{1}{8}$. Since this is a negative number, change the direction of the inequality symbol.

$$-\frac{1}{8}(-8k) \ge -\frac{1}{8}(32) \Rightarrow k \ge -4$$

The solution is $[-4, \infty)$.

−4

5. $-2b > 0$

Multiply both sides by $-\frac{1}{2}$.

$$-2b > 0 \Rightarrow -\frac{1}{2}(-2b) < -\frac{1}{2}(0) \Rightarrow b < 0$$

The solution is $(-\infty, 0)$. To graph this solution, put a parenthesis at 0 and draw an arrow extending to the left.

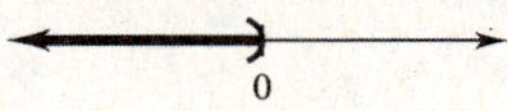

7. $3x + 4 \le 14$

Subtract 4 from both sides.

$$3x + 4 - 4 \le 14 - 4 \Rightarrow 3x \le 10$$

Multiply each side by $\frac{1}{3}$.

$$\frac{1}{3}(3x) \le \frac{1}{3}(10) \Rightarrow x \le \frac{10}{3}$$

The solution is $\left(-\infty, \frac{10}{3}\right]$.

$\frac{10}{3}$

For exercises 9–26, we give the solutions without additional explanation.

9.
$$-5 - p \geq 3$$
$$-5 + 5 - p \geq 3 + 5$$
$$-p \geq 8$$
$$(-1)(-p) \leq (-1)(8)$$
$$p \leq -8$$
The solution is $(-\infty, -8]$.

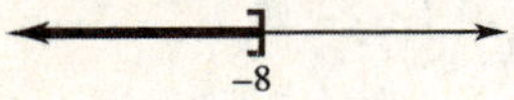

11.
$$7m - 5 < 2m + 10$$
$$5m - 5 < 10$$
$$5m < 15$$
$$\frac{1}{5}(5m) < \frac{1}{5}(15)$$
$$m < 3$$
The solution is $(-\infty, 3)$.

13.
$$m - (4 + 2m) + 3 < 2m + 2$$
$$m - 4 - 2m + 3 < 2m + 2$$
$$-1 - m < 2m + 2$$
$$-m - 2m < 2 + 1$$
$$-3m < 3$$
$$-\frac{1}{3}(-3m) > -\frac{1}{3}(3)$$
$$m > -1$$
The solution is $(-1, \infty)$.

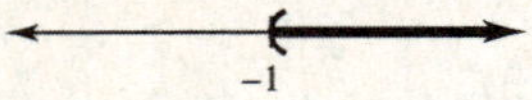

15.
$$-2(3y - 8) \geq 5(4y - 2)$$
$$-6y + 16 \geq 20y - 10$$
$$16 + 10 \geq 20y + 6y$$
$$26 \geq 26y$$
$$1 \geq y \text{ or } y \leq 1$$
The solution is $(-\infty, 1]$.

17.
$$3p - 1 < 6p + 2(p - 1)$$
$$3p - 1 < 6p + 2p - 2$$
$$-1 + 2 < 6p + 2p - 3p$$
$$1 < 5p$$
$$\frac{1}{5} < p \text{ or } p > \frac{1}{5}$$
The solution is $\left(\frac{1}{5}, \infty\right)$.

19.
$$-7 < y - 2 < 5$$
$$-7 + 2 < y - 2 + 2 < 5 + 2$$
$$-5 < y < 7$$
The solution is $(-5, 7)$.

21.
$$8 \leq 3r + 1 \leq 16$$
$$8 - 1 \leq 3r \leq 16 - 1$$
$$7 \leq 3r \leq 15$$
$$\frac{7}{3} \leq r \leq 5$$
The solution is $\left[\frac{7}{3}, 5\right]$.

23.
$$-4 \leq \frac{2k - 1}{3} \leq 2$$
$$-4(3) \leq 3\left(\frac{2k - 1}{3}\right) \leq 2(3)$$
$$-12 \leq 2k - 1 \leq 6$$
$$-12 + 1 \leq 2k \leq 6 + 1$$
$$-11 \leq 2k \leq 7$$
$$-\frac{11}{2} \leq k \leq \frac{7}{2}$$
The solution is $\left[-\frac{11}{2}, \frac{7}{2}\right]$.

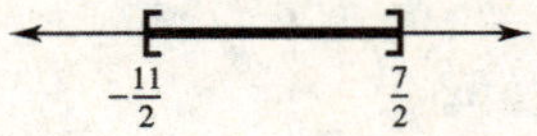

25.
$$\frac{3}{5}(2p+3) \geq \frac{1}{10}(5p+1)$$
$$10 \cdot \frac{3}{5}(2p+3) \geq 10 \cdot \frac{1}{10}(5p+1)$$
$$6(2p+3) \geq 5p+1$$
$$12p+18 \geq 5p+1$$
$$7p \geq -17$$
$$p \geq -\frac{17}{7}$$

The solution is $\left[-\frac{17}{7}, \infty\right)$.

27. $x \geq 2$

29. $-3 < x \leq 5$

31. $C = 50x + 6000$; $R = 65x$
To at least break even, $R \geq C$.
$65x \geq 50x + 6000$
$15x \geq 6000 \Rightarrow x \geq 400$
The number of units of wire must be in the interval $[400, \infty)$.

33. $C = 85x + 1000$; $R = 105x$
$R \geq C$
$105x \geq 85x + 1000$
$20x \geq 1000$
$$x \geq \frac{1000}{20} \Rightarrow x \geq 50$$
x must be in the interval $[50, \infty)$.

35. $C = 1000x + 5000$; $R = 900x$
$R \geq C$
$900x \geq 1000x + 5000$
$-100x \geq 5000$
$$x \leq \frac{5000}{-100} \Rightarrow x \leq -50$$
It is impossible to break even.

37. $|p| > 7$
$p < -7$ or $p > 7$
The solution is $(-\infty, -7)$ or $(7, \infty)$.

39. $|r| \leq 5 \Rightarrow -5 \leq r \leq 5$
The solution is $[-5, 5]$.

41. $|b| > -5$
The absolute value of a number is always nonnegative. Therefore, $|b| > -5$ is always true, so the solution is the set of all real numbers.

43.
$$\left|x - \frac{1}{2}\right| < 2$$
$$-2 < x - \frac{1}{2} < 2$$
$$-\frac{3}{2} < x < \frac{5}{2}$$

The solution is $\left(-\frac{3}{2}, \frac{5}{2}\right)$.

45. $|8b + 5| \geq 7$
$8b + 5 \leq -7$ or $8b + 5 \geq 7$
$8b \leq -12$ or $8b \geq 2$
$b \leq -\frac{3}{2}$ or $b \geq \frac{1}{4}$
The solution is $\left(-\infty, -\frac{3}{2}\right]$ or $\left[\frac{1}{4}, \infty\right)$.

47. $|T - 83| \leq 7$
$-7 \leq T - 83 \leq 7$
$76 \leq T \leq 90$

49. $|T - 61| \leq 21$
$-21 \leq T - 61 \leq 21$
$40 \leq T \leq 82$

51. $|R_L - 26.75| \le 1.42$
$|R_E - 38.75| \le 2.17$

a. $|R_L - 26.75| \le \pm 1.42 \Rightarrow$
$-1.42 \le R_L - 26.75 \le 1.42 \Rightarrow$
$25.33 \le R_L \le 28.17$
$|R_E - 38.75| \le 2.17 \Rightarrow$
$-2.17 \le R_E - 38.75 \le 2.17 \Rightarrow$
$36.58 \le R_E \le 40.92$

b. $225(25.33) \le T_L \le 225(28.17)$
$5699.25 \le T_L \le 6338.25$
$225(36.58) \le T_E \le 225(40.92)$
$8230.5 \le T_E \le 9207$

53. $35 \le B \le 43$

55. The six income ranges are:
$0 < x \le 8700$
$8700 < x \le 35,350$
$35,350 < x \le 85,650$
$85,650 < x \le 178,650$
$178,650 < x \le 388,350$
$x > 388,350$

Section 2.5 Polynomial and Rational Inequalities

1. $(x+4)(2x-3) \le 0$
Solve the corresponding equation.
$(x+4)(2x-3) = 0$
$x+4=0$ or $2x-3=0$
$x=-4$ $\quad x=\frac{3}{2}$
Note that because the inequality symbol is "≤," –4 and $\frac{3}{2}$ are solutions of the original inequality. These numbers separate the number line into three regions.

A B C
-5 -4 -3 -2 -1 0 1 2 3 4

In region A, let $x = -6$:
$(-6+4)[2(-6)-3] = 30 > 0$.
In region B, let $x = 0$:
$(0+4)[2(0)-3] = -12 < 0$.
In region C, let $x = 2$:
$(2+4)[2(2)-3] = 6 > 0$.
The only region where $(x+4)(2x-3)$ is negative is region B, so the solution is $\left[-4, \frac{3}{2}\right]$. To graph this solution, put brackets at –4 and $\frac{3}{2}$ and draw a line segment between these two endpoints.

–4 $\frac{3}{2}$

3. $r^2 + 4r > -3$
Solve the corresponding equation.
$r^2 + 4r = -3$
$r^2 + 4r + 3 = 0$
$(r+1)(r+3) = 0$
$r+1=0$ or $r+3=0$
$r=-1$ or $r=-3$
Note that because the inequality symbol is ">," –1 and –3 are not solutions of the original inequality.

A B C
-5 -4 -3 -2 -1 0 1 2 3 4

In region A, let $r = -4$:
$(-4)^2 + 4(-4) = 0 > -3$.
In region B, let $r = -2$:
$(-2)^2 + 4(-2) = -4 < -3$.
In region C, let $r = 0$:
$0^2 + 4(0) = 0 > -3$.
The solution is $(-\infty, -3)$ or $(-1, \infty)$.
To graph the solution, put a parenthesis at –3 and draw a ray extending to the left, and put a parenthesis at –1 and draw a ray extending to the right.

–3 –1

5. $4m^2 + 7m - 2 \le 0$
Solve the corresponding equation.

$$4m^2 + 7m - 2 = 0$$
$$(4m - 1)(m + 2) = 0$$
$$4m - 1 = 0 \quad \text{or} \quad m + 2 = 0$$
$$m = \frac{1}{4} \quad \text{or} \quad m = -2$$

Because the inequality symbol is "≤" $\frac{1}{4}$ and −2 are solutions of the original inequality.

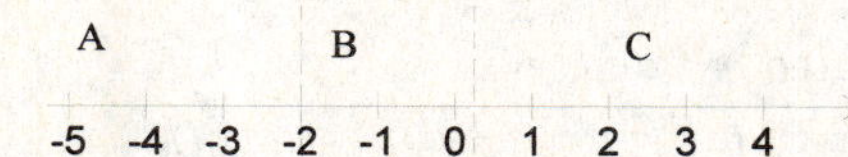

In region A, let $m = -3$:
$4(-3)^2 + 7(-3) - 2 = 13 > 0$.
In region B, let $m = 0$:
$4(0)^2 + 7(0) - 2 = -2 < 0$.
In region C, let $m = 1$:
$4(1)^2 + 7(1) - 2 = 9 > 0$.
The solution is $\left[-2, \frac{1}{4}\right]$.

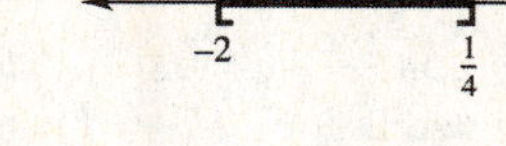

7. $4x^2 + 3x - 1 > 0$
Solve the corresponding equation.

$$4x^2 + 3x - 1 = 0$$
$$(4x - 1)(x + 1) = 0$$
$$4x - 1 = 0 \quad \text{or} \quad x + 1 = 0$$
$$x = \frac{1}{4} \quad \text{or} \quad x = -1$$

Note that $\frac{1}{4}$ and −1 are not solutions of the original inequality.

A B C
-5 -4 -3 -2 -1 0 1 2 3 4

In region A, let $x = -2$:
$4(-2)^2 + 3(-2) - 1 = 9 > 0$.
In region B, let $x = 0$:
$4(0)^2 + 3(0) - 1 = -1 < 0$.
In region C, let $x = 1$:
$4(1)^2 + 3(1) - 1 = 6 > 0$.
The solution is $(-\infty, -1)$ or $\left(\frac{1}{4}, \infty\right)$.

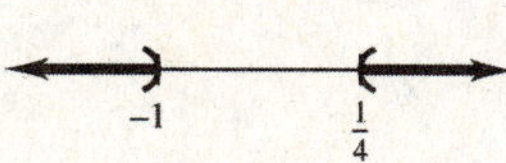

9. $x^2 \le 36$
Solve the corresponding equation.
$x^2 = 36 \Rightarrow x = \pm 6$

A B C
-7 -6 -5 -4 -3 -2 -1 0 1 2 3 4 5 6

For region A, let $x = -7$: $(-7)^2 = 49 > 36$.
For region B, let $x = 0$: $0^2 = 0 < 36$.
For region C, let $x = 7$: $7^2 = 49 > 36$.
Both endpoints are included. The solution is [−6, 6].

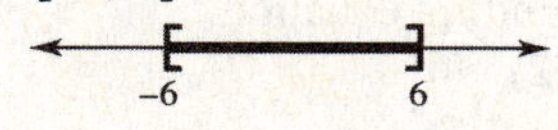

11. $p^2 - 16p > 0$
Solve the corresponding equation.
$p^2 - 16p = 0 \Rightarrow p(p - 16) = 0 \Rightarrow$
$p = 0$ or $p = 16$
Since the inequality is ">", 0 and 16 are not solutions of the original inequality.

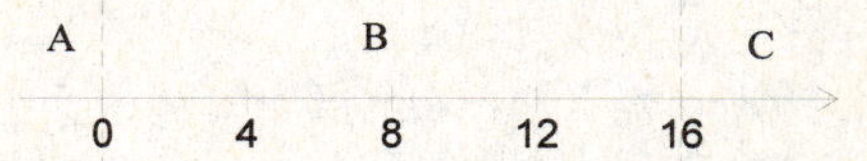

(continued next page)

For region A, let $p = -1$:
$(-1)^2 - 16(-1) = 17 > 0$.
For region B, let $p = 1$:
$1^2 - 16(1) = -15 < 0$.
For region C, let $p = 17$:
$17^2 - 16(17) = 17 > 0$.
The solution is $(-\infty, 0)$ or $(16, \infty)$.

13. $x^3 - 9x \geq 0$

Solve the corresponding equation.

$$x^3 - 9x = 0$$
$$x\left(x^2 - 9\right) = 0$$
$$x(x+3)(x-3) = 0$$

$x = 0$ or $x = -3$ or $x = 3$

Note that 0, –3, and 3 are all solutions of the original inequality.

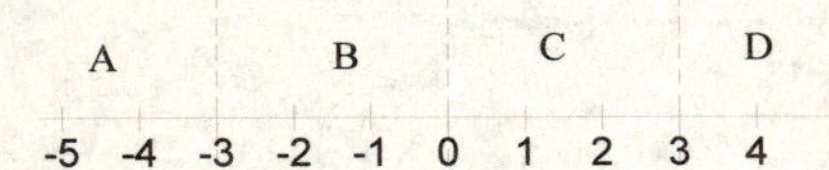

In region A, let $x = -4$:
$(-4)^3 - 9(-4) = -28 < 0$.
In region B, let $x = -1$:
$(-1)^3 - 9(-1) = 8 > 0$.
In region C, let $x = 1$:
$(1)^3 - 9(1) = -8 < 0$
In region D, let $x = 4$:
$4^3 - 9(4) = 28 > 0$.
The solution is $[-3, 0]$ or $[3, \infty)$.

15. $(x + 7)(x + 2)(x - 2) \geq 0$

Solve the corresponding equation.

$(x + 7)(x + 2)(x - 2) = 0$

$x + 7 = 0$ or $x + 2 = 0$ or $x - 2 = 0$

$x = -7$ or $x = -2$ or $x = 2$

Note that –7, –2 and 2 are all solutions of the original inequality.

(*continued next page*)

A B C D
–8 –7 –6 –5 –4 –3 –2 –1 0 1 2 3

In region A, let $x = -8$:
$(-8 + 7)(-8 + 2)(-8 - 2) = -60 < 0$
In region B, let $x = -4$:
$(-4 + 7)(-4 + 2)(-4 - 2) = 36 > 0$
In region C, let $x = 0$:
$(0 + 7)(0 + 2)(0 - 2) = -28 < 0$
In region D, let $x = 3$:
$(3 + 7)(3 + 2)(3 - 2) = 50 > 0$
The solution is $[-7, -2]$ or $[2, \infty)$.

17. $(x+5)\left(x^2 - 2x - 3\right) < 0$

Solve the corresponding equation.

$(x+5)\left(x^2 - 2x - 3\right) = 0$

$(x + 5)(x + 1)(x - 3) = 0$

$x + 5 = 0$ or $x + 1 = 0$ or $x - 3 = 0$

$x = -5$ or $x = -1$ or $x = 3$

Note that –5, –1 and 3 are not solutions of the original inequality.

A B C D
–7 –6 –5 –4 –3 –2 –1 0 1 2 3 4

In region A, let $x = -6$:
$(-6+5)\left[(-6)^2 - 2(-6) - 3\right] = (-1)(45)$
$= -45 < 0$
In region B, let $x = -2$:
$(-2+5)\left[(-2)^2 - 2(-2) - 3\right] = 3(5) = 15 > 0$
In region C, let $x = 0$:
$(0+5)\left[(0)^2 - 2(0) - 3\right] = 5(-3) = -15 < 0$
In region D, let $x = 4$:
$(4+5)\left[(4)^2 - 2(4) - 3\right] = 9(5) = 45 > 0$
The solution is $(\infty, -5)$ or $(-1, 3)$.

19. $6k^3 - 5k^2 < 4k \Rightarrow 6k^3 - 5k^2 - 4k < 0$

Solve the corresponding equation.

$$6k^3 - 5k^2 - 4k = 0$$
$$k\left(6k^2 - 5k - 4\right) = 0$$
$$k(3k - 4)(2k + 1) = 0$$

$k = 0$ or $k = \frac{4}{3}$ or $k = -\frac{1}{2}$

Note that 0, $\frac{4}{3}$, and $-\frac{1}{2}$ are not solutions of the original inequality.

A B C D
-5 -4 -3 -2 -1 0 1 2 3 4

(*continued next page*)

In region A, let $k = -1$:
$6(-1)^3 - 5(-1)^2 - 4(-1) = -7 < 0$

In region B, let $k = -\frac{1}{4}$:

$6\left(-\frac{1}{4}\right)^3 - 5\left(-\frac{1}{4}\right)^2 - 4\left(-\frac{1}{4}\right) = \frac{19}{32} > 0;$

In region C, let $k = 1$:
$6(1)^3 - 5(1)^2 - 4(1) = -3 < 0$

In region D, let $k = 10$:
$6(10)^3 - 5(10)^2 - 4(10) = 5460$

The given inequality is true in regions A and C.

The solution is $\left(-\infty, -\frac{1}{2}\right)$ or $\left(0, \frac{4}{3}\right)$.

21. The inequality $p^2 < 16$ should be rewritten as $p^2 - 16 < 0$ and solved by the method shown in this section for solving quadratic inequalities. This method will lead to the correct solution (–4, 4). The student's method and solution are incorrect.

23. To solve $.5x^2 - 1.2x < .2$, write the inequality as $.5x^2 - 1.2x - .2 < 0$. Graph the equation $y = .5x^2 - 1.2x - .2$. Enter this equation as y_1 and use $-4 \le x \le 6$ and $-5 \le y \le 5$. On the CALC menu, use "zero" to find the x-values where the graph crosses the x-axis. These values are $x = -.1565$ and $x = 2.5565$. The graph is below the x-axis between these two values. The solution of the inequality is (–.1565, 2.5565).

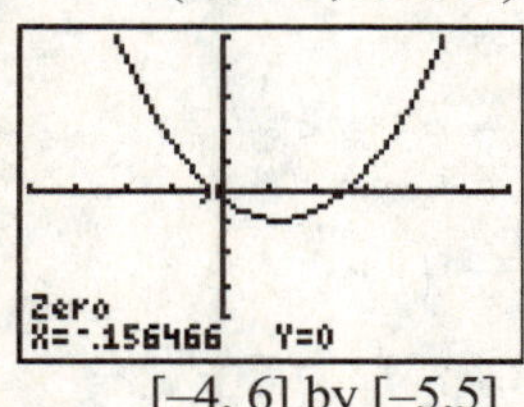

Zero
X=2.556466 Y=1E-13

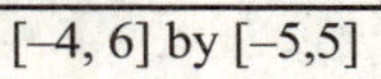
[–4, 6] by [–5,5]

25. To solve $x^3 - 2x^2 - 5x + 7 \ge 2x + 1$, graph $y_1 = x^3 - 2x^2 - 5x + 7$ and $y_2 = 2x + 1$ in the window [–5, 5] by[–10, 10]. On the CALC menu, use "intersect" to find the x-values where the graphs intersect. These values are $x = -2.2635$, $x = .7556$ and $x = 3.5079$. The graph of y_1 is above the graph of y_2 for [–2.2635, .7556] or [3.5079, ∞).

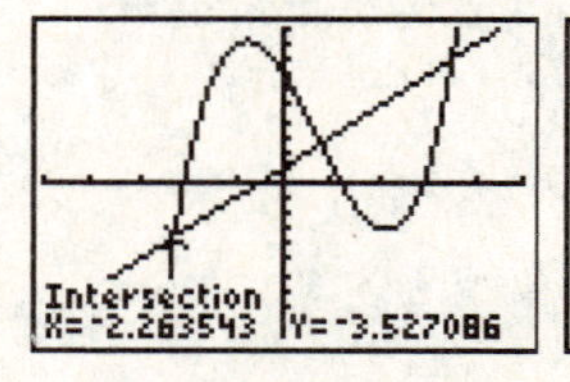

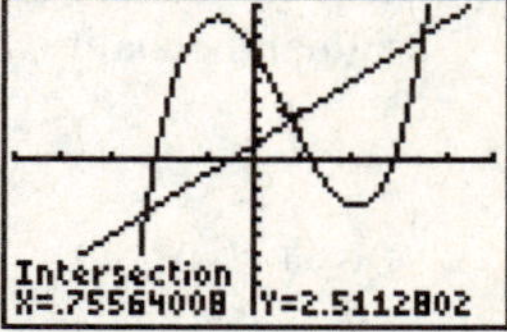

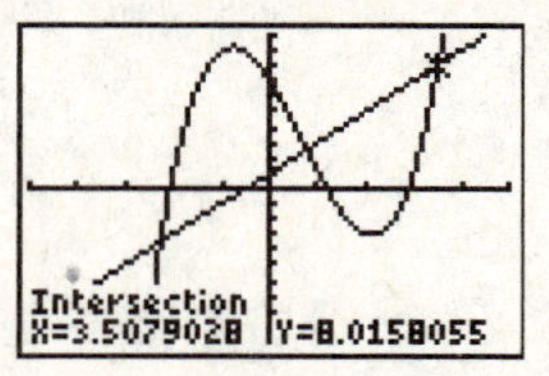

27. To solve $2x^4 + 3x^3 < 2x^2 + 4x - 2$, graph $y_1 = 2x^4 + 3x^3$ and $y_2 = 2x^2 + 4x - 2$ in the window [–2, 2] by[–5, 5]. On the CALC menu, use "intersect" to find the x-values where the graphs intersect. These values are $x = .5$ and $x = .8393$. The graph of y_1 is below the graph of y_2 to the right of .5 and to the left of .8393. The solution of the inequality is (.5, .8393).

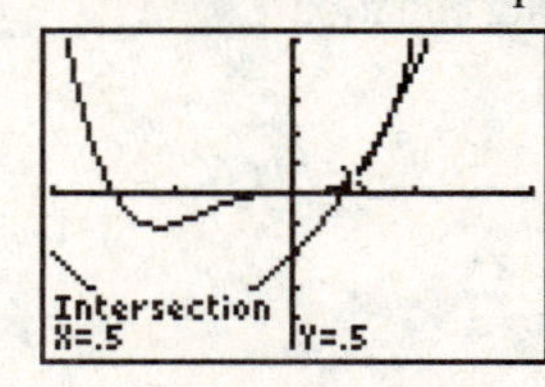

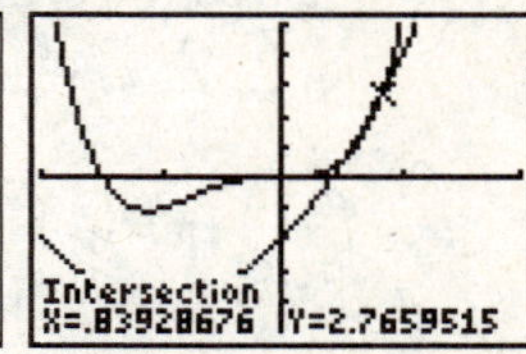

29. $\dfrac{r-4}{r-1} \geq 0$

Solve the corresponding equation.

$\dfrac{r-4}{r-1} = 0$

The quotient can change sign only when the numerator is 0 or the denominator is 0. The numerator is 0 when $r = 4$. The denominator is 0 when $r = 1$. Note that 4 is a solution of the original inequality, but 1 is not.

A B C

-2 -1 0 1 2 3 4 5

In region A, let $r = 0$:

$\dfrac{0-4}{0-1} = 4 > 0$.

In region B, let $r = 2$:

$\dfrac{2-4}{2-1} = -2 < 0$.

In region C, let $r = 5$:

$\dfrac{5-3}{5-1} = \dfrac{1}{4} > 0$.

The given inequality is true in regions A and C, so the solution is $(-\infty, 1)$ or $[4, \infty)$.

31. $\dfrac{a-2}{a-5} < -1$

Solve the corresponding equation.

$$\frac{a-2}{a-5} = -1$$

$$\frac{a-2}{a-5} + 1 = 0$$

$$\frac{a-2}{a-5} + \frac{a-5}{a-5} = 0$$

$$\frac{2a-7}{a-5} = 0$$

The numerator is 0 when $a = \dfrac{7}{2}$. The denominator is 0 when $a = 5$. Note that $\dfrac{7}{2}$ and 5 are not solutions of the original inequality.

A B C

-3 -2 -1 0 1 2 3 4 5 6 7 8

In region A, let $a = 0$:

$\dfrac{0-2}{0-5} = \dfrac{2}{5} > -1$.

In region B, let $a = 4$:

$\dfrac{4-2}{4-5} = \dfrac{2}{-1} = -2 < -1$.

In region C, let $a = 10$:

$\dfrac{10-2}{10-5} = \dfrac{8}{5} > -1$.

The solution is $\left(\dfrac{7}{2}, 5\right)$.

33. $\dfrac{1}{p-2} < \dfrac{1}{3}$

Solve the corresponding equation.

$$\frac{1}{p-2} = \frac{1}{3}$$

$$\frac{1}{p-2} - \frac{1}{3} = 0$$

$$\frac{3-(p-2)}{3(p-2)} = 0$$

$$\frac{3-p+2}{3(p-2)} = 0$$

$$\frac{5-p}{3(p-2)} = 0$$

The numerator is 0 when $p = 5$. The denominator is 0 when $p = 2$. Note that 2 and 5 are not solutions of the original inequality.

A B C

-2 -1 0 1 2 3 4 5 6

In region A, let $p = 0$: $\dfrac{1}{0-2} = -\dfrac{1}{2} < \dfrac{1}{3}$.

In region B, let $p = 3$: $\dfrac{1}{3-2} = 1 > \dfrac{1}{3}$.

In region C, let $p = 6$: $\dfrac{1}{6-2} = \dfrac{1}{4} < \dfrac{1}{3}$.

The solution is $(-\infty, 2)$ or $(5, \infty)$.

35. $\dfrac{5}{p+1} > \dfrac{12}{p+1}$

Solve the corresponding equation.

$$\frac{5}{p+1} = \frac{12}{p+1}$$

$$\frac{5}{p+1} - \frac{12}{p+1} = 0$$

$$\frac{-7}{p+1} = 0$$

The numerator is never 0. The denominator is 0 when $p = -1$. Therefore, in this case, we separate the number line into only two regions.

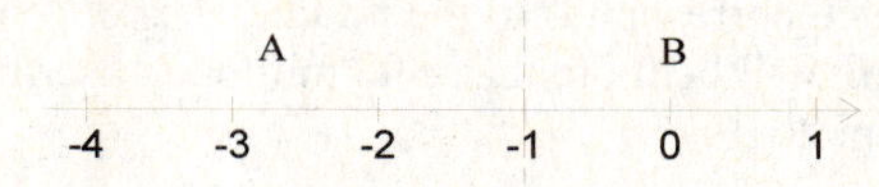

In region A, let $p = -2$:

$$\frac{5}{-2+1} = -5$$

$$\frac{12}{-2+1} = -12$$

$$-5 > -12$$

In region B, let $p = 0$:

$$\frac{5}{0+1} = 5$$

$$\frac{12}{0+1} = 12$$

$$12 > 5$$

Therefore, the given inequality is true in region A. The only endpoint, –1, is not included because the symbol is ">." Therefore, the solution is $(-\infty, -1)$.

37. $\dfrac{x^2 - x - 6}{x} < 0$

Solve the corresponding equation.

$$\frac{x^2 - x - 6}{x} = 0$$

$$x^2 - x - 6 = 0 \quad \text{or} \quad x = 0$$

$$(x-3)(x+2) = 0 \quad \text{or} \quad x = 0$$

$$x - 3 = 0 \quad \text{or} \quad x + 2 = 0 \quad \text{or} \quad x = 0$$

$$x = 3 \quad \text{or} \quad x = -2 \quad \text{or} \quad x = 0$$

Note that –2, 0 and 3 are not solutions of the original inequality.

In region A, let $x = -3$:

$$\frac{(-3)^2 - (-3) - 6}{-3} = \frac{9+3-6}{-3} = \frac{6}{-3} = -2 < 0.$$

In region B, let $x = -1$:

$$\frac{(-1)^2 - (-1) - 6}{-1} = \frac{1+1-6}{-1} = \frac{-4}{-1} = 4 > 0$$

In region C, let $x = 1$:

$$\frac{1^2 - 1 - 6}{1} = -6 < 0.$$

In region D, let $x = 4$:

$$\frac{4^2 - 4 - 6}{4} = \frac{16-10}{4} = \frac{6}{4} = \frac{3}{2} > 0.$$

The solution is $(-\infty, -2)$ or $(0, 3)$.

39. To solve $\dfrac{2x^2 + x - 1}{x^2 - 4x + 4} \le 0$, break the inequality into two inequalities $2x^2 + x - 1 \le 0$ and $x^2 - 4x + 4 \le 0$. Graph the equations $y = 2x^2 + x - 1$ and $y = x^2 - 4x + 4$. Enter these equations as y_1 and y_2, and use $-3 < x < 3$ and $-2 < y < 2$. On the CALC menu, use "zero" to find the x-values where the graphs cross the x-axis. These values for y_1 are $x = -1$ and $x = .5$. The graph of y_1 is below the x-axis to the right of –1 and to the left of .5. The graph of y_2 is never below the x-axis. The solution of the inequality is $[-1, .5]$.

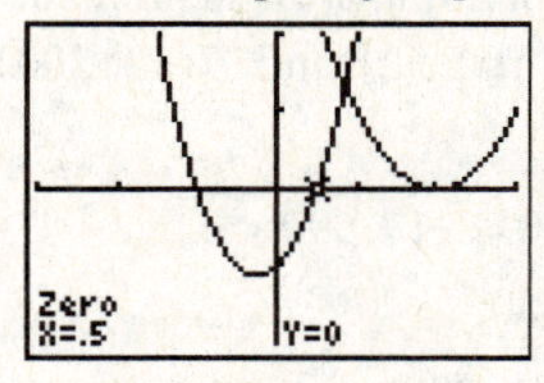

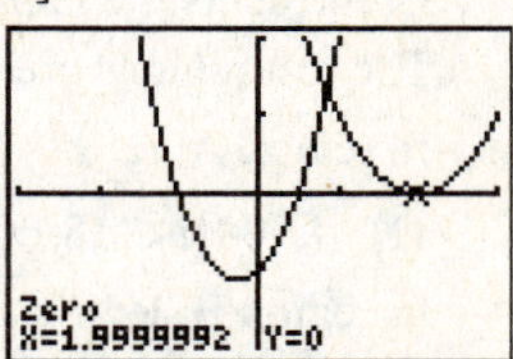

41. $P = 2x^2 - 12x - 32$
The company makes a profit when
$2x^2 - 12x - 32 > 0$.
Solve the corresponding equation.

$$2x^2 - 12x - 32 = 0$$
$$2(x^2 - 6x - 16) = 0$$
$$(x+2)(x-8) = 0 \Rightarrow x = -2 \text{ or } x = 8$$

The test regions are $A(-\infty, -2)$, $B(-2, 8)$, and $C(8, \infty)$. Region A makes no sense in this context, so we ignore this. Test a number from regions B and C in the original inequality.
For region B, let $x = 0$.

$$2(0)^2 - 12(0) - 32 = -32 < 0$$

For region C, let $x = 10$.

$$2(10)^2 - 12(10) - 32 = 48 > 0$$

The numbers in region C satisfy the inequality. The company makes a profit when the amount spent on advertising in hundreds of thousands of dollars is in the interval $(8, \infty)$.

43. $P = x^2 + 300x - 18{,}000$
The complex makes a profit when
$x^2 + 300x - 18{,}000 > 0$.
Solve the corresponding equation.

$$0 = x^2 + 300x - 18{,}000$$
$$x = \frac{-300 \pm \sqrt{(300)^2 - 4(1)(-18{,}000)}}{2(1)}$$
$$x \approx 51.25 \text{ or } x \approx -351.25$$

We only consider positive values of x because x represents the number of apartments rented.
The test regions are $A(0, 52)$ and $B(52, 200)$.
In region A, let $x = 1$:

$$(1)^2 + 300(1) - 18{,}000 = -17{,}699 < 0.$$

In region B, let $x = 100$:

$$(100)^2 + 300(100) - 18{,}000 = 22{,}000 > 0.$$

The complex makes a profit when the number of units rented is between 52 and 200, inclusive, or when x is in the interval [52, 200].

45. $.79x^2 + 5.4x + 178 > 300$
Use a graphing calculator to solve
$.79x^2 + 5.4x + 178 = 300 \Rightarrow$
$.79x^2 + 5.4x - 122 = 0$

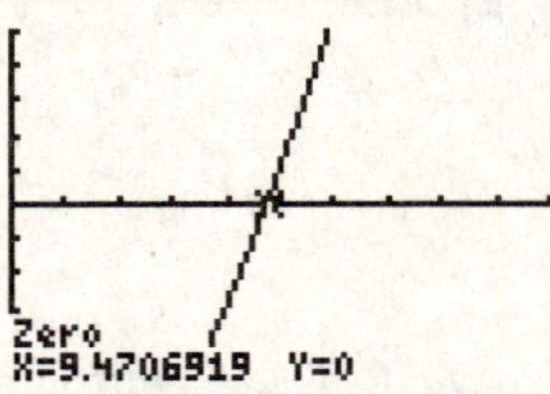

[0, 20] by [–50, 50]

The graph lies above the x-axis for $x > 9.47$, which corresponds to the middle of 2009. Thus, there will be more than 300 million subscribers from 2010.

47. $-.2x^2 + 3.44x + .16 > 13$
Use a graphing calculator to solve
$-.2x^2 + 3.44x + .16 > 13$
$-.2x^2 + 3.44x - 12.84 > 0$

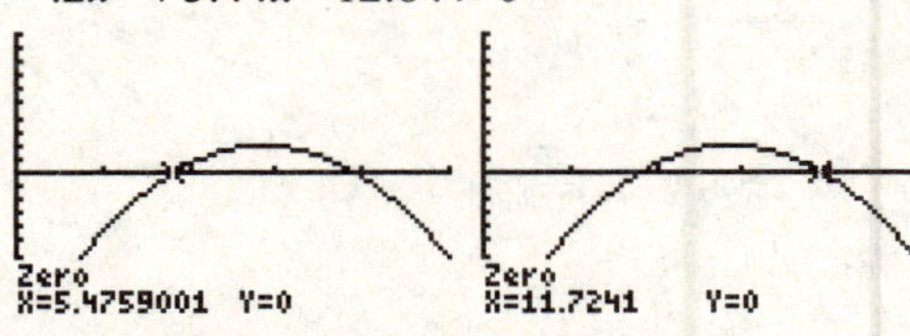

[0, 15] by [–10, 10]

The graph is above the x-axis for $5.48 \le x \le 11.72$. These values correspond to the years 2006 and 2011. There were greater than \$13 trillion of outstanding mortgage debt in the years 2006–2011, inclusive or [2006, 2011].

Chapter 2 Review Exercises

1. $y = x^2 - 2x - 5$

$(-2, 3)$:

$(-2)^2 - 2(-2) - 5 = 4 + 4 - 5 = 3$

$(0, -5)$

$(0)^2 - 2(0) - 5 = 0 - 0 - 5 = -5$

$(2, -3)$:

$(2)^2 - 2(2) - 5 = 4 - 4 - 5 = -5 \neq -3$

$(3, -2)$:

$(3)^2 - 2(3) - 5 = 9 - 6 - 5 = -2$

$(4, 3)$:

$(4)^2 - 2(4) - 5 = 16 - 8 - 5 = 3$

$(7, 2)$:

$(7)^2 - 2(7) - 5 = 49 - 14 - 5 = 30 \neq 2$

Solutions are (–2, 3), (0, –5), (3, –2), (4, 3).

2. $x - y = 5$

$(-2, 3): -2 - 3 = -5 \neq 5$

$(0, -5): 0 - (-5) = 0 + 5 = 5$

$(2, -3): 2 - (-3) = 2 + 3 = 5$

$(3, -2): 3 - (-2) = 3 + 2 = 5$

$(4, 3): 4 - 3 = 1 \neq 5$

$(7, 2): 7 - 2 = 5$

Solutions are (0, –5), (2, –3), (3, –2), (7, 2).

3. $5x - 3y = 15$

First, we find the y-intercept. If $x = 0$, $y = -5$, so the y-intercept is –5. Next we find the x-intercept. If $y = 0$, $x = 3$, so the x-intercept is 3. Using these intercepts, we graph the line.

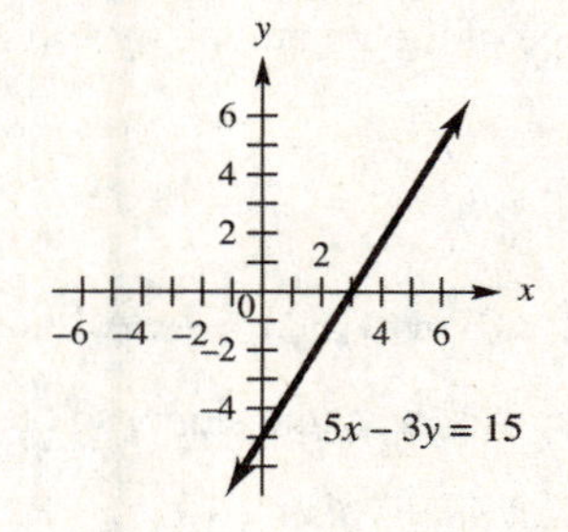

4. $2x + 7y - 21 = 0$

First we find the y-intercept. If $x = 0$, $y = 3$, so the y-intercept is 3. Next we find the x-intercept. If $y = 0$, $x = \frac{21}{2}$, so the x-intercept is $\frac{21}{2}$. Using these intercepts, we graph the line.

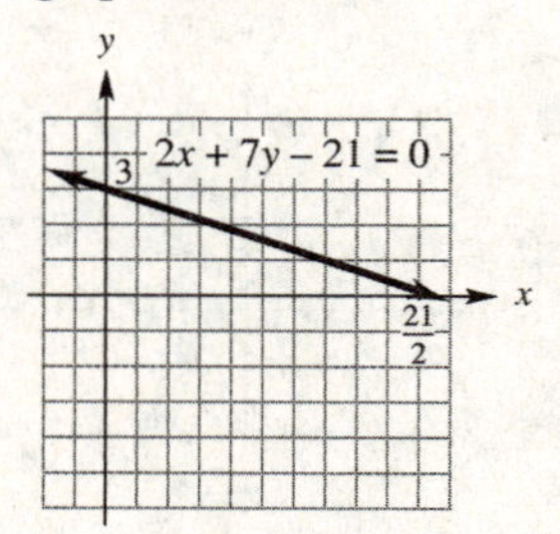

5. $y + 3 = 0$

The equation may be rewritten as $y = -3$. The graph of $y = -3$ is a horizontal line with y-intercept of –3.

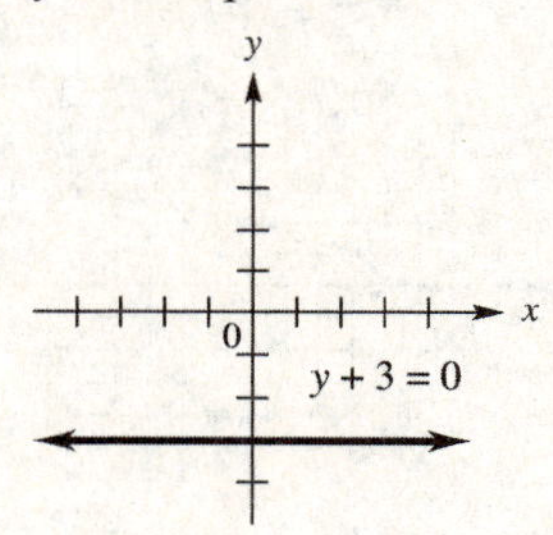

6. $y - 2x = 0$

First, we find the y-intercept. If $x = 0$, $y = 0$, so the y-intercept is 0. Since the line passes through the origin, the x-intercept is also 0. We find another point on the line by arbitrarily choosing a value for x. Let $x = 2$. Then $y - 2(2) = 0$, or $y = 4$. The point with coordinates (2, 4) is on the line. Using this point and the origin, we graph the line.

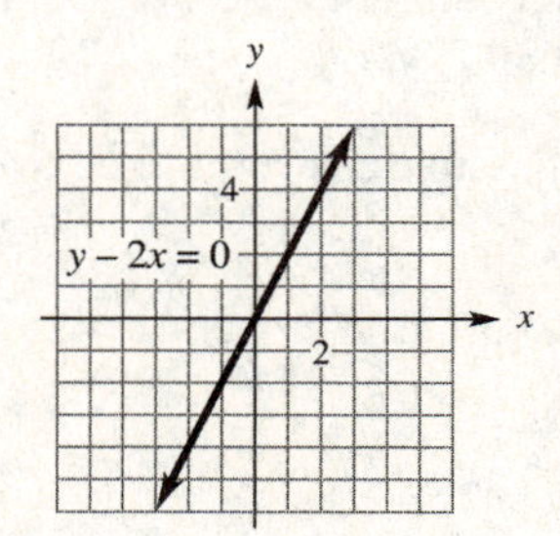

7. $y = .25x^2 + 1$
First we find the y-intercept. If $x = 0$, $y = .25(0)^2 + 1 = 1$, so the y-intercept is 1. Next we find the x-intercepts. If $y = 0$, $0 = .25x^2 + 1 \Rightarrow .25x^2 = -1 \Rightarrow x = \sqrt{-4}$, not a real number. There are no x-intercepts.
Make a table of points and plot them.

x	$.25x^2 + 1$
–4	5
–2	2
0	1
2	2
4	5

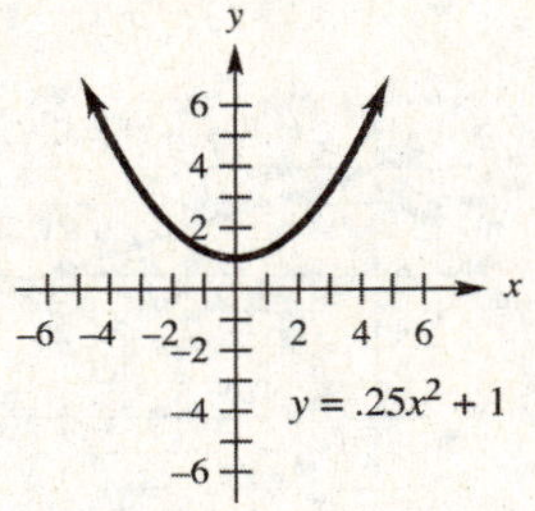

$y = .25x^2 + 1$

8. $y = \sqrt{x+4}$
Make a table of points and plot them.

x	$\sqrt{x+4}$
–4	0
–3	1
0	2
5	3

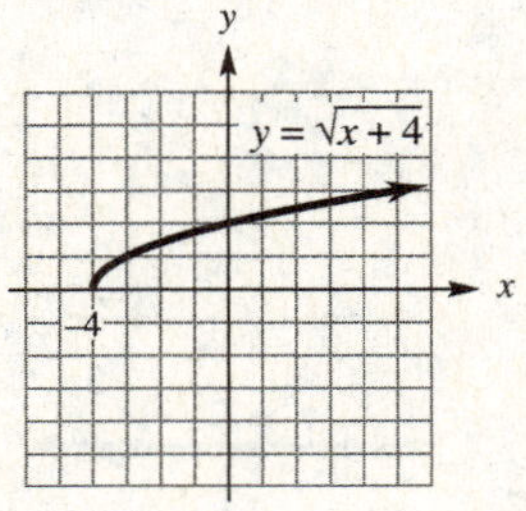

$y = \sqrt{x+4}$

9. a. The temperature was over 55° from about 11:30 A.M. to about 7:30 P.M.

b. The temperature was below 40° from midnight until about 5 A.M., and after about 10:30 P.M.

10. At noon in Bratenahl the temperature was about 57°. The temperature in Greenville is 57° when the temperature in Bratenahl is 50°, or at about 10:30 A.M. and 8:30 P.M.

11. Answers vary. A possible answer is "rise over run".

12. Through (–1, 3) and (2, 6)
$$\text{slope} = \frac{\Delta y}{\Delta x} = \frac{6-3}{2-(-1)} = \frac{3}{3} = 1$$

13. Through (4, –5) and (1, 4)
$$\text{slope} = \frac{-5-4}{4-1} = \frac{-9}{3} = -3$$

14. Through (8, –3) and the origin
The coordinates of the origin are (0, 0).
$$\text{slope} = \frac{-3-0}{8-0} = -\frac{3}{8}$$

15. Through (8, 2) and (0, 4)
$$\text{slope} = \frac{4-2}{0-8} = \frac{2}{-8} = -\frac{1}{4}$$

In exercises 16 and 17, we give the solution by rewriting the equation in slope-intercept form. Alternatively, the solution can be obtained by determining two points on the line and then using the definition of slope.

16. $3x + 5y = 25$
First we solve for y.
$$5y = -3x + 25 \Rightarrow y = -\frac{3}{5}x + 5$$
When the equation is written in slope-intercept form, the coefficient of x gives the slope. The slope is $-\frac{3}{5}$.

17. $6x - 2y = 7$
First we solve for y.
$$6x - 2y = 7 \Rightarrow 6x - 7 = 2y \Rightarrow 3x - \frac{7}{2} = y$$
The coefficient of x gives the slope, so the slope is 3.

18. $x - 2 = 0$
The graph of $x - 2 = 0$ is a vertical line. Therefore, the slope is undefined.

19. $y = -4$
The graph of $y = -4$ is a horizontal line. Therefore, the slope is 0.

20. Parallel to $3x + 8y = 0$
First, find the slope of the given line by solving for y.
$$8y = -3x \Rightarrow y = -\frac{3}{8}x$$
The slope is the coefficient of x, $-\frac{3}{8}$. A line parallel to this line has the same slope, so the slope of the parallel line is also $-\frac{3}{8}$.

21. Perpendicular to $x = 3y$
First, find the slope of the given line by solving for y: $y = \frac{1}{3}x$

The slope of this line is the coefficient of x, $\frac{1}{3}$.
The slope of a line perpendicular to this line is the negative reciprocal of this slope, so the slope of the perpendicular line is –3.

22. Through (0, 5) with $m = -\frac{2}{3}$

Since $m = -\frac{2}{3} = \frac{-2}{3}$, we start at the point with coordinates (0, 5) and move 2 units down and 3 units to the right to obtain a second point on the line. Using these two points, we graph the line.

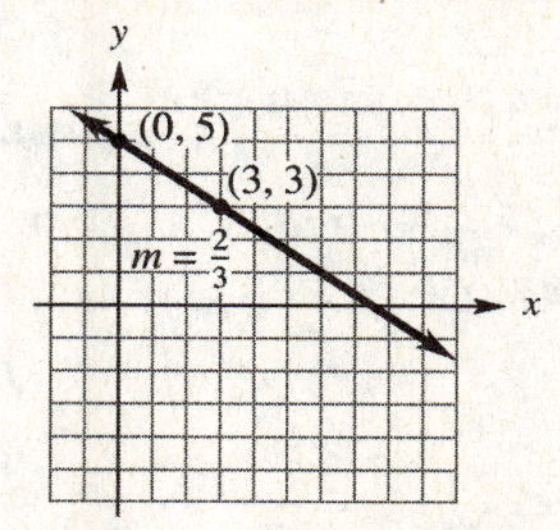

23. Through (–4, 1) with m = 3
Since $m = 3 = \frac{3}{1}$, we start at the point with coordinates (–4, 1) and move 3 units up and 1 unit to the right to obtain a second point on the line. Using these two points, we graph the line.

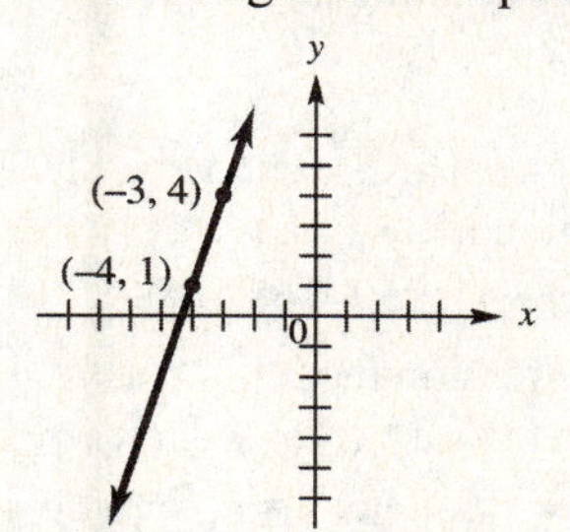

24. Answers vary. One example is:
You need two points; one point and the slope; the y-intercept and the slope.

25. Through (5, –1), slope $\frac{2}{3}$
Use the point slope form with $x_1 = 5$, $y_1 = -1$, and $m = \frac{2}{3}$.

$$y - y_1 = m(x - x_1)$$
$$y - (-1) = \frac{2}{3}(x - 5)$$
$$y + 1 = \frac{2}{3}x - \frac{10}{3}$$

Multiplying by 3 gives

$$3y + 3 = 2x - 10$$
$$3y = 2x - 13$$

26. Through (8, 0), $m = -\frac{1}{4}$

$$y - 0 = -\frac{1}{4}(x - 8)$$
$$4y = -1(x - 8)$$
$$4y = -x + 8$$

27. Through (5, –2) and (1, 3)

$$m = \frac{3 - (-2)}{1 - 5} = \frac{5}{-4} = -\frac{5}{4}$$
$$y - 3 = -\frac{5}{4}(x - 1)$$
$$4(y - 3) = -5(x - 1)$$
$$4y - 12 = -5x + 5$$
$$4y = -5x + 17$$

28. (2, –3) and (–3, 4)

$$m = \frac{-3 - 4}{2 - (-3)} = -\frac{7}{5}$$
$$y - (-3) = -\frac{7}{5}(x - 2)$$
$$5(y + 3) = -7(x - 2)$$
$$5y + 15 = -7x + 14$$
$$5y = -7x - 1$$

29. Undefined slope, through (–1, 4)
This is a vertical line. Its equation is $x = -1$.

30. Slope 0, (–2, 5)
This is a horizontal line. Its equation is $y = 5$.

31. x-intercept –3, y-intercept 5
Use the points (–3, 0) and (0, 5).

$$m = \frac{5-0}{0-(-3)} = \frac{5}{3}$$
$$y = \frac{5}{3}x + 5$$
$$3y = 3(\tfrac{5}{3}x + 5)$$
$$3y = 5x + 15$$

32. x-intercept 3, y-intercept 2.
Use the points $(3, 0)$ and $(0, 2)$.

$$m = \frac{2-0}{0-3} = -\frac{2}{3}$$
$$y = -\frac{2}{3}x + 2$$
$$3y = 3\left(-\frac{2}{3}x + 2\right)$$
$$3y = -2x + 6$$
$$2x + 3y = 6$$

The answer is (d).

33. a. Let (x_1, y_1) be (5, 14.0) and (x_2, y_2) be (11, 17.3). Find the slope.

$$m = \frac{17.3 - 14.0}{11-5} = \frac{3.3}{6} = .55$$
$$y - 14 = .55(x-5)$$
$$y - 14 = .55x - 2.75$$
$$y = .55x + 11.25$$

b. The slope is positive because the amount of wheat exported is increasing.

c. The year 2014 corresponds to $x = 14$.
$$y = .55(14) + 11.25 = 18.95$$
If the linear trend continues, there will be 18.95 million hectoliters of fruit juice and wine exported in 2014.

34. a. Let (x_1, y_1) be (5, 9.7) and (x_2, y_2) be (10, 8.2). Find the slope.

$$m = \frac{8.2 - 9.7}{10-5} = \frac{-1.5}{5} \approx -.3$$
$$y - 9.7 = -.3(x-5)$$
$$y - 9.7 = -.3x + 1.5$$
$$y = -.3x + 11.2$$

b.

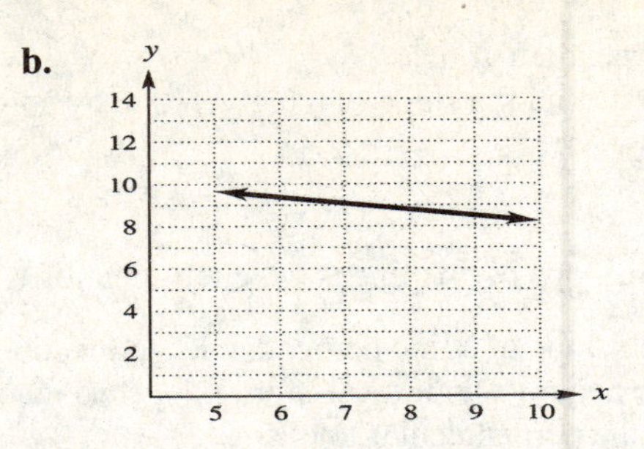

c. The year 2013 corresponds to $x = 13$.
$$y = -.3(13) + 11.2 = 7.3$$
If the linear trend continues, there will be about 7.3 billion pounds of fish and shellfish caught in 2013.

35. a. Let (x_1, y_1) be (0, 47059) and (x_2, y_2) be (10, 66249).
Find the slope.

$$m = \frac{66,249 - 47,059}{10-0} = \frac{19,190}{10} = 1919$$

Use the y intercept (0, 47059)
$$y = 1919x + 47,059$$

b.

```
LinReg
 y=ax+b
 a=1917.383117
 b=47051.25974
 r²=.9999959607
 r=.9999979803
```

Using a graphing calculator, the least squares regression line is
$$y = 1917.38x + 47,051.26.$$

c. The year 2011 corresponds to $x = 11$. Using the two-point model, we have
$$y = 1919(11) + 47,059 = 68,168.$$
Using the regression model, we have
$$y = 1917.38(11) + 47,051.26 \approx 68,142.44.$$
The two-point model is off by \$39, while the regression model is off by \$13.44, therefore the least-squares approximation is a better estimate.

d. The year 2015 corresponds to $x = 15$.
Regression model:
$$y = 1917.38(15) + 47,051.26 \approx 75,811.96$$
Thus, the compensation per full-time employee in the year 2015 will be about \$75,811.96.

36. a. $900 = 17.4x + 639 \Rightarrow 261 = 17.4x \Rightarrow 15 = x$
The weekly median wages for men will earn \$900 per week in the year 2000 + 15 = 2015.

b. $900 = 17.3x + 495 \Rightarrow 405 = 17.3x \Rightarrow$
$23.4 \approx x$
The weekly median wages for women will earn \$900 per week in the year 2000 + 23 = 2023.

37. a.

```
LinReg
 y=ax+b
 a=10.72
 b=98.8
 r²=.9088769377
 r=.9533503751
```

The least-squares regression line is $y = 10.72x + 98.8$.

b.

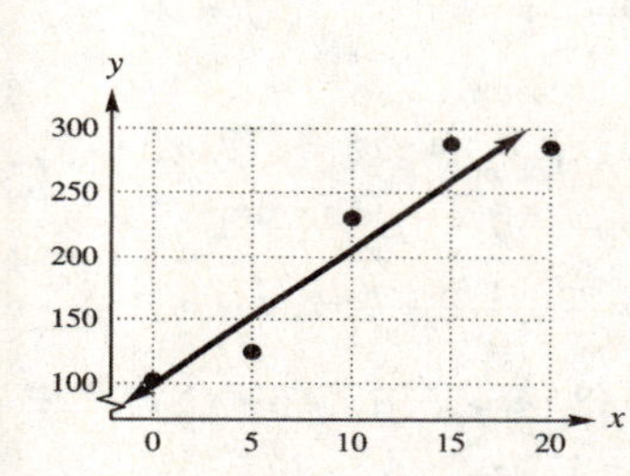

c. Yes; the line appears to fit.

d. The correlation coefficient is .953. This indicates that the line is a good fit.

38. a.

```
LinReg
 y=ax+b
 a=131.3722334
 b=2850.498994
 r²=.9760896701
 r=.9879725047
```

The least-squares regression line is $y = 131.4x + 2850$.

b.

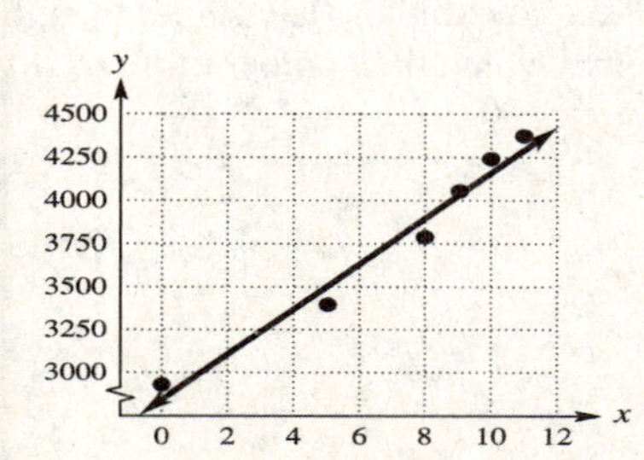

c. Yes; the line appears to fit.

d. The correlation coefficient is .988. This indicates that the line is a good fit.

39.

$$
\begin{aligned}
-6x + 3 &< 2x \\
-6x + 6x + 3 &< 2x + 6x \\
3 &< 8x \\
\frac{3}{8} &< \frac{8x}{8} \\
\frac{3}{8} < x &\text{ or } x > \frac{3}{8}
\end{aligned}
$$

The solution is $\left(\frac{3}{8}, \infty\right)$.

40.

$$
\begin{aligned}
12z &\ge 5z - 7 \\
12z - 5z &\ge 5z - 5z - 7 \\
7z &\ge -7 \\
\frac{7z}{7} &\ge \frac{-7}{7} \\
z &\ge -1
\end{aligned}
$$

The solution is $[-1, \infty)$.

41.

$$
\begin{aligned}
2(3 - 2m) &\ge 8m + 3 \\
6 - 4m &\ge 8m + 3 \\
6 - 4m - 8m &\ge 8m - 8m + 3 \\
6 - 12m &\ge 3 \\
6 - 6 - 12m &\ge 3 - 6 \\
-12m &\ge -3 \\
\frac{-12m}{-12} &\le \frac{-3}{-12} \\
m &\le \frac{1}{4}
\end{aligned}
$$

The solution is $\left(-\infty, \frac{1}{4}\right]$.

42.

$$
\begin{aligned}
6p - 5 &> -(2p + 3) \\
6p - 5 &> -2p - 3 \\
8p - 5 &> -3 \\
8p &> 2 \\
\frac{8p}{8} &> \frac{2}{8} \\
p &> \frac{1}{4}
\end{aligned}
$$

The solution is $\left(\frac{1}{4}, \infty\right)$.

43. $-3 \le 4x - 1 \le 7$

$-2 \le 4x \le 8$

$-\frac{1}{2} \le x \le 2$

The solution is $\left[-\frac{1}{2}, 2\right]$.

44. $0 \le 3 - 2a \le 15$

$0 - 3 \le 3 - 3 - 2a \le 15 - 3$

$-3 \le -2a \le 12$

$\frac{-3}{-2} \ge \frac{-2a}{-2} \ge \frac{12}{-2}$

$\frac{3}{2} \ge a \ge -6$

The solution is $\left[-6, \frac{3}{2}\right]$.

45. $|b| \le 8 \Rightarrow -8 \le b \le 8$
The solution is [–8, 8].

46. $|a| > 7 \Rightarrow a < -7$ or $a > 7$
The solution is (–∞, –7) or (7, ∞).

47. $|2x - 7| \ge 3$

$2x - 7 \le -3$ or $2x - 7 \ge 3$

$2x \le 4$ or $2x \ge 10$

$x \le 2$ or $x \ge 5$

The solution is (–∞, 2] or [5, ∞).

48. $|4m + 9| \le 16$

$-16 \le 4m + 9 \le 16$

$-25 \le 4m \le 7$

$-\frac{25}{4} \le m \le \frac{7}{4}$

The solution is $\left[-\frac{25}{4}, \frac{7}{4}\right]$.

49. $|5k + 2| - 3 \le 4$

$|5k + 2| \le 7$

$-7 \le 5k + 2 \le 7$

$-9 \le 5k \le 5$

$-\frac{9}{5} \le k \le 1$

The solution is $\left[-\frac{9}{5}, 1\right]$.

50. $|3z - 5| + 2 \ge 10$

$|3x - 5| \ge 8$

$3z - 5 \le -8$ or $3z - 5 \ge 8$

$3z \le -3$ or $3z \ge 13$

$z \le -1$ or $z \ge \frac{13}{3}$

The solution is (–∞, –1] or $\left[\frac{13}{3}, \infty\right)$.

51. The inequalities that represent the weight of pumpkin that he will not use are $x < 2$ or $x > 10$. This is equivalent to the following inequalities:
$x - 6 < 2 - 6$ or $x - 6 > 10 - 6$

$x - 6 < -4$ or $x - 6 > 4$

$|x - 6| > 4$

Choose answer option (d).

52. Let x = the price of the snow thrower
$|x - 600| \le 55$

53. **a.** Let (x_1, y_1) be (5, 1873) and (x_2, y_2) be (10, 2250). Find the slope

$m = \frac{2250 - 1873}{10 - 5} = \frac{377}{5} = 75.4$

$y - 1873 = 75.4(x - 5)$

$y - 1873 = 75.4x - 377$

$y = 75.4x + 1496.$

b. $75.4x + 1496 > 2500 \Rightarrow 75.4x > 1004 \Rightarrow$
$x > 13.32$

Assuming the linear trend continues, the amount of energy consumed will exceed 2500 trillion BTU's sometime during 2013 and after.

54. Let m = number of miles driven. The rate for the second rental company is $95 + .2m$. We want to determine when the second company is cheaper than the first.
$125 > 95 + .2m \Rightarrow 30 > .2m \Rightarrow 150 > m$
The second company is cheaper than the first company when the number of miles driven is less than 150.

55. $r^2 + r - 6 < 0$
Solve the corresponding equation.
$r^2 + r - 6 = 0 \Rightarrow (r+3)(r-2) = 0 \Rightarrow$
$r = -3$ or $r = 2$

For region A, test –4:
$(-4)^2 + (-4) - 6 = 6 > 0$.
For region B, test 0:
$0^2 + 0 - 6 = -6 < 0$.
For region C, test 3:
$3^2 + 3 - 6 = 6 > 0$.
The solution is (–3, 2).

56. $y^2 + 4y - 5 \ge 0$
Solve the corresponding equation.
$y^2 + 4y - 5 = 0 \Rightarrow (y+5)(y-1) = 0$
$y = -5$ or $y = 1$

For region A, test –6:
$(-6)^2 + 4(-6) - 5 = 7 > 0$.
For region B, test 0:
$0^2 + 4(0) - 5 = -5 < 0$.
For region C, test 2:
$2^2 + 4(2) - 5 = 7 > 0$.
Both endpoints are included because the inequality symbol is "≥." The solution is $(-\infty, -5]$ or $[1, \infty)$.

57. $2z^2 + 7z \ge 15$
Solve the corresponding equation.

$$2z^2 + 7z = 15$$
$$2z^2 + 7z - 15 = 0$$
$$(2z-3)(z+5) = 0$$
$$z = \frac{3}{2} \text{ or } z = -5$$

These numbers are solutions of the inequality because the inequality symbol is "≥."

For region A, test –6:
$2(-6)^2 + 7(-6) = 30 > 15$.
For region B, test 0:
$2 \cdot 0^2 + 7 \cdot 0 = 0 < 15$.
For region C, test 2:
$2 \cdot 2^2 + 7 \cdot 2 = 22 > 15$.
The solution is $(-\infty, -5]$ or $\left[\frac{3}{2}, \infty\right)$.

58. $3k^2 \le k + 14$
Solve the corresponding equation.

$$3k^2 = k + 14$$
$$3k^2 - k - 14 = 0$$
$$(3k-7)(k+2) = 0$$
$$k = \frac{7}{3} \text{ or } k = -2$$

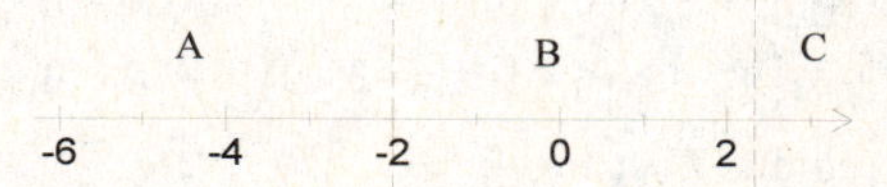

For region A, test –3:
$3(-3)^2 = 27,\ -3 + 14 = 11 \Rightarrow 27 > 11$
For region B, test 0:
$3(0)^2 = 0,\ 0 + 14 = 14 \Rightarrow 0 < 14$
For region C, test 3:
$3(3)^2 = 27,\ 3 + 14 = 17 \Rightarrow 27 > 17$
The given inequality is true in region B and at both endpoints, so the solution is $\left[-2, \frac{7}{3}\right]$.

59. $(x-3)\left(x^2+7x+10\right)\leq 0$

Solve the corresponding equation.

$(x-3)\left(x^2+7x+10\right)=0$

$(x-3)(x+2)(x+5)=0$

$x-3=0$ or $x+2=0$ or $x+5=0$

$x=3$ or $x=-2$ or $x=-5$

Note that –5, –2, and 3 are solutions of the original inequality.

A B C D

-6 -4 -2 0 2 4

In region A, let $x = -6$:

$(-6-3)\left((-6)^2+7(-6)+10\right)$

$= -9(36 - 42 + 10) = -9(4) = -36 < 0$

In region B, let $x = -3$:

$(-3-3)\left((-3)^2+7(-3)+10\right)$

$= -6(9 - 21 + 10) = -6(-2) = 12 > 0.$

In region C, let $x = 0$:

$(0-3)\left(0^2+7(0)+10\right)=-3(10)=-30<0$.

In region D, let $x = 4$:

$(4-3)\left(4^2+7(4)+10\right)$

$=1(16+28+10)=54>0$.

The solution is $(-\infty, -5]$ or $[-2, 3]$.

60. $(x+4)\left(x^2-1\right)\geq 0$

Solve the corresponding equation.

$(x+4)\left(x^2-1\right)=0$

$(x+4)(x+1)(x-1)=0$

$x+4=0$ or $x+1=0$ or $x-1=0$

$x=-4$ or $x=-1$ or $x=1$

Note that –4, –1, and 1 are solutions of the original inequality.

A B C D

-6 -4 -2 0 2 4

In region A, let $x = -5$:

$(-5+4)\left((-5)^2-1\right)=-1(24)=-24\leq 0$.

In region B, let $x = -2$:

$(-2+4)\left((-2)^2-1\right)=2(3)=6>0$.

In region C, let $x = 0$:

$(0+4)\left(0^2-1\right)=4(-1)=-4<0$.

In region D, let $x = 2$:

$(2+4)\left(2^2-1\right)=6(3)=18>0$.

The solution is $[-4, -1]$ or $[1, \infty)$.

61. $\frac{m+2}{m}\leq 0$

Solve the corresponding equation $\frac{m+2}{m}=0$.

The quotient changes sign when

$m+2=0$ or $m=0$

$m=-2$ or $m=0$

–2 is a solution of the inequality, but the inequality is undefined when $m = 0$, so the endpoint 0 must be excluded.

A B C

-6 -4 -2 0 2 4

For region A, test –3:

$\frac{-3+2}{-3}=\frac{1}{3}>0$.

For region B, test –1:

$\frac{-1+2}{-1}=-1<0$.

For region C, test 1:

$\frac{1+2}{1}=3>0$.

The solution is $[-2, 0)$.

62. $\frac{q-4}{q+3} > 0$

Solve the corresponding equation $\frac{q-4}{q+3} = 0$.

The numerator is 0 when $q = 4$. The denominator is 0 when $q = -3$.

A B C
-4 -2 0 2 4

For region A, test –4:
$\frac{-4-4}{-4+3} = \frac{-8}{-1} = 8 > 0$.
For region B, test 0:
$\frac{0-4}{0+3} = -\frac{4}{3} < 0$.
For region C, test 5:
$\frac{5-4}{5+3} = \frac{1}{8} > 0$.
The inequality is true in regions A and C, and both endpoints are excluded. Therefore, the solution is $(-\infty, -3)$ or $(4, \infty)$.

63. $\frac{5}{p+1} > 2$
Solve the corresponding equation.

$$\frac{5}{p+1} = 2$$
$$\frac{5}{p+1} - 2 = 0$$
$$\frac{5-2(p+1)}{p+1} = 0$$
$$\frac{3-2p}{p+1} = 0$$

The numerator is 0 when $p = \frac{3}{2}$. The denominator is 0 when $p = -1$.
Neither of these numbers is a solution of the inequality.

A B C
-4 -2 0 2 4

In region A, test –2:
$\frac{5}{-2+1} = -5 < 2$.
In region B, test 0:
$\frac{5}{0+1} = 5 > 2$.
In region C, test 2:
$\frac{5}{2+1} = \frac{5}{3} < 2$.

The solution is $\left(-1, \frac{3}{2}\right)$.

64. $\frac{6}{a-2} \le -3$
Solve the corresponding equation.

$$\frac{6}{a-2} = -3$$
$$\frac{6}{a-2} + 3 = 0$$
$$\frac{6+3(a-2)}{a-2} = 0$$
$$\frac{3a}{a-2} = 0$$

The numerator is 0 when $a = 0$. The denominator is 0 when $a = 2$.

A B C
-4 -2 0 2 4

For region A, test –1:
$\frac{6}{-1-2} = -2 \ge -3$.
For region B, test 1:
$\frac{6}{1-2} = -6 \le -3$.
For region C, test 3:
$\frac{6}{3-2} = 6 \ge -3$.
The given inequality is true in region B. The endpoint 0 is included because the inequality symbol is "≤." However, the endpoint 2 must be excluded because it makes the denominator 0. The solution is $[0, 2)$.

65. $\frac{2}{r+5} \le \frac{3}{r-2}$

Write the corresponding equation and then set one side equal to zero.

$$\frac{2}{r+5} = \frac{3}{r-2}$$

$$\frac{2}{r+5} - \frac{3}{r-2} = 0$$

$$\frac{2(r-2)-3(r+5)}{(r+5)(r-2)} = 0$$

$$\frac{2r-4-3r-15}{(r+5)(r-2)} = 0$$

$$\frac{-r-19}{(r+5)(r-2)} = 0$$

The numerator is 0 when $r = -19$. The denominator is 0 when $r = -5$ or $r = 2$. -19 is a solution of the inequality, but the inequality is undefined when $r = -5$ or $r = 2$.

A B C D

-20 -16 -12 -8 -4 0 4

For region A, test -20:

$\frac{2}{-20+5} = -\frac{2}{15} \approx -.13$ and

$\frac{3}{-20-2} = -\frac{3}{22} \approx -.14$

Since $-.13 > -.14$, -20 is not a solution of the inequality.

For region B, test -6:

$\frac{2}{-6+5} = -2$ and $\frac{3}{-6-2} = -\frac{3}{8}$.

Since $-2 < -\frac{3}{8}$, -6 is a solution.

For region C, test 0: $\frac{2}{0+5} = \frac{2}{5}$ and $\frac{3}{0-2} = -\frac{3}{2}$.

Since $\frac{2}{5} > -\frac{3}{2}$, 0 is not a solution.

For region D, test 3: $\frac{2}{3+5} = \frac{1}{4}$ and $\frac{3}{3-2} = 3$.

Since $\frac{1}{4} < 3$, 3 is a solution. The solution is $[-19, -5)$ or $(2, \infty)$.

66. $\frac{1}{z-1} > \frac{2}{z+1}$

Write the corresponding equation and then set one side equal to zero.

$$\frac{1}{z-1} = \frac{2}{z+1}$$

$$\frac{1}{z-1} - \frac{2}{z+1} = 0$$

$$\frac{(z+1)-2(z-1)}{(z-1)(z+1)} = 0$$

$$\frac{3-z}{(z-1)(z+1)} = 0$$

The numerator is 0 when $z = 3$. The denominator is 0 when $z = 1$ and when $z = -1$. These three numbers, -1, 1, and 3, separate the number line into four regions.

A B C D

-4 -2 0 2 4

For region A, test -3.

$\frac{1}{-3-1} > \frac{2}{-3+1} \Rightarrow -\frac{1}{4} > -1$, which is true.

For region B, test 0.

$\frac{1}{0-1} > \frac{2}{0+1} \Rightarrow -1 > 2$, which is false.

For region C, test 2.

$\frac{1}{2-1} > \frac{2}{2+1} \Rightarrow 1 > \frac{2}{3}$, which is true.

For region D, test 4.

$\frac{1}{4-1} > \frac{2}{4+1} \Rightarrow \frac{1}{3} > \frac{2}{5}$, which is false.

Thus, the solution is $(-\infty, -1)$ or $(1, 3)$.

67. $r = 340.1x^2 - 5360x + 18{,}834$
We want to determine when
$340.1x^2 - 5360x + 18{,}834 > 0$ for $6 \le x \le 12$.
Using a graphing calculator, plot
$Y_1 = 340.1x^2 - 5360x + 18{,}834$ on
[5, 13] by [–10, 10]. Then determine where the graph of Y_1 lies above the x axis.

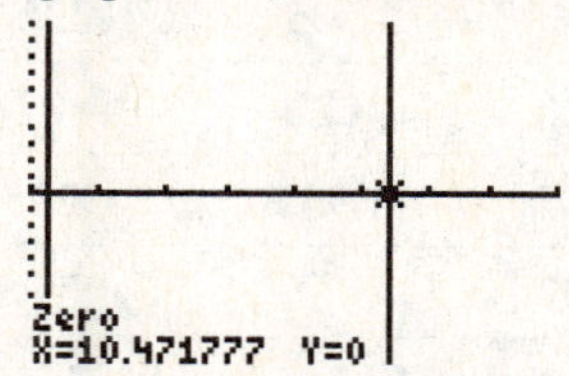

The profit was positive in the years 2011 through 2012.

68. $r = 89.29x^2 - 1517x + 7505$
We want to determine when
$89.29x^2 - 1577x + 7505 > 1000$ for $6 \le x \le 12$.
Using a graphing calculator, plot
$Y_1 = 89.29x^2 - 1577x + 7505$ and $Y_2 = 1000$ on
[0, 6] by [900, 1100]. Then determine where the graph of Y_1 lies above the graph of Y_2.

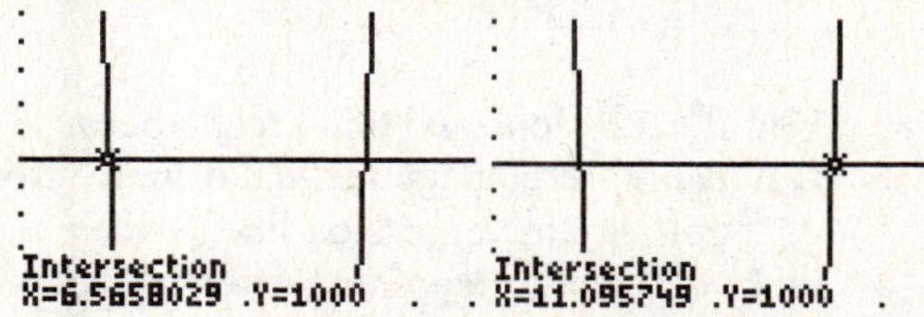

The net income exceeded $1000 million in 2006 and then again in 2012.

Case 2 Using Extrapolation for Prediction

1.

```
LinReg
 y=ax+b
 a=-146.0818182
 b=2330.118182
 r²=.7571345586
 r=-.8701347933
```

This verifies the regression equation.

2.

Year ($x = 0$ is 2000)	Table value	Predicted value	Residual
2	1648	2037.8	–389.8
3	1679	1891.7	–212.7
4	1842	1745.6	96.4
5	1931	1599.5	331.5
6	1979	1453.4	525.6
7	1503	1307.3	195.7
8	1120	1161.2	–41.2
9	794	1015.1	–221.1
10	652	869.0	–217.0
11	585	722.9	–137.9
12	650	576.8	73.2

3. See the table in problem 2 for residual values.

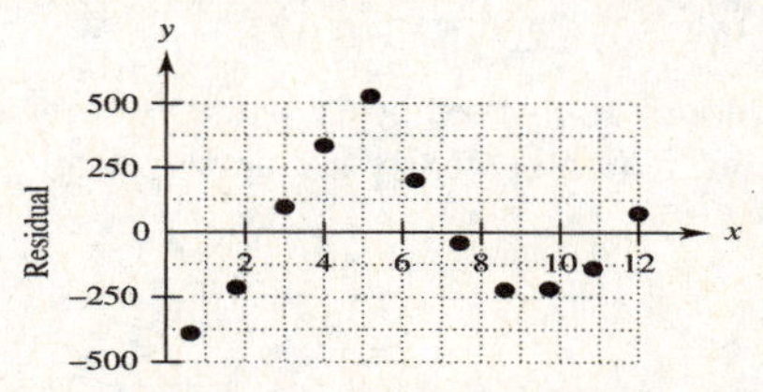

4. No; because the residuals show over fitting, under fitting, and then over fitting.

5. Since $x = 0$ corresponds to the year 1900, enter the following data into a computing device.

x	y
70	3.40
75	4.73
80	6.85
85	8.74
90	10.20
95	11.65
100	14.02
105	16.13
110	16.26

Then determine the least squares regression line.

```
LinReg
 y=ax+b
 a=.3429666667
 b=-20.647
 r²=.9907038368
 r=.9953410655
```

The model is verified.

6. The year 2002 corresponds to $x = 102$.

$y = .343(102) - 20.65 \approx 14.34.$

According to the model, the hourly wage in 2002 was about \$14.34, about 63¢ too low.

7. The year 1960 corresponds to $x = 60$.

$y = .343(60) - 20.65 \approx -0.07.$

The model gives the hourly wage as a negative amount, which is clearly not appropriate.

8.

Year ($x = 0$ is 1900)	Table value	Predicted value	Residual
70	3.40	3.36	.04
75	4.73	5.075	–.345
80	6.85	6.79	.06
85	8.74	8.505	.235
90	10.20	10.22	–.02
95	11.65	11.935	–.285
100	14.02	13.65	.37
105	16.13	15.365	.765
110	16.26	17.08	–.82

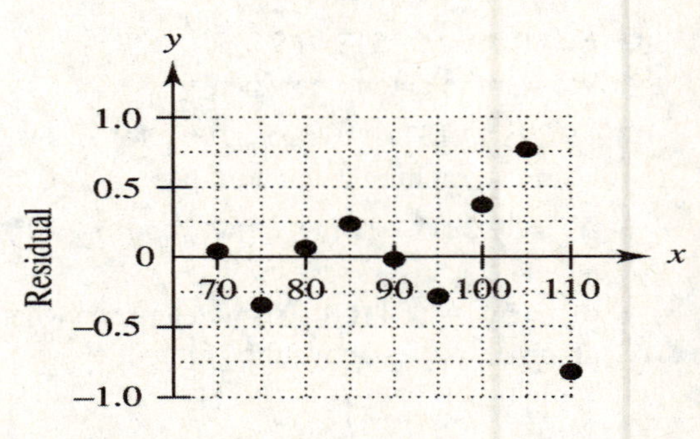

9. You'll get 0 slope and 0 intercept, because the residual represents the vertical distance from the data point to the regression line. Since r is very close to 1, the data points lie very close to the regression line.

```
LinReg
 y=ax+b
 a=-3.333333E-5
 b=.003
 r²=1.006684E-6
 r=-.0010033366
```

Chapter 3 Functions and Graphs

Section 3.1 Functions

1.

x	3	2	1	0	−1	−2	−3
y	9	4	1	0	1	4	9

This rule defines y as a function of x because each value of x determines one and only one value of y.

3. The rule $y = x^3$ defines y as a function of x because each value of x determines one and only one value of y.

5. The rule $x = |y+2|$ does not define y as a function of x because some values of x determine two values for y. For example, if $x = 4,\ 4 = |y+2| \Rightarrow$

$y + 2 = 4$ or $y + 2 = -4$

$y = 2$ or $y = -6$

7. The rule $y = \frac{-1}{x-1}$ defines y as a function of x because each value of x determines one and only one value of y (or none).

9. $f(x) = 4x - 1$
The domain of f is all real numbers since x may take on any real-number value. Therefore, the domain is $(-\infty, \infty)$.

11. $f(x) = x^4 - 1$
The domain of f is all real numbers since x may take on any real-number value. Therefore, the domain is $(-\infty, \infty)$.

13. $f(x) = \sqrt{-x} + 3$

In order to have $\sqrt{-x}$ be a real number, we must have $-x \geq 0$ or $x \leq 0$. Thus, the domain is all nonpositive real numbers, or $(-\infty, 0]$.

15. $g(x) = \frac{1}{x-2}$
Since the denominator cannot be zero, $x \neq 2$. Thus, the domain is all real numbers except 2, which in interval notation is written $(-\infty, 2)$ or $(2, \infty)$.

17. $g(x) = \frac{x^2+4}{x^2-4}$

Solve $x^2 - 4 = 0$ and exclude the solutions from the domain because the solutions make the denominator equal to 0.

$x^2 - 4 = 0$

$(x-2)(x+2) = 0$

$x - 2 = 0$ or $x + 2 = 0$

$x = 2$ or $x = -2$

The domain of g is all real numbers except $x = \pm 2$ since x cannot take on $x = \pm 2$. Therefore, the domain is $(-\infty, -2)$ or $(-2, 2)$ or $(2, \infty)$.

19. $h(x) = \frac{\sqrt{x+4}}{x^2+x-12}$

Solve $x^2 + x - 12 = 0$ and exclude the solutions from the domain since x cannot take on these numbers.

$x^2 + x - 12 = 0$

$(x-3)(x+4) = 0$

$x - 3 = 0$ or $x + 4 = 0$

$x = 3$ or $x = -4$

For $\sqrt{x+4}$, $x + 4$ must be positive or 0. That is $x + 4 \geq 0$ or $x \geq -4$. Therefore, the domain is x such that $x \neq 3$ and $x \geq -4$.

21. $g(x) = \begin{cases} \frac{1}{x} & \text{if } x < 0 \\ \sqrt{x^2+1} & \text{if } x \geq 0 \end{cases}$

For $\frac{1}{x}$, if $x < 0$, x can take on any real number.

For $\sqrt{x^2+1}$, x can take on any real number. The domain is all real numbers or $(-\infty, \infty)$.

23. $f(x) = 8$
For any value of x, the value of $f(x)$ will always be 8. (This is a constant function).

a. $f(4) = 8$

b. $f(-3) = 8$

c. $f(2.7) = 8$

d. $f(-4.9) = 8$

25. $f(x) = 2x^2 + 4x$

a. $f(4) = 2(4^2) + 4(4) = 2(16) + 16 = 32 + 16 = 48$

b. $f(-3) = 2(-3)^2 + 4(-3) = 2(9) + (-12) = 18 - 12 = 6$

c. $f(2.7) = 2(2.7)^2 + 4(2.7) = 2(7.29) + 10.8 = 14.58 + 10.8 = 25.38$

d. $f(-4.9) = 2(-4.9)^2 + 4(-4.9) = 2(24.01) + (-19.6) = 48.02 - 19.6 = 28.42$

27. $f(x) = \sqrt{x+3}$

a. $f(4) = \sqrt{4+3} = \sqrt{7}$

b. $f(-3) = \sqrt{-3+3} = \sqrt{0} = 0$

c. $f(2.7) = \sqrt{2.7+3} = \sqrt{5.7}$

d. $f(-4.9) = \sqrt{-4.9+3} = \sqrt{-1.9}$ not defined

29. $f(x) = |x^2 - 6x - 4|$

a. $f(4) = |4^2 - 6(4) - 4|$
$f(4) = |16 - 24 - 4| = |-12| = 12$

b. $f(-3) = |(-3)^2 - 6(-3) - 4| = |9 + 18 - 4| = 23$

c. $f(2.7) = |(2.7)^2 - 6(2.7) - 4| = |7.29 - 16.2 - 4| = 12.91$

d. $f(-4.9) = |(-4.9)^2 - 6(-4.9) - 4| = |24.01 + 29.4 - 4| = |49.41| = 49.41$

31. $f(x) = \dfrac{\sqrt{x-1}}{x^2-1}$

a. $f(4) = \dfrac{\sqrt{4-1}}{4^2-1} = \dfrac{\sqrt{3}}{15}$

b. $f(-3) = \dfrac{\sqrt{-3-1}}{(-3)^2-1} = \dfrac{\sqrt{-4}}{8}$ not defined

c. $f(2.7) = \dfrac{\sqrt{2.7-1}}{2.7^2-1} = \dfrac{\sqrt{1.7}}{6.29}$

d. $f(-4.9) = \dfrac{\sqrt{-4.9-1}}{(-4.9)^2-1} = \dfrac{\sqrt{-5.9}}{23.01}$ not defined

33. $f(x) = \begin{cases} x^2 & \text{if } x < 2 \\ 5x-7 & \text{if } x \geq 2 \end{cases}$

a. $f(4) = 5(4) - 7 = 13$

b. $f(-3) = (-3)^2 = 9$

c. $f(2.7) = 5(2.7) - 7 = 6.5$

d. $f(-4.9) = (-4.9)^2 = 24.01$

35. $f(x) = 6 - x$

a. $f(p) = 6 - p$

b. $f(-r) = 6 - (-r) = 6 + r$

c. $f(m+3) = 6 - (m+3) = 6 - m - 3 = 3 - m$

37. $f(x)=\sqrt{4-x}$

a. $f(p)=\sqrt{4-p}\ \ (p\le 4)$

b. $f(-r)=\sqrt{4-(-r)}$
$=\sqrt{4+r}\ \ (r\ge -4)$

c. $f(m+3)=\sqrt{4-(m+3)}$
$=\sqrt{4-m-3}$
$=\sqrt{1-m}\ \ (m\le 1)$

39. $f(x)=x^3+1$

a. $f(p)=p^3+1$

b. $f(-r)=(-r)^3+1=-r^3+1$

c. $f(m+3)=(m+3)^3+1$
$=m^3+9m^2+27m+27+1$
$=m^3+9m^2+27m+28$

41. $f(x)=\dfrac{3}{x-1}$

a. $f(p)=\dfrac{3}{p-1}\ \ (p\ne 1)$

b. $f(-r)=\dfrac{3}{-r-1}\ \ (r\ne -1)$

c. $f(m+3)=\dfrac{3}{(m+3)-1}=\dfrac{3}{m+2}\ \ (m\ne -2)$

43. $f(x)=2x-4$

$$\frac{f(x+h)-f(x)}{h}=\frac{[2(x+h)-4]-(2x-4)}{h}$$
$$=\frac{2x+2h-4-2x+4}{h}$$
$$=\frac{2h}{h}=2$$

45. $f(x)=x^2+1$

$$\frac{f(x+h)-f(x)}{h}=\frac{(x+h)^2+1-\left(x^2+1\right)}{h}$$
$$=\frac{x^2+2hx+h^2+1-x^2-1}{h}$$
$$=\frac{2hx+h^2}{h}=2x+h$$

47. $g(x)=3x^4-x^3+2x$

X	Y1
3.5	414.31
3.9	642.51
4.3	954.73
4.7	1369.5
5.1	1907.1
5.5	2589.8

Y1=3X^4-X^3+2X

49. $T(x)=$

$$\begin{cases} .0535x & \text{if } 0\le x\le 23{,}100 \\ 1235.85+.0705(x-23{,}100) & \text{if } 23{,}100<x\le 75{,}891 \\ 4957.85+.0785(x-75{,}891) & \text{if } x>75{,}891 \end{cases}$$

a. $.0535(20{,}000)=\$1070.00$

b. $1235.85+.0705(70{,}000-23{,}100)$
$=\$4542.30$

c. $4957.62+.0785(120{,}000-75{,}891)$
$=\$8420.18$

51. $R(x)=-.722x^3+19.23x^2-161.2x+421.8$

a. The year 2008 corresponds to $x=8$.
$R(8)=$
$-.722(8)^3+19.23(8)^2-161.2(8)+421.8$
$=-6.744$
In 2008, the revenue was \$–6.744 billion.

b. The year 2011 corresponds to $x=11$.
$R(11)=$
$-.722(11)^3+19.23(11)^2-161.2(11)+421.8$
$=14.448$
In 2011, the revenue was about \$14.448 billion.

53. $g(x) = -21.1x^2 + 205x + 2164$

a. The year 2001 corresponds to $x = 1$.

$g(1) = -21.1(1)^2 + 205(1) + 2164 \approx 2347.9$

In 2001, the value of electric household ranges and ovens shipped was about 2347.9 million.

b. The year 2009 corresponds to $x = 9$.

$g(9) = -21.1(9)^2 + 205(9) + 2164 \approx 2299.9$

In 2009, the value of electric household ranges and ovens shipped was about 2299.9 million.

55. $f(t) = 2050 - 500t$

57. Fixed costs = $1800
Each bag costs $.50 to produce, and sells for $1.20.

a. $c(x) = 1800 + .5x$

b. $r(x) = 1.2x$

c. $p(x) = r(x) - c(x)$
$p(x) = 1.2x - (1800 + .5x)$
$p(x) = .7x - 1800$

59. $h(x) = -4.83x^3 + 50.3x^2 + 25.5x + 4149$

The years 2005–2010 correspond to $x = 5$ through $x = 10$.

X	Y1
5	4930.3
6	5069.5
7	5135.5
8	5099.2
9	4931.7
10	4604
11	4087.1

Y1=-4.83X^3+50....

Section 3.2 Graphs of Functions

1. $f(x) = -.5x + 2$
The graph is a straight line with slope –.5 and y-intercept 2.

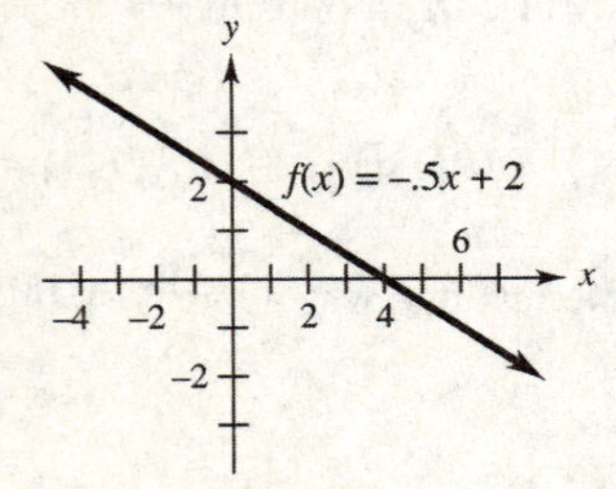

3. $f(x) = \begin{cases} x+3 & \text{if } x \le 1 \\ 4 & \text{if } x > 1 \end{cases}$

Graph the line $y = x + 3$ for $x \le 1$.
Graph the horizontal line $y = 4$ for $x > 1$.

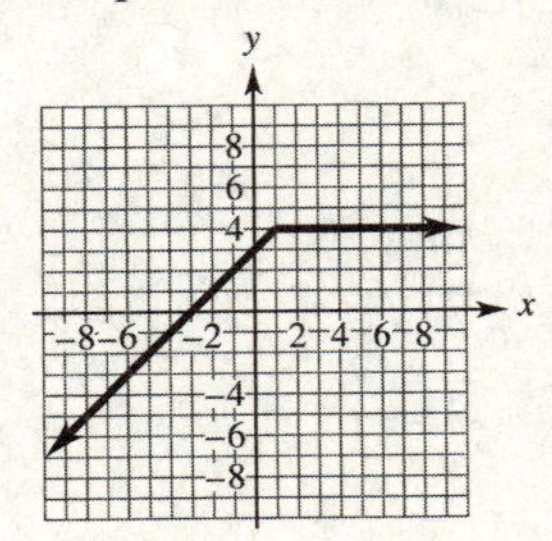

$f(x) = \begin{cases} x+3 & \text{if } x \le 1 \\ 4 & \text{if } x > 1 \end{cases}$

5. $y = \begin{cases} 4-x & \text{if } x \le 0 \\ 3x+4 & \text{if } x > 0 \end{cases}$

Graph the line $y = 4 - x$ for $x \le 0$.
Graph the line $y = 3x + 3$ for $x > 0$.

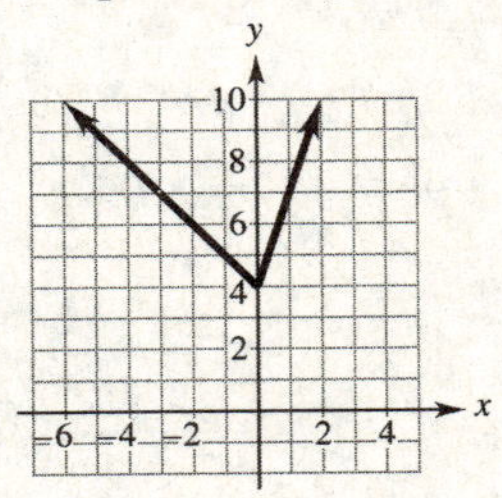

$y = \begin{cases} 4-x & \text{if } x \le 0 \\ 3x+4 & \text{if } x > 0 \end{cases}$

7. $f(x) = \begin{cases} |x| & \text{if } x < 2 \\ -2x & \text{if } x \ge 2 \end{cases}$

Rewrite the function as

$f(x) = \begin{cases} -x & \text{if } x < 0 \\ x & \text{if } 0 \le x < 2 \\ -2x & \text{if } x \ge 2 \end{cases}$

Graph the line $y = -x$ for $x \le 0$.
Graph the line $y = x$ for $0 < x \le 2$.
Graph the line $y = -2x$ for $x \ge 2$.

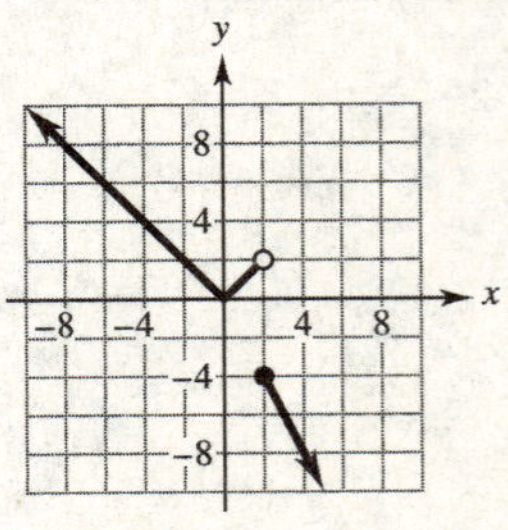

$f(x) = \begin{cases} |x| & \text{if } x < 2 \\ -2x & \text{if } x \ge 2 \end{cases}$

9. $f(x) = |x - 4|$

Using the definition of absolute values gives

$$f(x) = \begin{cases} x - 4 & \text{if } x - 4 \geq 0 \\ -(x - 4) & \text{if } x - 4 < 0 \end{cases} \text{ or}$$

$$f(x) = \begin{cases} x - 4 & \text{if } x \geq 4 \\ -x + 4 & \text{if } x < 4 \end{cases}$$

We graph the line $y = x - 4$ with slope 1 and y-intercept -4 for $x \geq 4$. We graph the line $y = -x + 4$ with slope -1 and y-intercept 4 for $x < 4$. Note that these partial lines meet at the point (4, 0).

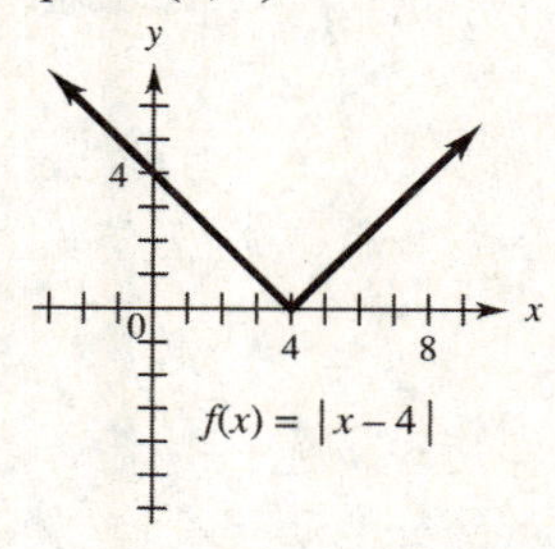

11. $f(x) = |3 - 3x|$

$$f(x) = \begin{cases} 3 - 3x & \text{if } 3 - 3x \geq 0 \\ -(3 - 3x) & \text{if } 3 - 3x < 0 \end{cases} \text{ or}$$

$$f(x) = \begin{cases} 3 - 3x & \text{if } x \leq 1 \\ -3 + 3x & \text{if } x > 1 \end{cases}$$

Graph the line $y = 3 - 3x$ for $x \leq 1$.
Graph the line $y = 3x - 3$ for $x > 1$.
The graph consists of two lines that meet at (1, 0).

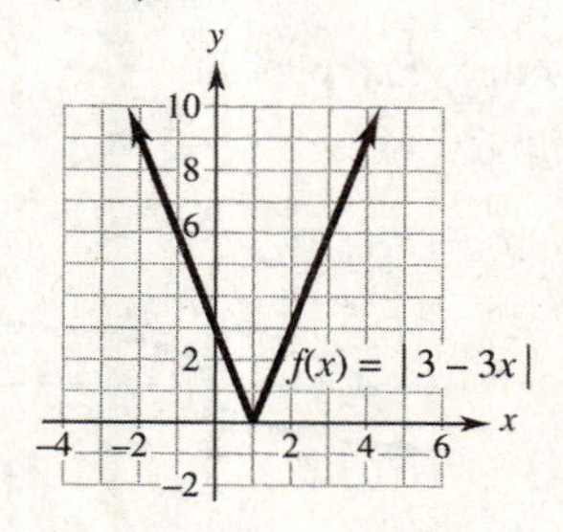

13. $y = -|x - 1|$

$$y = \begin{cases} -(x - 1) & \text{if } x - 1 \geq 0 \\ -[-(x - 1)] & \text{if } x - 1 < 0 \end{cases} \text{ or}$$

$$y = \begin{cases} -x + 1 & \text{if } x \geq 1 \\ x - 1 & \text{if } x < 1 \end{cases}$$

Graph the line $y = -x + 1$ for $x \geq 1$.
Graph the line $y = x - 1$ for $x < 1$.
The graph consists of two lines that meet at (1, 0).

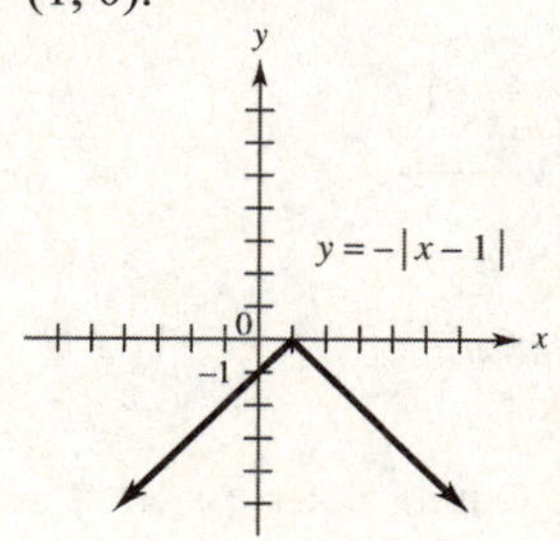

15. $y = |x - 2| + 3$

$$y = \begin{cases} x + 1 & \text{if } x \geq 2 \\ -x + 5 & \text{if } x < 2 \end{cases}$$

Graph the line $y = x + 1$ for $x \geq 2$.
Graph the line $y = -x + 5$ for $x < 2$.
The graph consists of two lines that meet at (2, 3).

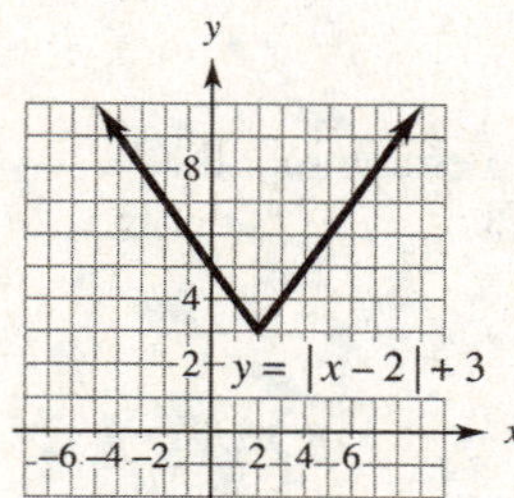

17. $f(x) = [x - 3]$

For x in the interval [0, 1), the value of $[x - 3] = -3$. For x in the interval [1, 2), the value of $[x - 3] = -2$. For x in the interval [2, 3), the value of $[x - 3] = -1$, and so on. The graph consists of a series of line segments. In each case, the left endpoint is included, and the right endpoint is excluded.

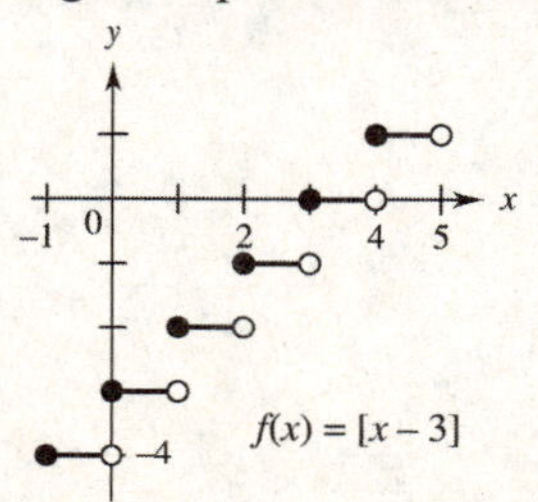

19. $g(x) = [-x]$
For x in the interval (0, 1], the value of $[-x] = -1$. For x in the interval (1, 2], the value of $[-x] = -2$. For x in the interval (2, 3], the value of $[-x] = -3$, and so on. The graph consists of a series of line segments. In each case, the left endpoint is excluded, and the right endpoint is included.

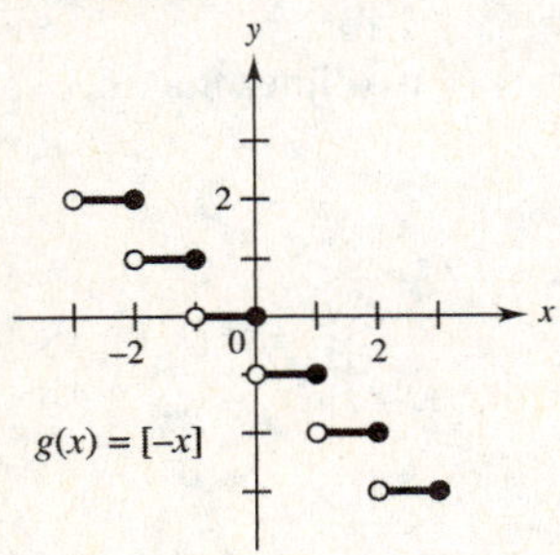

21. $f(x)$ = the price to mail a first-class letter weighing x ounces.

x	$p(x)$
(0, 1]	\$0.46
(1, 2]	\$0.66
(2, 3]	\$0.86
(3, 3.5]	\$1.06

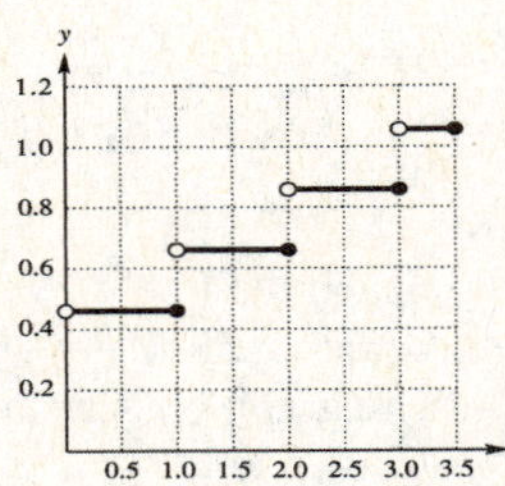

23. $f(x) = 3 - 2x^2$
Make a table of values and plot the corresponding points.

x	$f(x) = 3 - 2x^2$
-3	-15
-2	-5
-1	1
0	3
1	1
2	-5
3	-15

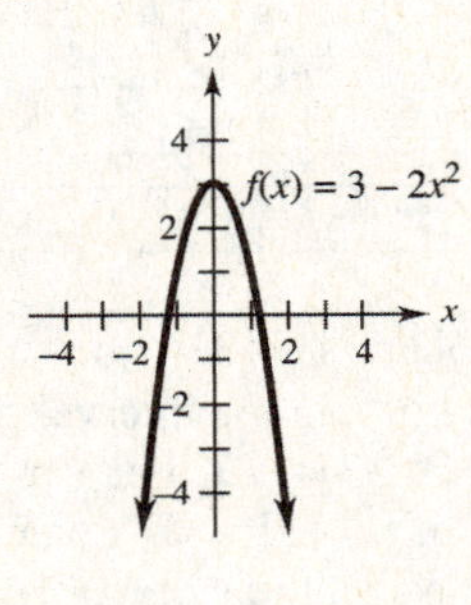

25. $h(x) = \dfrac{x^3}{10} + 2$
Make a table of values and plot the corresponding points.

x	$h(x) = \frac{x^3}{10} + 2$
-4	-4.4
-3	-0.7
-2	1.2
-1	1.9
0	2
1	2.1
2	2.8
3	4.7

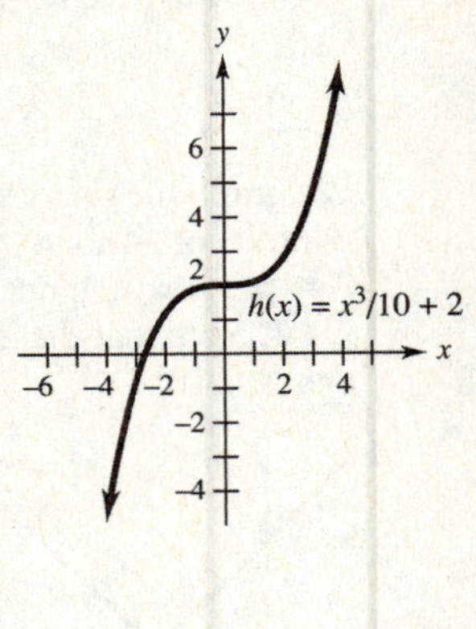

27. $g(x) = \sqrt{-x}$
Make a table of values and plot the corresponding points. The function is not defined for $x > 0$.

x	$g(x) = \sqrt{-x}$
-9	3
-4	2
-1	1
0	0

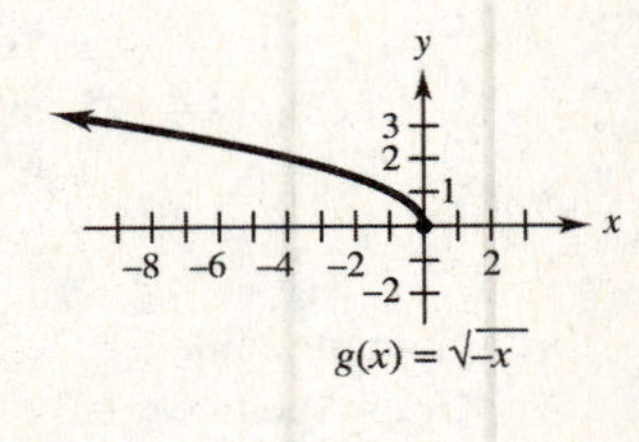

29. $f(x) = \sqrt[3]{x}$
Make a table of values and plot the corresponding points.

x	$f(x) = \sqrt[3]{x}$
-8	-2
-1	-1
0	0
1	1
8	2

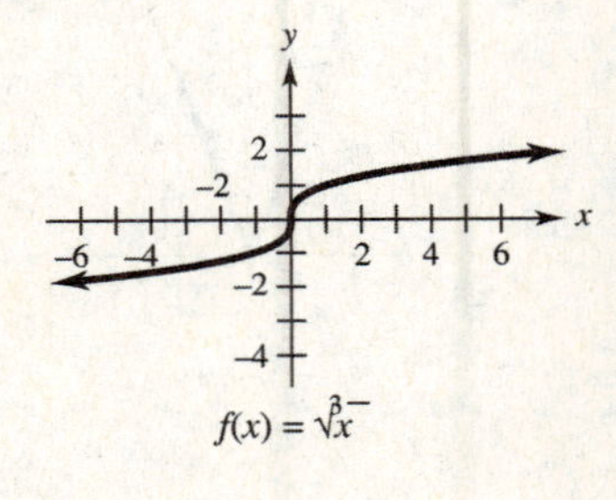

31. $f(x) = \begin{cases} x^2 & \text{if } x < 2 \\ -2x + 2 & \text{if } x \geq 2 \end{cases}$

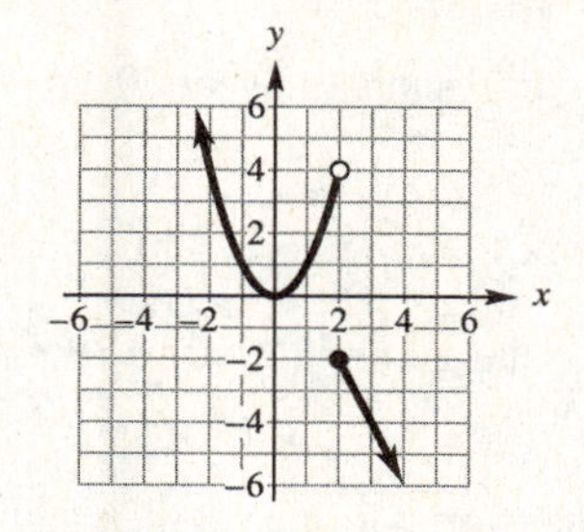

33. Every vertical line intersects this graph in at most one point, so this is the graph of a function.

35. A vertical line intersects the graph in more than one point, so this is not the graph of a function.

37. Every vertical line intersects this graph in at most one point, so this is the graph of a function.

39.

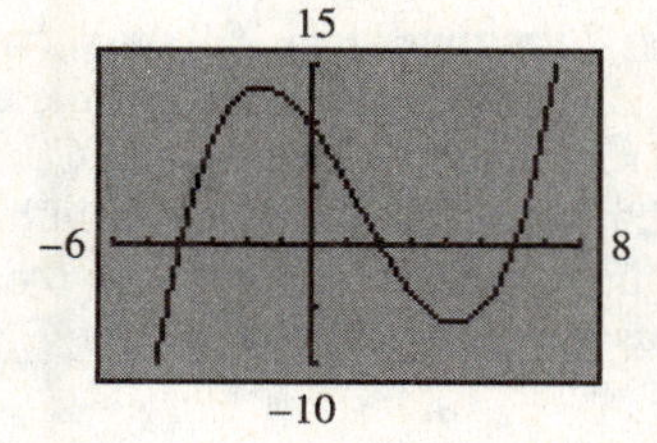

41.

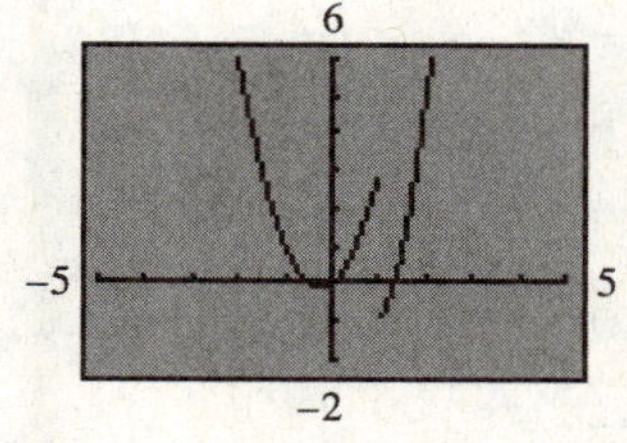

The endpoints are (1, –1), which is on the graph, and (1, 3), which is not on the graph.

43. Draw the graph in the window [–10, 10] by [–10, 15] and locate the x-intercepts.

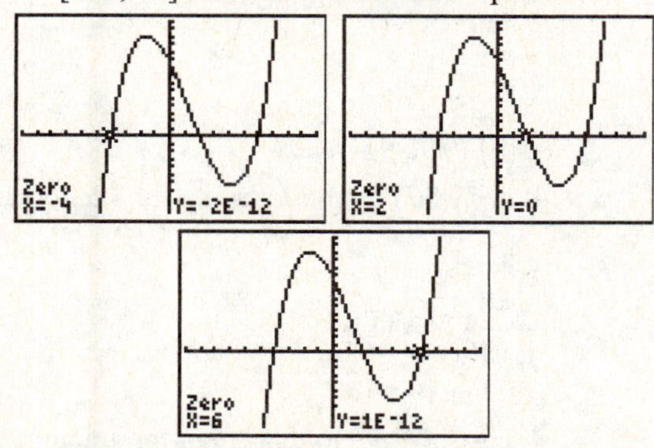

The x-intercepts are –4, 2, 6.

45. Draw the graph in the window [–10, 10] by [–10, 5].

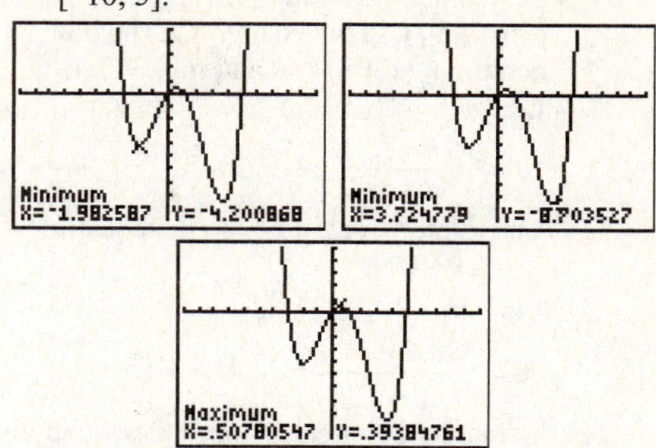

There is a maximum at (.5078, .3938), and there are minima at (– 1.9826, –4.2009) and (3.7248, –8.7035).

47.

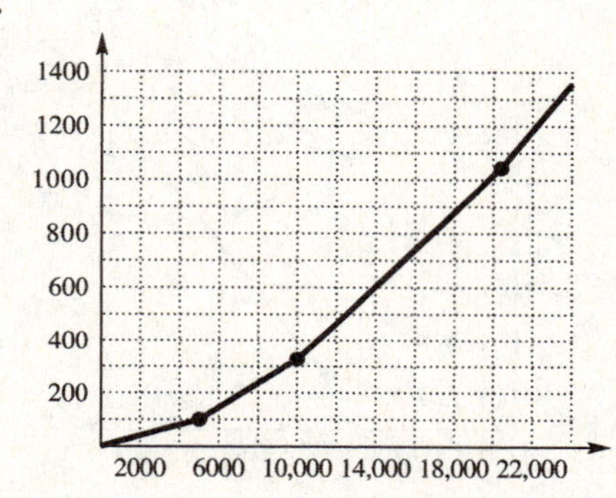

49. $f(x) = \begin{cases} 9.04 - .027x & \text{if } 0 \leq x \leq 43 \\ 7.879 + .032(x - 43) & \text{if } 43 < x \leq 113 \end{cases}$

a.

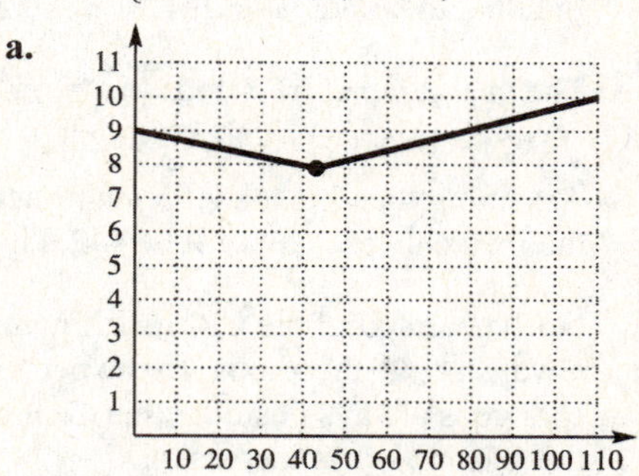

b. The lowest point on the graph occurs when $x = 43$.

$f(43) = 9.04 - .027(43) = 7.88$

The lowest stock price during the period defined by $f(x)$ is $7.88.

51. a. We must find the equation of the line containing the points (0, 15.1) and (30, 74.9) and the equation of the line containing (30, 74.9) and (60, 388.4).

Line 1:

$m = \frac{74.9 - 15.1}{30 - 0} = 1.993 \approx 2$

The y-intercept is 15.1, so the equation is $y = 2x + 15.1$.

Line 2:

$m = \frac{388.4 - 74.9}{60 - 30} = 10.45 \approx 10.5$

Using the point-slope form, the equation is

$y - 74.9 = 10.5(x - 30) \Rightarrow$

$y - 74.9 = 10.5x - 315 \Rightarrow y = 10.5x - 240.1.$

$$f(x) = \begin{cases} 2x + 15.1 & \text{if } 0 \le x \le 30 \\ 10.5x - 240.1 & \text{if } x > 30 \end{cases}$$

b.

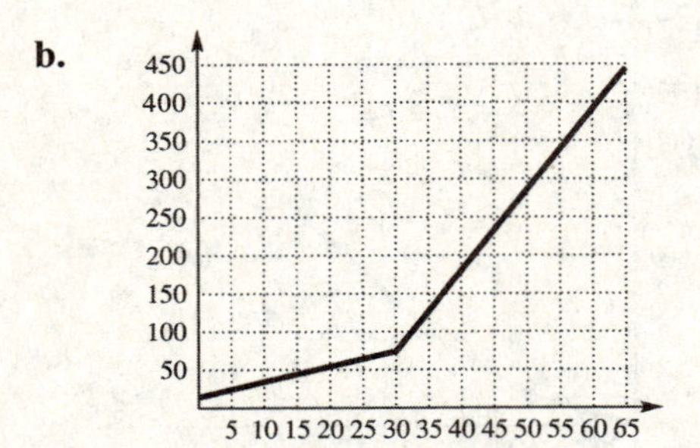

c. The year 2004 corresponds to $x = 54$.

$f(54) = 10.5(54) - 240.1 = 326.9$

In 2004, the medical-care CPI was 326.9.

d. The year 2015 corresponds to $x = 65$.

$f(65) = 10.5(65) - 240.1 = 442.4$

Assuming that this model remains accurate, the medical-care CPI will be 442.4 in 2015.

53. a. No. The graph of the CPI for all items is always above the x-axis, so the percent of change is always positive; this means that the CPI is always increasing.

b. The CPI for energy was decreasing in the years 1998, 2002, and 2009.

c. The CPI for energy was decreasing at the fastest rate in the year 2009.

55. a. Yes, this is the graph of a function because no vertical line intersects the graph in more than one point.

b. The domain represents the years from 1990 to 2010.

c. An estimate of the range is [7750, 15400].

57. a. According to the graph, $f(2000) = 33$ and $f(2011) = 44$.

b. The figure has vertical line segments, which can't be part of the graph of a function. To make the figure into the graph of f, delete the vertical line segments, then for each horizontal segment of the graph, put a closed dot on the left end and an open-circle dot on the right end (or vice versa, depending on when the cost change occurred).

59. First convert 75 minutes to 1.25 hours, and 15 minutes to .25 hours. Let x represent the number of hours to rent the van. To calculate $C(x)$, subtract 1.25 from x, then divide this number by .25 and round to the next integer. Finally, multiply this result by 5 and add $19.99.

a. 2 hours

$\frac{2 - 1.25}{.25} = 3$

$5(3) + 19.99 = 34.99$

It costs $34.99 to rent a van for 2 hours.

b. 1.5 hours

$\frac{1.5 - 1.25}{.25} = 1$

$5(1) + 19.99 = 24.99$

It costs $24.99 to rent a van for 1.5 hours.

c. 3.5 hours

$\frac{3.5 - 1.25}{.25} = 9$

$5(9) + 19.99 = 64.99$

It costs $64.99 to rent a van for 3.5 hours.

d. 4 hours

$\frac{4 - 1.25}{.25} = 11$

$5(11) + 19.99 = 74.99$

It costs $74.99 to rent a van for 4 hours.

e.

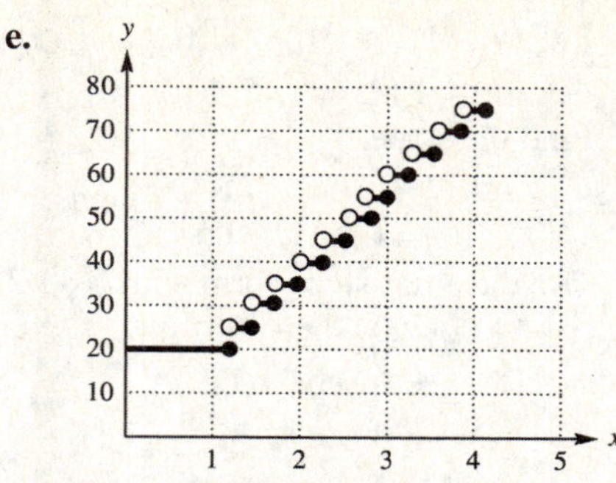

Note that for each step, the left endpoint is not included and the right point is included.

61. Make table of values:

Time	t	$g(t)$	Slope
midnight	0	1,000,000	increasing
noon	12	?	decreasing
4:00 P.M.	16	?	increasing
9:00 P.M.	21	?	vertical

There are many correct answers, including:

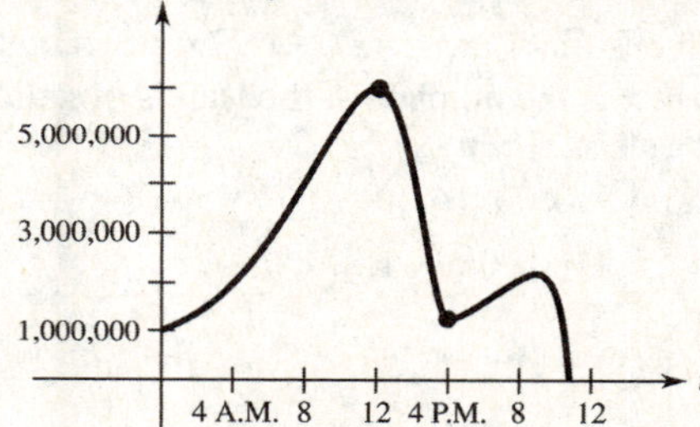

Section 3.3 Applications of Linear Functions

1. The marginal cost is \$5, while the fixed cost is \$25. Let $C(x)$ be the cost of renting a saw for x hours. Then $C(x) = 5x + 25$.

3. Let x = the number of half-hours and $C(x)$ = the total cost in dollar for x half-hours. Then $C(x) = 2.50x + 8$

5. Fixed cost, \$200, 50 items cost \$2000 to produce. Since the fixed cost is \$200,

$C(x) = mx + 200$

$C(50) = 2000.$

Therefore,

$2000 = m(50) + 200 \Rightarrow 50m = 1800 \Rightarrow m = 36$

$C(x) = 36x + 200$

7. Marginal cost, \$120; 100 items cost \$15,800 to produce.

$C(x) = mx + b,\ m = 120$

$C(100) = 15{,}800 = 120(100) + b$

$b = 15{,}800 - 12{,}000 = 3800$

$C(x) = 120x + 3800$

9. $\overline{C}(50) = \dfrac{C(50)}{50} = \dfrac{12(50)+1800}{50} = 48$

The average cost per item when 50 items are produced is \$48.

$\overline{C}(500) = \dfrac{C(500)}{500} = \dfrac{12(500)+1800}{500} = 15.60$

The average cost per item when 500 items are produced is \$15.60.

$\overline{C}(1000) = \dfrac{C(1000)}{1000} = \dfrac{12(1000)+1800}{1000} = 13.80$

The average cost per item when 1000 items are produced is \$13.80.

11. $\overline{C}(200) = \dfrac{C(200)}{200} = \dfrac{6.5(200)+9800}{200} = 55.50$

The average cost per item when 200 items are produced is \$55.50.

$\overline{C}(2000) = \dfrac{C(2000)}{2000} = \dfrac{6.5(2000)+9800}{2000}$

$= \$11.40$

The average cost per item when 2000 items are produced is \$11.40.

$\overline{C}(5000) = \dfrac{C(5000)}{5000} = \dfrac{6.5(5000)+9800}{5000}$

$= \$8.46$

The average cost per item when 5000 items are produced is \$8.46.

13. a. Let $(x_1, y_1) = (0, 16615)$ and let $(x_2, y_2) = (4, 8950)$.

$m = \dfrac{16{,}615 - 8950}{0 - 4} \approx -1916$

The y-intercept is 16,615, so the equation is $y = -1916x + 16{,}615$.

b. $f(5) = -1916(5) + 16{,}615 = 7035$

The car will be worthe \$7035.

c. Because the slope is –1916, the car is depreciating at a rate of \$1916 per year.

15. a. Let $(x_1, y_1) = (0, 120,000)$ and $(x_2, y_2) = (8, 25,000)$.

$$m = \frac{25,000 - 120,000}{8 - 0} = -11,875$$
$$y - y_1 = m(x - x_1)$$
$$y - 120,000 = -11,875(x - 0)$$
$$y - 120,000 = -11,875x$$
$$y = -11,875x + 120,000$$
$$f(x) = -11,875x + 120,000$$

b. The domain ranges from 0 to 8 years, that is [0, 8].

c. $f(6) = -11,875(6) + 120,000 = 48,750$
The machine will be worth $48,750 in 6 years.

17. a. The fixed costs are
$C(0) = \$42.5(0) + 80,000 = \$80,000$

b. The slope of $C(x) = 42.5x + 80,000$ is 42.5, so the marginal cost is $42.50.

c. The cost of producing 1000 books is
$C(1000) = 42.5(1000) + 80,000 = \$122,500.$
The cost of producing 32,000 books is
$$C(32,000) = 42.5(32,000) + 80,000$$
$$= \$1,440,000.$$

d. The average cost per book when 1000 are produced is:
$$\overline{C}(1000) = \frac{C(1000)}{1000}$$
$$= \frac{42.5(1000) + 80,000}{1000}$$
$$= \$122.50$$
The average cost per book when 32,000 are produced is,
$$\overline{C}(32,000) = \frac{C(32,000)}{32,000}$$
$$= \frac{42.5(32,000) + 80,000}{32,000}$$
$$= \$45$$

19. a. Let (x_1, y_1) be (100, 11.02) and (x_2, y_2) be (400, 40.12).
Find the slope.
$$m = \frac{40.12 - 11.02}{400 - 100} = \frac{29.1}{300} = .097$$
Use the point-slope form with (100, 11.02).
$$y - 11.02 = .097(x - 100)$$
$$y = .097x + 1.32$$
$$C(x) = .097x + 1.32$$

b. The total cost of producing 1000 cups is
$C(1000) = .097(1000) + 1.32 = \$98.32.$

c. The total cost of producing 1001 cups is
$C(1001) = .097(1001) + 1.32 = \$98.42.$

d. The marginal cost of the 1001st cup is
$C(1001) - C(1000) = 98.417 - 98.32 =$ $.097, which is about 9.7¢.

e. The slope of $C(x) = .097x + 1.32$ is .097, so the marginal cost is $.097 or 9.7 cents.

21. Each customer pays $4.20 + 1.77x$ per month, where x is the number of thousands of gallons of water used. Then,
$$R(x) = 550,000(4.20) + 1.77x$$
$$= 2,310,000 + 1.77x.$$

23. a. $C(x) = 10x + 750$

b. $R(x) = 35x$

c. $P(x) = R(x) - C(x) = 35x - (10x + 750)$
$= 25x - 750$

d. The profit on 100 items is,
$P(100) = 25(100) - 750 = \$1750$

25. a. $C(x) = 18x + 300$

b. $R(x) = 28x$

c. $P(x) = R(x) - C(x) = 28x - (18x + 300)$
$= 10x - 300$

d. The profit on 100 items is,
$P(100) = 10(100) - 300 = \$700$

27. a. $C(x) = 12.5x + 20{,}000$

b. $R(x) = 30x$

c. $P(x) = R(x) - C(x)$
$= 30x - (12.5x + 20{,}000)$
$= 17.5x - 20{,}000$

d. The profit on 100 items is,
$P(100) = 17.5(100) - 20{,}000 = -\$18{,}250$ (a loss of $18,250)

29. $2x - y = 7$ and $y = 8 - 3x$
Solve the first equation for y:
$2x - y = 7 \Rightarrow y = 2x - 7.$
Set the two equations equal and solve for x.
$2x - 7 = 8 - 3x \Rightarrow 5x = 15 \Rightarrow x = 3$
Substitute x into one equation to find y.
$y = 8 - 3(3) = -1$
The lines intersect at $(3, -1)$.

31. $y = 3x - 7$ and $y = 7x + 4$
Set the two equations equal and solve for x.
$3x - 7 = 7x + 4 \Rightarrow -4x = 11 \Rightarrow x = -\frac{11}{4}$
Substitute x into one equation to find y.
$y = 3\left(-\frac{11}{4}\right) - 7 = -\frac{33}{4} - \frac{28}{4} = -\frac{61}{4}$
The lines intersect at $\left(-\frac{11}{4}, -\frac{61}{4}\right)$.

33. a. The break-even point occurs when $R(x) = C(x)$.
$125x = 100x + 5000 \Rightarrow 25x = 5000 \Rightarrow$
$x = 200$
The break-even point is $x = 200$ or 200,000 policies.

b. To graph the revenue function, graph $y = 125x$. If $x = 0$, $y = 0$. If $x = 20$, $y = 2500$. Use the points (0, 0) and (20, 2500) to graph the line. To graph the cost function, graph $y = 100x + 5000$. If $x = 0$, $y = 5000$. If $x = 30$, $y = 8000$. Use the points (0, 5000) and (30, 8000) to graph the line.

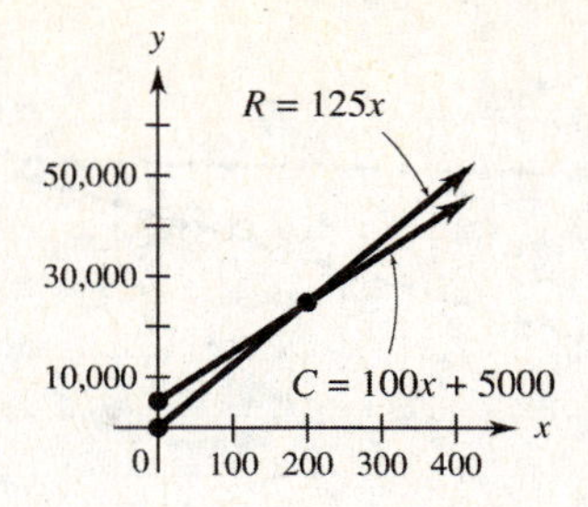

c. From the graph, when $x = 100$, cost is $15,000 and revenue is $12,500.

35. Given $R(x) = .21x$ and $P(x) = .084x - 1.5$

a. Cost equals revenue – profit, so
$C(x) = R(x) - P(x) = .21x - (.084x - 1.5)$
$= .21x - .084x + 1.5 = .126x + 1.5$

b. $C(7) = .126(7) + 1.5 = 2.382$
The cost of producing 7 units is $2.382 million.

c. At the break-even point, profit $P(x) = 0$, so $C(x) = R(x)$
$.126x + 1.5 = .21x \Rightarrow 1.5 = .084x \Rightarrow$
$x \approx 17.857$
The break-even point occurs at about 17.857 units.

37. $C(x) = 80x + 7000$; $R(x) = 95x$
$95x = 80x + 7000 \Rightarrow 15x = 7000 \Rightarrow x \approx 467$
The break-even point is about 467 units. Do not produce the item since $467 > 400$.

39. $C(x) = 125x + 42{,}000$; $R(x) = 165.5x$
$125x + 42{,}000 = 165.5x \Rightarrow 42{,}000 = 40.5x \Rightarrow$
$x \approx 1037$
The break-even point is about 1037 units. Produce the item since $1037 < 2000$.

41. The percentage of female workers in both countires was the same in 1998 at about 53%.

43. a.

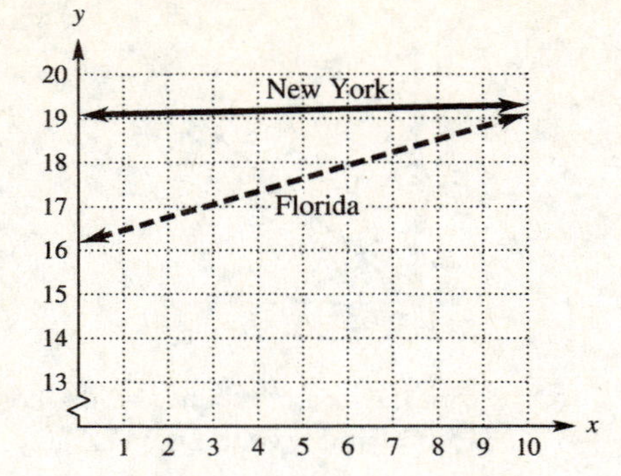

b. No, the graphs do not intersect in this window.

c. Yes, Florida will eventually overtake New York in population.

To find the year when Florida will overtake New York in population, equate the two equations and solve for x.

$.292x + 16.17 = .026x + 19.04 \Rightarrow$
$.266x = 2.87 \Rightarrow x \approx 10.79.$
Florida's population will overtake New York's sometime in late 2010.

45. On the supply curve, when $q = 20$, $p = 140$. The point (20, 140) is on the graph. When 20 items are supplied, the price is \$140.

47. The two curves intersect at the point (10, 120). The equilibrium supply and equilibrium demand are both $q = 10$ or 10 items.

49. $p = 16 - \frac{5}{4}q$

a. If $q = 0$, $p = 16$, so for a demand of 0 units, the price is \$16.

b. If $q = 4,\ p = 16 - \frac{5}{4}(4) = 16 - 5 = 11$, so for a demand of 4 units, the price is \$11.

c. If $q = 8,\ p = 16 - \frac{5}{4}(8) = 16 - 10 = 6$, so for a demand of 8 units, the price is \$6.

d. From (c), if $p = 6$, $q = 8$, so at a price of \$6, the demand is 8 units.

e. From (b), if $p = 11$, $q = 4$, so at a price of \$11, the demand is 4 units.

f. From (a), if $p = 16$, $q = 0$, so at a price of \$16, the demand is 0 units.

g.

h. $p = \frac{3}{4}q$

If $p = 0$, $q = 0$, so at a price of \$0, the supply is 0 units.

i. If $p = 10$,

$$10 = \frac{3}{4}q \Rightarrow \frac{4}{3}(10) = \frac{4}{3}\left(\frac{3}{4}\right)q \Rightarrow \frac{40}{3} = q$$

When the price is \$10, the supply is $\frac{40}{3}$ units.

j. If $p = 20$,

$$20 = \frac{3}{4}q \Rightarrow \frac{4}{3}(20) = \frac{4}{3}\left(\frac{3}{4}q\right) \Rightarrow \frac{80}{3} = q$$

When the price is \$20, the supply is $\frac{80}{3}$ units.

k. See(g).

l. The two graphs intersect at the point (8, 6). The equilibrium supply is 8 units.

m. The equilibrium price is \$6.

51. a.

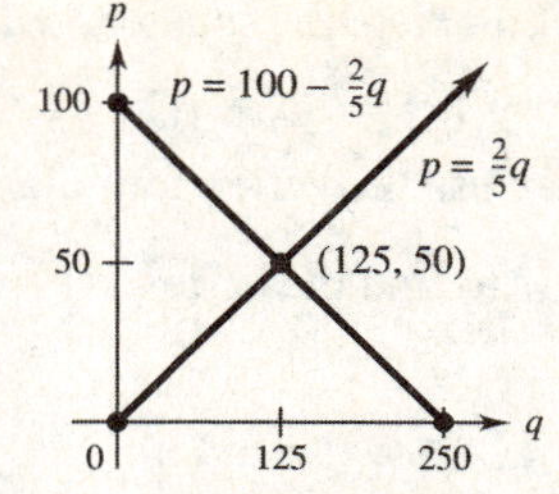

b. The two graphs intersect at the point (125, 50). The equilibrium demand is 125 units.

c. The equilibrium price is 50 cents.

d. $100-\frac{2}{5}q>\frac{2}{5}q \Rightarrow 100>\frac{2}{5}q+\frac{2}{5}q \Rightarrow$

$100>\frac{4}{5}q \Rightarrow \frac{5}{4}(100)>\frac{5}{4}\left(\frac{4}{5}q\right) \Rightarrow$

$125>q$ or $q<125$

Demand exceeds supply when q is in the interval [0, 125).

53. Total cost increases when more items are made (because it includes the cost of all previously made items), so the graph cannot move downward. No; the average cost can decrease as more items are made, so its graph may move downward.

Section 3.4 Quadratic Functions and Applications

1. $f(x)=x^2-3x-12$

This function is written in the form $y=ax^2+bx+c$ with a = 1, b = –3 and the parabola opens upward since $a = 1 > 0$.

3. $h(x)=-3x^2+14x+1$

The parabola opens downward since $a = -3 < 0$.

5. $f(x)=-2(x-5)^2+7$

This function is written in the form $y=a(x-h)^2+k$ with $a = -2$, $h = 5$, and $k = 7$. The vertex of the parabola, (h, k), is (5, 7). The parabola opens downward since $a = -2 < 0$.

7. $h(x)=4(x+1)^2-9$

This function is written in the form $y=a(x-h)^2+k$ with $a = 4$, $h = -1$, and $k = -9$. The vertex of the parabola, (h, k), is (–1, –9). The parabola opens upward since $a = 4 > 0$.

9. I **11.** K

13. J **15.** F

17. vertex (1, 2); point (5, 6)

$f(x)=a(x-h)^2+k$

$f(x)=a(x-1)^2+2$

Use (5, 6) to find a.

$6=a(5-1)^2+2 \Rightarrow 4=16a \Rightarrow a=\frac{1}{4}$

$f(x)=\frac{1}{4}(x-1)^2+2$

19. vertex (–1, –2); point (1, 2)

$f(x)=a(x-h)^2+k$

$f(x)=a(x+1)^2-2$

Use (1, 2) to find a.

$2=a(1+1)^2-2 \Rightarrow 4=4a \Rightarrow a=1$

$f(x)=(x+1)^2-2$

21. $f(x)=-x^2-6x+3$

$a=-1,\ b=-6$

To find the vertex of the function, use the vertex formula $x=-\frac{b}{2a}$ and $y=f\left(-\frac{b}{2a}\right)$.

$x=-\frac{-6}{2(-1)}=-3$

$f(-3)=-(-3)^2-6(-3)+3=12$

The vertex is (–3, 12).

23. $f(x)=3x^2-12x+5$

$a=3,\ b=-12$

To find the vertex of the function, use the vertex formula $x=-\frac{b}{2a}$ and $y=f\left(-\frac{b}{2a}\right)$.

$x=-\frac{-12}{2(3)}=2$

$f(2)=3(2)^2-12(2)+5=-7$

The vertex is (2, –7).

25. $f(x) = 3(x-2)^2 - 3$

To find the x-intercepts, set $f(x) = 0$ and then solve for x.

$0 = 3(x-2)^2 - 3 \Rightarrow 3(x-2)^2 = 3 \Rightarrow$

$(x-2)^2 = 1 \Rightarrow x-2 = \pm 1 \Rightarrow x = 1 \text{ or } x = 3.$

The x-intercepts are (1, 0) and (3, 0).

Let $x = 0$ to find the y-intercept.

$f(0) = 3(0-2)^2 - 3 = 9$

The y-intercept is 9.

27. $g(x) = 2x^2 + 8x + 6$

To find the intercepts, set each variable equal to 0. If $x = 0$, $g(0) = 2(0)^2 + 8(0) + 6 = 6$.

The y-intercept is 6.

If $y = 0$, $f(x) = y = 0$, so

$2x^2 + 8x + 6 = 0 \Rightarrow x^2 + 4x + 3 = 0 \Rightarrow$

$(x+3)(x+1) = 0 \Rightarrow x = -3 \text{ or } x = -1$

The x-intercepts are –3 and –1.

29. $f(x) = (x+2)^2 \Rightarrow y = 1(x+2)^2 + 0$

The vertex is (–2, 0). The axis is $x = -2$.

x	–1	0	1
y	1	4	9

Plot these points and use the axis of symmetry to find corresponding points on the other side of the axis. Connect the points with a smooth curve.

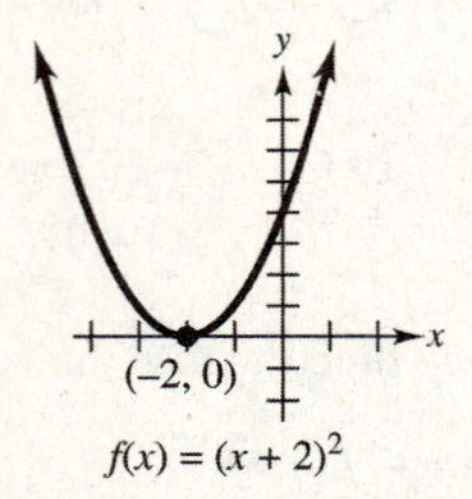

$f(x) = (x+2)^2$

In exercises 31, we complete the square to write the equation in the form $f(x) = a(x-h)^2 + k$. Alternatively, we can use the vertex formula, $h = -\frac{b}{2a}$, $k = f(h)$, to find the vertex.

31. $f(x) = x^2 - 4x + 6$

$= (x^2 - 4x + 4) + (6 - 4)$

$= (x-2)^2 + 2$

The vertex is (2, 2). The axis is $x = 2$.

x	3	4	5
y	3	6	11

Use the axis of symmetry to find corresponding points on the other side of the axis.

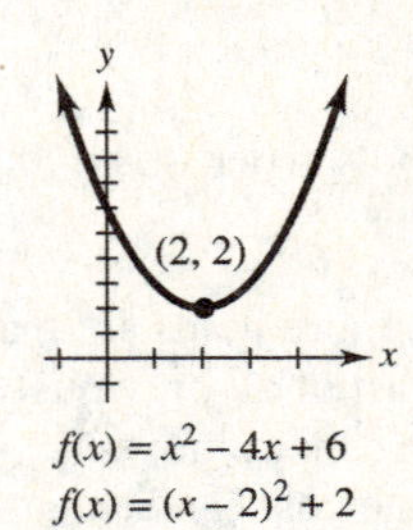

$f(x) = x^2 - 4x + 6$
$f(x) = (x-2)^2 + 2$

33. $f(x) = .0328x^2 - 3.55x + 115$

Let x be the age of the driver. Because the graph is a parabola which opens upward, the rate is lowest at the vertex. We must find the x-coordinate of the vertex.

$a = .0328, b = -3.55$

$$x = -\frac{b}{2a} = -\frac{-3.55}{2(.0328)} \approx 54.12$$

The rate is lowest at about age 54.

35. **a.** $y = -x^2 + 20x - 60 = -(x^2 - 20x) - 60$

$= -(x^2 - 20x + 100) - 60 + 100$

$= -(x-10)^2 + 40$

The graph of this parabola opens downward, so the maximum occurs at the vertex, (10, 40). The maximum firing rate will be reached in 10 milliseconds.

b. When $x = 10$, $y = 40$, so the maximum firing rate is 40 responses/millisec.

37. a. If $P(x) = 0$, then

$$0 = -2x^2 + 60x - 120 \Rightarrow 0 = x^2 - 30x + 60$$

$$x = \frac{-(-30) \pm \sqrt{(-30)^2 - 4(1)(60)}}{2(1)}$$

$$= \frac{30 \pm \sqrt{900 - 240}}{2} = \frac{30 \pm \sqrt{660}}{2}$$

$$x = \frac{30 + \sqrt{660}}{2} \approx 27.8 \text{ or}$$

$$x = \frac{30 - \sqrt{660}}{2} \approx 2.2$$

Therefore, $P(x) > 0$ if $2.2 < x < 27.8$. The largest number of cases she can sell and still make a profit is 27.

b. There are many possibilities. For example, she might have to buy additional machinery or pay high overtime wages in order to increase product, thus increasing costs and decreasing profits.

c. Find the vertex to determine the maximum.

$$P(x) = -2x^2 + 60x - 120$$
$$= -2\left(x^2 - 30x\right) - 120$$
$$= -2\left(x^2 - 30x + 225 - 225\right) - 120$$
$$= -2\left(x^2 - 30x + 225\right) + 450 - 120$$
$$= -2(x - 15)^2 + 330$$

Since the vertex is at (15, 330), she should sell 15 cases to maximize her profit.

39. Given supply: $p = \frac{1}{5}q^2$, and

demand: $p = -\frac{1}{5}q^2 + 40$.

a. $10 = -\frac{1}{5}q^2 + 40 \Rightarrow -30 = -\frac{1}{5}q^2 \Rightarrow$

$150 = q^2 \Rightarrow q \approx 12$

About 12 books are demanded.

b. $20 = -\frac{1}{5}q^2 + 40 \Rightarrow -20 = -\frac{1}{5}q^2 \Rightarrow$

$100 = q^2 \Rightarrow q = 10$

Ten books are demanded.

c. $30 = -\frac{1}{5}q^2 + 40 \Rightarrow -10 = -\frac{1}{5}q^2 \Rightarrow$

$50 = q^2 \Rightarrow q \approx 7$

About 7 books are demanded.

d. $40 = -\frac{1}{5}q^2 + 40 \Rightarrow 0 = -\frac{1}{5}q^2 \Rightarrow q = 0$

No books are demanded.

e. $5 = \frac{1}{5}q^2 \Rightarrow 25 = q^2 \Rightarrow 5 = q$

5 books are supplied.

f. $10 = \frac{1}{5}q^2 \Rightarrow 50 = q^2 \Rightarrow 7 \approx q$

About 7 books are supplied.

g. $20 = \frac{1}{5}q^2 \Rightarrow 100 = q^2 \Rightarrow 10 = q$

10 books are supplied.

h. $30 = \frac{1}{5}q^2 \Rightarrow 150 = q^2 \Rightarrow 12 \approx q$

About 12 books are supplied.

i. Using the values from the tables, plot points and graph the functions.

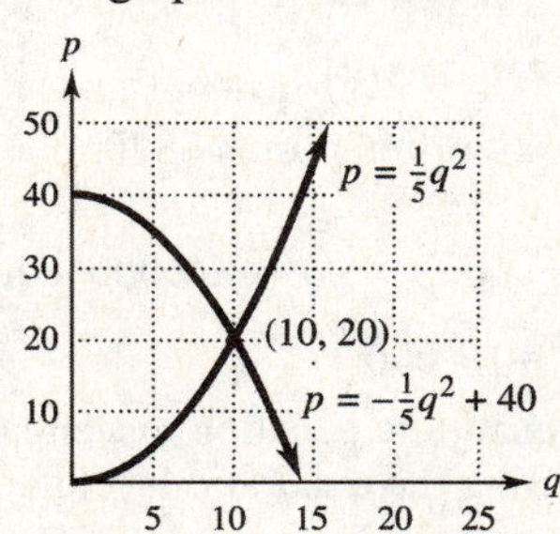

41. a. $p = 640 - 5(0)^2 = 640$

When 0 widgets are demanded, the price is \$640.

b. $p = 640 - 5(5)^2 = 515$

When 5 hundred widgets are demanded, the price is \$515.

c. $p = 640 - 5(10)^2 = 140$

When 10 hundred widgets are demanded, the price is \$140.

d. $p = 640 - 5q^2$

q	0	4	8	10
p	640	560	320	140

$p = 5q^2$

q	0	4	8	10
p	0	80	320	500

Graph these two parabolas on the same axes.

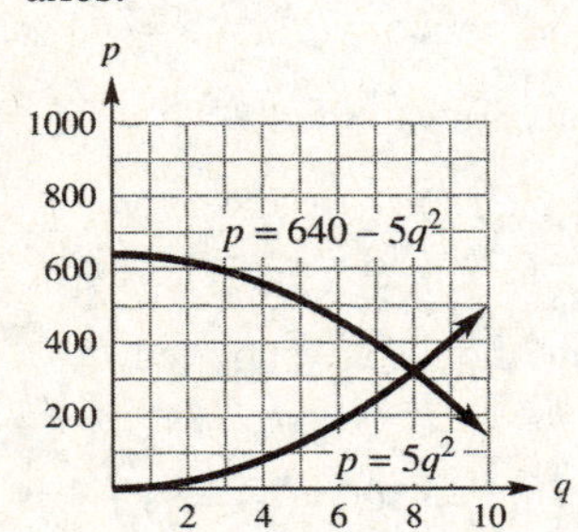

e. supply = demand ⇒

$5q^2 = 640 - 5q^2 \Rightarrow 10q^2 = 640 \Rightarrow$

$q^2 = 640 \Rightarrow q = \pm 8$

A negative value of q is not meaningful. The equilibrium supply is 8 hundreds or 800 units.

f. If $q = 8,\ p = 5(8)^2 = 320.$

The equilibrium cost is \$320.

43. Set $p = 45q$ and $p = -q^2 + 10{,}000$ equal.

$45q = -q^2 + 10{,}000$

Write this quadratic equation in standard form and solve using the quadratic formula.

$q^2 + 45q - 10{,}000 = 0$

$$q = \frac{-45 \pm \sqrt{45^2 - 4(1)(-10{,}000)}}{2(1)}$$

$$= \frac{-45 \pm \sqrt{42025}}{2} = \frac{-45 \pm 205}{2}$$

$$q = \frac{-45 + 205}{2} = 80 \text{ or } q = \frac{-45 - 205}{2} = -125$$

Since q cannot be negative, $q = 80$.
Since $p = 45(80) = 3600$ and

$p = -(80)^2 + 10{,}000 = 3600,$ the equilibrium quantity is $q = 80$ units and the equilibrium price is $p = \$3600$.

45. Set $p = q^2 + 20q$ and $p = -2q^2 + 10q + 3000$ equal.

$q^2 + 20q = -2q^2 + 10q + 3000$

Write this quadratic equation in standard form and solve using the quadratic formula.

$3q^2 + 10q - 3000 = 0$

$$q = \frac{-10 \pm \sqrt{10^2 - 4(3)(-3000)}}{2(3)}$$

$$= \frac{-10 \pm \sqrt{36{,}100}}{6} = \frac{-10 \pm 190}{6}$$

$$q = \frac{-10 - 190}{6} = \frac{-200}{6} = -\frac{100}{3} \text{ or}$$

$$q = \frac{-10 + 190}{6} = \frac{180}{6} = 30$$

Since q cannot be negative, $q = 30$.
Since $p = 30^2 + 20(30) = 1500$ and

$p = -2(30)^2 + 10(30) + 3000 = 1500,$ the equilibrium quantity is $q = 30$ units and the equilibrium price is $p = \$1500$.

47. Set the revenue function equal to the cost function and solve for x.

$200x - x^2 = 70x + 2200$

$0 = x^2 - 130x + 2200$

By the quadratic formula,

$$x = \frac{-(-130) \pm \sqrt{(-130)^2 - 4(1)(2200)}}{2(1)}$$

$$= \frac{130 \pm \sqrt{8100}}{2}$$

$x = 20$ or $x = 110$

Since these equations are only valid for $0 \le x \le 100$, 20 is the number of units needed to break even.

49. Set the revenue function equal to the cost function and solve for x.

$$400x - 2x^2 = -x^2 + 200x + 1900$$

$$-x^2 + 200x - 1900 = 0$$

By the quadratic formula,

$$x = \frac{-200 \pm \sqrt{(200)^2 - 4(-1)(-1900)}}{2(-1)} = \frac{-200 \pm \sqrt{32400}}{-2}$$

$x = 10$ or $x = 190$

Since these equations are only valid for $0 \le x \le 100$, 10 is the number of units needed to break even.

51. a. Let x be the number of unsold seats. Then the number of people flying is $100 - x$, and the price per ticket is $200 + 4x$. The total revenue is

$$R(x) = (200 + 4x)(100 - x) = 20{,}000 - 200x + 400x - 4x^2$$

$$R(x) = 20{,}000 + 200x - 4x^2$$

b.

$$R(x) = -4\left(x^2 - 50x\right) + 20{,}000 = -4\left(x^2 - 50x + 625 - 625\right) + 20{,}000 = -4(x-25)^2 + 2500 + 20{,}000$$

$$R(x) = -4(x-25)^2 + 22{,}500$$

The vertex is (25, 22,500).

x	5	15	25
$R(x)$	20,900	22,100	22,500
x	35	45	
$R(x)$	22,100	20,900	

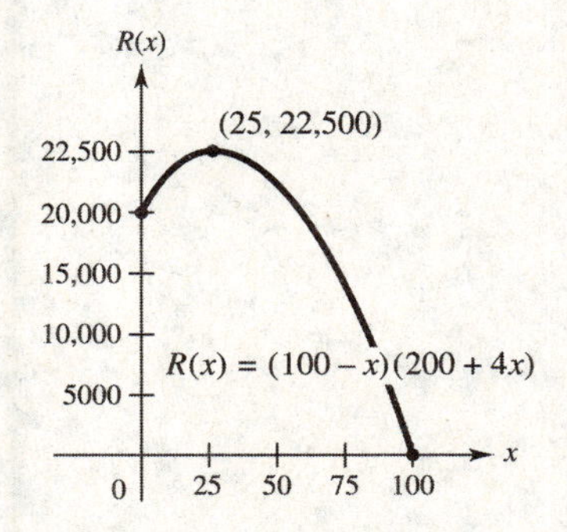

c. The maximum revenue occurs at the vertex, when 25 seats are unsold.

d. $R(25) = 22{,}500$, so the maximum revenue is \$22,500.

53. Let x represent the number of weeks she should wait. Let R represent her revenue in dollars per hog.

$$R(x) = (90 + 5x)(.88 - .02x) = 79.2 - 1.8x + 4.4x - .1x^2 = \left(-1.x^2 + 2.6x\right) + 79.2 = -1.\left(x^2 - 26x\right) + 79.2 = -.1\left(x^2 - 26x + 169 - 169\right) + 79.2$$

$$R(x) = -.1\left(x^2 - 26x + 169\right) + 16.9 + 79.2 = -1.(x-13)^2 + 96.1$$

The vertex is (13, 96.1). She should wait 13 weeks and will receive \$96.10/hog.

55. a. The vertex is given as (20, 0), so the equation is of the form $g(x) = a(x-20)^2$. The year 2005 corresponds to $x = 35$. Using the point (35, 3), we have

$3 = a(35-20)^2 \Rightarrow 3 = 225a \Rightarrow$

$a = \frac{3}{225} \approx .013.$

The equation is $g(x) = .013(x-20)^2$.

b. The year 2008 corresponds to $x = 38$.

$g(38) = .013(38-20)^2 \approx 4.$

According to the model, there were about 4 deaths in 2008.

57. a. The vertex is given as (4, 57.8), so the equation is of the form

$f(x) = a(x-4)^2 + 57.8.$ The year 2008 corresponds to $x = 8$. Using the point (8, 120.8), we have

$120.8 = a(8-4)^2 + 57.8 \Rightarrow 63 = 16a \Rightarrow$

$a = \frac{63}{16} = 3.9375.$

The equation is

$f(x) = 3.9375(x-4)^2 + 57.8.$

b. The year 2012 corresponds to $x = 12$.

$f(12) = 3.9375(12-4)^2 + 57.8 = 309.8.$

According to the model, there were about \$309.8 billion in China's gross domestic expeditures on research and development in 2012.

59. QuadReg
y=ax²+bx+c
a=.0344155844
b=-1.866623377
c=26.16363636

The quadratic regression is
$f(x) = .034x^2 - 1.87x + 26.16.$

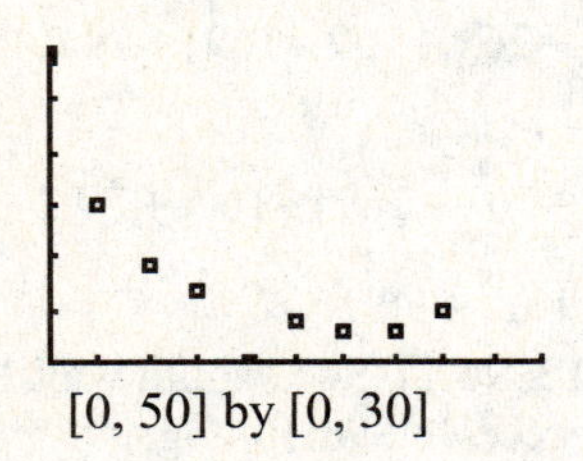

[0, 50] by [0, 30]

$g(38) = .034(38)^2 - 1.87(38) + 26.16 \approx 4$

This is the same as the model in exercise 55 estimates.

61.

The quadratic regression is
$f(x) = 1.808x^2 - 5.185x + 50.72.$

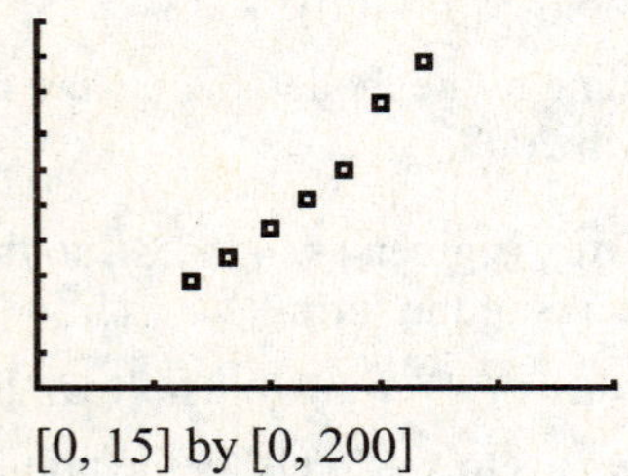

[0, 15] by [0, 200]

$f(12) = 1.808(12)^2 - 5.185(12) + 50.72 \approx 248.85$

The quadratic regression estimates that there will be $248.85 billion in expendatures. This is about $60.95 billion less than the model in exercise 57.

63. $R(x) = 400x - 2x^2$; $C(x) = 200x + 2000$

a.

$$R(x) = C(x)$$
$$400x - 2x^2 = 200x + 2000$$
$$0 = 2x^2 - 200x + 2000$$
$$0 = x^2 - 100x + 1000$$
$$x = \frac{100 \pm \sqrt{(-100)^2 - 4(1)(1000)}}{2(1)}$$
$$= \frac{100 \pm \sqrt{6000}}{2} = \frac{100 \pm 20\sqrt{15}}{2}$$
$$= 50 \pm 10\sqrt{15}$$

$x \approx 88.7$ and 11.3

The break-even point occurs when $x \approx 11.3$ and $x \approx 88.7$.

b.

$$P(x) = R(x) - C(x)$$
$$= (400x - 2x^2) - (200x + 2000)$$
$$= -2x^2 + 200x - 2000$$
$$= -2(x^2 - 100x + 2500) - 2000 + 5000$$
$$= -2(x - 50)^2 + 3000$$

The vertex is (50, 3000). Profit is maximum when $x = 50$.

c. From the vertex, the maximum profit is $3000.

d. Since the break-even points are 88.7 and 11.3, a loss will occur if $x < 11.3$ or $x > 88.7$.

e. A profit will occur for $11.3 < x < 88.7$.

Section 3.5 Polynomial Functions

1. $f(x) = x^4$

First we find several ordered pairs.

x	−3	−2	−1	0	1	2	3
y	81	16	1	0	1	16	81

Plot these ordered pairs and draw a smooth curve through them.

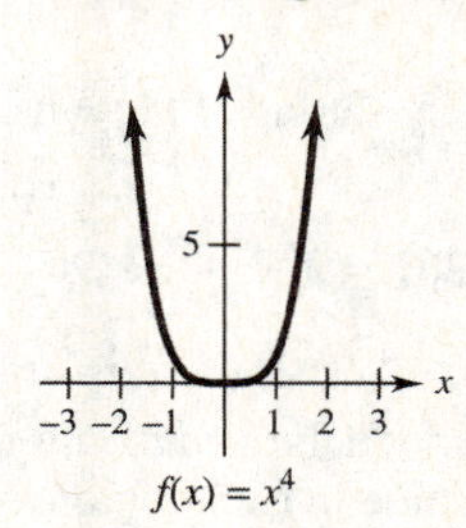

3. $h(x) = -.2x^5$

First find several ordered pairs

x	−3	−2	−1	0	1	2	3
y	48.6	6.4	.2	0	−.2	−6.4	−48.6

To graph the function, plot these ordered pairs, and draw a smooth curve through them.

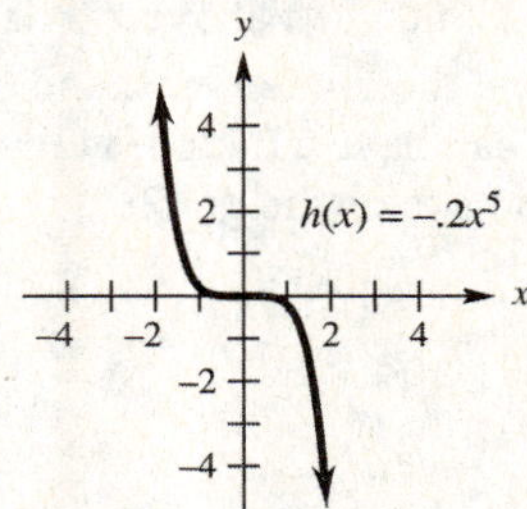

5. a. Yes. The graph could possibly be the graph of some polynomial function because it moves sharply away from the x-axis at the far left and far right.

b. No. The total number of peaks and valleys on the graph of a polynomial function of degree n is at most $n - 1$. Since the graph has 2 peaks and 2 valleys, this could not be the graph of a polynomial function of degree 3.

c. No. Since the graph has 2 peaks and 2 valleys, it could not be the graph of a polynomial function of degree 4.

d. Yes. Since the graph has 2 peaks and 2 valleys and the opposite ends of the graph move sharply away from the x-axis in the opposite direction, it could be the graph of a polynomial function of degree 5.

7. a. Yes. The graph could possibly be the graph of some polynomial function because it moves sharply away from the x-axis as $|x|$ gets large.

b. No. Since it has 1 peak and 2 valleys, it could not be the graph of a polynomial function of degree 3 (since $n - 1 = 3 - 1 = 2$).

c. Yes. Since it has 1 peak and 2 valleys, it could be the graph of a polynomial function of degree 4 (since $n - 1 = 4 - 1 = 3$).

d. No. The graph could not be the graph of a polynomial function of degree 5 because the opposite ends of the graph move sharply away from the x-axis in the same direction, indicating an even degree.

9. $f(x) = x^3 - 7x - 9$

Since $f(x)$ is degree 3, it has at most 2 peaks and valleys. When $|x|$ is large, the graph resembles the graph of x^3. Therefore, since the y-intercept is −9, D is the graph of the function.

11. $f(x) = x^4 - 5x^2 + 7.$

Since $f(x)$ is degree 4, it has at most 3 peaks and valleys. When $|x|$ is large, the graph resembles the graph of x^4. Therefore, since the y-intercept is 7, B is the graph of the function.

13. $f(x) = .7x^5 - 2.5x^4 - x^3 + 8x^2 + x + 2$

Since $f(x)$ is degree 5, it has at most 4 peaks and valleys. When $|x|$ is large, the graph resembles the graph of $.7x^5$. Therefore, since the y-intercept is 2, E is the graph of the function.

15. $f(x) = (x+3)(x-4)(x+1)$

First, find x-intercepts by setting $f(x) = 0$ and solving for x.

$f(x) = (x+3)(x-4)(x+1) = 0 \Rightarrow$

$x = -3, 4, -1$

These three numbers divide the x-axis into four regions. Choose an x-value in each region as a test number and compute the function value.

Region	Test No.	Value of $f(x)$	Graph
$x < -3$	–5	–72	below x-axis
$-3 < x < -1$	–2	6	above x-axis
$-1 < x < 4$	2	–30	below x-axis
$4 < x$	5	48	above x-axis

Use this information, the x-intercepts, and the fact that the graph can have a total of at most 2 peaks and valleys to sketch the graph.

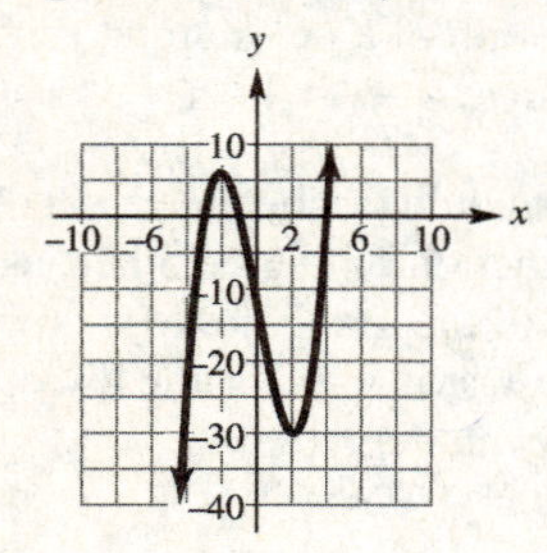

17. $f(x) = x^2(x+3)(x-1)$

First, find x-intercepts by setting $f(x) = 0$ and solving for x.

$f(x) = x^2(x+3)(x-1) = 0 \Rightarrow$

$x = 0, -3, 1$

These three numbers divide the x-axis into four regions. Choose an x-value in each region as a test number and compute the function value.

Region	Test No.	Value of $f(x)$	Graph
$x < -3$	–4	80	above x-axis
$-3 < x < 0$	–1	–4	below x-axis
$0 < x < 1$	0.5	–.4375	below x-axis
$1 < x$	2	20	above x-axis

Use this information, the x-intercepts, and the fact that the graph can have a total of at most 3 peaks and valleys to sketch the graph.

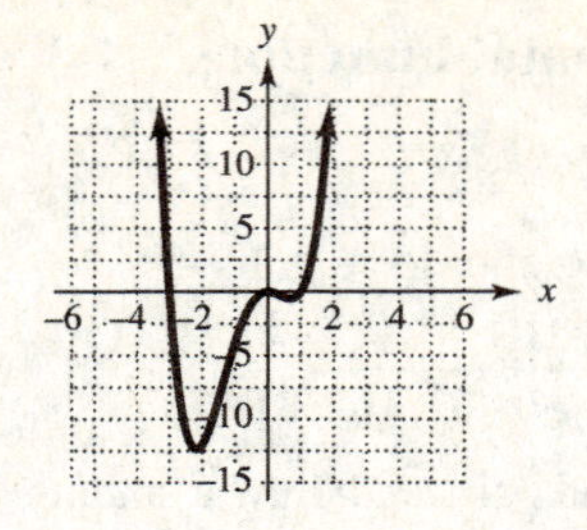

19. $f(x) = x^3 - x^2 - 20x$

First, find x-intercepts by setting $f(x) = 0$ and solving for x.

$f(x) = 0 = x^3 - x^2 - 20x = x(x^2 - x - 20)$

$= x(x-5)(x+4) \Rightarrow x = 0, 5, -4$

These three numbers divide the x-axis into four regions. Choose an x-value in each region as a test number and compute the function value.

Region	Test No.	Value of $f(x)$	Graph
$x < -4$	–5	–50	below x-axis
$-4 < x < 0$	–1	18	above x-axis
$0 < x < 5$	1	–20	below x-axis
$5 < x$	10	700	above x-axis

Use this information, the x-intercepts, and the fact that the graph can have a total of at most 2 peaks and valleys to sketch the graph.

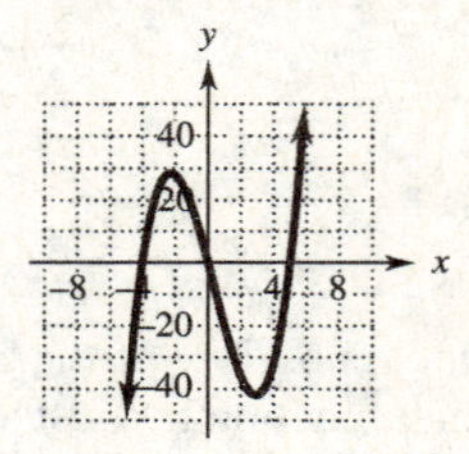

21. $f(x) = x^3 + 4x^2 - 7x$

First, find x-intercepts by setting $f(x) = 0$ and solving for x.

$f(x) = 0 = x^3 + 4x^2 - 7x = x(x^2 + 4x - 7) \Rightarrow$

$x = 0$ or $x^2 + 4x - 7 = 0$.

Use the quadratic formula to solve

$x^2 + 4x - 7 = 0$.

$$x = \frac{-4 \pm \sqrt{4^2 - 4(1)(-7)}}{2(1)} = \frac{-4 \pm \sqrt{44}}{2} \Rightarrow$$

$x \approx 1.3$ or $x \approx -5.3$

These three numbers divide the x-axis into four regions. Choose an x-value in each region as a test number and compute the function value.

Region	Test No.	Value of $f(x)$	Graph
$x < -5.3$	−6	−30	below x-axis
$-5.3 < x < 0$	−1	10	above x-axis
$0 < x < 1.3$	1	−2	below x-axis
$1.3 < x$	2	10	above x-axis

Use this information, the x-intercepts, and the fact that the graph can have a total of at most 2 peaks and valleys to sketch the graph.

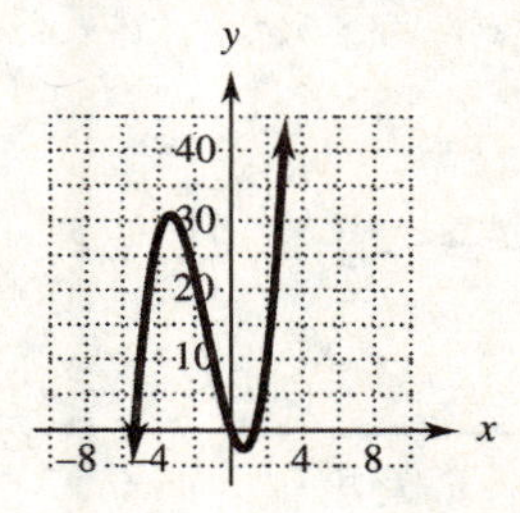

23. To graph $g(x) = x^3 - 3x^2 - 4x - 5$, enter the function as Y_1. There are many possible viewing windows that will show the complete graph. One such window is $-3 \le x \le 5$ and $-20 \le y \le 5$.

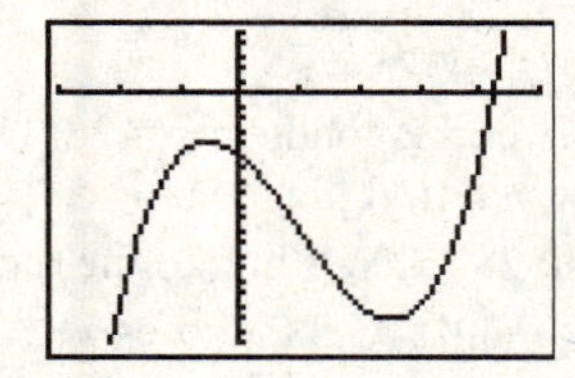

25. To graph $f(x) = 2x^5 - 3.5x^4 - 10x^3 + 5x^2 + 12x + 6$, enter the function as Y_1. There are many possible viewing windows that will show the complete graph. One such window is $-3 \le x \le 4$ and $-35 \le y \le 20$.

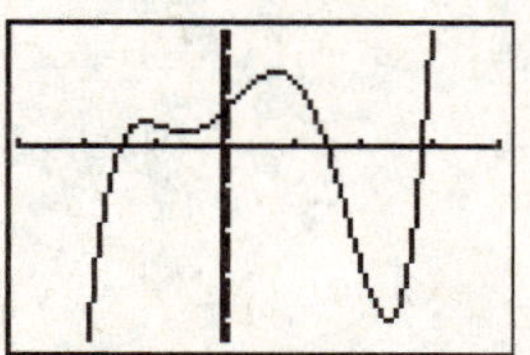

27. a. For a 20% tax rate the revenue is,

$$f(20) = \frac{20(20-100)(20-160)}{240} = \frac{224,000}{240} \approx \$933.33 \text{ billion}$$

b. For a 40% tax rate the revenue is,

$$f(40) = \frac{40(40-100)(40-160)}{240} = \frac{288,000}{240} = \$1200 \text{ billion}$$

c. For a 50% tax rate the revenue is,

$$f(50) = \frac{50(50-100)(50-160)}{240} = \frac{275,000}{240} \approx \$1145.8 \text{ billion}$$

d. For a 70% tax rate the revenue is,

$$f(70) = \frac{70(70-100)(70-160)}{240} = \frac{189,000}{240} = \$787.5 \text{ billion}$$

e.

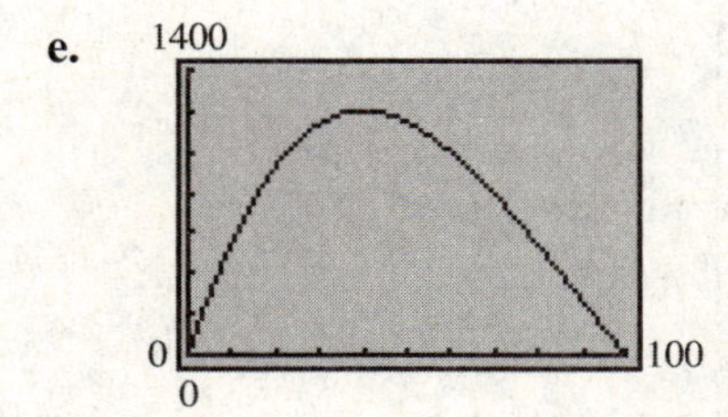

29. a. $R(x) = .141x^3 - 3.11x^2 + 26.98x - 19.99$

$R(x) = .141(5)^3 - 3.11(5)^2 + 26.98(5) - 19.99 = 54.785$

$R(x) = .141(7)^3 - 3.11(7)^2 + 26.98(7) - 19.99 = 64.843$

$R(x) = .141(12)^3 - 3.11(12)^2 + 26.98(12) - 19.99 = 99.578$

b.

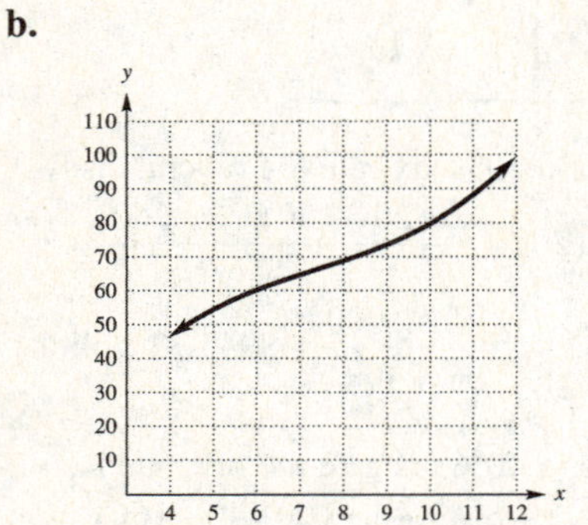

c. From the graph, the revenue is always increasing because the slope is always positive.

31. a. $C(x) = .135x^3 - 2.98x^2 + 25.99x - 18.71$

$C(x) = .135(5)^3 - 2.98(5)^2 + 25.99(5) - 18.71 = 53.615$

$C(x) = .135(7)^3 - 2.98(7)^2 + 25.99(7) - 18.71 = 63.505$

$C(x) = .135(12)^3 - 2.98(12)^2 + 25.99(12) - 18.71 = 97.33$

b.

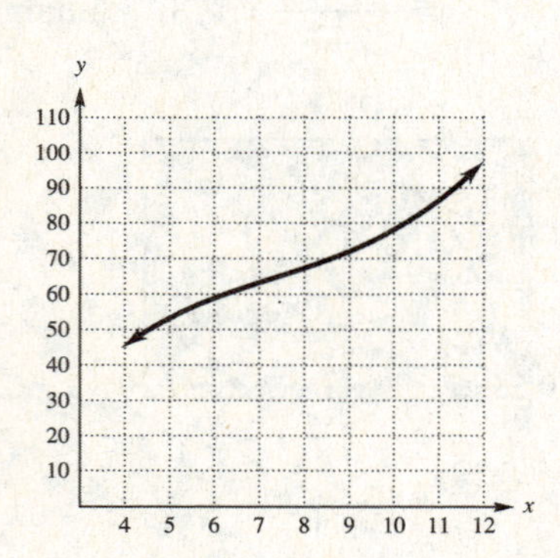

c. From the graph, the cost is always increasing because the slope is always positive.

33. $P(x) = R(x) - C(x)$

$= .141x^3 - 3.11x^2 + 26.98x + 19.99 - (.135x^3 - 2.98x^2 + 25.99x + 18.71)$

$= .006x^3 - .13x^2 + .99x - 1.28$

$P(12) = .006(12)^3 - .13(12)^2 + .99(12) - 1.28 = 2.248$

According to the models, the profit was \$2.248 million in 2012.

35. a.

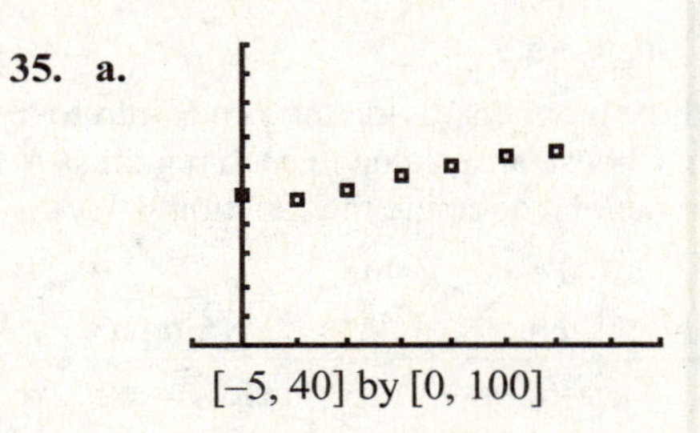

[–5, 40] by [0, 100]

b.

```
CubicReg
 y=ax³+bx²+cx+d
 a=-.0015555556
 b=.0764761905
 c=-.4249206349
 d=50.01190476
```

$g(x) = -.0016x^3 + .0765x^2 - .425x + 50.0$

c.

[–5, 40] by [0, 100]

Yes; the graph fits reasonably well.

d.

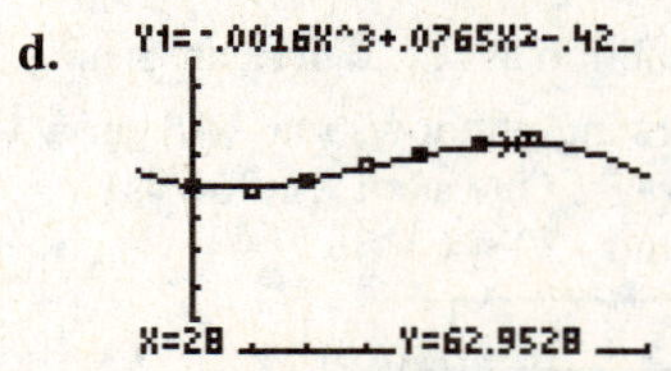

The year 2008 corresponds to $x = 28$. Using the calculator, we find that $f(28) \approx 62.9528$. According to the model, the enrollment will be about 62.9528 million in 2008.

37. a.

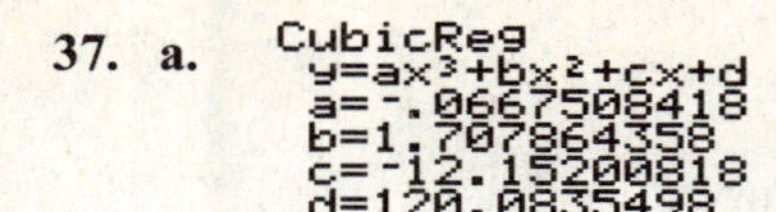

$$R(x) = -.0668x^3 + 1.709x^2 - 12.15x + 120.1$$

b.

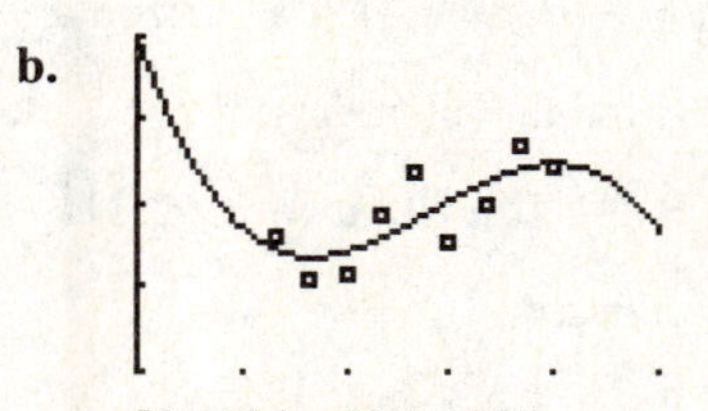

[0,15] by [80, 120]

Yes; the graph fits reasonably well.

39. $P(x) = R(x) - C(x)$

$$= .0668x^3 + 1.709x^2 - 12.15x + 120.1 - \left(-.0355x^3 + .935x^2 .7.32x + 103.0\right)$$

$$= -.0313x^3 + .774x^2 - 4.83x + 17.1$$

Section 3.6 Rational Functions

1. $f(x) = \dfrac{1}{x+5}$

Vertical asymptote:
If $x + 5 = 0$, then $x = -5$, so the line $x = -5$ is a vertical asymptote.

Horizontal asymptote: $y = \dfrac{1}{x+5} = \dfrac{0x+1}{1x+5}$

The line $y = \dfrac{a}{c} = \dfrac{0}{1}$ or $y = 0$ is a horizontal asymptote. There are no x-intercepts.

y-intercept: $y = \dfrac{1}{0+5} = \dfrac{1}{5}$

Make a table of values.

x	−8	−7	−6	−5.5	−4.5	−4	−3	−2
y	$-\frac{1}{3}$	$-\frac{1}{2}$	−1	−2	2	1	$\frac{1}{2}$	$\frac{1}{3}$

Using these points and the asymptotes, we graph the function.

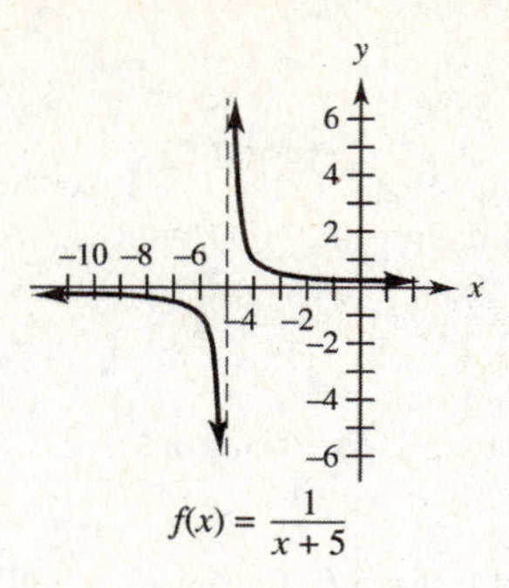

$f(x) = \dfrac{1}{x+5}$

3. $f(x) = \dfrac{-3}{2x+5}$

Vertical asymptote:

If $2x + 5 = 0$, $x = -\dfrac{5}{2}$, so the line $x = -\dfrac{5}{2}$ is a vertical asymptote.

Horizontal asymptote: $y = -\dfrac{3}{2x+5} = \dfrac{0x-3}{2x+5}$

The line $y = \dfrac{0}{2}$ or $y = 0$ is a horizontal asymptote. There are no x-intercepts. If $x = 0$, then $y = \dfrac{-3}{2(0)+5} = -\dfrac{3}{5}$.

Make a table of values.

x	−5	−4	−3	−2	−1	0
y	$\frac{3}{5}$	1	3	−3	−1	$-\frac{3}{5}$

Using these points and the information above, we graph the function.

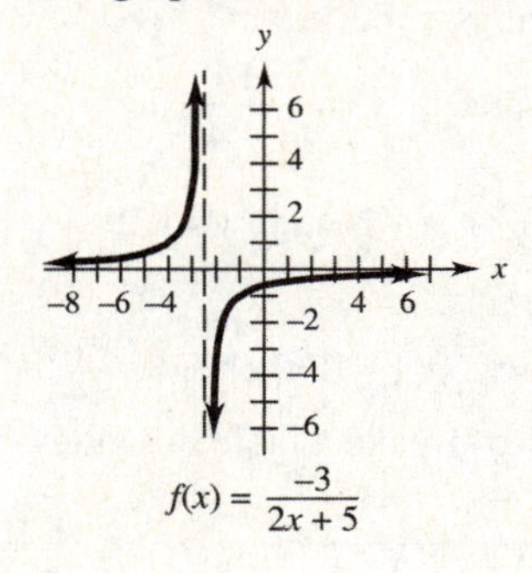

$f(x) = \dfrac{-3}{2x+5}$

5. $f(x)=\dfrac{3x}{x-1}$

Vertical asymptote: If $x-1=0$, $x=1$, so the line $x=1$ is a vertical asymptote.

Horizontal asymptote: $y=\dfrac{3x+0}{1x-1}$

The line $y=\dfrac{3}{1}$ or $y=3$ is a horizontal asymptote.

Find the intercepts. If $x=0$, then $y=f(x)=0$. Next, make a table of values.

x	–3	–2	–1	0	2	3	4	5
y	$\frac{9}{4}$	2	$\frac{3}{2}$	0	6	$\frac{9}{2}$	4	$\frac{15}{4}$

Using these points and the information above, we graph the function.

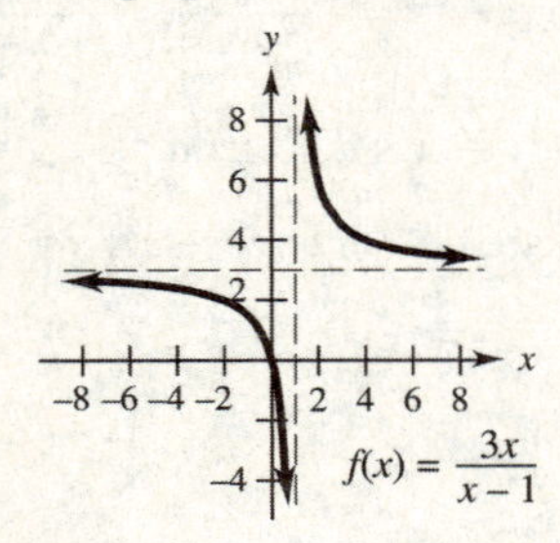

7. $f(x)=\dfrac{x+1}{x-4}$

Vertical asymptote: If $x-4=0$, $x=4$, so the line $x=4$ is a vertical asymptote.

Horizontal asymptote: $y=\dfrac{1x+1}{1x-4}$

The line $y=\dfrac{1}{1}$ or $y=1$ is a horizontal asymptote.

Find the intercepts. If $x=0$, then $y=-\frac{1}{4}$. If $y=0$, then $x=-1$. Now make a table of values.

x	0	1	2	3	5	6	7	8
y	$-\frac{1}{4}$	$-\frac{2}{3}$	$-\frac{3}{2}$	–4	6	$\frac{7}{2}$	$\frac{8}{3}$	$\frac{9}{4}$

Using these points and the information above, we graph the function.

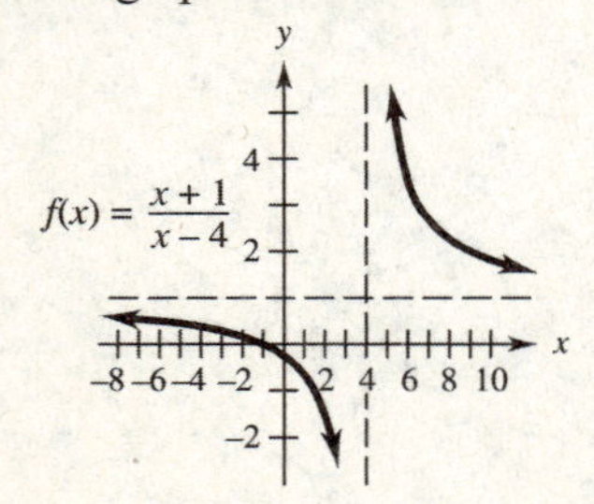

9. $f(x)=\dfrac{2-x}{x-3}$

Vertical asymptote:
If $x-3=0$, $x=3$, so the line $x=3$ is a vertical asymptote. Horizontal asymptote: $y=\dfrac{-1x+2}{1x-3}$

The line $y=-\dfrac{1}{1}$ or $y=-1$ is a horizontal asymptote. Find the intercepts. If $x=0$, then $f(x)=y=-\dfrac{2}{3}$. If $y=0$, then $x=2$. Now, make a table of values.

x	–1	0	1	2	4	5	6	7
y	$-\frac{3}{4}$	$-\frac{2}{3}$	$-\frac{1}{2}$	0	–2	$-\frac{3}{2}$	$-\frac{4}{3}$	$-\frac{5}{4}$

Using these points and the information above, we graph the function.

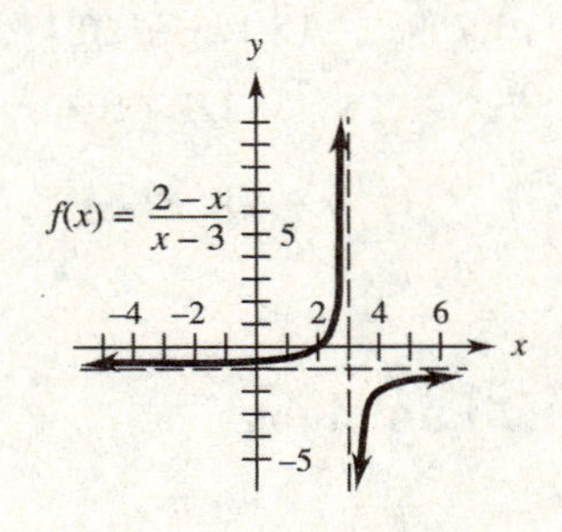

11. $f(x)=\dfrac{3x+2}{2x+4}$

Vertical asymptote: Set the denominator equal to 0 and solve for x. $2x+4=0 \Rightarrow x=-2$.
The line $x=-2$ is a vertical asymptote.
Horizontal asymptote:

$y=\dfrac{3x+2}{2x+4} \Rightarrow y=\dfrac{3}{2}$ is a horizontal asymptote.

Find the intercepts. If $x=0$, then

$y=\dfrac{3(0)+2}{2(0)+4}=\dfrac{1}{2}$. If $y=0$, then $0=\dfrac{3x+2}{2x+4} \Rightarrow$

$3x+2=0 \Rightarrow x=-\dfrac{2}{3}$.

(*continued next page*)

Now make a table of values.

x	−10	−6	−4	−3	−1	0	2	8
y	$\frac{7}{4}$	2	$\frac{5}{2}$	$\frac{7}{2}$	$-\frac{1}{2}$	$\frac{1}{2}$	1	$\frac{13}{10}$

Using these points and the information above, we graph the function.

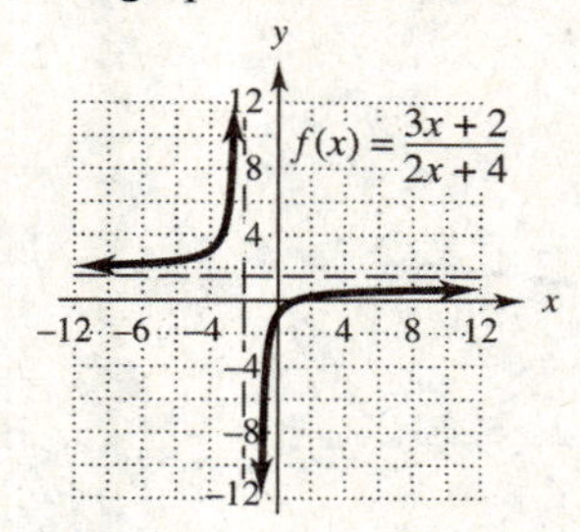

13. $h(x)=\frac{x+1}{x^2+2x-8}=\frac{x+1}{(x-4)(x+2)}$

Set the factors in the denominator equal to 0 and solve. $x-4=0 \Rightarrow x=4$; $x+2=0 \Rightarrow x=-2$.

To find horizontal asymptote, divide the numerator and the denominator by x^2.

$$\frac{\frac{x}{x^2}+\frac{1}{x^2}}{\frac{x^2}{x^2}+\frac{2x}{x^2}-\frac{8}{x^2}}=\frac{\frac{1}{x}+\frac{1}{x^2}}{1+\frac{2x}{x^2}-\frac{8}{x^2}}$$

As $|x|$ gets very large, the numerator gets close to 0 and the denominator gets close to 1, so the function has a horizontal asymptote at $y = 0$.

Find the intercepts. If $x = 0$, $y=-\frac{1}{8}$.

If $y = 0$, $0=\frac{x+1}{(x-4)(x+2)} \Rightarrow$

$x+1=0 \Rightarrow x=-1$. Now, make a table of values.

x	−10	−8	−6	−4
y	$-\frac{1}{8}$	$-\frac{7}{40}$	$-\frac{5}{16}$	undefined

x	−3	−2	−1	0	1
y	$\frac{2}{5}$	$\frac{1}{8}$	0	$-\frac{1}{8}$	$-\frac{2}{5}$

x	2	4	6	8	10
y	undefined	$\frac{5}{16}$	$\frac{7}{40}$	$\frac{1}{8}$	$\frac{11}{112}$

Using these points and the asymptotes, graph the function.

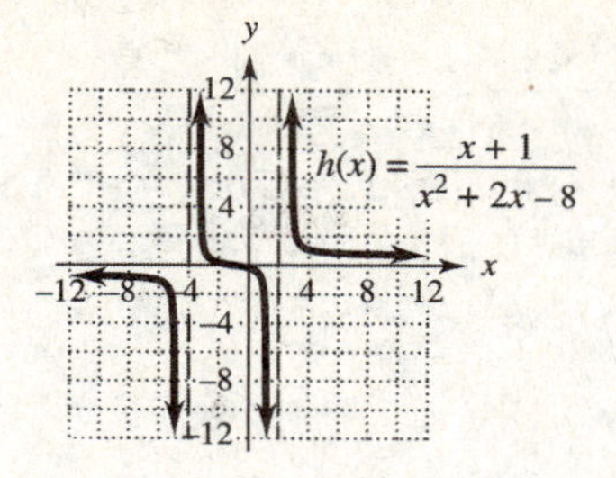

15. $f(x)=\frac{x^2+4}{x^2-4}$

To find the vertical asymptotes, set the denominator equal to 0 and solve.

$x^2-4=0 \Rightarrow (x-2)(x+2)=0 \Rightarrow x=\pm 2$

To find horizontal asymptotes, divide the numerator and the denominator by x^2.

$$\frac{\frac{x^2}{x^2}+\frac{4}{x^2}}{\frac{x^2}{x^2}-\frac{4}{x^2}}=\frac{1+\frac{4}{x^2}}{1-\frac{4}{x^2}}$$

As $|x|$ gets very large, the function approaches 1. So the function has a horizontal asymptote at $y = 1$. Find the intercepts.

If $x = 0$, $\frac{0+4}{0-4}=-1$, so the y-intercept is −1.

If $y = 0$, $0=\frac{x^2+4}{x^2-4} \Rightarrow x^2=-4$ is undefined, so there is no x-intercept. Now make a table of values.

x	−6	−4	−3	−1	0	1	3	4
y	$1\frac{1}{4}$	$1\frac{2}{3}$	2.6	$-1\frac{2}{3}$	−1	$-1\frac{2}{3}$	$2\frac{3}{5}$	$1\frac{2}{3}$

Use these points and the information above, graph the function.

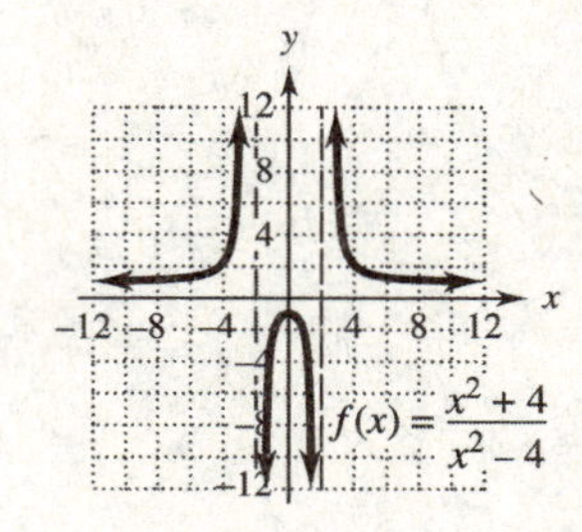

17. $f(x)=\frac{x-3}{x^2+x-2}=\frac{(x-3)}{(x+2)(x-1)}$

If $(x + 2)(x - 1) = 0$, then $x = -2$ or $x = 1$. The lines $x = -2$ and $x = 1$ are vertical asymptotes.

19. $g(x) = \dfrac{x^2 + 2x}{x^2 - 4x - 5} = \dfrac{x(x+2)}{(x-5)(x+1)}$

If $(x - 5)(x + 1) = 0$, then $x = 5$ or $x = -1$. The lines $x = 5$ and $x = -1$ are vertical asymptotes.

21. $f(x) = \dfrac{4.3x}{100 - x}$

a. $f(50) = \dfrac{4.3(50)}{100 - 50} \Rightarrow = 4.3$ or \$4300

b. $f(70) = \dfrac{4.3(70)}{100 - 70} = 10.03333$ or \$10,033.33

c. $f(80) = \dfrac{4.3(80)}{100 - 80} = 17.2$ or \$17,200

d. $f(90) = \dfrac{4.3(90)}{100 - 90} = 38.7$ or \$38,700

e. $f(95) = \dfrac{4.3(95)}{100 - 95} = 81.7$ or \$81,700

f. $f(98) = \dfrac{4.3(98)}{100 - 98} = 210.7$ or \$210,700

g. $f(99) = \dfrac{4.3(99)}{100 - 99} = 425.7$ or \$425,700

h. Since $f(100)$ is undefined, all the pollutant cannot be removed according to his model.

i.

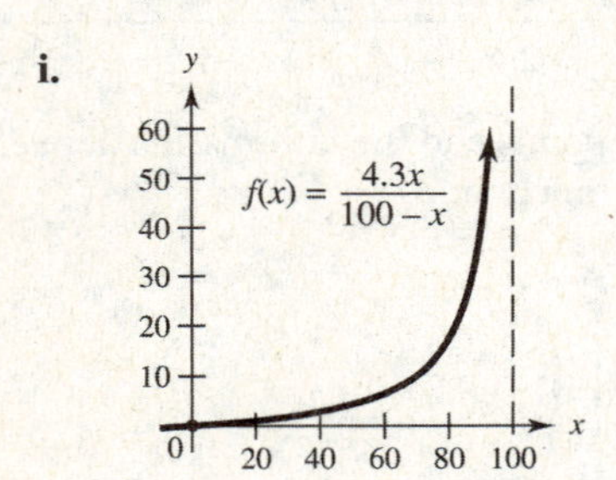

23. a. Since x represents the size of a generation, x cannot be a negative number. The domain of x is $[0, \infty)$.

b. Let $\lambda = a = b = 1$ and $x \geq 0$.

$f(x) = \dfrac{x}{1 + x}$

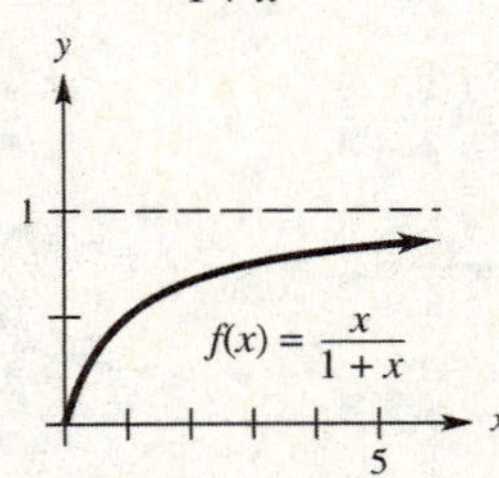

c. Let $\lambda = a = 1$, $b = 2$, and $x \geq 0$

$f(x) = \dfrac{x}{1 + x^2}$

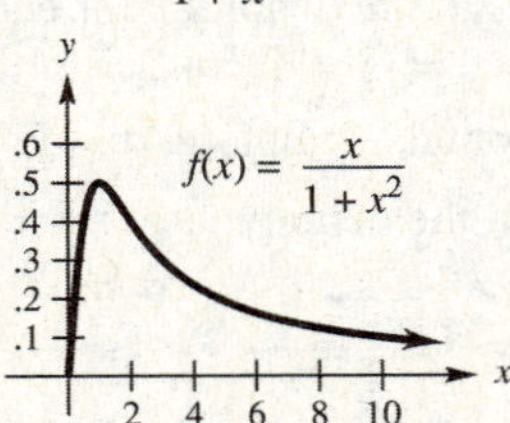

d. Increasing b makes the next generation smaller when this generation is larger.

25. $W = \dfrac{S(S - A)}{A} = \dfrac{3(3 - A)}{A}$

a. If $A = 1$, $W = \dfrac{3(3-1)}{1} = 6$

The waiting time is 6 min.

b. If $A = 2$, $W = \dfrac{3(3-2)}{2} = 1.5$

The waiting time is 1.5 min.

c. If $A = 2.5$, $W = \dfrac{3(3-2.5)}{2.5} = .6$

The waiting time is .6 min.

d. The vertical asymptote occurs when the denominator is zero or A = 0.

e. Use the vertical asymptote and the values found in (a), (b), and (c) to graph the function for $0 < A \le 3$.

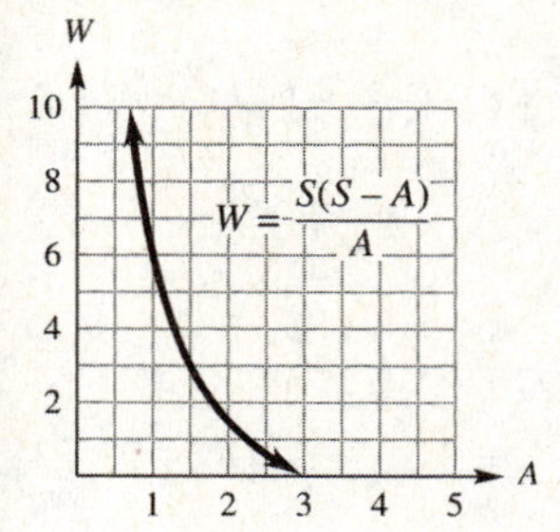

f. When $A < 3$, W is decreasing. The waiting time approaches 0 as A approaches 3. The formula does not apply for $A > 3$ because there will be no waiting if people arrive more than 3 minutes apart

27. $y = \dfrac{900,000,000 - 30,000x}{x + 90,000}$

x	0	10,000	20,000	30,000
y	10,000	6000	2727	0

The maximum number of red tranquilizers is 30,000. The maximum number of blue ones is 10,000.

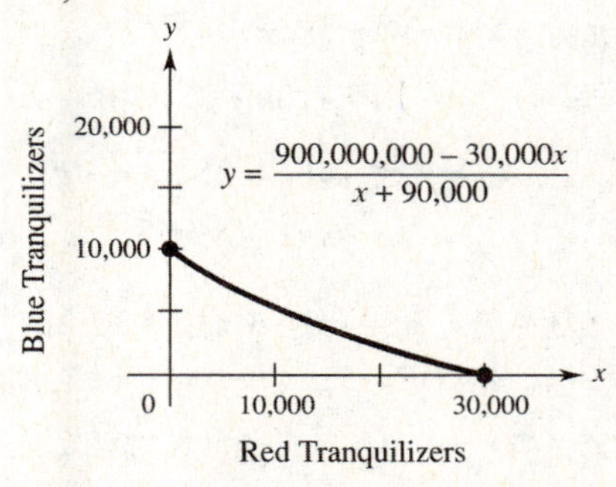

29. Fixed costs = \$40,000
Marginal cost = \$2.60/unit

a. $C(x) = 2.6x + 40,000$

b. Average cost per item:

$$\overline{C}(x) = \frac{C(x)}{x} = \frac{2.6x + 40,000}{x}$$
$$= 2.6 + \frac{40,000}{x}$$

c. As $|x|$ becomes large, $C(x)$ approaches 2.6. The horizontal asymptote is at $y = 2.6$. This means the average cost per item may get close to \$2.60, but never quite reach it.

31. $C(x) = .2x^3 - 25x^2 + 1531x + 25,000$

$$\overline{C}(x) = \frac{C(x)}{x} = .2x^2 - 25x + 1531 + \frac{25000}{x}$$

Graph the average cost function in the window [0, 150] by [0, 2000], and use the minimum finder to find the point with the smallest y-coordinate.

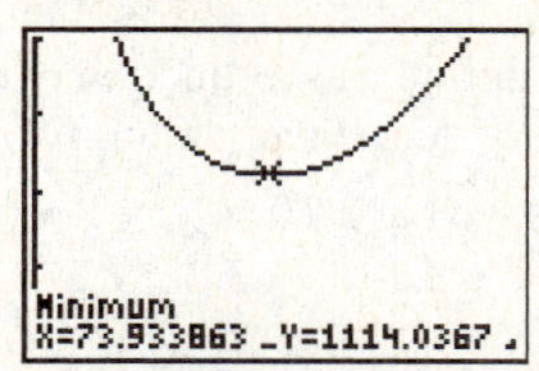

About 73.9 units should be produced to have the lowest possible average cost.

33. a.

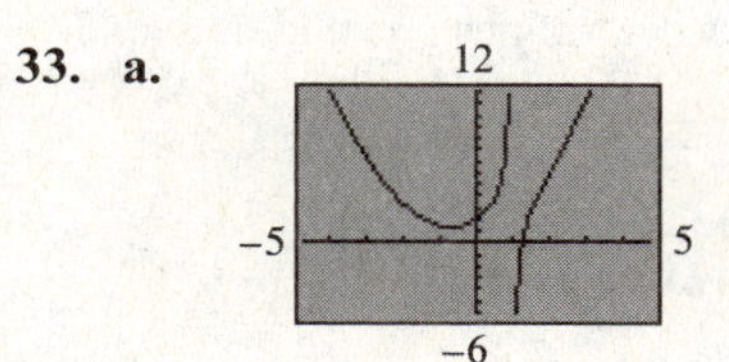

b.

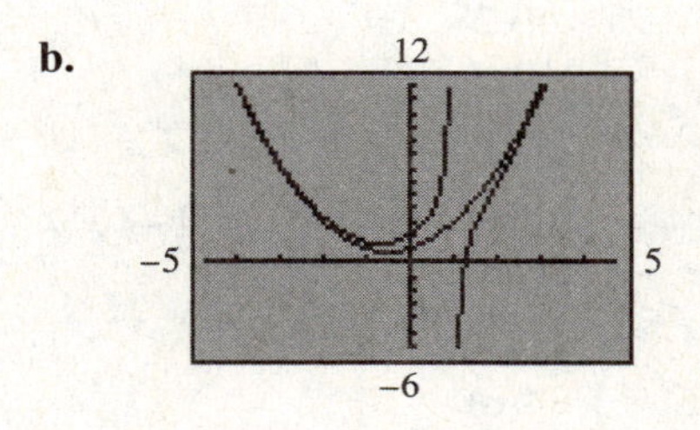

For $|x| \ge 2$, the two graphs appear almost identical because the parabola is an asymptote of the graph.

Chapter 3 Review Exercises

1. This rule does not define y as a function of x because the x-value of 2 corresponds to both 5 and −5.

2. This rule defines y as a function of x because each value of x determines one and only one value of y.

3. $y = \sqrt{x}$
   This rule defines y as a function of x because each value of x determines one and only one value of y.

4. $x = |y|$
This rule does not define y as a function of x. A given value of x may define two values for y. For example, if $x = 4$, we have $4 = |y|$, so $y = 4$ and $y = -4$.

5. $x = y^2 + 1$
This rule does not define y as a function of x. A given value of x may define two values for y. For example, if $x = 10$, we have $10 = y^2 + 1$, or $y^2 = 9$, so $y = \pm 3$.

6. $y = 5x - 2$
This rule defines y as a function of x because each value of x determines one and only one value of y.

7. $f(x) = 4x - 1$

a. $f(6) = 4(6) - 1 = 24 - 1 = 23$

b. $f(-2) = 4(-2) - 1 = -8 - 1 = -9$

c. $f(p) = 4p - 1$

d. $f(r+1) = 4(r+1) - 1 = 4r + 4 - 1 = 4r + 3$

8. $f(x) = 3 - 4x$

a. $f(6) = 3 - 4(6) = -21$

b. $f(-2) = 3 - 4(-2) = 3 + 8 = 11$

c. $f(p) = 3 - 4p$

d. $f(r+1) = 3 - 4(r+1) = -1 - 4r$

9. $f(x) = -x^2 + 2x - 4$

a. $f(6) = -6^2 + 2(6) - 4 = -36 + 12 - 4 = -28$

b. $f(-2) = -(-2)^2 + 2(-2) - 4$
$= -4 - 4 - 4 = -12$

c. $f(p) = -p^2 + 2p - 4$

d. $f(r+1) = -(r+1)^2 + 2(r+1) - 4$
$= -(r^2 + 2r + 1) + 2r + 2 - 4$
$= -r^2 - 2r - 1 + 2r + 2 - 4$
$= -r^2 - 3$

10. $f(x) = 8 - x - x^2$

a. $f(6) = 8 - 6 - 36 = -34$

b. $f(-2) = 8 - (-2) - (-2)^2 = 8 + 2 - 4 = 6$

c. $f(p) = 8 - p - p^2$

d. $f(r+1) = 8 - (r+1) - (r+1)^2$
$= 8 - r - 1 - (r^2 + 2r + 1)$
$= 8 - r - 1 - r^2 - 2r - 1$
$= -r^2 - 3r + 6$

11. $f(x) = 5x - 3$ and $g(x) = -x^2 + 4x$

a. $f(-2) = 5(-2) - 3 = -10 - 3 = -13$

b. $g(3) = -3^2 + 4(3) = -9 + 12 = 3$

c. $g(-k) = -(-k)^2 + 4(-k) = -k^2 - 4k$

d. $g(3m) = -(3m)^2 + 4(3m) = -9m^2 + 12m$

e. $g(k-5) = -(k-5)^2 + 4(k-5)$
$= -(k^2 - 10k + 25) + 4k - 20$
$= -k^2 + 10k - 25 + 4k - 20$
$= -k^2 + 14k - 45$

f. $f(3-p) = 5(3-p) - 3 = 15 - 5p - 3$
$= 12 - 5p$

12. $f(x) = x^2 + x + 1$

a. $f(3) = 3^2 + 3 + 1 = 13$

b. $f(1) = 1^2 + 1 + 1 = 3$

c. $f(4) = 4^2 + 4 + 1 = 21$

d. No, $f(3) + f(1) = 13 + 3 = 16$, while $f(3 + 1) = f(4) = 21$.

13. $f(x) = |x| - 3$

$$y = \begin{cases} x - 3 & \text{if } x \geq 0 \\ -x - 3 & \text{if } x < 0 \end{cases}$$

We graph the line $y = x - 3$ for $x \geq 0$. We graph the line $y = -x - 3$ for $x < 0$.

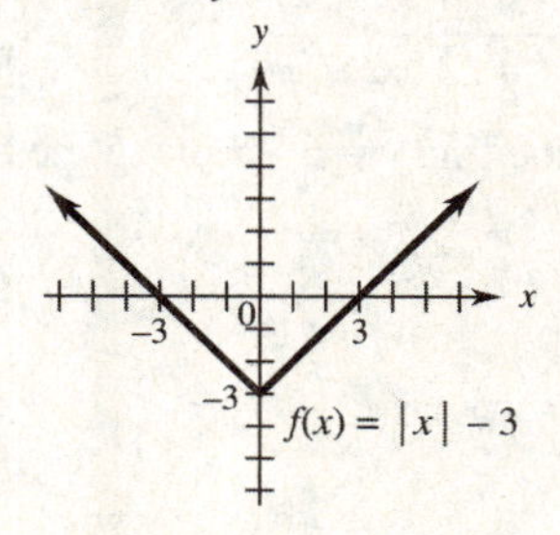

14. $f(x) = -|x| - 2$

$$y = \begin{cases} -x - 2 & \text{for } x \geq 0 \\ -(-x) - 2 & \text{for } x < 0 \end{cases}$$

or

$$y = \begin{cases} -x - 2 & \text{for } x \geq 0 \\ x - 2 & \text{for } x < 0 \end{cases}$$

We graph the line $y = -x - 2$ for $x \geq 0$. We graph the line $y = x - 2$ for $x < 0$.

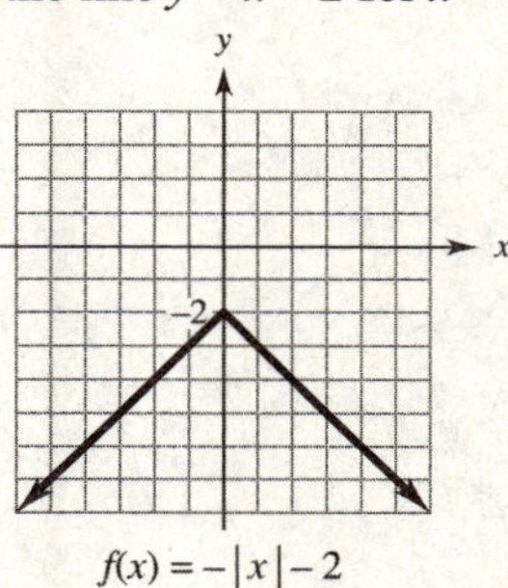

15. $f(x) = -|x + 1| + 3$

$$y = \begin{cases} -(x + 1) + 3 & \text{if } x + 1 \geq 0 \\ -[-(x + 1)] + 3 & \text{if } x + 1 < 0 \end{cases}$$

or

$$y = \begin{cases} -x + 2 & \text{if } x \geq -1 \\ x + 4 & \text{if } x < -1 \end{cases}$$

We graph the line $y = -x + 2$ for $x \geq -1$. We graph the line $y = x + 4$ for $x < -1$.

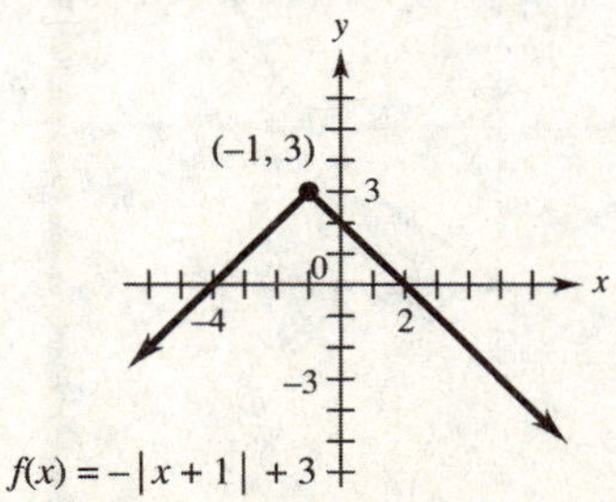

16. $f(x) = 2|x - 3| - 4$

$$y = \begin{cases} 2(x - 3) - 4 & \text{if } x - 3 \geq 0 \\ -2(x - 3) - 4 & \text{if } x - 3 < 0 \end{cases}$$

or

$$y = \begin{cases} 2x - 10 & \text{if } x \geq 3 \\ -2x + 2 & \text{if } x < 3 \end{cases}$$

We graph the line $y = 2x - 10$ for $x \geq 3$. We graph the line $y = -2x + 2$ for $x < 3$.

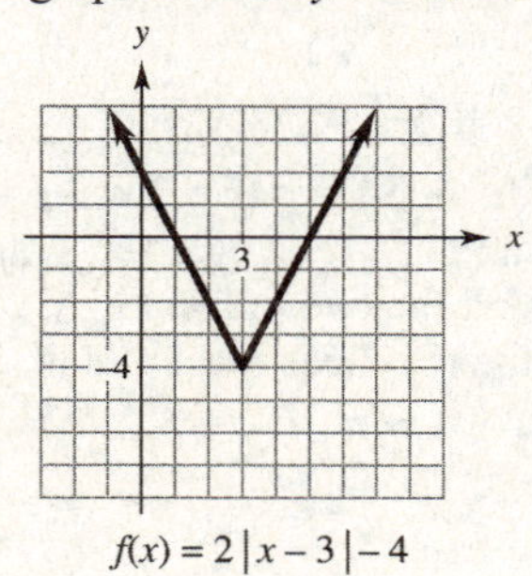

17. $f(x) = [x - 3]$

For x in the interval $[0, 1)$, $[x - 3] = -3$.
For x in the interval $[1, 2)$, $[x - 3] = -2$.
For x in the interval $[2, 3)$, $[x - 3) = -1$.
Continue in this pattern.
The graph consists of a series of line segments. In each case the left endpoint is included, and the right endpoint is excluded.

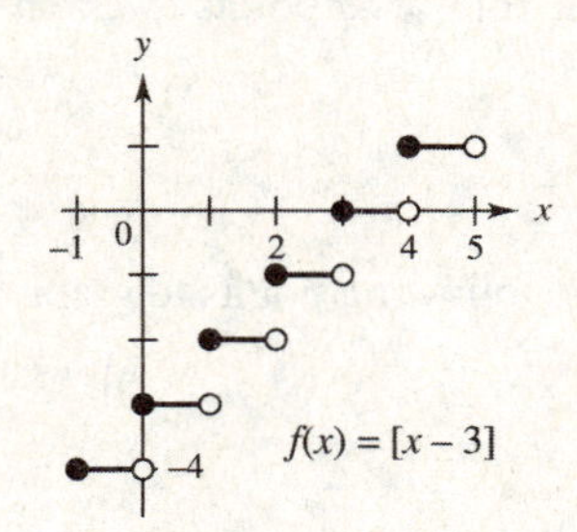

18. $f(x) = \left[\frac{1}{2}x - 2\right]$

$f(x)$ is a step function with breaks in the graph whenever x is an even integer.
If $x \in [-2, 0)$, $f(x) = -3$.
If $x \in [0, 2)$, $f(x) = -2$.
If $x \in [2, 4)$, $f(x) = -1$.
If $x \in [4, 6)$, $f(x) = 0$.
If $x \in [6, 8)$, $f(x) = 1$.

(*continued next page*)

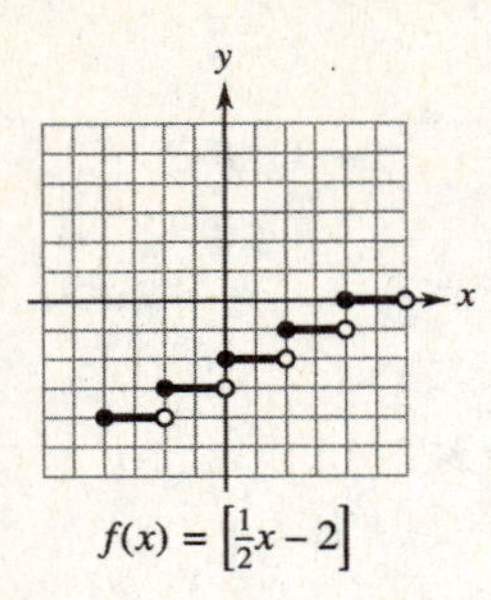

19. $f(x) = \begin{cases} -4x+2 & \text{if } x \le 1 \\ 3x-5 & \text{if } x > 1 \end{cases}$

For $x \le 1$, graph the line $y = -4x + 2$ using the two points (0, 2) and (1, –2). For $x > 1$, graph the line $y = 3x - 5$ using the two points (1, –2) and (4, 7). Note that the two lines meet at their common endpoint (1, –2).

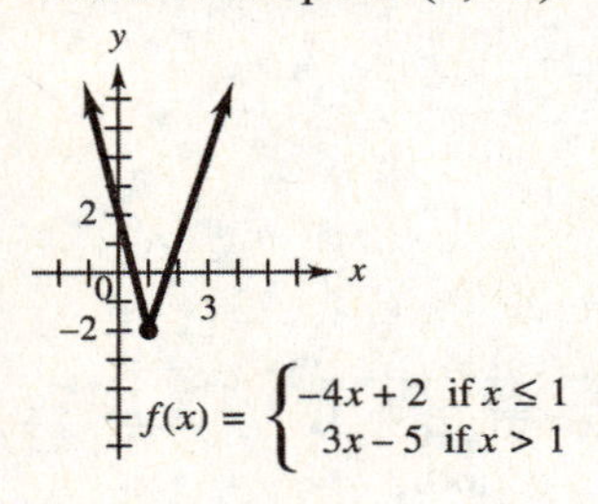

20. $f(x) = \begin{cases} 3x+1 & \text{if } x < 2 \\ -x+4 & \text{if } x \ge 2 \end{cases}$

For $x < 2$, plot the following points to graph the line $y = 3x + 1$.

x	0	1	2
y	1	4	7

For $x \ge 2$, plot the following points to graph the line $y = -x + 4$.

x	2	3	4
y	2	1	0

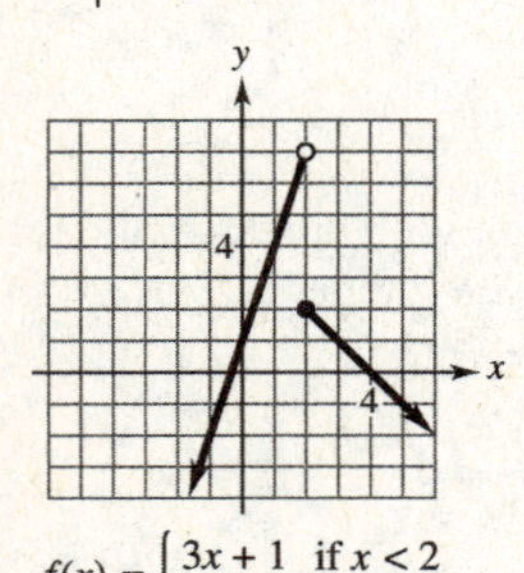

21. $f(x) = \begin{cases} |x| & \text{if } x < 3 \\ 6-x & \text{if } x \ge 3 \end{cases}$

For $x < 3$, graph $y = |x|$ using points from the following table.

x	–3	–2	–1	0	1	2	2.9
y	3	2	1	0	1	2	2.9

For $x \ge 3$, graph the line $y = 6 - x$ using the two points (3, 3) and (6, 0).

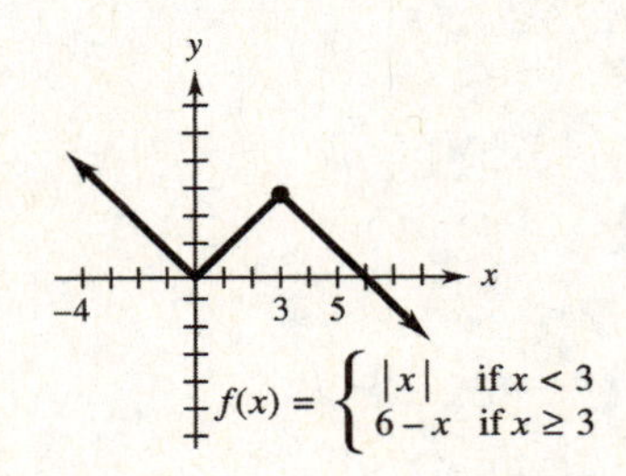

22. $f(x) = \sqrt{x^2}$

Use the table of values to draw the graph.

x	–4	–3	–2	–1	0	1	2	3	4	5
$f(x)$	4	3	2	1	0	1	2	3	4	5

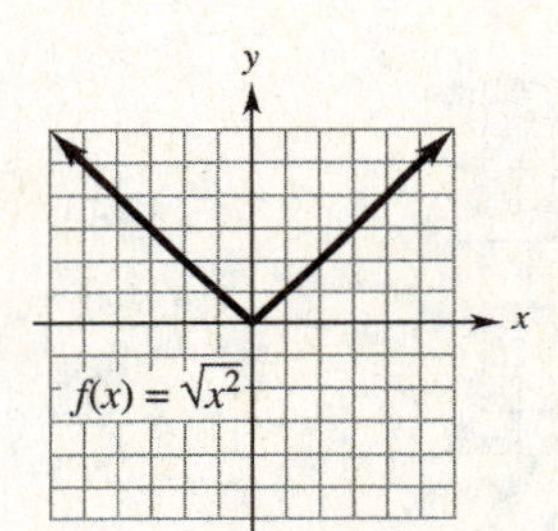

23. $g(x) = \dfrac{x^2}{8} - 3$

Use the table of values to draw the graph.

x	–8	–4	–2	0	2	4	8
$g(x)$	5	–1	$-2\frac{1}{2}$	–3	$-2\frac{1}{2}$	–1	5

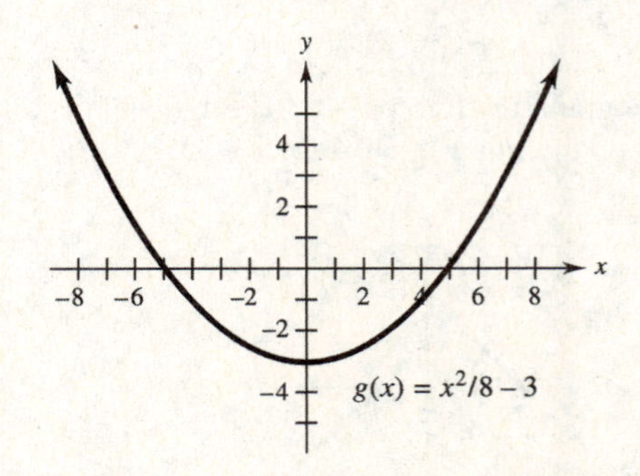

24. $h(x) = \sqrt{x} + 2$

Note that x must be greater than or equal to 0. Use the table of values to draw the graph.

x	0	1	4	9
$h(x)$	2	3	4	5

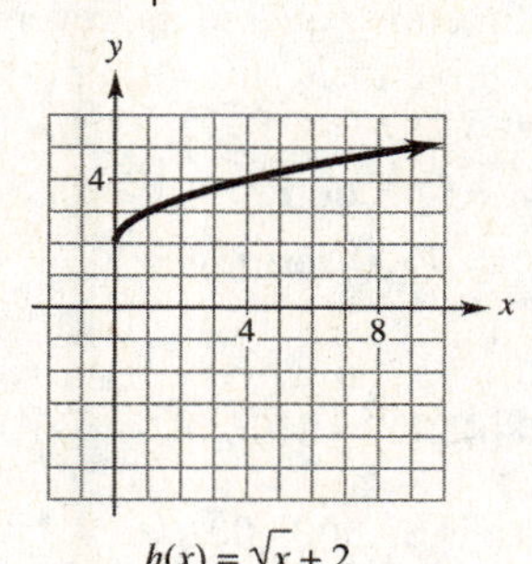

$h(x) = \sqrt{x} + 2$

25. a.

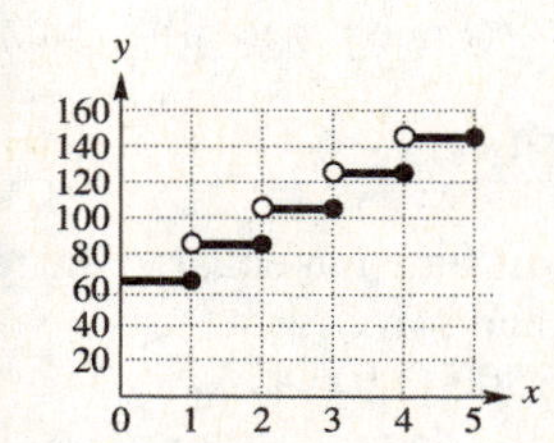

b. The domain is $(0, \infty)$.
The range is $\{65, 85, 105, 125, \ldots\}$.

c. If he can spend no more than \$90, he can rent the washer for at most 2 days.

26. a. The charge for 5 hours is
$400 + (80)(5) = 800$.
It costs \$800 for 5 hours work, so \$750 is not enough.

b.

c. The domain is $(0, \infty)$
The range is $\{400, 480, 560, 640, \ldots\}$.

27. a. For female high school seniors, the general trend is decreasing. For females 65 years and older, the general trend is relatively flat.

b. Using the points (0, 30) and (10, 16), we have $m = \dfrac{16-30}{10-0} = -1.4$.

$y - 30 = -1.4(x - 0) \Rightarrow$
$y - 30 = -1.4x \Rightarrow y = -1.4x + 30$

c. The year 2012 corresponds to $x = 12$.
$f(12) = -1.4(12) + 30 = 13.2$
If the trend continues, about 13.2% of female high school seniors smoked cigarettes in 2012.

28. a. For youths aged 15–19 years, the general trend is decreasing.

b. Using the points (10, 22) and (20, 11), we have $m = \dfrac{11-22}{20-10} = -1.1$.

$y - 22 = -1.1(x - 10) \Rightarrow$
$y - 22 = -1.1x + 11 \Rightarrow y = -1.1x + 33$

c. The year 2012 corresponds to $x = 22$.
$f(22) = -1.1(22) + 33 = 8.8$
If the trend continues, there will be about 8.8 deaths per 100,000 people in 2012.

29. Eight units cost \$300; fixed cost is \$60.

a. $C(x) = mx + b,\ b = 60$ and $C(8) = 300$
$C(x) = mx + 60 \Rightarrow 300 = m(8) + 60 \Rightarrow$
$8m = 240 \Rightarrow m = 30$
$C(x) = 30x + 60$

b. The marginal cost is given by the slope, so the marginal cost is \$30.

c.
$$\text{Average cost} = \frac{C(100)}{100} = \frac{30(100)+60}{100} = \frac{3060}{100} = 30.6$$
The average cost per unit to produce 100 units is \$30.60.

30. Fixed cost is \$2000; 36 units cost \$8480.

a. $C(x) = mx + 2000$
$C(36) = 8480 = 36m + 2000$
$m = \dfrac{8480 - 2000}{36} = 180$
$C(x) = 180x + 2000$

b. Since $m = 180$, the marginal cost is \$180.

c. Average cost $= \dfrac{C(100)}{100} = \dfrac{180(100) + 2000}{100}$
$= \dfrac{20,000}{100} = 200$
The average cost per unit to produce 100 units is \$200.

31. Twelve units cost \$445; 50 units cost \$1585.

a. $C(x) = mx + b$
Use the two points (12, 445) and (50, 1585).
$m = \dfrac{1585 - 445}{50 - 12} = \dfrac{1140}{38} = 30$
$y - 1585 = 30(x - 50) \Rightarrow$
$y - 1585 = 30x - 1500 \Rightarrow y = 30x + 85$
$C(x) = 30x + 85$

b. The slope is 30, so the marginal cost is \$30.

c. Average cost $= \dfrac{C(100)}{100} = \dfrac{30(100) + 85}{100}$
$= \dfrac{3085}{100} = 30.85$
The average cost per unit to produce 100 units is \$30.85.

32. Thirty units cost \$1500; 120 units cost \$5640. Use the points (30, 1500) and (120, 5640).

a. $m = \dfrac{5640 - 1500}{120 - 30} = \dfrac{4140}{90} = 46$
$C(x) = 46x + b \Rightarrow C(30) = 1500 \Rightarrow$
$46(30) + b = 1500 \Rightarrow b = 120$
$C(x) = 46x + 120$

b. The slope is 46, so the marginal cost is \$46.

c. Average cost $= \dfrac{C(100)}{100} = \dfrac{46(100) + 120}{100}$
$= \dfrac{4720}{100} = 47.20$
The average cost per item to produce 100 units is \$47.20.

33. a. The fixed cost is \$18,000.

b. Revenue = (Price per item) × (Number of items)
$R(x) = 28x$

c. Set the cost function equal to the revenue function and solve for x.
$24x + 18,000 = 28x$
$-4x = -18,000$
$x = 4500$ cartridges

d. Let $x = 4500$.
$R(4500) = 28(4500) = \$126,000$

34. a. The fixed cost is \$2,300,000.

b. Revenue = (Price per item) × (Number of items)
$R(x) = 450x$

c. Set the cost function equal to the revenue function and solve for x.
$325x + 2,300,000 = 450x$
$-125x = -2,300,000$
$x = 18,400$ printers

d. Let $x = 18,400$.
$R(18,400) = 450(18,400) = \$8,280,000$

35. $-.5q + 30.95 = .3q + 2.15 \Rightarrow -.8q = -28.8 \Rightarrow$
$q = 36$
$-.5(36) + 30.95 = \$12.95$ per month
The equilibrium quantity is 36 million subscribers at an equilibrium price of \$12.95 per month.

36. $.0015q + 1 = -.0025q + 64.36 \Rightarrow$
$.004q = 63.36 \Rightarrow q = 15,840$
$p = .0015(15,840) + 1 = 24.76$
The equilibrium quantity is 15,840 prescriptions at an equilibrium price of \$24.76.

37. $f(x) = 3(x - 2)^2 + 6$
Here $a = 3$, $h = 2$, and $k = 6$. The vertex is (2, 6). The parabola opens upward since $a = 3 > 0$.

38. $f(x) = 2(x + 3)^2 - 5$
Here $a = 2$, $h = -3$, and $k = -5$. The vertex is (–3, –5). The parabola opens upward since $a = 2 > 0$.

39. $g(x) = -4(x+1)^2 + 8$
$a = -4$, $h = -1$, $k = 8$. The vertex is $(-1, 8)$.
The parabola opens downward since $a = -4 < 0$.

40. $g(x) = -5(x-4)^2 - 6$
$a = -5$, $h = 4$, $k = -6$. The vertex is $(4, -6)$. The parabola opens downward since $a = -5 < 0$.

41. $f(x) = x^2 - 9$
First, locate the vertex of the parabola and find the axis of the parabola.
$y = 1(x-0)^2 - 9$
The vertex is at $(0, -9)$. The axis is the line $x = 0$. Make a table of values to find points on one side of the axis and then use the axis of symmetry to find the corresponding points on the other side.

x	1	2	3
y	−8	−5	0

Connect the points with a smooth curve.

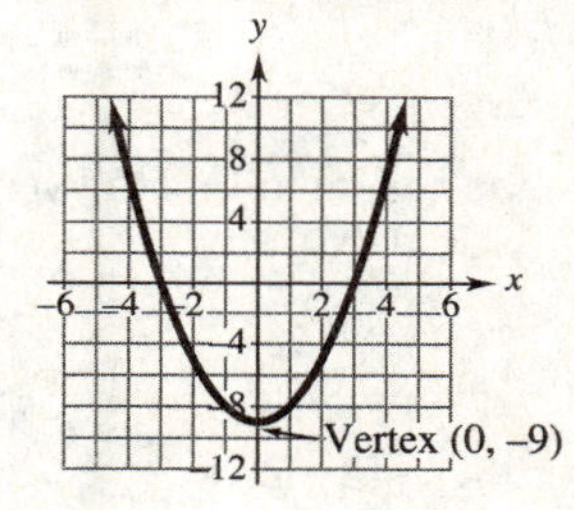

42. $f(x) = 5 - 2x^2 = -2(x-0)^2 + 5$
The vertex is $(0, 5)$. The axis is the line $x = 0$. Make a table of values to find points on one side of the axis and then use the axis of symmetry to find the corresponding points on the other side.

x	−3	−2	−1	0
y	−13	−3	3	5

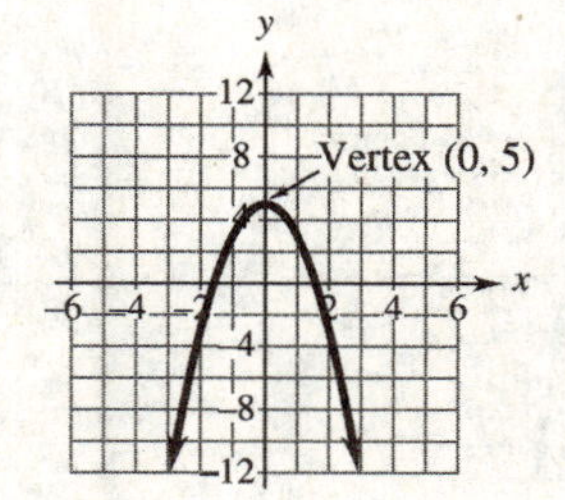

43. $f(x) = x^2 + 2x - 6$
We must complete the square to find the vertex.
$$f(x) = \left(x^2 + 2x + 1\right) - 6 - 1 = (x+1)^2 - 7$$
The vertex is $(-1, -7)$. The axis is the line $x = -1$. Make a table of values to find points on one side of the axis and then use the axis of symmetry to find the corresponding points on the other side.

x	−4	−3	−2	−1
y	2	−3	−6	−7

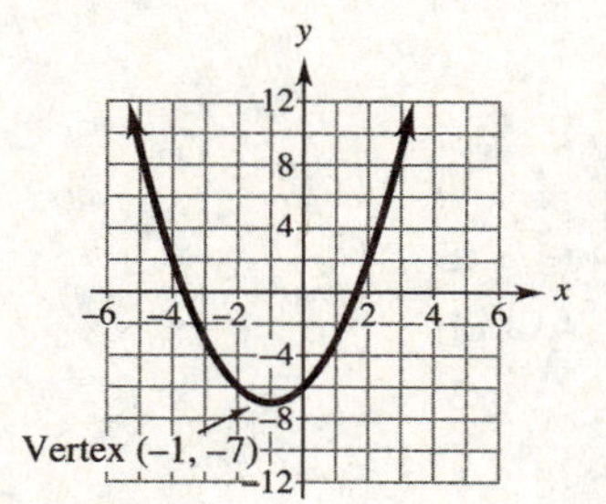

44. $f(x) = -x^2 + 8x - 1$
We must complete the square to find the vertex.
$$\begin{aligned} f(x) &= -\left(x^2 - 8x\right) - 1 \\ &= -\left(x^2 - 8x + 16 - 16\right) - 1 \\ &= -\left(x^2 - 8x + 16\right) - 1 + 16 \\ &= -(x-4)^2 + 15 \end{aligned}$$
The vertex is $(4, 15)$. The axis is the line $x = 4$. Make a table of values to find points on one side of the axis and then use the axis of symmetry to find the corresponding points on the other side.

x	0	1	2	3	4
y	−1	6	11	14	15

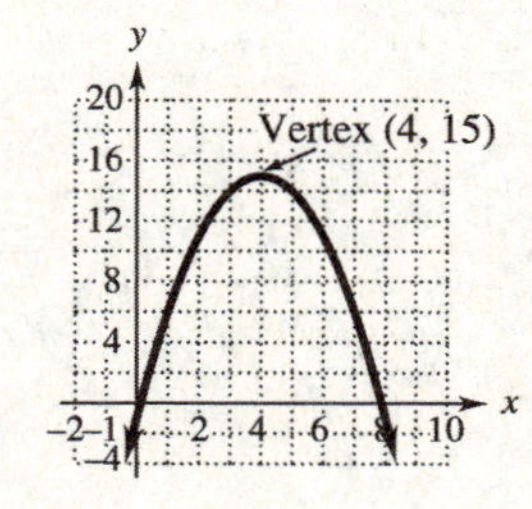

45. $f(x) = -x^2 - 6x + 5$

We must complete the square to find the vertex.

$$f(x) = -\left(x^2 + 6x\right) + 5 = -\left(x^2 + 6x + 9 - 9\right) + 5$$
$$= -\left(x^2 + 6x + 9\right) + 5 + 9 = -(x+3)^2 + 14$$

The vertex is (–3, 14). The axis is the line $x = -3$. Make a table of values to find points on one side of the axis and then use the axis of symmetry to find the corresponding points on the other side.

x	–6	–5	–4	–3
y	5	10	13	14

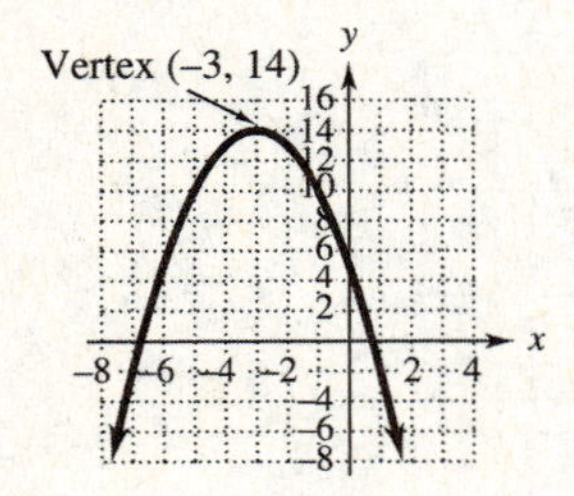

46. $f(x) = 5x^2 + 20x - 2$

We must complete the square to find the vertex.

$$f(x) = 5\left(x^2 + 4x\right) - 2 = 5\left(x^2 + 4x + 4 - 4\right) - 2$$
$$= 5\left(x^2 + 4x + 4\right) - 20 - 2 = 5(x+2)^2 - 22$$

The vertex is (–2, –22). The axis is the line $x = -2$. Make a table of values to find points on one side of the axis and then use the axis of symmetry to find the corresponding points on the other side.

x	–5	–4	–3	–2
y	23	–2	–17	–22

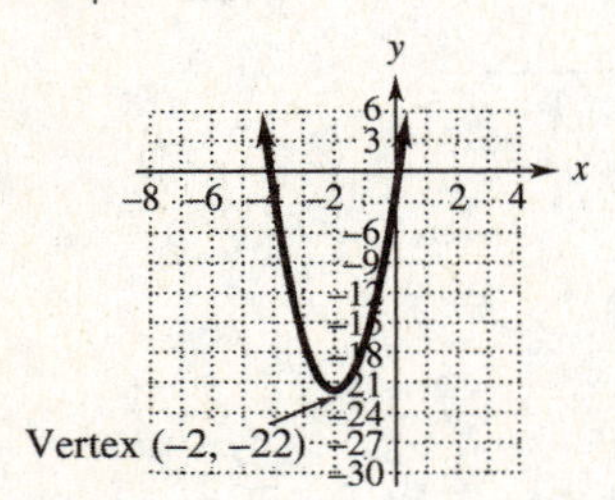

47. $f(x) = 2x^2 - 12x + 10$

We must complete the square to find the vertex.

$$f(x) = 2\left(x^2 - 6x\right) + 10 = 2\left(x^2 - 6x + 9 - 9\right) + 10$$
$$= 2\left(x^2 - 6x + 9\right) - 18 + 10 = 2(x-3)^2 - 8$$

The vertex is (3, –8). The axis is the line $x = 3$. Make a table of values to find points on one side of the axis and then use the axis of symmetry to find the corresponding points on the other side.

x	0	1	2	3
y	10	0	–6	–8

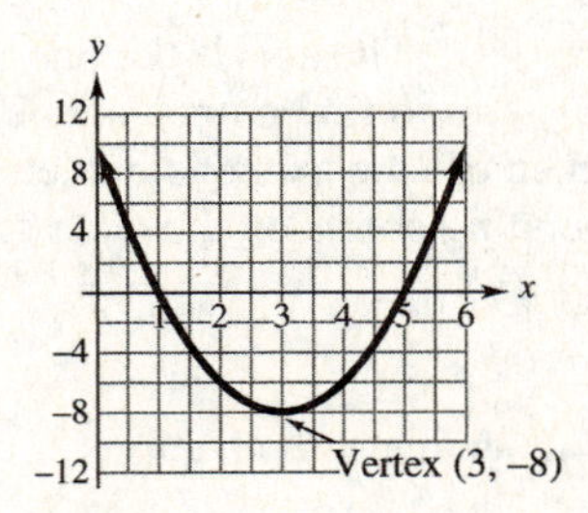

48. $f(x) = -3x^2 - 12x - 2$

We must complete the square to find the vertex.

$$f(x) = -3\left(x^2 + 4x\right) - 2$$
$$= -3\left(x^2 + 4x + 4 - 4\right) - 2$$
$$= -3\left(x^2 + 4x + 4\right) + 12 - 2$$
$$= -3(x+2)^2 + 10$$

The vertex is (–2, 10). The axis is the line $x = -2$. Make a table of values to find points on one side of the axis and then use the axis of symmetry to find the corresponding points on the other side.

x	–5	–4	–3	–2
y	–17	–2	7	10

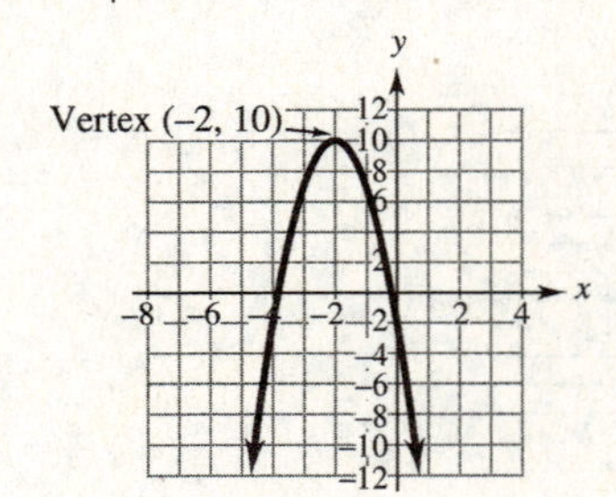

49. $f(x)=x^2+6x-2=\left(x^2+6x+9-9\right)-2$
$=\left(x^2+6x+9\right)-9-2=(x+3)^2-11$

The vertex is (–3, –11). Because $a=1>0$, the parabola opens upward and the function has a minimum value. This is the y-value of the vertex, which is –11.

50. $f(x)=x^2+4x+5=\left(x^2+4x+4\right)-4+5$
$==(x+2)^2+1$

The vertex is (–2, 1). Because $a=1>0$, the parabola opens upward and the function has a minimum value, which is 1.

51. $g(x)=-4x^2+8x+3=\left(-4x^2+8x\right)+3$
$=-4\left(x^2-2x\right)+3=-4\left(x^2-2x+1\right)+3+4$
$=-4(x-1)^2+7$

The vertex is (1, 7). Because $a=-4<0$, the parabola opens downward and the function has a maximum value, which is 7.

52. $g(x)=-3x^2-6x+3=\left(-3x^2-6x\right)+3$
$=-3\left(x^2+2x\right)+3$
$=-3\left(x^2+2x+1\right)+3+3=-3(x+1)^2+6$

The vertex is (–1, 6). The parabola opens downward and the function has a maximum value, which is 6.

53. The peak will occur at the vertex of the quadratic function. Use a formula to find the x coordinate:
$x=-\frac{b}{2a}=-\frac{.644}{2(-.002)}=161$. The peak will occur 161 weeks after the beginning of 2009. Now use the 161 weeks to find the peak price.
Thus, $f(161)=-.002(161)^2+.644(161)+45.32$
≈ 97.16

The peak price will be \$97.16 and occur 161 week after the beginning of 2009.

54. The lowest point will occur at the vertex of the quadratic function. Use a formula to find the x coordinate:
$x=-\frac{b}{2a}=-\frac{-2.40}{2(.096)}=12.5$. The lowest point will occur in the year 2012. Now use the year 2012 to find the lowest percentage. Thus,
$f(12.5)=.096(12.5)^2-2.40(12.5)+31.7$
≈ 16.7

The lowest percentage will be about 16.7% and occur in the year 2012.

55. The peak will occur at the vertex of the quadratic function. Use a formula to find the x coordinate:
$x=-\frac{b}{2a}=-\frac{12.86}{2(-.173)}\approx 37.16$. The peak will occur 37 years after 1990 or the year 2027. Now use the 37 to find the peak population. Thus,
$f(37.16)=-.173(37.16)^2+12.86(37.16)+1152$
≈ 1391

The peak population will be about 1391 million and occur in the year 2027.

56. The peak will occur at the vertex of the quadratic function. Use a formula to find the x coordinate:
$x=-\frac{b}{2a}=-\frac{220}{2(-1.44)}\approx 76.39$. The peak will occur 76 years after 1900 or the year 1976. Now use the 76 to find the peak amount of energy. Thus,
$f(76.39)=-1.44(76.39)^2+220(76.39)-6953$
≈ 1450

The peak amount of energy will be about 1450 kilowatt hours and occur in the year 1976.

57. a. The year 2005 corresponds to $x=35$.
$f(x)=a(x-5)^2+3672 \Rightarrow$
$26{,}908=a(35-5)^2+3672 \Rightarrow$
$26{,}908=900a+3672 \Rightarrow$
$900a=23{,}236 \Rightarrow a\approx 25.82$

The function is
$f(x)=25.82(x-5)^2+3672.$

b. The year 2015 corresponds to $x=45$.
$f(45)=25.82(45-5)^2+3672\approx 44{,}984$

According to the model, tuition and fees at private colleges will be about \$44,984 in 2015.

58. **a.** The year 2004 corresponds to $x = 4$.

$f(x) = a(x-6)^2 + 1418 \Rightarrow$

$1326 = a(9-6)^2 + 1418 \Rightarrow$

$1326 = 9a + 1418 \Rightarrow$

$9a = -92 \Rightarrow a \approx -10.22$

The function is

$f(x) = -10.22(x-6)^2 + 1418.$

b. The year 2012 corresponds to $x = 12$.

$f(12) = -10.22(12-6)^2 + 1418 \approx 1050$

According to the model, there were about 1050 thousand violent crimes in 2012.

59. **a.**

```
QuadReg
 y=ax²+bx+c
 a=9.738095238
 b=389
 c=1095.964286
```

$g(x) = 9.738x^2 + 389x + 1096$

b.

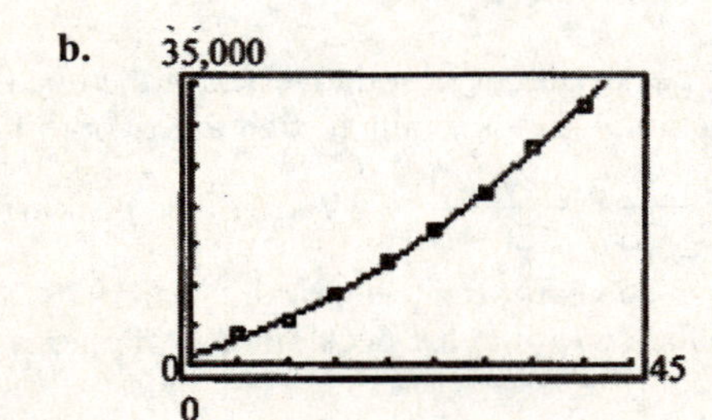

c. The year 2015 corresponds to $x = 45$.

$g(45) = 9.738(45)^2 + 389(45) + 1096$

$\approx 38{,}320.45$

According to the model, tuition and fees at private colleges will be about 38,320.45 in 2015. This is lower than the estimate in exercise 57.

60. **a.**

```
QuadReg
 y=ax²+bx+c
 a=-12.32142857
 b=154.75
 c=932.9285714
```

$g(x) = -12.321x^2 + 154.75x + 933$

b.

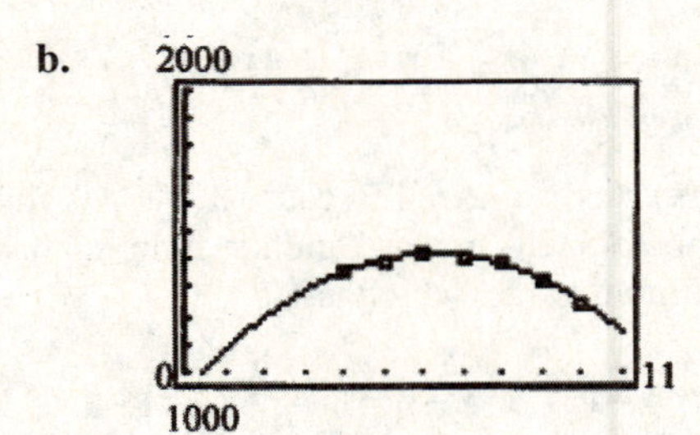

c. The year 2012 corresponds to $x = 12$.

$g(12) = -12.321(12)^2 + 154.75(12) + 933$

≈ 1016

According to the model, there were about 1016 thousand violent crimes in 2012. This is lower than the estimate in exercise 58.

61. $f(x) = x^4 - 5$

First we find several ordered pairs.

x	-2	-1.5	-1	0	1	1.5	2
y	11	$\frac{1}{16}$	-4	-5	-4	$\frac{1}{16}$	11

Plot these ordered pairs and draw a smooth curve through them.

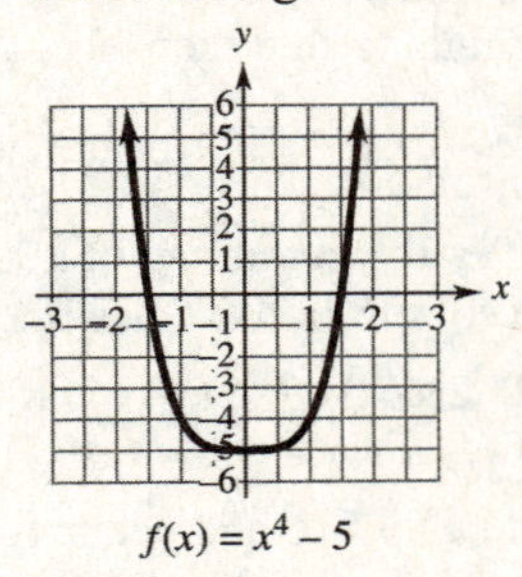

$f(x) = x^4 - 5$

62. $g(x) = x^3 - 4x = x(x^2 - 4) = x(x-2)(x+2)$

The x-intercepts are 0, –2, and 2.
These three numbers divide the x-axis into four regions. Choose an x-value in each region as a test number and compute the function value.

Region	Test No.	Value of $f(x)$	Graph
$x < -2$	–3	–15	below x-axis
$-2 < x < 0$	–1	3	above x-axis
$0 < x < 2$	1	–3	below x-axis
$2 < x$	3	15	above x-axis

Use this information, the x-intercepts, and the fact that the graph can have a total of at most 2 peaks and valleys to sketch the graph.

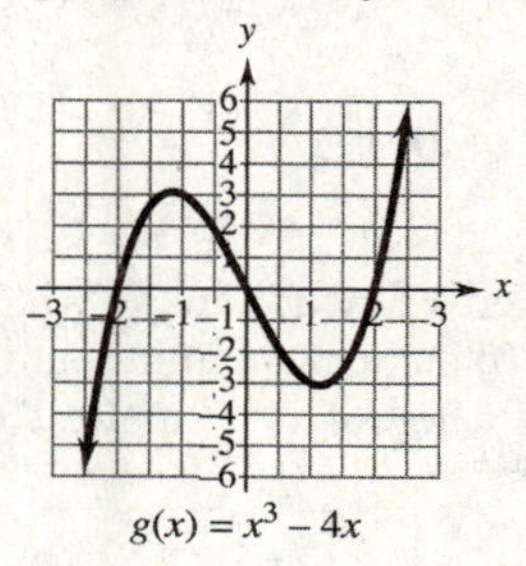

$g(x) = x^3 - 4x$

63. $f(x) = x(x-4)(x+1)$

The x-intercepts are 0, 4, and –1. These three numbers divide the x-axis into four regions. Choose an x-value in each region as a test number and compute the function value.

Region	Test No.	Value of $f(x)$	Graph
$x < -1$	–2	–12	below x-axis
$-1 < x < 0$	–.5	1.125	above x-axis
$0 < x < 4$	1	–6	below x-axis
$4 < x$	5	30	above x-axis

Use this information, the x-intercepts, and the fact that the graph can have a total of at most 2 peaks and valleys to sketch the graph.

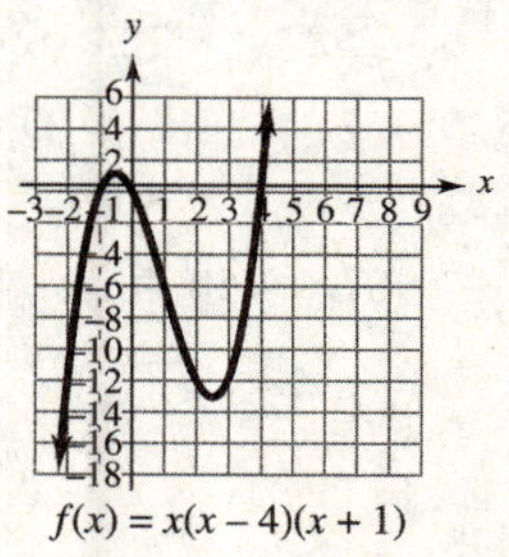

$f(x) = x(x - 4)(x + 1)$

64. $f(x) = (x-1)(x+2)(x-3)$

The x-intercepts are 1, –2, and 3.
These three numbers divide the x-axis into four regions. Choose an x-value in each region as a test number and compute the function value.

Region	Test No.	Value of $f(x)$	Graph
$x < -2$	–3	–24	below x-axis
$-2 < x < 1$	0	6	above x-axis
$1 < x < 3$	2	–4	below x-axis
$3 < x$	4	18	above x-axis

Use this information, the x-intercepts, and the fact that the graph can have a total of at most 2 peaks and valleys to sketch the graph.

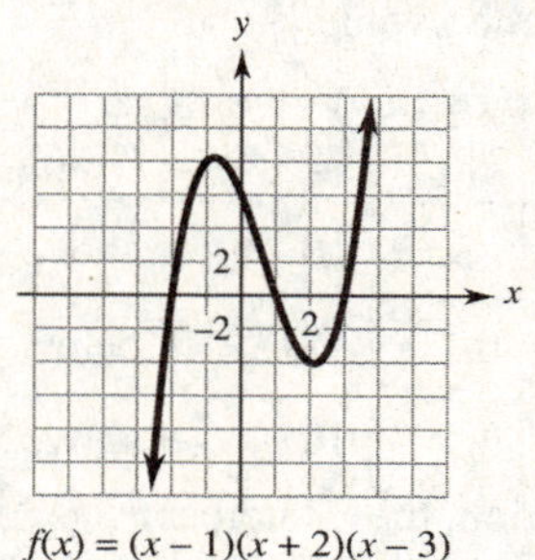

$f(x) = (x - 1)(x + 2)(x - 3)$

65. $f(x) = x^4 - 5x^2 - 6 = (x^2 - 6)(x^2 + 1)$

First, find x-intercepts by setting $f(x) = 0$ and solving for x. Note that there is no real solution for $x^2 + 1 = 0$.

$(x^2 - 6)(x^2 + 1) = 0 \Rightarrow x^2 = 6 \Rightarrow$

$x = \pm\sqrt{6} \approx \pm 2.4$

These two numbers divide the x-axis into three regions. Choose a point in each region as a test point and compute the function value.

Region	Test No.	Value of $f(x)$	Graph
$x < -\sqrt{6}$	–3	30	above x-axis
$-\sqrt{6} < x < \sqrt{6}$	0	–6	below x-axis
$\sqrt{6} < x$	3	30	above x-axis

Use this information, the x-intercepts, and the fact that the graph can have a total of at most 3 peaks and valleys to sketch the graph.

(*continued next page*)

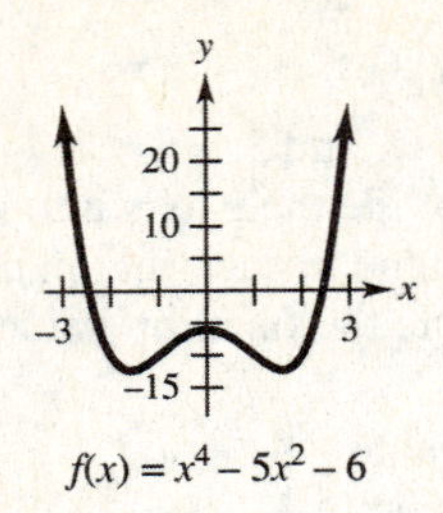

$f(x) = x^4 - 5x^2 - 6$

66. $f(x) = x^4 - 7x^2 - 8 = (x^2 - 8)(x^2 + 1)$

$= (x - \sqrt{8})(x + \sqrt{8})(x^2 + 1)$

The x-intercepts are $-\sqrt{8}$ and $\sqrt{8}$.

(Note that $x^2 + 1 \neq 0$ for any x.) These two numbers divide the x-axis into three regions. Choose a point in each region as a test point and compute the function value.

Region	Test No.	Value of $f(x)$	Graph
$x < -\sqrt{8}$	−3	10	above x-axis
$-\sqrt{8} < x < \sqrt{8}$	0	−8	below x-axis
$\sqrt{8} < x$	3	10	above x-axis

Use this information, the x-intercepts, and the fact that the graph can have a total of at most 3 peaks and valleys to sketch the graph.

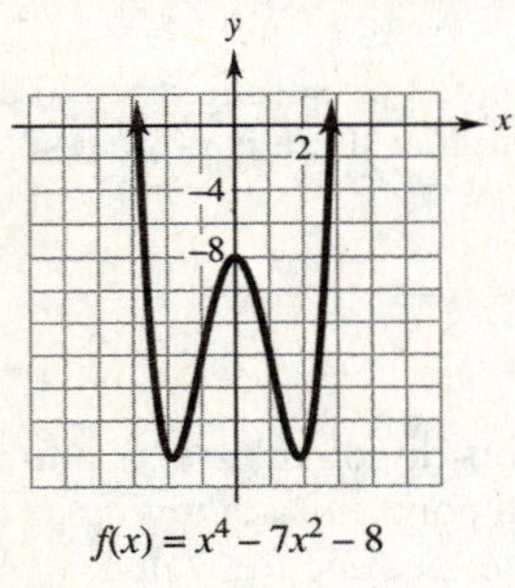

$f(x) = x^4 - 7x^2 - 8$

67. Demand equation:

$p = -.000012q^3 - .00498q^2 + .1264q + 1508$

Supply equation:

$p = -.000001q^3 + .00097q^2 + 2q$

Graph the supply and demand equations in the window [0, 500] by [0, 1500].. Use the intersection finder to determine the equilibrium point.

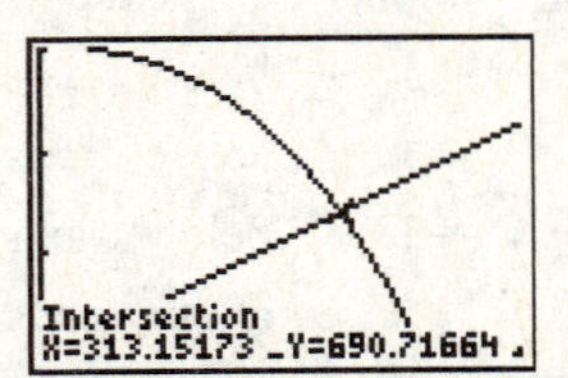

The equilibrium quantity is about 313,152 and the equilibrium price is about $690.72 per thousand.

68. $A(x) = -.000006x^4 + .0017x^3 + .03x^2 - 24x + 1110$

Graph the average cost function in the window [0, 150] by [−15, 80] and use the minimum finder to find point where $A(x)$ is smallest.

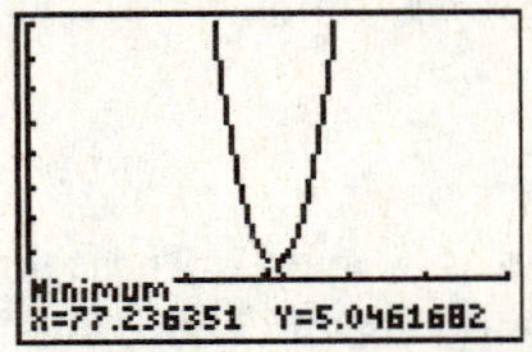

About 77,236 cans should be manufactured for an average cost of about $5.05 per can.

69. a. $R(10) = -1558(10)^3 - 36{,}587(10)^2 + 283{,}469(10) - 329{,}027 \approx 404{,}963$

b. $C(10) = 1261(10)^3 - 29{,}686(10)^2 + 233{,}069(10) - 249{,}868 \approx 373{,}222$

c. $P(x) = R(x) - C(x)$

$= 1558x^3 - 36{,}587x^2 + 283{,}469x - 329{,}027 - (1261x^3 - 29{,}686x^2 + 233{,}069x - 249{,}868)$

$= 297x^3 - 6901x^2 + 50{,}400x - 79{,}159$

d. $P(x) = 297x^3 - 6901x^2 + 50,400x - 79,159$

$P(10) = 297(10)^3 - 6901(10)^2 + 50,400(10) - 79,159 \approx 31,741$

$P(12) = 297(12)^3 - 6901(12)^2 + 50,400(12) - 79,159 \approx 45,113$

Thus, the profit for the years 2010 and 2012 were about \$31,741 million and 45,113 million respectively.

70. a.

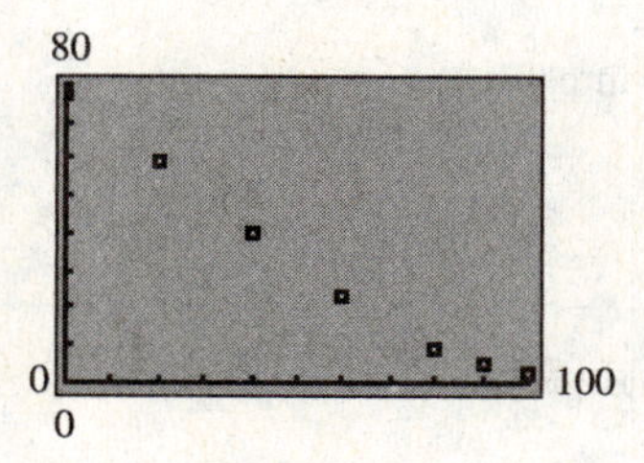

b.

$f(x) = .000000143x^4 + .0000178x^3 - .00145x^2 - .9279x + 77.8$

c. $f(25) \approx 54.0;\ f(35) \approx 44.5;\ f(50) \approx 30.9$

d. Answers will vary.

71. $f(x) = \dfrac{1}{x-3}$

Find the vertical asymptote by setting the denominator equal to zero and solving for x.

$x - 3 = 0 \Rightarrow x = 3$

The line $x = 3$ is a vertical asymptote. The horizontal asymptote of the function in the form $y = \dfrac{0x+1}{1x-3}$ is $y = \dfrac{0}{1}$ or $y = 0$.

If $x = 0$, $y = -\dfrac{1}{3}$.

If $y = 0$, there is no solution for x.

x	0	1	2	4	5	6
y	$-\frac{1}{3}$	$-\frac{1}{2}$	-1	1	$\frac{1}{2}$	$\frac{1}{3}$

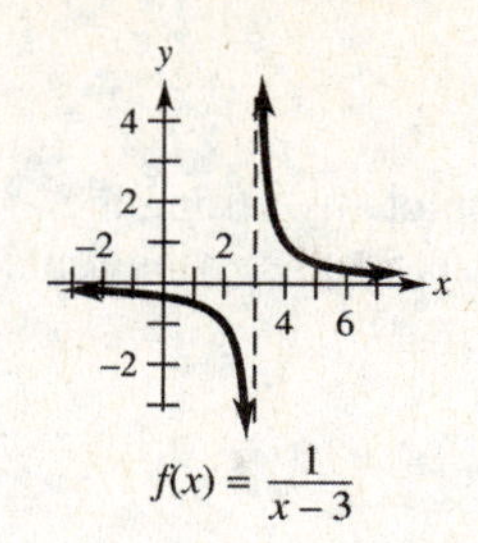

72. $f(x) = \dfrac{-2}{x+4}$

Vertical asymptote: $x = -4$

Horizontal asymptote: $y = 0$

x	-8	-6	-5	$-\frac{9}{2}$	$-\frac{7}{2}$	-3	-2	0
y	$\frac{1}{2}$	1	2	4	-4	-2	-1	$-\frac{1}{2}$

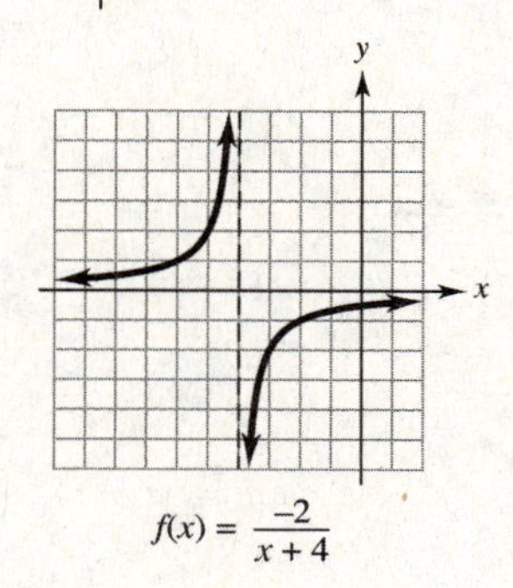

73. $f(x) = \dfrac{-3}{2x-4}$

Find the vertical asymptote by setting the denominator equal to zero and solving for x.

$2x - 4 = 0 \Rightarrow x = 2$

The line $x = 2$ is a vertical asymptote. The horizontal asymptote of the function in the form $y = \dfrac{0x-3}{2x-4}$ is $y = \dfrac{0}{2}$ or $y = 0$.

If $x = 0$, $y = \dfrac{3}{4}$.

If $y = 0$, there is no solution for x.

x	-1	0	1	3	4	5
y	$\frac{1}{2}$	$\frac{3}{4}$	$\frac{3}{2}$	$-\frac{3}{2}$	$-\frac{3}{4}$	$-\frac{1}{2}$

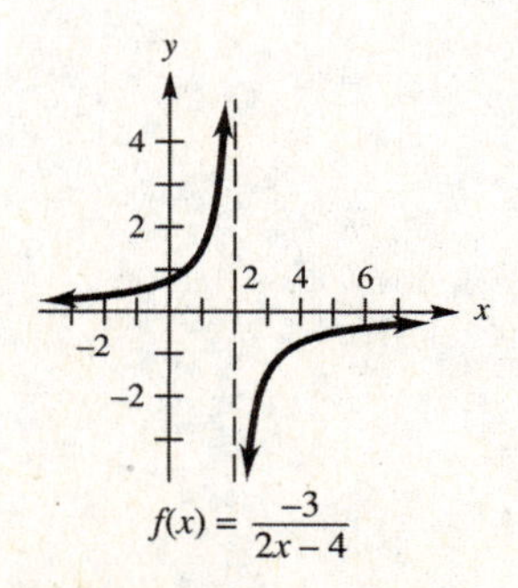

74. $f(x)=\dfrac{5}{3x+7}$

Find the vertical asymptote by setting the denominator equal to zero and solving for x.

$3x+7=0 \Rightarrow x=-\dfrac{7}{3}$

Vertical asymptote: $x=-\dfrac{7}{3}$

The horizontal asymptote of the function in the form $y=\dfrac{0x+5}{3x+7}$ is $y=\dfrac{0}{3}$ or $y=0$.

x	−6	−5	−4	−3	−2	−1	0	1
y	$-\frac{5}{11}$	$-\frac{5}{8}$	−1	$-\frac{5}{2}$	5	$\frac{5}{4}$	$\frac{5}{7}$	$\frac{1}{2}$

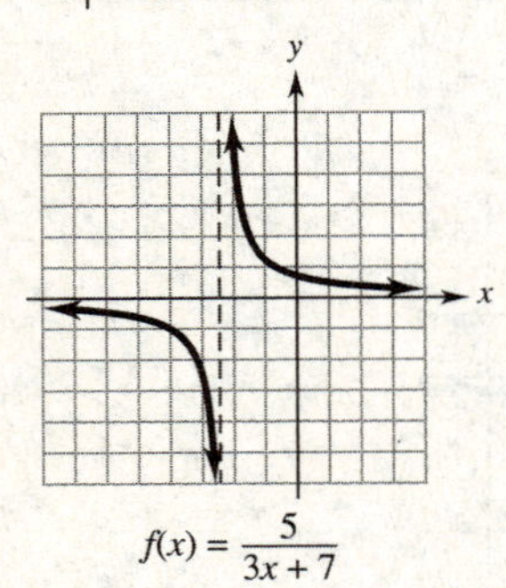

$f(x)=\dfrac{5}{3x+7}$

75. $g(x)=\dfrac{5x-2}{4x^2-4x-3}$

Find the vertical asymptotes by setting the denominator equal to zero and solving for x.

$4x^2-4x-3=0 \Rightarrow (2x-3)(2x+1)=0 \Rightarrow$

$2x-3=0$ or $2x+1=0$

$2x=3$ or $2x=-1$

$x=\dfrac{3}{2}$ or $x=-\dfrac{1}{2}$

The lines $x=\dfrac{3}{2}$ and $x=-\dfrac{1}{2}$ are vertical asymptotes.

Divide the numerator and denominator by x^2 to find the horizontal asymptotes.

$$\frac{\frac{5x}{x^2}-\frac{2}{x^2}}{\frac{4x^2}{x^2}-\frac{4x}{x^2}-\frac{3}{x^2}}=\frac{\frac{5}{x}-\frac{2}{x^2}}{4-\frac{4}{x}-\frac{3}{x^2}}$$

As $|x|$ gets very large, the numerator gets close to 0. The function has a horizontal asymptote at $y=0$. Find the intercepts.

If $x=0$, $y=\dfrac{-2}{-3}=\dfrac{2}{3}$. The y-intercept is $\dfrac{2}{3}$.

If $y=0$, $0=\dfrac{5x-2}{4x^2-4x-3} \Rightarrow 5x=2 \Rightarrow x=\dfrac{2}{5}$

The x-intercept is $\dfrac{2}{5}$.

x	−5	−3	−1	−.3	0	$\frac{2}{5}$
y	−.23	−.38	−1.4	2.4	$\frac{2}{3}$	0

x	1	$1\frac{1}{4}$	$1\frac{3}{4}$	2	4
y	−1	−2.4	3	1.6	.4

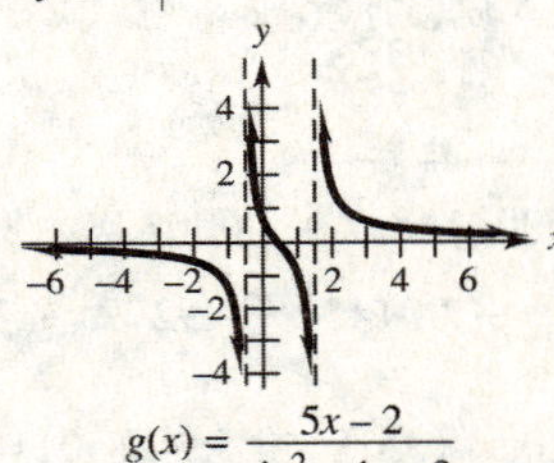

$g(x)=\dfrac{5x-2}{4x^2-4x-3}$

76. $g(x)=\dfrac{x^2}{x^2-1}$

Find the vertical asymptotes by setting the denominator equal to zero and solving for x.

$x^2-1=0 \Rightarrow x^2=1 \Rightarrow x=\pm 1$

The lines $x=1$ and $x=-1$ are vertical asymptotes.

Divide the numerator and denominator by x^2 to find the horizontal asymptotes.

$$\frac{\frac{x^2}{x^2}}{\frac{x^2}{x^2}-\frac{1}{x^2}}=\frac{1}{1-\frac{1}{x^2}}$$

As $|x|$ gets very large, the function approaches 1. The function has a horizontal asymptote at $y=1$.

(*continued next page*)

Find the intercepts.
If $x = 0$, $y = 0$. The y-intercept is 0.

If $y = 0$, $0 = \dfrac{x^2}{x^2 - 1} \Rightarrow x = 0$

The x-intercept is 0.

x	-3	-2	$-\frac{3}{2}$	$-\frac{2}{3}$	$-\frac{1}{2}$	0
y	$\frac{9}{8}$	$\frac{4}{3}$	$\frac{9}{5}$	$-\frac{4}{5}$	$-\frac{1}{3}$	0

x	$\frac{1}{2}$	$\frac{2}{3}$	$\frac{3}{2}$	2	3
y	$-\frac{1}{3}$	$-\frac{4}{5}$	$\frac{9}{5}$	$\frac{4}{3}$	$\frac{9}{8}$

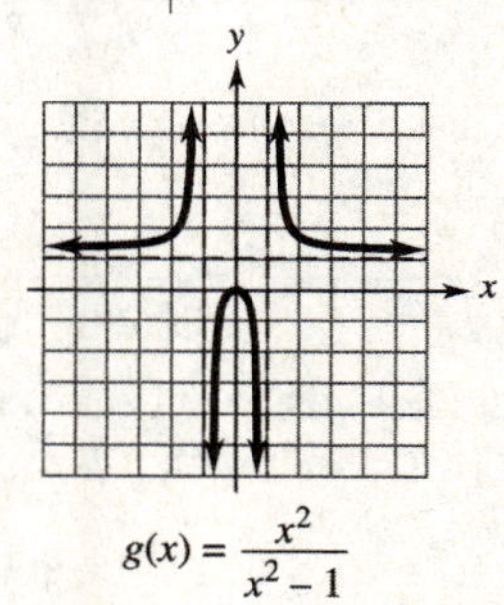

77. $A(x) = \dfrac{650}{2x + 40}$

a. $A(10) = \dfrac{650}{2(10) + 40} \approx 10.83$

The average cost per carton to produce 10 cartons is about \$10.83.

b. $A(50) = \dfrac{650}{2(50) + 40} \approx 4.64$

The average cost per carton to produce 50 cartons is about \$4.64.

c. $A(70) = \dfrac{650}{2(70) + 40} \approx 3.61$

The average cost per carton to produce 70 cartons is about \$3.61.

d. $A(100) = \dfrac{650}{2(100) + 40} \approx 2.71$

The average cost per carton to produce 100 cartons is about \$2.71.

e. We see that $y = 0$ (the x-axis) is a horizontal asymptote.

x	0	10	30	50	70	100
y	16.25	10.83	6.50	4.64	3.61	2.71

Using these points and the horizontal asymptote, we graph the function.

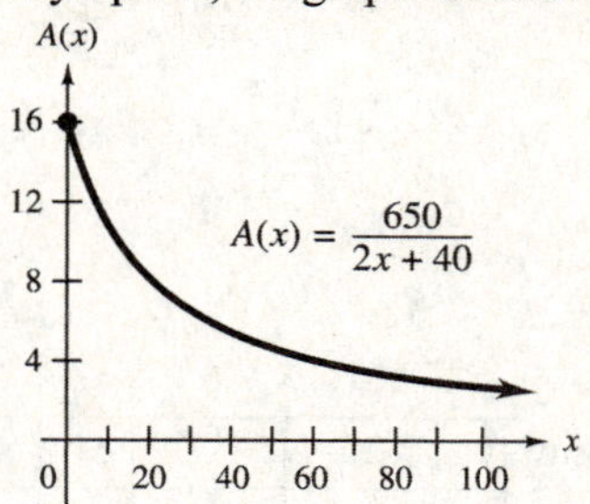

78. $C(x) = \dfrac{400x + 400}{x + 4} = \dfrac{400(x + 1)}{x + 4}$ and $R(x) = 100x$

a. In graphing the function $C(x)$ we see that $y = 400$ is a horizontal asymptote.

x	0	1	2	3	4	5
y	100	160	200	228.6	250	266.7

Using the asymptote and these points we graph the function $C(x)$. The graph of the function $R(x)$ is a line with a slope of 100 and a y-intercept of 0.

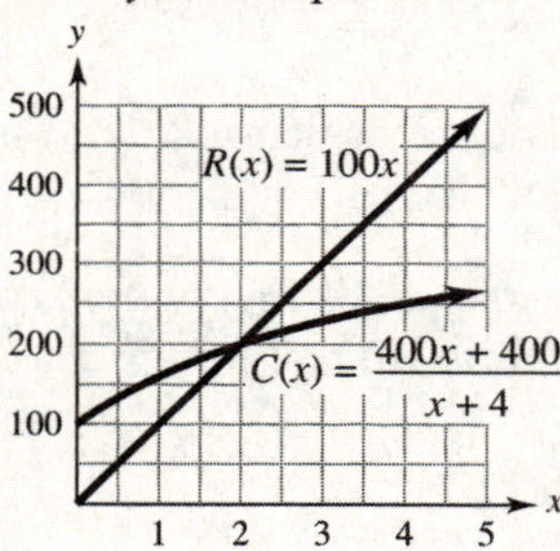

b. To find the break-even point, set $C(x) = R(x)$ and solve for x.

$$\frac{400(x + 1)}{x + 4} = 100x$$

Divide both sides by 100.

$$\frac{4(x + 1)}{x + 4} = x \Rightarrow 4x + 4 = x^2 + 4x \Rightarrow$$

$$x^2 = 4 \Rightarrow x = \pm 2$$

Reject -2 as a solution because the number of hundreds of units cannot be negative. $x = 2$ represents 200 units, so the break-even point is 200 units.

c. $P(1)$ represents a loss. From the graph we see that $C(1) > R(1)$. If the cost is greater than revenue, there is a loss.

d. $P(4)$ represents a profit. From the graph we see that $R(4) > C(4)$. If revenue is greater than cost, there is a profit.

79. Supply: $p = \frac{q^2}{4} + 25$

Demand: $p = \frac{500}{q}$

a.

q	0	5	10	15
Supply	25	31.25	50	81.25
Demand	No value	100	50	33.3

The equilibrium point is (10, 50).

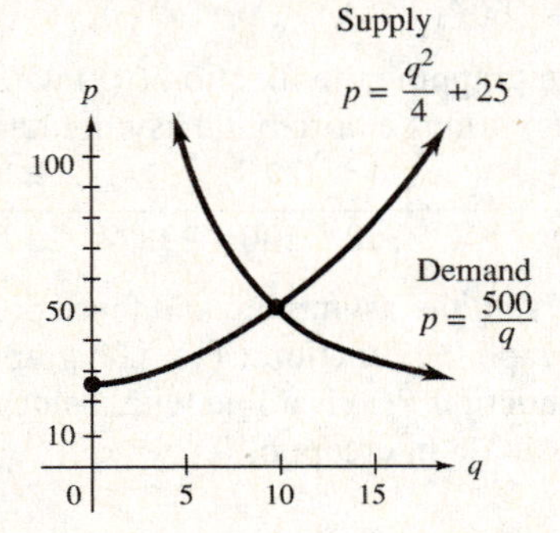

b. Supply exceeds demand if q is in the interval $(10, \infty)$.

c. Demand exceeds supply if q is in the interval (0, 10).

80. $y = \frac{9.2x}{106 - x}$

a. If $x = 50$, $y = \frac{9.2(50)}{106 - 50} \approx 8.2$ or about \$8200.

b. If $x = 98$, $y = \frac{9.2(98)}{106 - 98} \approx 112.7$ or about \$112,700

c. If $y = \$22{,}000$, then we have

$22 = \frac{9.2x}{106 - x} \Rightarrow 22(106 - x) = 9.2x \Rightarrow$

$2332 - 22x = 9.2x \Rightarrow 2332 = 31.2x \Rightarrow$

$x \approx 74.7$

About 75% of the pollutant can be removed for \$22,000.

Case 3 Architectural Arches

1. $f(x) = \frac{-k}{c^2}x^2 + k$

If $k = 20$, and $c = 7$, then $f(x) = \frac{-20}{49}x^2 + 20$

2. $f(x) = \frac{-k}{c^2}x^2 + k$

If $k = 25$, and $c = 15$ then

$f(x) = \frac{-25}{225}x^2 + 25 = \frac{-1}{9}x^2 + 25$

3. $g(x) = \sqrt{r^2 - x^2} = \sqrt{25^2 - x^2} = \sqrt{625 - x^2}$ The arch is 50 feet wide at its base because the width of the base is twice the radius.

4. $g(x) = \sqrt{r^2 - x^2} = \sqrt{40^2 - x^2} = \sqrt{1600 - x^2}$

The arch is 80 feet wide at its base because the width of the base is twice the radius.

5. The arch is 20 feet tall and 24 feet wide at the base. Thus, $r = 12$ and the vertical sides of the arch are $20 - 12 = 8$ feet tall.

$h(x) = \sqrt{r^2 - x^2} + 8 = \sqrt{12^2 - x^2} + 8$

$= \sqrt{144 - x^2} + 8$

6. The arch is 32 feet tall and 44 feet wide at the base. Thus, $r = 22$ and the vertical sides of the arch are $32 - 22 = 10$ feet tall.

$h(x) = \sqrt{r^2 - x^2} + 10 = \sqrt{22^2 - x^2} + 10$

$= \sqrt{484 - x^2} + 10$

7. It would fit through the semicircular and Norman arches, but not the parabolic arch. To allow it to fit through the parabolic arch, increase the width of the arch to 15 ft.

8. Yes; it would fit through all of the arches

Chapter 4 Exponential and Logarithmic Functions

Section 4.1 Exponential Functions

1. $f(x) = 6^x$

This function is exponential, because the variable is in the exponent and the base is a positive constant other than 1.

3. $h(x) = 4x^2 - x + 5$

This function is quadratic because it is a polynomial function of degree 2.

5. $f(x) = 675\left(1.055^x\right)$

This function is exponential because the variable is in the exponent and the base is a positive constant other than 1.

7. $f(x) = .6^x$

a. The equation describing this function has the format $f(x) = a^x$ with $0 < a < 1$, so the graph lies entirely above the x-axis and falls from left to right. It falls relatively steeply until it reaches the y-intercept 1 and then falls slowly, with the positive x-axis as a horizontal asymptote.

b. $f(0) = .8^0 = 1$; (0, 1)

$f(1) = .8^1 = .8$; (1, .8)

9. $h(x) = 2^{.5x}$

a. The base of this exponential function is greater than 1, so the graph lies entirely above the x-axis and rises from left to right. The negative x-axis is a horizontal asymptote. The graph rises slowly until it reaches the y-intercept 1 and then rises quite steeply.

b. $h(0) = 2^{.5(0)} = 2^0 = 1$; (0, 1)

$h(1) = 2^{.5(1)} = 2^{.5}$; $(1, 2^{.5})$

11. $f(x) = e^{-x}$

a. $e^{-x} = \left(\frac{1}{e}\right)^x$, so this exponential function is of the form $f(x) = a^x$ with $0 < a < 1$. The graph lies entirely above the x-axis and falls from left to right. It falls very steeply until it reaches the y-intercept 1 and then falls slowly, with the positive x-axis as a horizontal asymptote.

b. $f(0) = e^{-0} = 1$; (0,1)

$f(1) = e^{-1} = \frac{1}{e}$; $\left(1, \frac{1}{e}\right) \approx (1, .367879)$

13. $f(x) = 3^x$

Construct a table of ordered pairs, plot the points, and connect them to form a smooth curve.

x	–2	–1	0	1	2
y	$\frac{1}{9}$	$\frac{1}{3}$	1	3	9

This graph crosses the y-axis at 1 and it is always above the x-axis.

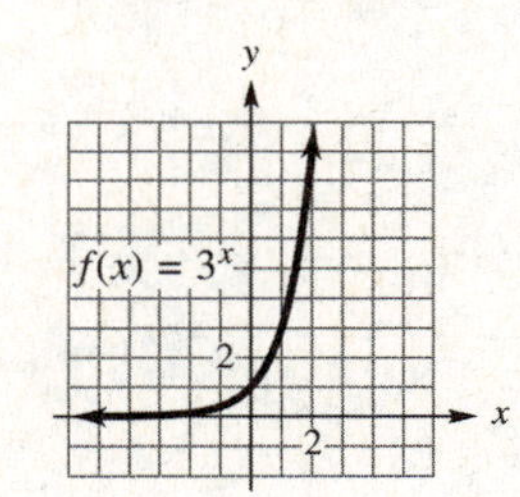

15. $f(x) = 2^{x/2}$

Construct a table of ordered pairs.

x	–4	–2	0	2	4
y	$\frac{1}{4}$	$\frac{1}{2}$	1	2	4

This graph crosses the y-axis at 1 and it is always above the x-axis.

(*continued next page*)

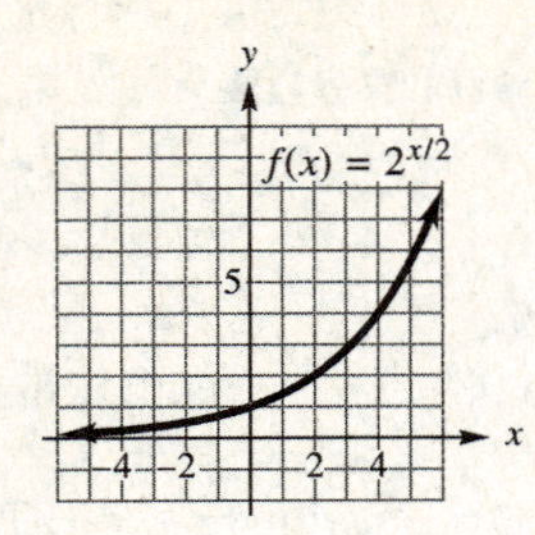

17. $f(x) = \left(\frac{1}{5}\right)^x$

x	-2	-1	0	1	2
y	25	5	1	$\frac{1}{5}$	$\frac{1}{25}$

This graph crosses the y-axis at 1 and it is always above the x-axis.

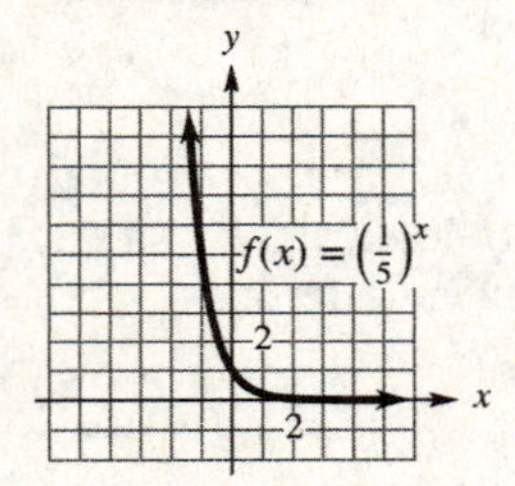

19. **a.** $f(x) = 2^x$

x	-2	-1	0	1	2
y	$\frac{1}{4}$	$\frac{1}{2}$	1	2	4

b. $g(x) = 2^{x+3}$

x	-2	-1	0	1
y	2	4	8	16

c. $h(x) = 2^{x-4}$

x	-1	0	1	2	3	4
y	$\frac{1}{32}$	$\frac{1}{16}$	$\frac{1}{8}$	$\frac{1}{4}$	$\frac{1}{2}$	1

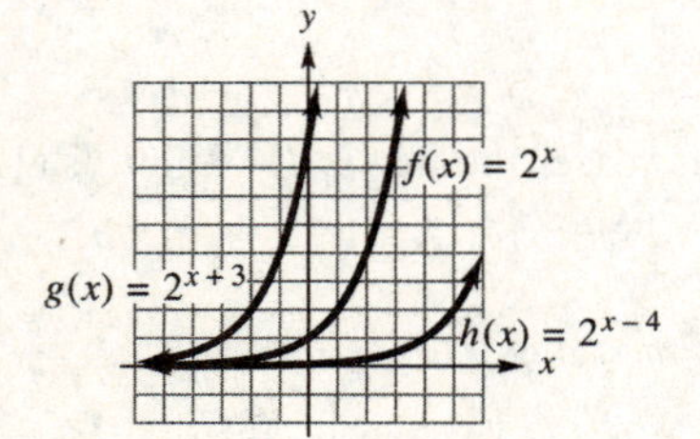

d. The graph of $y = 2^{x+c}$ is the graph of $f(x) = 2^x$ shifted c units to the left. The graph of $y = 2^{x-c}$ is the graph of $f(x) = 2^x$ shifted c units to the right.

21. $y = a^x$

$y = 2.3^x$ is A

23. $y = a^x$

$y = .75^x$ is C

25. $y = a^x$

$y = .31^x$ is E

27. **a.** $a > 1$

b. Domain: $(-\infty, \infty)$
Range: $(0, \infty)$

c.

y
(0, −1)
x
$g(x) = -a^x$

d. Domain: $(-\infty, \infty)$
Range: $(-\infty, 0)$

e.

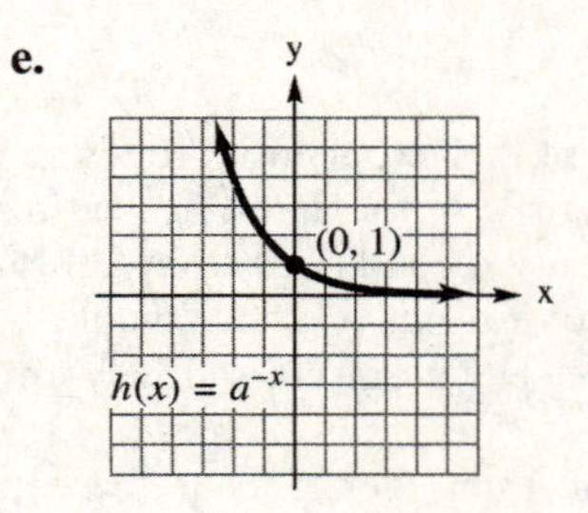

f. Domain: $(-\infty, \infty)$
Range: $(0, \infty)$

29. Since $f(x) = a^x$ and $f(3) = 27$,

$a^3 = 27 \Rightarrow a^3 = 3^3 \Rightarrow a = 3$

a. $f(x) = 3^x$

$f(1) = 3^1 = 3$

b. $f(x) = 3^x$

$f(-1) = 3^{-1} = \frac{1}{3}$

c. $f(x) = 3^x$

$f(2) = 3^2 = 9$

d. $f(x) = 3^x$

$f(0) = 3^0 = 1$

31. $f(x) = 2^{-x^2+2}$

x	-2	-1	0	1	2
y	$\frac{1}{4}$	2	4	2	$\frac{1}{4}$

For this graph, the x-axis is an asymptote on the left and on the right.

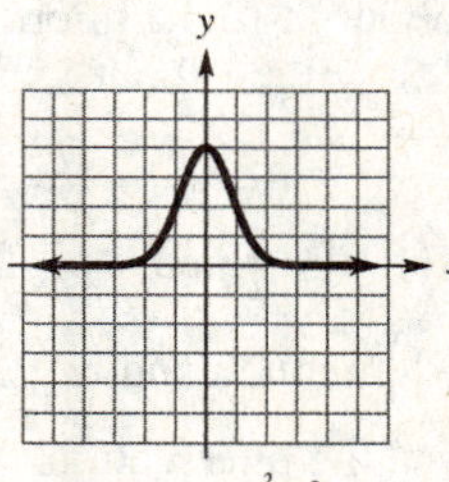

33. $f(x) = x \cdot 2^x$

x	-4	-2	0	1	2
y	$-\frac{1}{4}$	$-\frac{1}{2}$	0	2	8

This graph passes through the origin, and it has the negative x-axis as a horizontal asymptote towards the left.

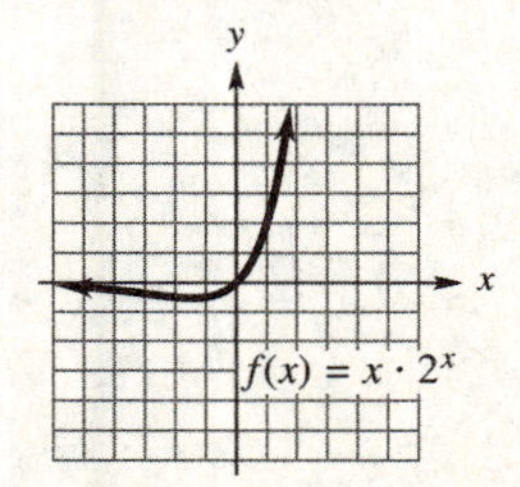

35. $y = (1.06)^t$

a.

t	0	1	2	3	4	5
y	1	1.06	1.12	1.19	1.26	1.34

t	6	7	8	9	10
y	1.42	1.50	1.59	1.69	1.79

b. Plot the eleven points in the table and draw a smooth curve connecting them. (Realize that negative values are not acceptable for t, which represents time.)

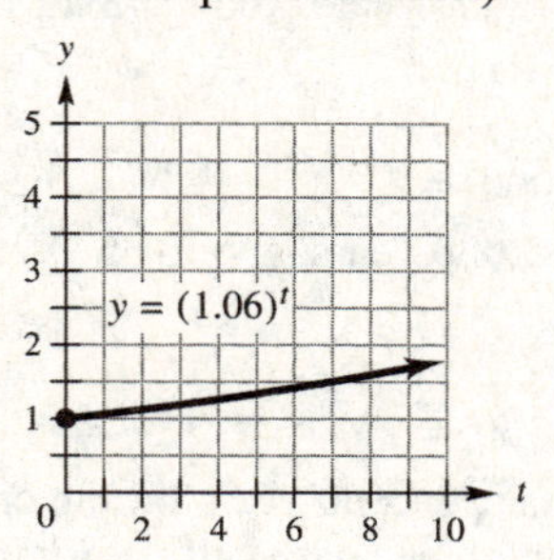

37. $y = (.97)^t$

a. When $t = 10, y = (.97)^{10} \approx .74$.

Let x represent the unknown cost.

$.74x = 105{,}000$

$x = \frac{105{,}000}{.74}$

$x \approx 141{,}892$

The house will cost about \$141,892.

b. When $t = 8$, $y = (.97)^8 \approx .78$.

Let x represent the unknown cost.

$.78x = 50 \Rightarrow x = \frac{50}{.78} \Rightarrow x \approx 64.10$

The book will cost about \$64.10.

39. $f(x) = 116.75e^{0.101x}$

$x = 0$ corresponds to 2000.

a. $x = 2004 - 2000 = 4$

$f(4) = 116.75e^{0.101(4)} \approx 175$

There were approximately 175,000,000 cell phone subscribers in 2004.

b. $x = 2010 - 2000 = 10$

$f(10) = 116.75e^{0.101(10)} \approx 321$

There were approximately 321,000,000 cell phone subscribers in 2010.

c. $x = 2011 - 2000 = 11$

$f(11) = 116.75e^{0.101(11)} \approx 355$

There were approximately 355,000,000 cell phone subscribers in 2011.

41. $W(x) = 2^{-x/24,360}$

a. $W(1000) = 2^{-1000/24,360} \approx .97$

After 1000 years, about .97 kg will be left.

b. $W(10,000) = 2^{-10,000/24,360} \approx .75$

After 10,000 years, about .75 kg will be left.

c. $W(15,000) = 2^{-15,000/24,360} \approx .65$

After 15,000 years, about .65 kg will be left.

d. $W(24,360) = 2^{-24,360/24,360} \approx .5$

It will take 24,360 years for the one kilogram to decay to half its original weight.

43. If $C = \$244,000$, $n = 12$, and $r = .15$, then

$$S = C(1-r)^n = 244,000(1-.15)^{12}$$
$$= 244,000(.85)^{12} \approx \$34,706.99$$

The scrap value is about \$34,706.99.

45. $P(t) = 4.834\left(1.01^{(t-1980)}\right)$

a. $t = 2005$

$$P(2005) = 4.834\left(1.01^{(2005-1980)}\right)$$
$$= 4.834(1.01^{25}) \approx 6.2$$

In 2005, the population was about 6.2 billion.

b. $t = 2010$

$$P(2010) = 4.834\left(1.01^{(2010-1980)}\right)$$
$$= 4.834(1.01^{30}) \approx 6.5$$

In 2010, the population is about 6.5 billion.

c. $t = 2030$

$$P(2010) = 4.834\left(1.01^{(2030-1980)}\right)$$
$$= 4.834(1.01^{50}) \approx 8.0$$

In 2030, the population will be about 8.0 billion.

d. Answers will vary.

47. $f(x) = 10.21\left(1.103^x\right),\ g(x) = 14.28\left(1.046^x\right)$

$x = 0$ corresponds to 2010.

a. $x = 2012 - 2010 = 2$

$$f(2) = 10.21\left(1.103^2\right) \approx 12.4$$
$$g(2) = 14.28\left(1.046^2\right) \approx 15.6$$

According to the model, in 2012, the GDP of China will be about \$12.4 trillion, and the GDP of the U.S. will be about \$15.6 trillion.

b. $x = 2020 - 2010 = 10$

$$f(10) = 10.21\left(1.103^{10}\right) \approx 27.2$$
$$g(10) = 14.28\left(1.046^{10}\right) \approx 22.4$$

According to the model, in 2020, the GDP of China will be about \$27.2 trillion, and the GDP of the U.S. will be about \$22.4 trillion.

c. $x = 2050 - 2010 = 40$

$$f(40) = 10.21\left(1.103^{40}\right) \approx 515.3$$
$$g(40) = 14.28\left(1.046^{40}\right) \approx 86.3$$

According to the model, in 2050, the GDP of China will be about \$515.3 trillion, and the GDP of the U.S. will be about \$86.3 trillion.

d. Graph $Y_1 = 10.21\left(1.103^x\right)$ and $Y_2 = 14.28\left(1.046^x\right)$ in the window [0, 10] by [0, 30], then find the intersection.

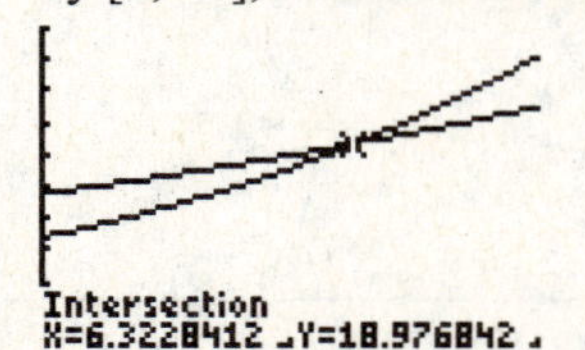

The Chinese GDP will surpass the U.S. GDP sometime during 2016.

49. $g(x) = 9.78(1.07^x)$

$x = 1$ corresponds to first quarter of 2009.

a. $x = 5$

$g(5) = 9.78(1.07^5) \approx 13.72$

According to the model, the number of total subscribers to Netflix was about 13.72 million in the first quarter of 2010.

b. $x = 16$

$g(16) = 9.78(1.07^{16}) \approx 28.9$

According to the model, the number of total subscribers to Netflix was about 28.9 million in the fourth quarter of 2012.

c. Graph $Y_1 = 9.78(1.07^x)$ and $Y_2 = 25$ in the window [0, 20] by [0, 30], then find the intersection.

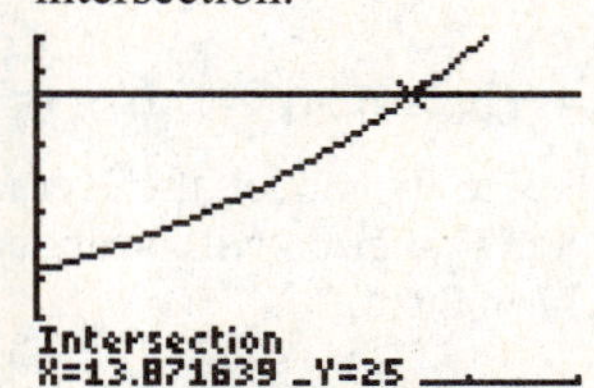

The number of total subscribers to Netflix surpassed 25 million in the first quarter of 2012.

51. $f(x) = .54e^{.191x}$

$x = 1$ corresponds to first quarter of 2011.

a. $x = 4$

$f(4) = .54e^{.191(4)} \approx 1.2$

According to the model, the amount of music listened to online streaming site Pandora was about 1.2 billion hours in the fourth quarter of 2011.

b. $x = 7$

$f(7) = .54e^{.191(7)} \approx 2.1$

According to the model, the amount of music listened to online streaming site Pandora was about 2.1 billion hours in the third quarter of 2012.

c. Graph $Y_1 = .54e^{.191x}$ and $Y_2 = 2.5$ in the window [0, 10] by [0, 5], then find the intersection.

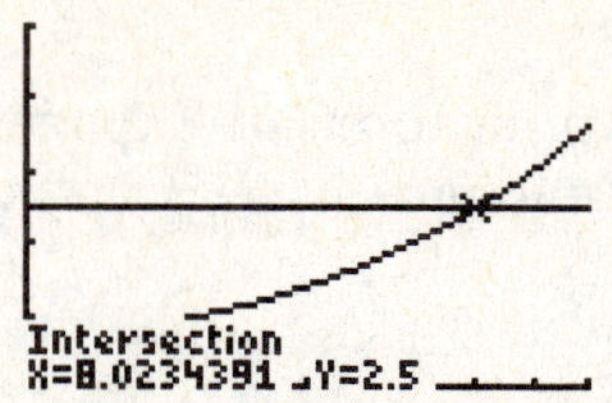

There was 2.5 billion hours of music listened to online through Pandaora in the fourth quarter of 2012.

53. $C(t) = 20e^{-.1155t}$

a. $C(4) = 20e^{-.1155(4)} \approx 12.6$

According to the model, there will be about 12.6 mg of aminophylline remaining in the bloodstream after 4 hours.

b. $C(8) = 20e^{-.1155(8)} \approx 7.9$

According to the model, there will be about 7.9 mg of aminophylline remaining in the bloodstream after 8 hours.

c. Graph $Y_1 = 20e^{-.1155x}$ and $Y_2 = 3.5$ in the window [0, 20] by [–5, 20], then find the intersection.

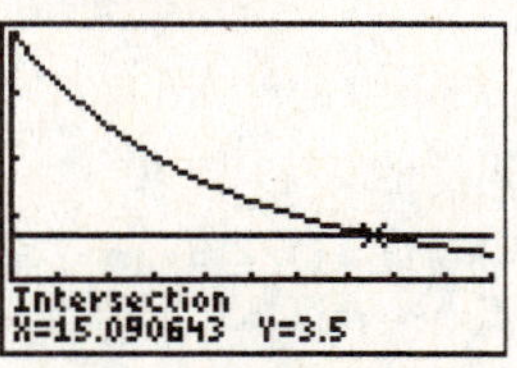

According to the model, there will be 3.59 mg of aminophylline remaining in the bloodstream after about 15.1 hours.

Section 4.2 Applications of Exponential Functions

1. $B(t) = 800(.9898^t)$

a. $B(6) = 800(.9898^6) \approx 752.27$

Your balance after 6 months is $752.27.

b. $B(12) = 800(.9898^{12}) \approx 707.39$

Your balance after 1 year is $707.39.

c. $B(60) = 800(.9898^{60}) \approx 432.45$

Your balance after 5 years is $732.45.

d. $B(96) = 800(.9898^{96}) \approx 298.98$

Your balance after 8 years is $298.98.

e. Answers vary. Your balance will take years to reach 0, so never pay the minimum amount.

3. $W(t) = w_0(1.30^t)$, where $t = 0$ corresponds to 2000.

a. In 2000, $t = 0$. Therefore $W(t) = 2540$.

$$W(0) = w_0(1.30^0) = 2540$$
$$= w_0(1) = w_0 = 2540$$

b. $t = 2005 - 2000 = 5$

$$W(5) = 2540(1.30^5) \approx 9431$$

According to the model, in 2005, the amount of energy generated in the United States from wind power was about 941 megawatts.

$t = 2012 - 2000 = 12$

$$W(12) = 2540(1.30^{12}) \approx 59{,}177$$

According to the model, in 2012, the amount of energy generated in the United States from wind power was about 59,177 megawatts.

5. a. Let $f(t)$ represent the average asking price for rent in San Francisco in the year t after 1990.

The values of $f(t)$ at $t = 0$ and $t = 2012 - 1990 = 22$ are given, that is, $f(0) = 1000$ and $f(22) = 2663$. Solving the first of these equations for y_0 in

$f(t) = y_0 b^t$:

$$f(0) = 1000 \Rightarrow y_0 b^0 = 1000 \Rightarrow y_0 = 1000$$

The model has the form $f(t) = 1000b^t$.

Solving the second equation, $f(22) = 2663$, for b:

$$f(22) = 2663 \Rightarrow 1000b^{22} = 2663 \Rightarrow$$
$$b^{22} = \frac{2663}{1000} \Rightarrow b = \left(\frac{2663}{1000}\right)^{1/22} \approx 1.046$$

The model is $f(t) = 1000(1.046)^t$.

b. $t = 2015 - 1990 = 25$

$$f(25) = 1000(1.046)^{25} \approx 3078.17$$

According to the model, the average asking rent in San Francisco will be about $3078.17 in 2015.

c. Graph $Y_1 = 1000(1.046)^x$ and $Y_2 = 2000$ in the window [0, 20] by [1000, 3000], then find the intersection.

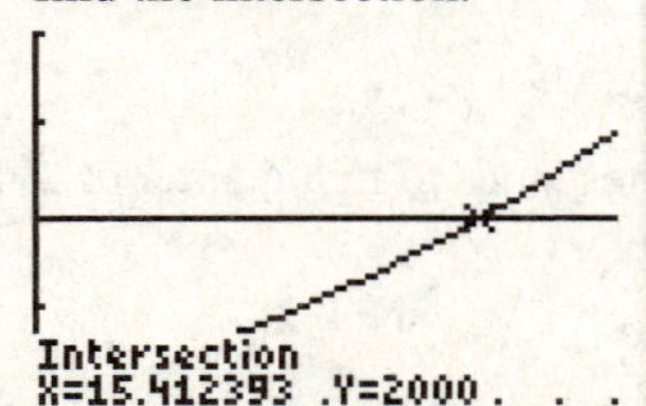

According to the model, the average asking pice for rent in San Francisco was $2000 in 2005.

7. a. Let $f(t)$ represent the sales of video and audio equipment, computers, and related services in billions in year t after 1990. The values of $f(t)$ at $t = 0$ and $t = 2010 - 1990 = 20$ are given, that is, $f(0) = 81.1$ and $f(20) = 296$. Solve the first of these equations for y_0 in $f(t) = y_0 b^t$.

$f(0) = 81.1 \Rightarrow y_0 b^0 = 81.1 \Rightarrow \ y_0 = 81.1$

Now solve the second equation $f(20) = 296$ for b. $f(20) = 296 \Rightarrow 81.1b^{20} = 296 \Rightarrow$

$$b^{20} = \frac{296}{81.1} \Rightarrow b = \left(\frac{296}{81.1}\right)^{1/20} \approx 1.067$$

The model is $f(t) = 81.1\left(1.067^t\right)$.

b. $t = 2008 - 1990 = 18$

$f(18) = 81.1\left(1.067^{18}\right) \approx 260.6$

According to the model, sales were about $260.6 billion in 2008.

c. $t = 2015 - 1990 = 25$

$f(25) = 81.1\left(1.067^{25}\right) \approx 410.3$

According to the model, sales will be about $410.3 billion in 2015.

9. a. Let $t = 0$ correspond to 2000. Using the data for 2000 and 2012 to find a function of the form $f(t) = y_0 b^t$, first solve for y_0:

$f(0) = 1.00 \Rightarrow y_0 b^0 = 1.00 \Rightarrow y_0 = 1.00$

The function has the form $f(t) = 1.00b^t$.

$t = 2012 - 2000 = 12$

Now solve for b.

$f(12) = 0.75 \Rightarrow 1.00b^{12} = 0.75 \Rightarrow$

$b = 0.75^{1/12} \approx 0.9763$

Thus the function is $f(t) = 0.9763^t$.

Using exponential regression on a graphing calculator, the function produced is $g(t) = .998\left(0.976^t\right)$.

```
ExpReg
 y=a*b^x
 a=.9980409002
 b=.9756564469
 r²=.9786678372
 r=-.989276421
```

b. Two-point model:

$t = 2015 - 2000 = 15$

$f(15) = 0.9763^{15} \approx 0.70$

In 2015, a dollar will buy what about 70 cents did in 2000.

$t = 2018 - 2000 = 18$

$f(18) = 0.9763^{18} \approx 0.65$

In 2018, a dollar will buy what about 65 cents did in 2000.

Exponential regression model:

$g(15) = .998\left(0.976^{15}\right) \approx 0.69$

In 2015, a dollar will buy what about 69 cents did in 2000.

$g(18) = .998\left(0.976^{18}\right) \approx 0.64$

In 2018, a dollar will buy what about 64 cents did in 2000.

c. Two point model:

Graph $Y_1 = 0.9763^x$ and $Y_2 = 0.40$ in the window [0, 50] by [0, 1], then find the intersection.

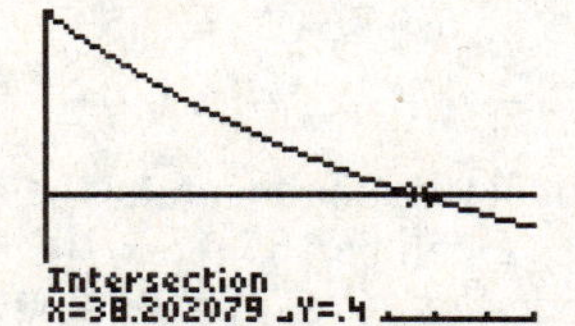

According to the two-point model, the purchasing power of a dollar will drop to 40 cents sometime during 2038.

Exponential regression model:

Graph $Y_1 = .998\left(0.976^x\right)$ and $Y_2 = 0.40$ in the window [0, 50] by [0, 1], then find the intersection.

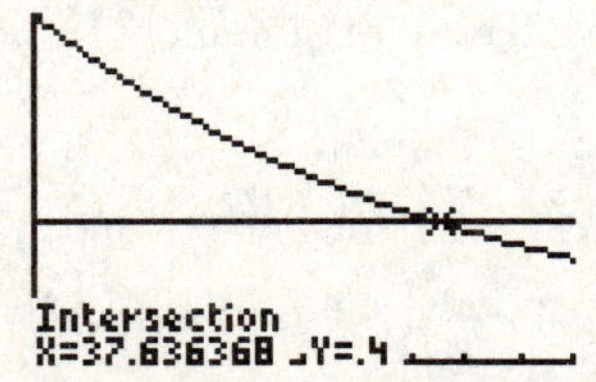

According to the exponential regression model, the purchasing power of a dollar will drop to 40 cents sometime during 2037.

11. a. Let $f(t)$ be the death rate per 100,000 population in the year t after 2000.
Two-point model:

$f(0) = 257.6 \Rightarrow y_0 b^0 = 257.6 \Rightarrow$
$y_0 = 257.6$
$t = 2010 - 2000 = 10$
$f(10) = 178.5 \Rightarrow 257.6b^{10} = 178.5 \Rightarrow$

$$b^{10} = \frac{178.5}{257.6} \Rightarrow b = \left(\frac{178.5}{257.6}\right)^{1/10} \approx 0.964$$

$f(t) = 257.6(0.964)^t$, where $t = 0$ corresponds to 2000.
Exponential regression model:

```
ExpReg
 y=a*b^x
 a=257.4341321
 b=.9630906559
 r²=.9843477455
 r=-.9921430066
```

$g(t) = 257.4(0.963)^t$

b. $t = 2012 - 2000 = 12$

$f(12) = 257.6(0.964)^{12} \approx 165.9$

$g(12) = 257.4(0.963)^{12} \approx 163.7$

If the model remains accurate, the death-rate in 2012 will be 165.9 per 100,000 (two-point model) or 163.7 per 100,000 (exponential regression model).
$t = 2016 - 2000 = 16$

$f(16) = 257.6(0.964)^{16} \approx 143.3$

$g(16) = 257.4(0.963)^{16} \approx 140.8$

If the model remains accurate, the death-rate in 2016 will be 143.3 per 100,000 (two-point model) or 140.8 per 100,000 (exponential regression model).

c. Two point model:

Graph $Y_1 = 257.6(0.964)^x$ and $Y_2 = 100$ in the window [0, 40] by [0, 150], then find the intersection.

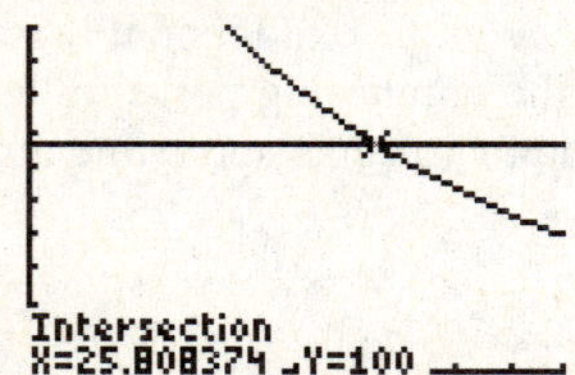

According to the two-point model, the death rate from heart disease will drop to 100 per 100,000 population sometime during 2025.

Exponential regression model:

Graph $Y_1 = 257.4(0.963)^x$ and $Y_2 = 100$ in the window [0, 40] by [0, 150], then find the intersection.

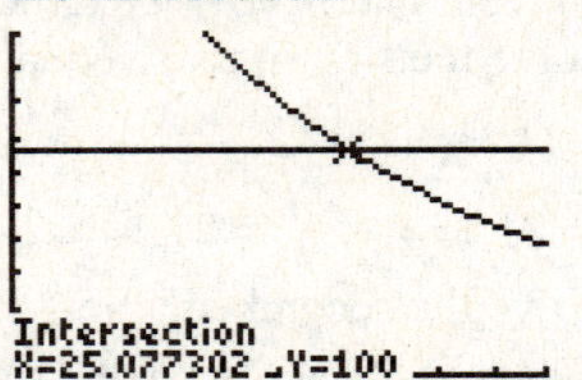

According to the exponential regression model, the death rate from heart disease will drop to 100 per 100,000 population sometime during 2025.

13. $P(t) = 25 - 25e^{-.3t}$

a. $P(1) = 25 - 25e^{-.3(1)} \approx 6$

b. $P(8) = 25 - 25e^{-.3(8)} \approx 23$

c. The maximum number of items that can be produced is 25.

15. $F(t) = T_0 + Cb^t$, where T_0 is the temperature of the constant environment. At $t = 0$, $F(0) = 100$. So,
$100 = T_0 + Cb^0 \Rightarrow 100 = -18 + Cb^0 \Rightarrow$
$100 = -18 + C(1) \Rightarrow C = 118$. The function then becomes $F(t) = -18 + 118b^t$. At $t = 24$, $F(24) = 50$.

So $50 = -18 + 118b^{24} \Rightarrow$

$$68 = 118b^{24} \Rightarrow b^{24} = \frac{68}{118} \Rightarrow$$

$$b = \left(\frac{68}{118}\right)^{1/24} \approx 0.9773$$

The function now becomes
$F(t) = -18 + 118\left(0.9773^t\right)$. At $t = 76$,
$F(76) = -18 + 118\left(0.9773^{76}\right) \approx 2.6$. So after 76 minutes, the temperature of the water will be about 26° C.

17. $y(t) = \dfrac{y_0 e^{kt}}{1 - y_0(1 - e^{kt})}$

a. Let $k = 0.1$ and $y_0 = 0.05$. Then

$$y(10) = \frac{0.05e^{.1(10)}}{1 - 0.05(1 - e^{.1(10)})} = \frac{0.05e}{1 - 0.05(1 - e)} \approx 0.13.$$

b. Let $k = 0.2$ and $y_0 = 0.1$. Then

$$y(5) = \frac{0.1e^{.2(5)}}{1 - 0.1\left(1 - e^{.2(5)}\right)} = \frac{0.1e}{1 - 0.1\left(1 - e\right)} \approx 0.23.$$

c. Plot $Y_1 = \dfrac{0.1e^{.2x}}{1 - 0.1\left(1 - e^{.2x}\right)}$ and $Y_2 = .65$ in the window [0, 60] by [–.1, 1.1], then find the intersection.

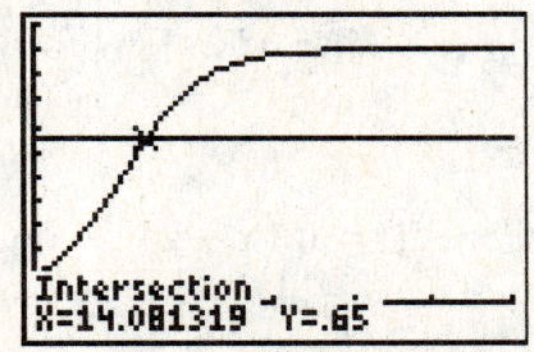

It will take about 14 days for 65% of the people to have heard the rumor.

19. $f(x) = \dfrac{1084}{1 + 1.94e^{-.171x}}$, $x = 0$ corresponds to 2000.

a. $x = 2005 - 2000 = 5$

$$f(5) = \frac{1084}{1 + 1.94e^{-.171(5)}} \approx 594.0$$

The function estimates expenditures of about $594.0 billion in 2005.

$x = 2010 - 2000 = 10$

$$f(10) = \frac{1084}{1 + 1.94e^{-.171(10)}} \approx 802.4$$

The function estimates expenditures of about $802.4 billion in 2010.

b. To graph $f(x) = \dfrac{1084}{1 + 1.94e^{-.171x}}$, enter this as y_1 and use [0, 20] by [0, 1100]

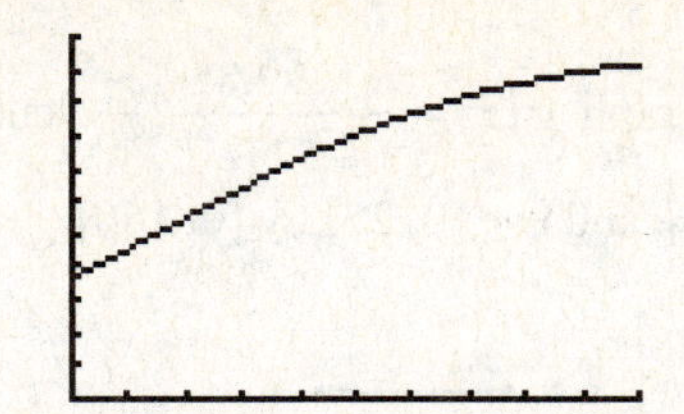

c. To determine the year in which revenues will reach $850 billion, graph y_1 from part *b* and $y_2 = 850$ on the same screen. On the CALC menu, select "intersect" and compute the point of intersection.

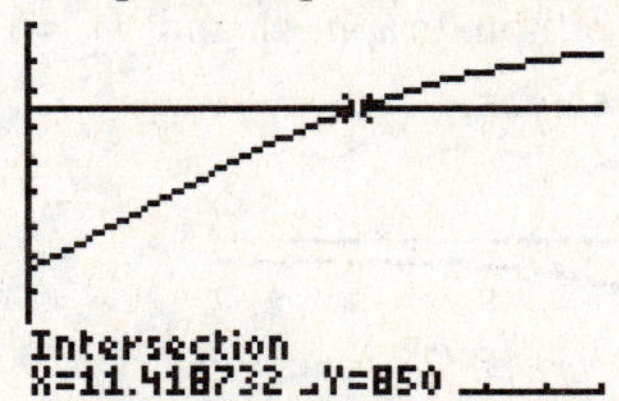

The coordinates of this point are (11.42, 850) so the revenues will reach $850 billion when $t \approx 11.42$ or sometime during 2011.

21. $g(x) = \dfrac{99.85}{1 + .527e^{-.258x}}$, $x = 0$ corresponds to 2000.

a. $x = 2007 - 2000 = 7$

$$g(7) = \frac{99.85}{1 + .527e^{-.258(7)}} \approx 91.89$$

The function estimates the amount spent on legal services to be $91.89 billion in 2007.

$x = 2010 - 2000 = 10$

$$g(10) = \frac{40.35}{1 + 6.39e^{-.0866(10)}} \approx 10.941$$

The function estimates the national debt to be $10.941 trillion in 2010.

$x = 2011 - 2000 = 11$

$$g(11) = \frac{99.85}{1 + .527e^{-.258(11)}} \approx 96.86$$

The function estimates the amount spent on legal services to be $96.86 billion in 2011.

b. To graph $g(x) = \frac{99.85}{1+.527e^{-.258x}}$, enter this as y_1 and use [0, 25] by [0, 150]

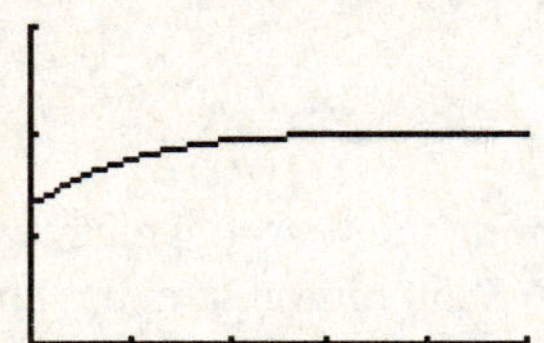

c. To determine the year (if any) when the amount spent on legal services will exceed \$110 billion, graph y_1 and $y_2 = 110$ on the same screen.

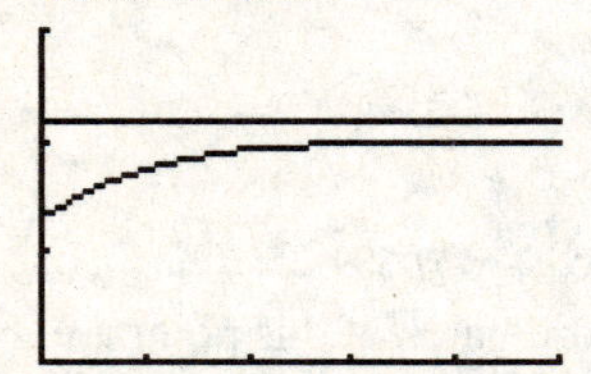

Based on the above graph, the amount spent on legal services will not exceed \$110 billion.

Section 4.3 Logarithmic Functions

1. $y = \log_a x$ means $\underline{x = a^y}$.

3. It is missing the value that equals b^y. If that value is *x*, it should read $y = \log_b x$.

5. log 100,000 = 5 is equivalent to $10^5 = 100{,}000$. (The base of the logarithm is understood to be 10.)

7. $\log_9 81 = 2$ is equivalent to $9^2 = 81$.

9. $10^{1.9823} = 96$ means log 96 = 1.9823.

11. $3^{-2} = \frac{1}{9}$ is equivalent to $\log_3\left(\frac{1}{9}\right) = -2$.

13. $\log 1000 = \log_{10} 10^3 = 3$

15. $\log_6 36 = \log_6 6^2 = 2$

17. $\log_4 64 = \log_4 4^3 = 3$

19. $\log_2 \frac{1}{4} = \log_2 2^{-2} = -2$

21. $\ln\sqrt{e} = \ln e^{1/2} = \frac{1}{2}$

23. $\ln e^{8.77} = 8.77$

25. $\log 53 \approx 1.724$

27. $\ln .0068 \approx -4.991$

29. Answers will vary. Possible answer: $\log_a 1 = 0$ because $a^0 = 1$ for any valid base *a*.

31. $\log 6 + \log 8 - \log 2 = \log\frac{6(8)}{2} = \log 24$

33. $2\ln 5 - \frac{1}{2}\ln 25 = \ln 5^2 - \ln 25^{1/2} = \ln 25 - \ln 5$

$$= \ln\frac{25}{5} = \ln 5$$

35. $2\log u + 3\log w - 6\log v$

$$= \log u^2 + \log w^3 - \log v^6 = \log\left(\frac{u^2w^3}{v^6}\right)$$

37. $2\ln(x+2) - \ln(x+3) = \ln(x+2)^2 - \ln(x+3)$

$$= \ln\left(\frac{(x+2)^2}{x+3}\right)$$

39. $\ln\sqrt{6m^4n^2} = \ln(6m^4n^2)^{1/2} = \frac{1}{2}\ln 6m^4n^2$

$$= \frac{1}{2}(\ln 6 + \ln m^4 + \ln n^2)$$
$$= \frac{1}{2}(\ln 6 + 4\ln m + 2\ln n)$$
$$= \frac{1}{2}\ln 6 + 2\ln m + \ln n$$

41. $\log\frac{\sqrt{xz}}{z^3} = \log\frac{(xz)^{1/2}}{z^3}$

$$= \log(xz)^{1/2} - \log z^3$$
$$= \frac{1}{2}\log(xz) - 3\log z$$
$$= \frac{1}{2}\log x + \frac{1}{2}\log z - 3\log z$$
$$= \frac{1}{2}\log x - \frac{5}{2}\log z$$

43. $\ln(x^2y^5) = \ln x^2 + \ln y^5$
$= 2\ln x + 5\ln y$
$= 2u + 5v$

45. $\ln\left(\dfrac{x^3}{y^2}\right) = \ln x^3 - \ln y^2 = 3\ln x - 2\ln y$
$= 3u - 2v$

47. $\log_6 384 = \dfrac{\ln 384}{\ln 6} \approx 3.32112$

49. $\log_{35} 5646 = \dfrac{\ln 5646}{\ln 35} \approx 2.429777$

51. $\log(b + c) = \log b + \log c$
Consider the values $b = 1$ and $c = 2$.
Then $\log(b + c)$ becomes
$\log(1 + 2) = \log 3 \approx .4771$,
while $\log b + \log c$ becomes
$\log 1 + \log 2 \approx .3010$.
Many other choices for the b and c values would demonstrate just as clearly that the statement $\log(b + c) = \log b + \log c$ is generally false.

53. $y = \ln(x + 2)$
We must have $x + 2 > 0$, so the domain is $x > -2$.

x	y
−1.99	−4.6
−1.5	−.7
−1	0
0	.7
2	1.4
4	1.8

Connect these points with a smooth curve.

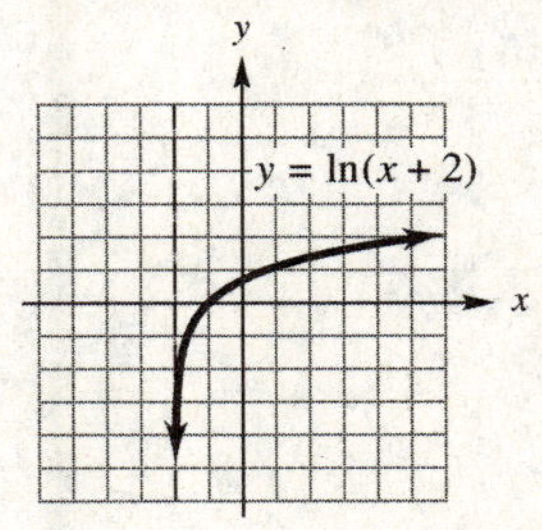

55. $y = \log(x - 3)$
We must have $x - 3 > 0$, so the domain is $x > 3$.

x	y
3.01	−2
3.5	−.30
4	0
6	.48
8	.70

Connect these points with a smooth curve.

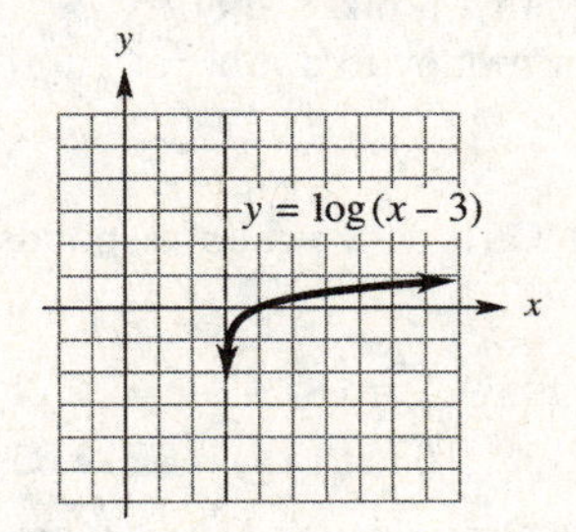

57. Answers will vary. Possible answer:

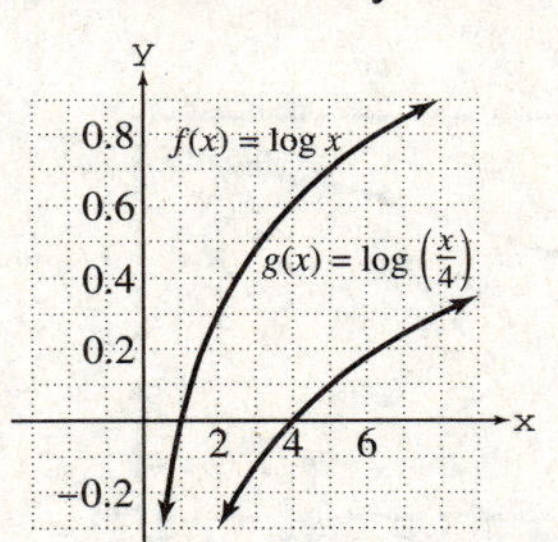

$\log\left(\dfrac{x}{4}\right) = \log x - \log 4$

$g(x)$ equals log 4 subtracted from $f(x)$

59. $\ln 2.75 = 1.0116009$
$e^{1.0116009} = 2.75$

61. 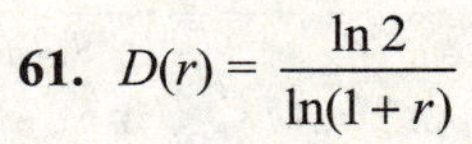 $D(r) = \dfrac{\ln 2}{\ln(1 + r)}$

a. 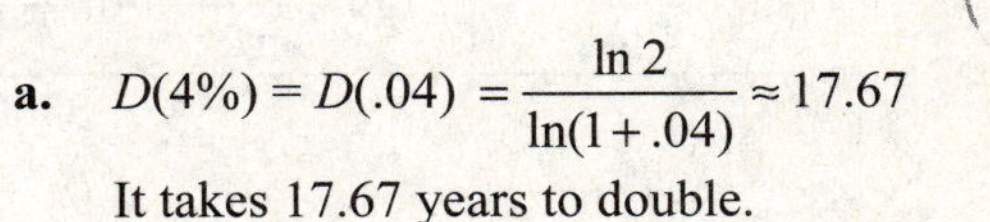 $D(4\%) = D(.04) = \dfrac{\ln 2}{\ln(1 + .04)} \approx 17.67$

It takes 17.67 years to double.

b. $D(8\%) = D(.08) = \dfrac{\ln 2}{\ln(1 + .08)} \approx 9.01$

It takes 9.01 years to double.

$3^2 = 4x - 3$
$9 = 4x - 3$
$9 + 3 = 4x$
$\dfrac{12}{4} = \dfrac{4x}{4}$
$3 = x$

c. $D(18\%) = D(.18) = \dfrac{\ln 2}{\ln(1+.18)} \approx 4.19$

It takes 4.19 years to double.

d. $D(36\%) = D(.36) = \dfrac{\ln 2}{\ln(1+.36)} \approx 2.25$

It takes 2.25 years to double.

e. $17.67 \approx 18 = \dfrac{72}{4};\ 9.01 \approx 9 = \dfrac{72}{8};$

$4.19 \approx 4 = \dfrac{72}{18};\ 2.25 \approx 2 = \dfrac{72}{36}$

The pattern is that it takes about $72/k$ years for money to double at k% interest.

63. $g(x) = -771.9 + 1035\ln x$, $x = 5$ corresponds to 1995.

a. $x = 2008 - 1995 + 5 = 18$

$g(18) = -771.9 + 1035\ln(18) \approx 2219.63$

In 2008, the average expenditures were about $2219.63.

b.

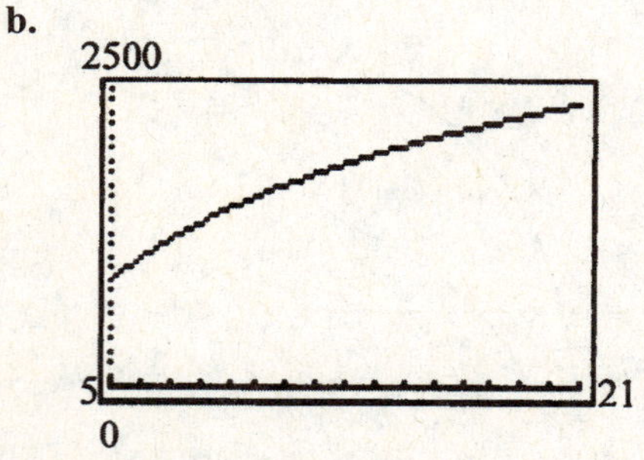

c. The graph indicates that the average expenditures on gasoline and motor oil will gradually increase as time goes on.

65. a. $x = 1990 - 1980 + 10 = 20$

$h(20) = 1.58 + 10.15\ln 20 \approx 31.99$

By 1990, approximately 31.99 million residents of the U.S. population will be age 65 or older.

$x = 2005 - 1980 + 10 = 35$

$g(35) = 1.58 + 10.15\ln 35 \approx 37.67$

By 2005, approximately 37.67 million residents of the U.S. population will be age 65 or older.

b.

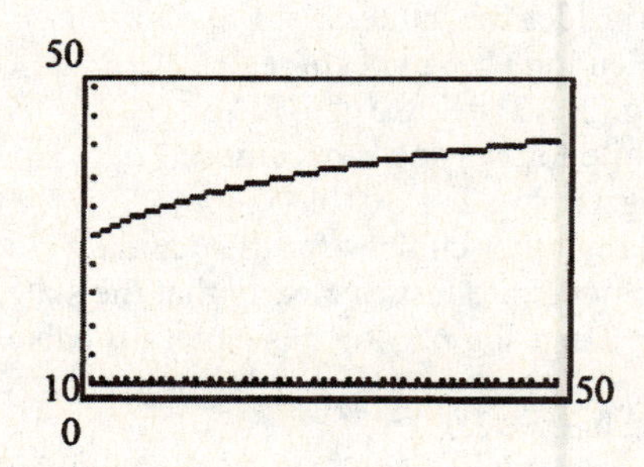

c. As time goes on, the number of residents of the U.S. population age 65 or over is gradually increasing.

67. $n_1 = 2754,\ n_2 = 689,\ n_3 = 4428$, and $n_4 = 629$

$N = n_1 + n_2 + n_3 + n_4 = 8500$

So, the index of diversity is

$$H = \frac{N\log_2 N - [n_1\log_2 n_1 + n_2\log_2 n_2 + n_3\log_2 n_3 + n_4\log_2 n_4]}{N}$$

$$= \frac{8500\log_2 8500 - [2754\log_2 2754 + 689\log_2 689 + 4428\log_2 4428 + 629\log_2 629]}{8500}$$

$$= \frac{8500\dfrac{\ln 8500}{\ln 2} - \left[2754\dfrac{\ln 2754}{\ln 2} + 689\dfrac{\ln 689}{\ln 2} + 4428\dfrac{\ln 4428}{\ln 2} + 629\dfrac{\ln 629}{\ln 2}\right]}{8500} \approx 1.5887$$

69. $f(x) = -1237 + 580.6\ln x$,
$x = 10$ corresponds to 1990.

a. $x = 1998 - 1990 + 10 = 18$
$f(18) = -1237 + 580.6\ln 18 \approx 441.15$
According to the model, in 1998, total assets held by credit unions were about \$441.15 billion.
$x = 2010 - 1990 + 10 = 30$
$f(30) = -1237 + 580.6\ln 30 \approx 737.74$
According to the model, in 2010, total assets held by credit unions were about \$737.74 billion.

b. Graph $Y_1 = -1237 + 580.6\ln x$ and $Y_2 = 1000$ in the window [0, 60] by [5, 2000], then find the intersection.

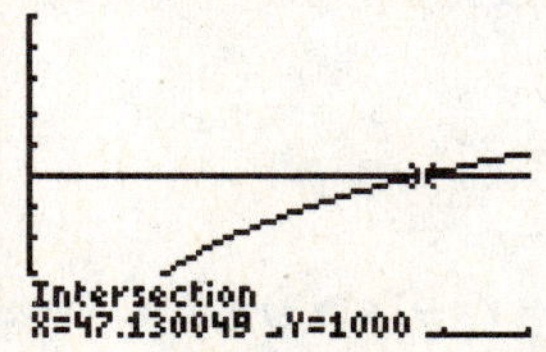

If the model remains accurate, total assets will reach \$1000 billion sometime during 2027.

71. $g(x) = 28.29 - 1.948\ln(x+1)$,
$x = 0$ corresponds to 1980.

a. $x = 1985 - 1980 = 5$
$g(5) = 28.29 - 1.948\ln(5+1)$
≈ 24.8
According to the model, in 1985, about 24.8 gallons of milk were consumed per person annually.
$x = 2005 - 1980 = 25$
$g(25) = 28.29 - 1.948\ln(25+1)$
≈ 21.9
According to the model, in 2005, about 21.9 gallons of milk were consumed per person annually.

b. Graph $Y_1 = 28.29 - 1.948\ln(x+1)$ and $Y_2 = 21$ in the window [0, 50] by [15, 25], then find the intersection.

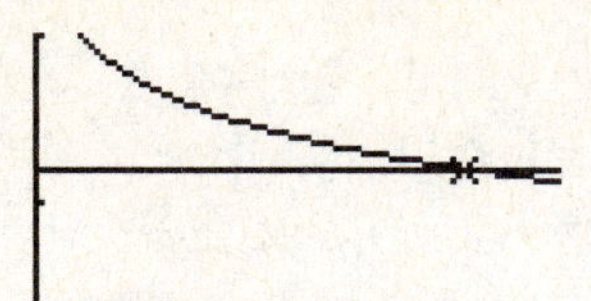

If the model remains accurate, 21 gallons of milk will be consumed annually sometime during 2021.

Section 4.4 Logarithmic and Exponential Equations

1. $\ln(x+3) = \ln(2x-5) \Rightarrow x+3 = 2x-5$
$x = 8$

3.
$$\ln(3x+1) - \ln(5+x) = \ln 2$$
$$\ln\frac{3x+1}{5+x} = \ln 2$$
$$\frac{3x+1}{5+x} = 2$$
$$3x+1 = 10+2x$$
$$x = 9$$

5.
$$2\ln(x-3) = \ln(x+5) + \ln 4$$
$$\ln(x-3)^2 = \ln[4(x+5)]$$
$$(x-3)^2 = 4(x+5)$$
$$x^2 - 6x + 9 = 4x + 20$$
$$x^2 - 10x - 11 = 0$$
$$(x-11)(x+1) = 0$$
$$x = 11 \text{ or } x = -1$$
Since -1 is not in the domain of $\ln(x-3)$, the only solution is 11.

7. $\log_3(6x-2) = 2 \Rightarrow 6x - 2 = 3^2 \Rightarrow$
$6x - 2 = 9 \Rightarrow 6x = 11 \Rightarrow x = \frac{11}{6}$

9. $\log x - \log(x+4) = -1 \Rightarrow \log\left(\frac{x}{x+4}\right) = -1 \Rightarrow$
$\frac{x}{x+4} = 10^{-1} \Rightarrow \frac{x}{x+4} = \frac{1}{10} \Rightarrow 10x = x + 4 \Rightarrow$
$9x = 4 \Rightarrow x = \frac{4}{9}$

11. $\log_3(y+2) = \log_3(y-7) + \log_3 4$

$\log_3(y+2) = \log_3[4(y-7)]$

$y+2 = 4(y-7)$

$y+2 = 4y-28$

$30 = 3y$

$y = 10$

13. $\ln(x+9) - \ln x = 1 \Rightarrow \ln\left(\dfrac{x+9}{x}\right) = 1 \Rightarrow$

$\dfrac{x+9}{x} = e^1 \Rightarrow x+9 = ex \Rightarrow x - ex = -9 \Rightarrow$

$(1-e)x = -9 \Rightarrow x = -\dfrac{-9}{e-1} \approx 5.2378$

15. $\log x + \log(x-9) = 1$

$\log[x(x-9)] = 1$

$x(x-9) = 10^1$

$x^2 - 9x = 10$

$x^2 - 9x - 10 = 0$

$(x-10)(x+1) = 0$

$x - 10 = 0$ or $x + 1 = 0$

$x = 10$ or $x = -1$

$x = -1$ is not in the domains of $\log x$ or $\log(x-9)$, so the only solution is $x = 10$.

17. $\log(3+b) = \log(4c-1) \Rightarrow 3+b = 4c-1 \Rightarrow$

$4+b = 4c \Rightarrow \dfrac{4+b}{4} = c \Rightarrow c = \dfrac{4+b}{4}$

19. $2 - b = \log(6c+5) \Rightarrow 6c+5 = 10^{2-b} \Rightarrow$

$6c = 10^{2-b} - 5 \Rightarrow c = \dfrac{10^{2-b}-5}{6}$

21. Answers will vary. Sample answer:
Not necessarily. For example, the solution to the example $\log(x+103) = 2$ is $x = -3$. Negative answers must be rejected when they are not in the domain of any of the terms in the equation.

23. $2^{x-1} = 8 \Rightarrow 2^{x-1} = 2^3 \Rightarrow x - 1 = 3 \Rightarrow x = 4$

25. $25^{-3x} = 3125 \Rightarrow (5^2)^{-3x} = 5^5 \Rightarrow$

$5^{-6x} = 5^5 \Rightarrow -6x = 5 \Rightarrow x = -\dfrac{5}{6}$

27. $6^{-x} \Rightarrow 36^{x+6} = 6^{-x} = \left(6^2\right)^{x+6} \Rightarrow$

$6^{-x} = 6^{2(x+6)} \Rightarrow -x = 2(x+6) \Rightarrow$

$-x = 2x + 12 \Rightarrow -3x = 12 \Rightarrow x = -4$

29. $\left(\dfrac{3}{4}\right)^x = \dfrac{16}{9} \Rightarrow \left(\dfrac{3}{4}\right)^x = \left(\dfrac{4}{3}\right)^2 \Rightarrow$

$\left(\dfrac{3}{4}\right)^x = \left(\dfrac{3}{4}\right)^{-2} \Rightarrow x = -2$

31. $2^x = 5$

Take natural logarithms of both sides.

$\ln 2^x = \ln 5$

$x \ln 2 = \ln 5$

$x = \dfrac{\ln 5}{\ln 2} \approx \dfrac{1.6094}{.6931} \approx 2.3219$

33. $2^x = 3^{x-1}$

$\ln 2^x = \ln 3^{x-1}$

$x \ln 2 = (x-1)(\ln 3)$

$x \ln 2 = x \ln 3 - 1 \ln 3$

$x \ln 2 - x \ln 3 = -\ln 3$

$(\ln 2 - \ln 3)x = -\ln 3$

$x = \dfrac{-\ln 3}{\ln 2 - \ln 3} \approx 2.710$

35. $3^{1-2x} = 5^{x+5}$

$\ln 3^{1-2x} = \ln 5^{x+5}$

$(1-2x)(\ln 3) = (x+5)(\ln 5)$

$\ln 3 - 2x \ln 3 = x \ln 5 + 5 \ln 5$

$\ln 3 - 5 \ln 5 = x \ln 5 + 2x \ln 3$

$\ln 3 - 5 \ln 5 = (\ln 5 + 2 \ln 3)x$

$\dfrac{\ln 3 - 5 \ln 5}{\ln 5 + 2 \ln 3} = x$

$x \approx -1.825$

37. $e^{3x} = 6 \Rightarrow \ln e^{3x} = \ln 6 \Rightarrow$

$3x = \ln 6 \Rightarrow x = \dfrac{\ln 6}{3} \approx .597253$

39.
$$2e^{5a+2}=8$$
$$e^{5a+2}=4$$
$$\ln e^{5a+2}=\ln 4$$
$$5a+2=\ln 4$$
$$5a=-2+\ln 4$$
$$a=\frac{-2+\ln 4}{5}\approx -.123$$

41.
$$10^{4c-3}=d$$
$$\log 10^{4c-3}=\log d$$
$$4c-3=\log d$$
$$4c=\log d+3$$
$$c=\frac{\log d+3}{4}$$

43. $e^{2c-1}=b\Rightarrow \ln e^{2c-1}=\ln b\Rightarrow$

$2c-1=\ln b\Rightarrow 2c=\ln b+1\Rightarrow c=\frac{\ln b+1}{2}$

45.
$$\log_7(r+3)+\log_7(r-3)=1$$
$$\log_7\left[(r+3)(r-3)\right]=1$$
$$(r+3)(r-3)=7^1$$
$$r^2-9=7$$
$$r^2=16\Rightarrow r=\pm 4$$

Since –4 is not in the domain of $\log_7(r-3)$, 4 is the only solution.

47. $\log_3(a-3)=1+\log_3(a+1)\Rightarrow$

$\log_3(a-3)-\log_3(a+1)=1\Rightarrow$

$\log_3\frac{a-3}{a+1}=1\Rightarrow\frac{a-3}{a+1}=3^1\Rightarrow$

$a-3=3a+3\Rightarrow -6=2a\Rightarrow a=-3$

Since –3 is not in the domain of $\log_3(a-3)$ nor in that of $\log_3(a+1)$, there is no solution.

49.
$$\log_2\sqrt{2y^2}-1=\frac{3}{2}$$
$$\log_2\sqrt{2y^2}=\frac{5}{2}$$
$$\sqrt{2y^2}=2^{5/2}$$
$$2y^2=2^5=32$$
$$y^2=16\Rightarrow y=\pm 4$$

51. $\log_2(\log_3 x)=1\Rightarrow \log_3 x=2^1=2\Rightarrow$

$x=3^2=9$

53. $5^{-2x}=\frac{1}{25}\Rightarrow 5^{-2x}=5^{-2}\Rightarrow -2x=-2\Rightarrow x=1$

55. $2^{|x|}=16\Rightarrow 2^{|x|}=2^4\Rightarrow |x|=4\Rightarrow x=\pm 4$

57.
$$2^{x^2-1}=10$$
$$\ln 2^{x^2-1}=\ln 10$$
$$(x^2-1)\ln 2=\ln 10$$
$$x^2-1=\frac{\ln 10}{\ln 2}$$
$$x^2-1\approx\frac{2.3026}{.6931}$$
$$x^2-1\approx 3.3219$$
$$x^2=4.3219$$
$$x=\pm\sqrt{4.3219}\approx\pm 2.0789$$

59. $2(e^x+1)=10\Rightarrow e^x+1=5\Rightarrow e^x=4\Rightarrow$

$\ln e^x=\ln 4\Rightarrow x=\ln 4\Rightarrow x\approx 1.386$

61. Answers will vary. Possible answer:
Since $x^2\ge 0$ for every x, $x^2+1\ge 1$ for every x.
Hence $4^{x^2+1}\ge 4^1=4$ for every x.

63. $g(x)=1.104(1.346)^x$,
$x=2$ corresponds to 2002.

a. $g(x)=20\Rightarrow 20=1.104(1.346)^x\Rightarrow$

$\frac{20}{1.104}=1.346^x\Rightarrow$

$\ln\left(\frac{20}{1.104}\right)=x\ln 1.346\Rightarrow$

$x=\frac{\ln\left(\frac{20}{1.104}\right)}{\ln 1.346}\approx 9.7$

Gambling revenue generated in Macau, China was \$20 billion sometime during 2009.

b. $g(x) = 45 \Rightarrow 45 = 1.104(1.346)^x \Rightarrow$
$\frac{45}{1.104} = 1.346^x \Rightarrow$
$\ln\left(\frac{45}{1.104}\right) = x\ln 1.346 \Rightarrow$
$x = \frac{\ln\left(\frac{45}{1.104}\right)}{\ln 1.346} \approx 12.5$
Gambling revenue generated in Macau, China was $45 billion sometime during 2012.

65. $f(x) = 17.6 + 12.8\ln x$,
$x = 10$ corresponds to 1910.

a. $f(x) = 75.5 \Rightarrow 75.5 = 17.6 + 12.8\ln x \Rightarrow$
$57.9 = 12.8\ln x \Rightarrow \frac{57.9}{12.8} = \ln x \Rightarrow$
$x = e^{57.9/12.8} \approx 92.1518$
A person whose life expectancy at birth is 75.5 years was born in 1992.

b. $f(x) = 77.5 \Rightarrow 77.5 = 17.6 + 12.8\ln x \Rightarrow$
$59.9 = 12.8\ln x \Rightarrow \frac{59.9}{12.8} = \ln x \Rightarrow$
$x = e^{59.9/12.8} \approx 107.7364$
A person whose life expectancy at birth is 77.5 years was born in 2007.

c. $f(x) = 81 \Rightarrow 81 = 17.6 + 12.8\ln x \Rightarrow$
$63.4 = 12.8\ln x \Rightarrow \frac{63.4}{12.8} = \ln x \Rightarrow$
$x = e^{63.4/12.8} \approx 141.62$
A person whose life expectancy at birth is 81 years will be born in 2041.

67. $f(x) = 7.9e^{.0254x}$, where $x = 0$ corresponds to 1970.

a. $f(x) = 15 \Rightarrow 7.9e^{.0254x} = 15 \Rightarrow$
$e^{.0254x} = \frac{15}{7.9} \Rightarrow .0254x = \ln\left(\frac{15}{7.9}\right) \Rightarrow$
$x = \frac{\ln\left(\frac{15}{7.9}\right)}{.0254} \approx 25.2$
The percent of U.S. income earned by the top one percent reached 15% sometime during 1995.

b. $f(x) = 30 \Rightarrow 7.9e^{.0254x} = 30 \Rightarrow$
$e^{.0254x} = \frac{30}{7.9} \Rightarrow .0254x = \ln\left(\frac{30}{7.9}\right) \Rightarrow$
$x = \frac{\ln\left(\frac{30}{7.9}\right)}{.0254} \approx 52.5$
The percent of U.S. income earned by the top one percent will reach 30% sometime during 2022.

69. $C(t) = 25e^{-.14t}$

a. $C(0) = 25e^{-.14(0)} = 25e^0 = 25$
Initially, 25 g of cobalt was present.

b. We determine the half-life by finding a value of t such that $C(t) = \left(\frac{1}{2}\right)(25) = 12.5.$

$12.5 = 25e^{-.14t} \Rightarrow \frac{1}{2} = e^{-.14t} \Rightarrow$

$\ln\frac{1}{2} = \ln e^{-.14t} \Rightarrow \ln\frac{1}{2} = -.14t \Rightarrow$

$t = \frac{\ln\frac{1}{2}}{-.14} = \frac{\ln .5}{-.14} \approx 4.95$
The half-life of cobalt is about 4.95 years.

71. $y = y_0\left(.5^{t/5730}\right)$
36% lost means 64% remains. Since 64% = .64, replace y with $.64y_0$ and solve for t.
$.64y_0 = y_0\left(.5^{t/5730}\right) \Rightarrow .64 = \left(.5^{t/5730}\right) \Rightarrow$
$\ln .64 = \ln\left(.5^{t/5730}\right) \Rightarrow \ln .64 = \frac{t}{5730}\ln .5 \Rightarrow$
$t = \frac{5730\ln .64}{\ln .5} \approx 3689.3$
The ivory is about 3689 years old.

73. Use the formula for the Richter scale given in Example 9 in this section of the textbook.
$R(i) = \log\left(\frac{i}{i_0}\right)$

a. $R(i) = 7.9 = \log\left(\frac{i}{i_0}\right) \Rightarrow$

$10^{7.9} = \frac{i}{i_0} \Rightarrow i = 10^{7.9}i_0 \approx 79{,}432{,}823i_0$

b. $R(i) = 5.4 = \log\left(\dfrac{i}{i_0}\right) \Rightarrow$

$$10^{5.4} = \frac{i}{i_0} \Rightarrow i = 10^{5.4} i_0 \approx 251{,}189 i_0$$

c. The China earthquake was

$$\frac{10^{7.9}}{10^{5.4}} = 10^{7.9-5.4} = 10^{2.5} \approx 316.23$$

times stronger than the Los Angeles earthquake.

75. $D(i) = 10 \cdot \log\left(\dfrac{i}{i_0}\right)$

a. $D(115 i_0) = 10\log\left(\dfrac{115 i_0}{i_0}\right)$

$$= 10\log 115 \approx 21$$

b. $D(10^{10} i_0) = 10\log\left(\dfrac{10^{10} i_0}{i_0}\right)$

$$= 10 \cdot (10\log 10)$$
$$= 10 \cdot (10 \cdot 1) = 100$$

c. $D(31{,}600{,}000{,}000 i_0)$

$$= 10\log\left(\frac{31{,}600{,}000{,}000 i_0}{i_0}\right)$$
$$= 10\log 31{,}600{,}000{,}000 \approx 105$$

d. $D(895{,}000{,}000{,}000 i_0)$

$$= 10\log\left(\frac{895{,}000{,}000{,}000 i_0}{i_0}\right)$$
$$= 10\log(895{,}000{,}000{,}000) \approx 120$$

e. $D(109{,}000{,}000{,}000{,}000 i_0)$

$$= 10\log\left(\frac{109{,}000{,}000{,}000{,}000 i_0}{i_0}\right)$$
$$= 10\log(109{,}000{,}000{,}000{,}000) \approx 140$$

77. a. $P(T) = 1 - e^{-.0034-.0053T}$

$$P(60) = 1 - e^{-.0034-.0053(60)} \approx .275$$

The reduction will be 27.5% when the tax is \$60.

b.

$$P(T) = 1 - e^{-.0034-.0053T}$$
$$.5 = 1 - e^{-.0034-.0053T}$$
$$e^{-.0034-.0053T} = 1 - .5$$
$$-.0034 - .0053T = \ln .5$$
$$T = \frac{\ln(.5) + .0034}{-.0053} \approx 130.14$$

The tax would be \$130.14

79. a. To graph $g(x) = 2.0 + 2.768\ln x$, enter this function as y_1 and use [2,22] by [0, 15]

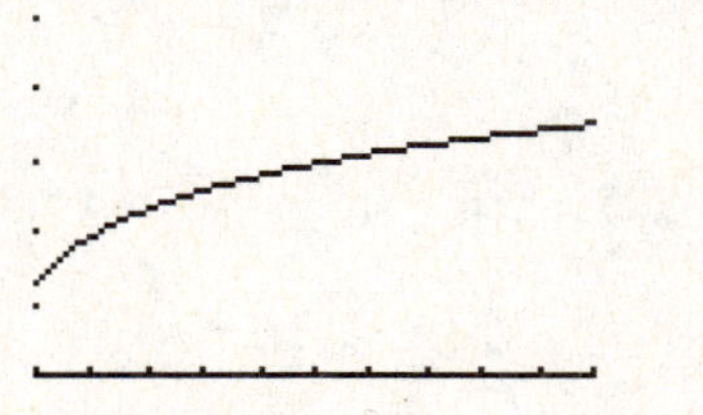

b. Graph $y = 9$, entering this function as y_2. On the CALC menu use "intersect."

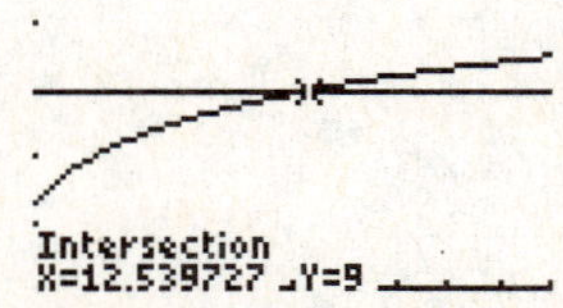

The point of intersection has approximate coordinates (12.54, 9). Sometime in 2002 the annual box office revenue reached \$9 billion.

Alternatively, solve

$$2.0 + 2.768\ln x = 9 \Rightarrow 2.768\ln x = 7 \Rightarrow$$
$$\ln x = \frac{7}{2.768} \Rightarrow x = e^{7/2.768} \approx 12.5.$$

Chapter 4 Review Exercises

1. $y = a^{x+2}$ is (c).

2. $y = a^x + 2$ is (a).

3. $y = -a^x + 2$ is (d).

4. $y = a^{-x} + 2$ is (b).

5. $0 < a < 1$

6. Domain of f: $(-\infty, \infty)$ or all real numbers.

7. Range of f: $(0, \infty)$ or all positive real numbers.

8. $f(0) = 1$

9. $f(x) = 4^x$

x	-2	-1	0	1	2
y	$\frac{1}{16}$	$\frac{1}{4}$	1	4	16

The negative x-axis is a horizontal asymptote for the graph.

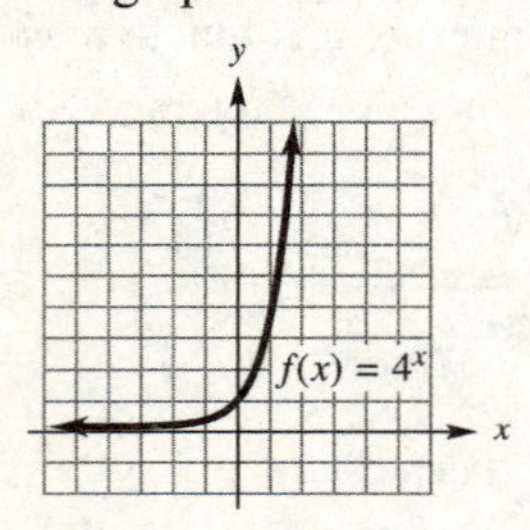

10. $g(x) = 4^{-x}$

x	-2	-1	0	1	2
y	16	4	1	$\frac{1}{4}$	$\frac{1}{16}$

The positive x-axis is a horizontal asymptote for the graph.

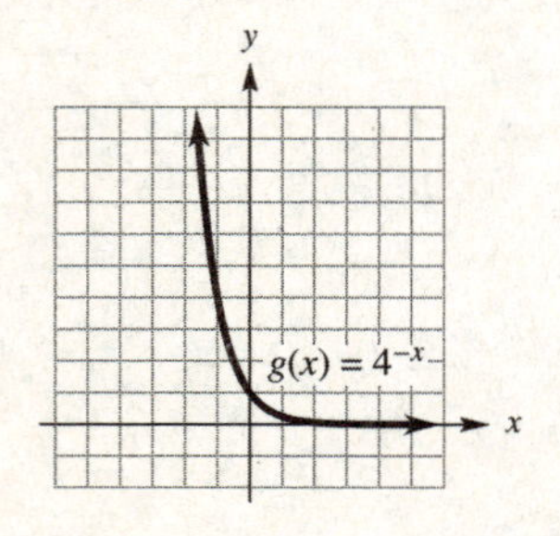

11. $f(x) = \ln x + 5$

x	.01	.1	1	2	4	6	8
y	.4	2.7	5	5.7	6.4	6.8	7.1

The negative y-axis is a vertical asymptote for the graph.

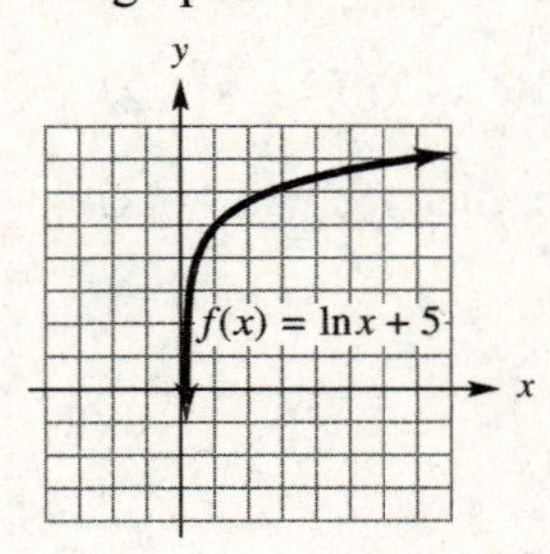

12. $g(x) = \log x - 3$

x	.01	.1	1	2	4	6	8
y	-5	-4	-3	-2.7	-2.4	-2.2	-2.1

The negative y-axis is a vertical asymptote for the graph.

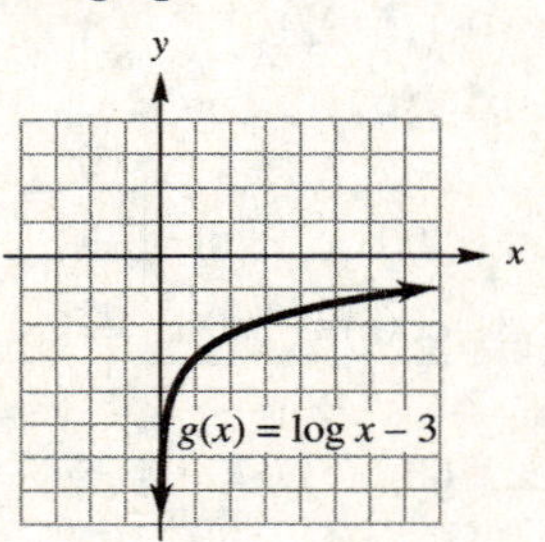

13. $f(x) = 39.61(1.1118)^x$,
$x = 1$ corresponds to January, 2011.

a. $x = 21$

$f(21) = 39.61(1.1118)^{21} \approx 412$

In September 2012, there were about 412 thousand retail jobs added.

b. To determine x for $f(x) = 300$, solve:

$$f(x) = 300 = 39.61(1.118)^x \Rightarrow$$

$$\frac{300}{39.61} = (1.118)^x \Rightarrow$$

$$\ln\left(\frac{300}{39.61}\right) = x \ln 1.118 \Rightarrow$$

$$x = \frac{\ln\left(\frac{300}{39.61}\right)}{\ln 1.118} \approx 18.2$$

Alternatively, graph $Y_1 = 39.61(1.118)^x$ and $Y_2 = 300$ in the window [0, 25] by [0, 400], then find the intersection.

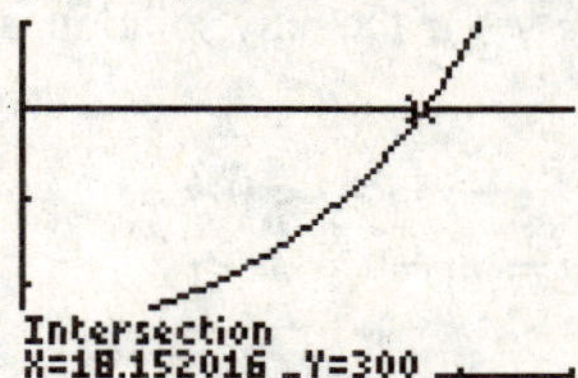

According to the model, the cumulative number of jobs hit 300,000 sometime during June 2012.

14. $f(x) = 28.7(1.120)^x$,
$x = 1$ corresponds to January, 2011.

a. $x = 18$

$$f(18) = 28.7(1.120)^{18} \approx 221$$

In June 2012, there were about 221 thousand construction jobs added.

b. To determine x for $f(x) = 250$, solve:

$$f(x) = 250 = 28.7(1.120)^x \Rightarrow$$

$$\frac{250}{28.7} = (1.120)^x \Rightarrow$$

$$\ln\left(\frac{250}{28.7}\right) = x \ln 1.120 \Rightarrow$$

$$x = \frac{\ln\left(\frac{250}{28.7}\right)}{\ln 1.120} \approx 19.1$$

Alternatively, graph $Y_1 = 28.7(1.120)^x$ and $Y_2 = 250$ in the window [0, 75] by [–100, 600], then find the intersection.

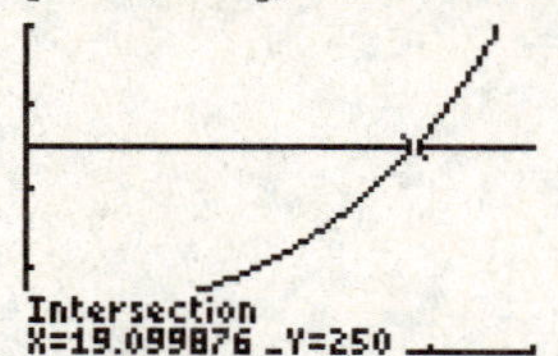

According to the model, the cumulative number of construction jobs hit 250,000 sometime during July 2012.

15. $10^{2.53148} = 340$ is equivalent to $\log 340 = 2.53148$.

16. $5^4 = 625$ is equivalent to $\log_5 625 = 4$.

17. $e^{3.8067} = 45$ is equivalent to $\ln 45 = 3.8067$.

18. $7^{1/2} = \sqrt{7}$ is equivalent to $\log_7 \sqrt{7} = \frac{1}{2}$.

19. $\log 10{,}000 = 4$ is equivalent to $10^4 = 10{,}000$.

20. $\log 26.3 = 1.4200$ is equivalent to $10^{1.4200} = 26.3$.

21. $\ln 81.1 = 4.3957$ is equivalent to $e^{4.3957} = 81.1$.

22. $\log_2 4096 = 12$ is equivalent to $2^{12} = 4096$.

23. $\ln e^5 = 5$ because $\ln e^k = k$ for every real number k.

24. $\log \sqrt[3]{10} = \log_{10} 10^{1/3} = \frac{1}{3}$ because $\log_a a^y = y$ for every positive real number y.

25. $10^{\log 8.9} = 8.9$ because $a^{\log_a x} = x$ for every positive real number x.

26. $\ln e^{3t^2} = 3t^2$ because $\ln e^j = j$ for every real number j.

27. Let $x = \log_8 2$.

$\log_8 2 = x \Rightarrow 8^x = 2 \Rightarrow (2^3)^x = 2 \Rightarrow 2^{3x} = 2 \Rightarrow$

$3x = 1 \Rightarrow x = \frac{1}{3}$

Therefore, $\log_8 2 = \frac{1}{3}$.

28. Let $x = \log_8 32$.

$\log_8 32 = x \Rightarrow 8^x = 32 \Rightarrow (2^3)^x = 2^5 \Rightarrow$

$2^{3x} = 2^5 \Rightarrow 3x = 5 \Rightarrow x = \frac{5}{3}$

Therefore, $\log_8 32 = \frac{5}{3}$.

29. $\log 4x + \log 5x^5 = \log(4x \cdot 5x^5) = \log(20x^6)$

30. $4\log u - 5\log u^6 = \log u^4 - \log(u^6)^5$

$= \log u^4 - \log u^{30}$

$= \log\left(\frac{u^4}{u^{30}}\right) = \log\left(\frac{1}{u^{26}}\right)$

31. $3\log b - 2\log c = \log b^3 - \log c^2 = \log\left(\frac{b^3}{c^2}\right)$

32. $7\ln x - 3(\ln x^3 + 5\ln x) = 7\ln x - 3\ln x^3 - 15\ln x$

$= -3\ln x^3 - 8\ln x$

$= \ln(x^3)^{-3} + \ln x^{-8}$

$= \ln(x^{-9}) + \ln x^{-8}$

$= \ln(x^{-9} \cdot x^{-8}) = \ln x^{-17}$

$= \ln\left(\frac{1}{x^{17}}\right)$

33. $\ln(m+8) - \ln m = \ln 3 \Rightarrow$

$\ln\frac{m+8}{m} = \ln 3 \Rightarrow \frac{m+8}{m} = 3 \Rightarrow 3m = m + 8 \Rightarrow$

$2m = 8 \Rightarrow m = 4$

34. $2\ln(y+1) = \ln(y^2-1) + \ln 5$

$\ln(y+1)^2 = \ln\left[(y^2-1)(5)\right]$

$(y+1)^2 = 5(y^2-1)$

$y^2 + 2y + 1 = 5y^2 - 5$

$0 = 4y^2 - 2y - 6$

$0 = 2(2y^2 - y - 3)$

$0 = 2(2y-3)(y+1)$

$y = \frac{3}{2}$ or $y = -1$

Since –1 is not in the domain of $\ln(y+1)$ nor in that of $\ln(y^2-1)$, the only solution is $\frac{3}{2}$.

35. $\log(m+3) = 2 \Rightarrow m + 3 = 10^2 \Rightarrow$

$m + 3 = 100 \Rightarrow m = 97$

36. $\log x^3 = 2 \Rightarrow x^3 = 10^2 \Rightarrow x^3 = 100 \Rightarrow$

$x = \sqrt[3]{100} \approx 4.642$

37. $\log_2(3k+1) = 4 \Rightarrow 3k + 1 = 2^4 \Rightarrow$

$3k + 1 = 16 \Rightarrow 3k = 15 \Rightarrow k = 5$

38. $\log_5\left(\frac{5z}{z-2}\right) = 2 \Rightarrow 5^2 = \frac{5z}{z-2} \Rightarrow$

$25(z-2) = 5z \Rightarrow 25z - 50 = 5z \Rightarrow$

$20z = 50 \Rightarrow z = \frac{50}{20} = 2.5$

39. $\log x + \log(x-3) = 1 \Rightarrow \log[x(x-3)] = 1 \Rightarrow$

$x(x-3) = 10^1 \Rightarrow x^2 - 3x = 10 \Rightarrow$

$x^2 - 3x - 10 = 0 \Rightarrow (x-5)(x+2) = 0 \Rightarrow$

$x = 5$ or $x = -2$

Since –2 is not in the domain of $\log x$ nor in that of $\log(x-3)$, the only solution is 5.

40. $\log_2 r + \log_2(r-2) = 3$

$\log_2[r(r-2)] = 3$

$r(r-2) = 2^3$

$r^2 - 2r - 8 = 0$

$(r-4)(r+2) = 0 \Rightarrow r = 4$ or $r = -2$

Since –2 is not in the domain of $\log_2 r$ nor in that of $\log_2(r-2)$, the only solution is 4.

41. $2^{3x} = \frac{1}{64} \Rightarrow 2^{3x} = 2^{-6} \Rightarrow 3x = -6 \Rightarrow x = -2$

42. $\left(\frac{9}{16}\right)^x = \frac{3}{4} \Rightarrow \left[\left(\frac{3}{4}\right)^2\right]^x = \left(\frac{3}{4}\right)^1 \Rightarrow$

$\left(\frac{3}{4}\right)^{2x} = \left(\frac{3}{4}\right)^1 \Rightarrow 2x = 1 \Rightarrow x = \frac{1}{2}$

43. $9^{2y+1} = 27^y \Rightarrow (3^2)^{2y+1} = (3^3)^y \Rightarrow$

$3^{4y+2} = 3^{3y} \Rightarrow 4y + 2 = 3y \Rightarrow y = -2$

44. $\frac{1}{2} = \left(\frac{b}{4}\right)^{1/4} \Rightarrow \left(\frac{1}{2}\right)^4 = \left[\left(\frac{b}{4}\right)^{1/4}\right]^4 \Rightarrow$

$\frac{1}{16} = \frac{b}{4} \Rightarrow 16b = 4 \Rightarrow b = \frac{4}{16} = \frac{1}{4}$

45. $8^p = 19 \Rightarrow \ln 8^p = \ln 19 \Rightarrow p \ln 8 = \ln 19 \Rightarrow$

$p = \frac{\ln 19}{\ln 8} \Rightarrow p \approx \frac{2.9444}{2.0794} \Rightarrow p \approx 1.416$

46. $3^z = 11 \Rightarrow \ln 3^z = \ln 11 \Rightarrow z \ln 3 = \ln 11 \Rightarrow$

$z = \frac{\ln 11}{\ln 3} \Rightarrow z \approx 2.183$

47. $5 \cdot 2^{-m} = 35 \Rightarrow 2^{-m} = 7 \Rightarrow \ln(2^{-m}) = \ln 7 \Rightarrow$

$-m \ln 2 = \ln 7 \Rightarrow -m = \frac{\ln 7}{\ln 2} \Rightarrow$

$-m = \frac{1.9459}{.6931} \Rightarrow m \approx -2.807$

48. $2 \cdot 15^{-k} = 18 \Rightarrow 15^{-k} = 9 \Rightarrow \ln 15^{-k} = \ln 9 \Rightarrow$

$-k \ln 15 = \ln 9 \Rightarrow k = -\frac{\ln 9}{\ln 15} \Rightarrow k \approx -.811$

49. $e^{-5-2x} = 5 \Rightarrow \ln e^{-5-2x} = \ln 5 \Rightarrow$

$-5 - 2x = \ln 5 \Rightarrow -5 - 2x \approx 1.6094 \Rightarrow$

$-2x = 6.6094 \Rightarrow x \approx -3.305$

50. $e^{3x-1} = 12 \Rightarrow \ln e^{3x-1} = \ln 12 \Rightarrow 3x - 1 = \ln 12 \Rightarrow$

$x = \frac{1 + \ln 12}{3} \approx 1.162$

51.

$$\begin{aligned} 6^{2-m} &= 2^{3m+1} \\ \ln 6^{2-m} &= \ln 2^{3m+1} \\ (2-m)\ln 6 &= (3m+1)\ln 2 \\ 2\ln 6 - m\ln 6 &= 3m\ln 2 + \ln 2 \\ 2\ln 6 - \ln 2 &= 3m\ln 2 + m\ln 6 \\ 2\ln 6 - \ln 2 &= (3\ln 2 + \ln 6)m \\ m &= \frac{2\ln 6 - \ln 2}{3\ln 2 + \ln 6} \approx .747 \end{aligned}$$

52.

$$\begin{aligned} 5^{3r-1} &= 6^{2r+5} \\ \ln 5^{3r-1} &= \ln 6^{2r+5} \\ (3r-1)\ln 5 &= (2r+5)\ln 6 \\ 3r\ln 5 - \ln 5 &= 2r\ln 6 + 5\ln 6 \\ (3\ln 5 - 2\ln 6)r &= 5\ln 6 + \ln 5 \\ r &= \frac{5\ln 6 + \ln 5}{3\ln 5 - 2\ln 6} \approx 8.490 \end{aligned}$$

53.

$$\begin{aligned} (1+.003)^k &= 1.089 \\ 1.003^k &= 1.089 \\ \ln 1.003^k &= \ln 1.089 \\ k\ln 1.003 &= \ln 1.089 \\ k &= \frac{\ln 1.089}{\ln 1.003} \approx 28.463 \end{aligned}$$

54.

$$\begin{aligned} (1+.094)^z &= 2.387 \\ 1.094^z &= 2.387 \\ \ln 1.094^z &= \ln 2.387 \\ z\ln 1.094 &= \ln 2.387 \\ z &= \frac{\ln 2.387}{\ln 1.094} \approx 9.684 \end{aligned}$$

55. $y = 4e^{.03t}$

a. The population will double, so $y = 2 \cdot 4$.
The answer is C.

b. The population will be 12 thousand, so $y = 12$.
The answer is A.

c. $t = 6$
The answer is D.

d. $t = 6(12) = 72$
The answer is B.

56. $y = 2e^{.02t}$

a. The population will triple, so $y = 3 \cdot 2$. The answer is B.

b. The population will be 3 million, so $y = 3$. The answer is D.

c. $t = 3$
The answer is C.

d. $t = \frac{4}{12} = \frac{1}{3}$
The answer is A.

57. $A(t) = 10e^{-.00495t}$

a. $A(0) = 10e^{-.00495(0)} = 10e^0 = 10$
The amount of polonium present initially was 10 g.

b. We determine the half-life by finding a value of t such that $A(t) = \left(\frac{1}{2}\right)(10) = 5$.

$$5 = 10e^{-.00495(t)}$$
$$\frac{1}{2} = e^{-.00495t}$$
$$\ln\frac{1}{2} = \ln e^{-.00495t}$$
$$\ln\frac{1}{2} = -.00495t$$
$$t = \frac{\ln .5}{-.00495} \approx 140$$

The half-life of polonium is about 140 days.

c. Find t such that $A(t) = 3$.

$$10e^{-.00495(t)} = 3$$
$$e^{-.00495(t)} = .3$$
$$\ln e^{-.00495(t)} = \ln .3$$
$$-.00495t = \ln .3$$
$$t = \frac{\ln .3}{-.00495} \approx 243$$

It will take about 243 days for the polonium to decay to 3 g.

58. $h(x) = 69.54e^{-.264x}$, $x = 8$ corresponds to 2008.

a. $x = 2010 - 2008 + 8 = 10$
$h(10) = 69.54e^{-.264(10)} \approx 4.96$
According to the model, the average price per million BTU's was about 4.96 in 2010.

b. $h(x) = 3 = 69.54e^{-.264x} \Rightarrow$
$\frac{3}{69.54} = e^{-.264x} \Rightarrow \ln\left(\frac{3}{69.54}\right) = -.264x \Rightarrow$

$$x = \frac{\ln\left(\frac{3}{69.54}\right)}{-.264} \approx 11.9$$

According to the model, the average price per million BTU's was \$3.00 late in 2011.

59. a. $i(7) = i_0 10^x = i_0 10^7 = 10{,}000{,}000 i_0$

b. $i(6.5) = i_0 10^{6.5} = 3{,}162{,}278 i_0$

c. $\frac{i(7)}{i(6.5)} = \frac{10{,}000{,}000 i_0}{3{,}162{,}278 i_0} \approx 3.2$

The earthquake in Indonesia was about 3.2 times more intense than the earthquake in Paupa New Guinea.

60. For earthquakes, increasing the ground motion by a factor of 10^k increases the Richter magnitude by k units. In this problem, the second earthquake, with ground motion $1000 = 10^3$ times greater than the first earthquake, will measure $4.6 + 3 = 7.6$ on the Richter scale.

61. $F(t) = T_0 + Ce^{-kt}$

$$T_0 = 50, F(0) = 50 + Ce^{-k(0)} = 300, C = 250$$
$$F(t) = 50 + 250e^{-kt}$$
$$F(4) = 175 = 50 + 250e^{-k(4)}$$
$$125 = 250e^{-k(4)}$$
$$\frac{125}{250} = e^{-k(4)} \Rightarrow \ln\left(\frac{125}{250}\right) = -k(4) \Rightarrow k \approx .1733$$
$$F(12) = 50 + 250e^{-.1733(12)} \approx 81.25$$

The temperature after 12 minutes is 81.25° C.

62. $F(t) = T_0 + Ce^{-kt}$, $T_0 = 18$

$F(0) = 3.4 = 18 + C \Rightarrow C = -14.6$

$F(30) = 18 - 14.6e^{-k(30)} = 7.2$

$-14.6e^{-k(30)} = 7.2 - 18$

$e^{-k(30)} = \dfrac{7.2-18}{-14.6} \approx 0.739726$

$k = \dfrac{\ln 0.739726}{-30} \approx 0.01$

$F(t) = 18 - 14.6e^{-(0.01)t} = 10 \Rightarrow$

$-14.6e^{-(0.01)t} = 10 - 18 \Rightarrow$

$e^{-(0.01)t} = \dfrac{-8}{-14.6} \Rightarrow t \approx \dfrac{\ln 0.5479}{-0.01} \approx 60$

It will thaw to 10°C in 1 hour.

63. a. $f(0) = 500,\ f(7) = 785$

$f(x) = a(b)^x \Rightarrow f(0) = 500 = a(b)^0 \Rightarrow$
$a = 500$

$f(7) = a(b)^7 \Rightarrow 785 = 500(b)^7 \Rightarrow$

$b^7 = \dfrac{785}{500} \Rightarrow b = \left(\dfrac{785}{500}\right)^{1/7} \Rightarrow b \approx 1.067$

The function is $f(x) = 500(1.067)^x$.

b.

```
ExpReg
 y=a*b^x
 a=474.8173496
 b=1.075351749
 r²=.96176873
 r=.980698083
```

$g(x) = 474.8(1.075)^x$

c. $x = 9$ for April, 2013

$f(9) = 500(1.067)^9 \approx 896$

$g(9) = 474.8(1.075)^9 \approx 910$

Using the two-point model, the average time spent on Facebook per unique visitor was about 896 minutes. Using the exponential regression equation, the average time was 910 minutes.

d. $f(x) = 1000 = 500(1.067)^x \Rightarrow$

$\dfrac{1000}{500} = (1.067)^x \Rightarrow$

$\ln(2) = x \ln 1.067 \Rightarrow$

$x = \dfrac{\ln(2)}{\ln 1.067} \approx 10.7$

$g(x) = 1000 = 474.8(1.075)^x \Rightarrow$

$\dfrac{1000}{474.8} = 1.075^x \Rightarrow$

$\ln\left(\dfrac{1000}{474.8}\right) = x \ln 1.075 \Rightarrow$

$x = \dfrac{\ln\left(\dfrac{1000}{474.8}\right)}{\ln 1.075} \approx 10.3$

Both models predict that the average times spent on Facebook per unique visitor reached 1000 minutes sometime during May 2013.

64. a. At $x = 0$, $f(0) = 1013$ and at $x = 10$, $f(10) = 265$.

To find $f(x) = a(b^x)$, first solve for a:

$f(0) = 1013 \Rightarrow a(b^0) = 1013 \Rightarrow a = 1013$

Now solve for b:

$f(10) = 265 \Rightarrow 1013b^{10} = 265 \Rightarrow$

$b^{10} = \dfrac{265}{1013} \Rightarrow b = \left(\dfrac{265}{1013}\right)^{1/10} \approx .8745$

The model becomes $f(x) = 1013(.8745^x)$

b.

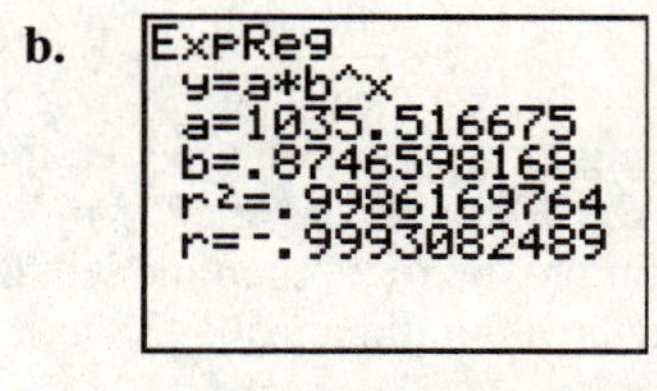
```
ExpReg
 y=a*b^x
 a=1035.516675
 b=.8746598168
 r²=.9986169764
 r=-.9993082489
```

$g(x) = 1035.52(.8747^x)$

c. $x = 1.5$

$f(1.5) = 1013(.8745^{1.5}) \approx 828.4$

$g(1.5) = 1035.52(.8747^{1.5}) \approx 847.1$

The models estimate the pressure at 1500m to be 828.4 and 847.1 millibars respectively. The models under and over estimate the actual value of 846 millibars respectively.

$x = 11$

$f(11) = 1013(.8745^{11}) \approx 231.7$

$g(11) = 1035.52(.8747^{11}) \approx 237.5$

The models estimate the pressure at 11,000 m to be 231.7 and 237.5 millibars respectively. Both models overestimate the actual pressure of 227 millibars.

d. Two-point model:

$1013(.8745^x) = 500 \Rightarrow .8745^x = \frac{500}{1013} \Rightarrow$

$x \ln .8745 = \ln\left(\frac{500}{1013}\right) \Rightarrow$

$x = \frac{\ln\left(\frac{500}{1013}\right)}{\ln .8745} \approx 5.265$

Regression model:

$1035.52(.8747^x) = 500$

$.8747^x = \frac{500}{1035.52}$

$x \ln(.8747) = \ln\left(\frac{500}{1035.52}\right)$

$x = \frac{\ln\left(\frac{500}{1035.52}\right)}{\ln .8747} \approx 5.438$

According to the two-point model, the pressure is 500 millibars at 5265 m. According to the regression model, the pressure is 500 millibars at 5438 m.

65. a. $f(1) = 18,\ f(7) = 59$

$f(x) = a + b\ln x \Rightarrow$

$f(1) = 18 = a + b\ln 1 \Rightarrow 18 = a$

$f(7) = 18 + b\ln 7 \Rightarrow 59 = 18 + b\ln 7 \Rightarrow$

$b\ln 7 = 41 \Rightarrow b = \frac{41}{\ln 7} \approx 21.07$

The function is $f(x) = 18 + 21.07\ln x$.

b.

```
LnReg
 y=a+blnx
 a=15.21212398
 b=20.47059577
 r²=.959533477
 r=.9795577967
```

$g(x) = 15.21 + 20.47\ln x$

c. 2014 corresponds to $x = 9$

$f(9) = 18 + 21.07\ln 9 \approx 64$

$g(9) = 15.21 + 20.47\ln 9 \approx 60$

According to the two-point model, the digital share of recorded music sales in the U.S. will be about 64% in 2014. According to the logarithmic regression, the digital share will be about 60% in 2014.

d. $f(x) = 70$

$18 + 21.07\ln x = 70 \Rightarrow$

$21.07\ln x = 52 \Rightarrow \ln x = \frac{52}{21.07} \Rightarrow$

$x = e^{52/21.07} \approx 11.8$

$g(x) = 70$

$15.21 + 20.47\ln x = 70 \Rightarrow$

$20.47\ln x = 54.79 \Rightarrow \ln x = \frac{54.79}{20.47} \Rightarrow$

$x = e^{54.79/20.47} \approx 14.5$

According to the two-point model, the share of digital music will reach 70% sometime during 2016. According to the logarithmic regression model, the share of digital music will reach 70% sometime during 2019.

66. a. $f(1) = 9.678,\ f(21) = 16.813$

$f(x) = a + b\ln x \Rightarrow$

$f(1) = 9.678 = a + b\ln 1 \Rightarrow 9.678 = a$

$f(21) = 9.678 + b\ln 21 \Rightarrow$

$16.813 = 9.678 + b\ln 21 \Rightarrow$

$b\ln 21 = 7.135 \Rightarrow b = \dfrac{7.135}{\ln 21} \approx 2.344$

The function is $f(x) = 9.678 + 2.344\ln x.$

b.

```
LnReg
 y=a+blnx
 a=8.811096985
 b=2.443197413
 r²=.8607540505
 r=.9277683173
```

$g(x) = 8.811 + 2.443\ln x$

c. 2013 corresponds to $x = 23$

$f(23) = 9.678 + 2.344\ln 23 \approx 17.028$

$g(20) = 8.811 + 2.443\ln 23 \approx 16.471$

According to the two-point model, there were about 17,028 kidney transplants in 2013. According to the logarithmic regression, there were about 16,471 kidney transplants in 2013.

d. $f(x) = 17.5$ thousand

$9.678 + 2.344\ln x = 17.5 \Rightarrow$

$2.344\ln x = 7.822 \Rightarrow \ln x = \dfrac{7.822}{2.344} \Rightarrow$

$x = e^{7.822/2.344} \approx 28.1$

$g(x) = 17.5$ thousand

$8.811 + 2.443\ln x = 17.5$

$2.443\ln x = 8.689 \Rightarrow$

$\ln x = \dfrac{8.689}{2.443} \Rightarrow$

$x = e^{8.689/2.443} \approx 35.0$

According to the two-point model, there will be 17,500 kidney transplants sometime during 2018. According to the logarithmic regression model, there will be 17,500 kidney transplants sometime during 2025.

Case 4 Gapminder.org

1.

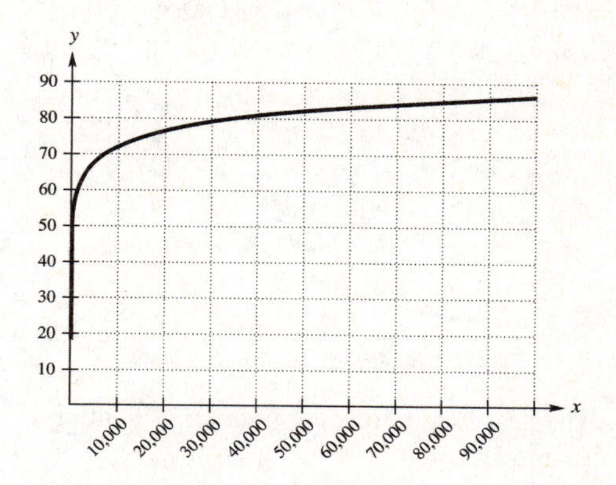

2. The graph is increasing and then leveling off.

3. $x = 4000$

$f(4000) = 15.32 + 6.178\ln 4000 \approx 66.6$

According to the model, the life expectancy in the Democratic Republic of the Congo if the per person income level was \$4000 would be about 66.6 years.

4. $x = 25{,}000$

$f(25{,}000) = 15.32 + 6.178\ln 25{,}000 \approx 77.9$

According to the model, the life expectancy in Russia if the per person income level was \$25,000 would be about 77.9 years.

5. $f(x) = 60 \Rightarrow 60 = 15.32 + 6.178\ln x \Rightarrow$

$44.68 = 6.178\ln x \Rightarrow \dfrac{44.68}{6.178} = \ln x \Rightarrow$

$x = e^{44.68/6.178} \approx 1383$

According to the model, if there is a life expectancy of 60 years, the per person income level would be about \$1383.

6. $f(x) = 75 \Rightarrow 75 = 15.32 + 6.178\ln x \Rightarrow$

$59.68 = 6.178\ln x \Rightarrow \dfrac{59.68}{6.178} = \ln x \Rightarrow$

$x = e^{59.68/6.178} \approx 15{,}679$

According to the model, if there is a life expectancy of 75 years, the per person income level would be about \$15,679.

7.

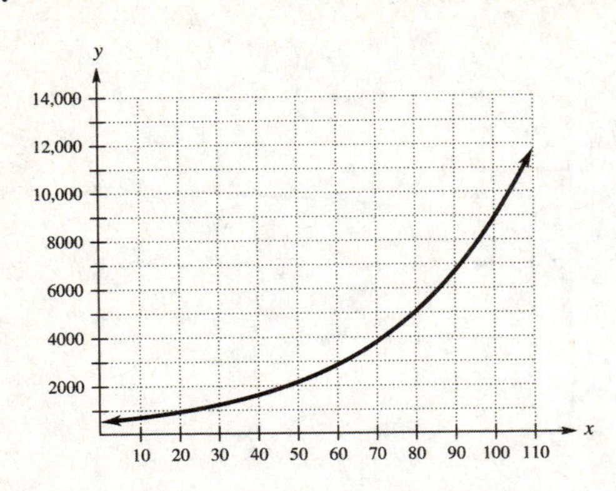

8. The graph is increasing in an exponential fashion.

9. $x = 1950 - 1900 = 50$

$g(50) = 510.3(1.029)^{50} \approx 2131$

According to the model, the per person income for Brazil in 1950 was about $2131.

$x = 2000 - 1900 = 100$

$g(100) = 510.3(1.029)^{100} \approx 8899$

According to the model, the per person income for Brazil in 2000 was about $8899.

10. $510.3(1.029^x) = 4000 \Rightarrow 1.029^x = \dfrac{4000}{510.3} \Rightarrow$

$x \ln 1.029 = \ln\left(\dfrac{4000}{510.3}\right) \Rightarrow$

$x = \dfrac{\ln\left(\frac{4000}{510.3}\right)}{\ln 1.029} \approx 72.0$

According to the model, the per person income for Brazil reached $4000 sometime in the year 1972.

11.

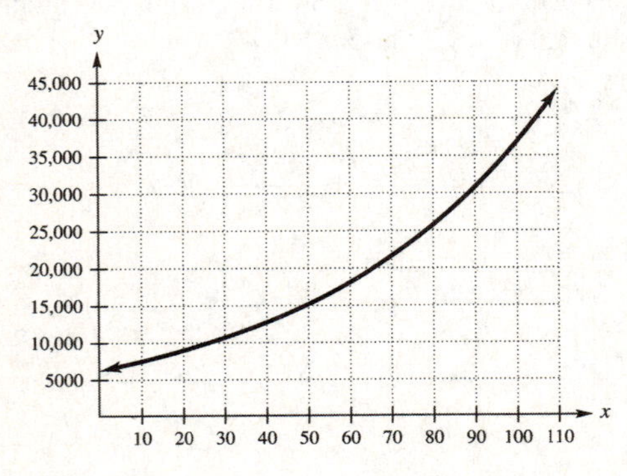

12. The graph is increasing in an exponential fashion.

13. $x = 1960 - 1900 = 60$

$h(60) = 6229.2(1.018)^{60} \approx 18{,}168$

According to the model, the per person income for the U.S. in 1960 was about $18,168.

$x = 2010 - 1900 = 110$

$h(110) = 6229.2(1.018)^{110} \approx 44{,}329$

According to the model, the per person income for the U.S. in 2010 was about $44,329.

14. $6229.2(1.018^x) = 20{,}000 \Rightarrow$

$1.018^x = \dfrac{20{,}000}{6229.2} \Rightarrow$

$x \ln 1.018 = \ln\left(\dfrac{20{,}000}{6229.2}\right) \Rightarrow$

$x = \dfrac{\ln\left(\frac{20{,}000}{6229.2}\right)}{\ln 1.018} \approx 65.4$

According to the model, the per person income for the U.S. reached $20,000 sometime in the year 1965.

Chapter 5 Mathematics of Finance

Section 5.1 Simple Interest and Discount

Note: Exercises in this chapter have been completed with a calculator. To ensure as much accuracy as possible, rounded values have been avoided in intermediate steps. When rounded values have been necessary, several decimal places have been carried throughout the exercise; only the final answer has been rounded to 1 or 2 decimal places. In most cases, answers involving money have been rounded to the nearest cent. Students who use rounded intermediate values should expect their final answers to differ from the answers given here. Depending on the magnitude of the numbers used in the exercise, the difference could be a few pennies or several thousand dollars.

1. The factors are time and interest rate.

3. \$2850 at 7% for 8 months

$P = 2850$, $r = .07$, and $t = \dfrac{8}{12}$.

$$I = Prt$$
$$= 2850(.07)\left(\frac{8}{12}\right) = 133.00$$

The simple interest is \$133.00.

5. \$3650 at 6.5% for 11 months

$P = 3650$, $r = .065$, $t = \dfrac{11}{12}$

$$I = Prt$$
$$= 3650(.065)\left(\frac{11}{12}\right) = 217.48$$

The simple interest is \$217.48.

7. \$2830 at 8.9% for 125 days

$P = 2830$, r = .089, $t = \dfrac{125}{365}$

$$I = Prt$$
$$= 2830(.089)\left(\frac{125}{365}\right) = 86.26$$

The simple interest is \$86.26.

9. \$5328 at 8%; loan made on August 16 is due December 30.
The duration of this loan is
$(31 - 16) + 30 + 31 + 30 + 30 = 136$ days.

$P = 5328$, $r = .08$, $t = \dfrac{136}{365}$

$$I = Prt$$
$$= 5328(.08)\left(\frac{136}{365}\right) = 158.82$$

The simple interest is \$158.82.

11. \$5,000, 3-year bond at 1.25%
To determine the semiannual interest payment, use the formula $I = P\,r\,t$, with $t = \frac{1}{2}$.

$$I = 5{,}000(0.0125)\left(\tfrac{1}{2}\right) = \$31.25.$$

Over the 3-year life of the bond, the total interest earned will be $\$31.25 \cdot 2 \cdot 3 = \$187.50$.

13. \$12,500, 10-year bond at 3.75%
To determine the semiannual interest payment, use the formula $I = P\,r\,t$, with $t = \frac{1}{2}$.

$$I = 12{,}500(0.0375)\left(\tfrac{1}{2}\right) = \$234.38.$$

Over the 10-year life of the bond, the total interest earned will be
$\$234.375 \cdot 2 \cdot 10 = \$4687.50$.

15. \$6,500, 10-year bond at 2.5%
To determine the semiannual interest payment, use the formula $I = P\,r\,t$, with $t = \frac{1}{2}$.

$$I = 6{,}500(0.025)\left(\tfrac{1}{2}\right) = \$81.25.$$

Over the 10-year life of the bond, the total interest earned will be $\$81.25 \cdot 2 \cdot 10 = \$1625.00$.

17. \$12,000 loan at 3.5% for three months
Three months is $\frac{1}{4}$ of a year.

$$A = P + I = P + Prt$$
$$= 12{,}000 + 12{,}000(.035)\left(\tfrac{1}{4}\right) = \$12{,}105$$

19. \$6500 loan at 5.25% for eight months.
Eight months is $\frac{2}{3}$ of a year.

$$A = P + I = P + Prt$$
$$= 6500 + 6500(.0525)\left(\tfrac{2}{3}\right) \approx \$6727.50$$

21. Answers vary. Sample answer: The present value is the amount of money that can be deposited today to yield some larger amount in the future.

23. \$15,000 for 9 months; money earns 6%. Use the formula for present value with

$A = 15{,}000,\ t = \dfrac{9}{12},\ r = .06$

$$P = \frac{A}{1+rt} = \frac{15{,}000}{1+(.06)\left(\frac{9}{12}\right)} = \frac{15{,}000}{1.045} = \$14{,}354.07$$

The present value is \$14,354.07

25. \$15,402 for 120 days; money earns 6.3%.

$A = 15{,}402,\ t = \dfrac{120}{365}\ r = .063$

$$P = \frac{A}{1+rt} = \frac{15{,}402}{1+(.063)\left(\frac{120}{365}\right)} = \frac{15{,}402}{1.0207} = \$15{,}089.46$$

The present value is \$15,089.46

27. Three-month \$20,000 T-bill with discount rate .075%

a. Three months is $\frac{1}{4}$ of a year.

Discount $= Prt = 20000 \cdot .00075 \cdot \frac{1}{4} = \$3.75$

The price of the T-bill is the face value minus the discount.
$20{,}000 - 3.75 = \$19{,}996.25$

b. $I = Prt \Rightarrow 3.75 = 19{,}996.25(r)\left(\frac{1}{4}\right) \Rightarrow$

$r \approx .0007501$
The actual interest rate is about .07501%.

29. Six-month \$15,500 T-bill with discount rate .105%

a. Six months is $\frac{1}{2}$ of a year.

Discount $= Prt$
$= 15{,}500 \cdot .00105 \cdot \frac{1}{2} = \$8.14$

The price of the T-bill is the face value minus the discount.
$15{,}500 - 8.14 = \$15{,}491.86$

b. $I = Prt \Rightarrow 8.14 = 15{,}491.86(r)\left(\frac{1}{2}\right) \Rightarrow$

$r \approx .00105087$
The actual interest rate is about .105087%.

31. Three-month \$20,000 T-bill with discount rate 4.96%

a. Three months is $\frac{1}{4}$ of a year.

Discount $= Prt$
$= 20000 \cdot .0496 \cdot \frac{1}{4}$
$= \$248.00$

The price of the T-bill is the face value minus the discount.
$20{,}000 - 248 = \$19{,}752.00$

b. $I = Prt \Rightarrow 248 = 19{,}752(r)\left(\frac{1}{4}\right) \Rightarrow$

$r \approx .050223$
The actual interest rate is about 5.0223%.

33. Six-month \$15,500 T-bill with discount rate 4.93%

a. Six months is $\frac{1}{2}$ of a year.

Discount $= Prt$
$= 15{,}500 \cdot .0493 \cdot \frac{1}{2} = \$382.07$

The price of the T-bill is the face value minus the discount.
$15{,}500 - 382.07 = \$15{,}117.93$

b. $I = Prt \Rightarrow 382.07 = 15{,}117.93(r)\left(\frac{1}{2}\right) \Rightarrow$

$r \approx .050545$
The actual interest rate is about 5.0545%.

35. a. New York City paid the interest from March 1874 until March 2009, 135 years.
$I = Prt = 1000 \cdot .07 \cdot 135 = \$9450$
As of March 2009, New York City paid \$9450 in interest on this bond.

b. The bond that matures in March 2147 will have been earning interest for 2147 – 1874 = 273 years.
$I = Prt = 1000 \cdot .07 \cdot 273 = \$19{,}110$
This bond will have earned \$19,110 in interest when it matures in March 2147.

37. $P = 3000,\ r = .025,\ t = \dfrac{9}{12}$

$$A = P(1+rt) = 3000\left[1+(.025)\left(\tfrac{9}{12}\right)\right] = 3056.25$$

After 9 months, the amount will be \$3056.25.

39. Interest = 67,359.39 – 67,081.20 = 278.19

$P = 67{,}081.20,\ t = \dfrac{1}{12}$

$I = Prt$

$r = \dfrac{I}{Pt} = \dfrac{278.19}{(67{,}081.20)\left(\frac{1}{12}\right)} = .050$

The interest rate was 5.0%.

41. Interest = 7675 – 7500 = 175, $P = 7500$, $r = .07$

$I = Prt \Rightarrow 175 = 7500(.07)t \Rightarrow$

$t = \dfrac{175}{7500(.07)} = \dfrac{1}{3}$

The time period for the loan was $\frac{1}{3}$ of a year, or four months.

43. Want present value of $1769 in 4 months at 6.25% interest.

$A = 1769,\ r = .0325,\ t = \dfrac{4}{12}$

$P = \dfrac{A}{1+rt} = \dfrac{1769}{1+(.0325)\left(\dfrac{4}{12}\right)} = 1750.04$

The student should deposit $1750.04.

45. $A = 6000,\ r = .036,\ t = \dfrac{10}{12} = \dfrac{5}{6}$

$P = \dfrac{A}{1+rt} = \dfrac{6000}{1+.036\left(\frac{5}{6}\right)} = 5825.24$

Yee should deposit $5825.24.

47. $P = 4000,\ t = \frac{1}{2}$

Discount = 4000 – 3930 = 70

Find the discount rate as follows:

$70 = 4000(r)\left(\frac{1}{2}\right) \Rightarrow r = .035$

The discount rate is 3.5%

49. Interest = (24 – 22) + .50 = 2.50, $P = 22$, $t = 1$

$I = Prt \Rightarrow 2.50 = 22(r)(1) \Rightarrow$

$r = \dfrac{2.50}{22(1)} \approx .114$

The simple interest rate is about 11.4%.

51. You will receive $700 from your tax preparer if you take the offer.

$60 = 760(r)\left(\frac{28}{365}\right) \Rightarrow r \approx 1.0291$

The actual interest rate is about 102.91%!

53. First find the maturity value of the loan, the amount that the contractor must repay.

$A = P(1+rt) \Rightarrow$

$A = 13{,}500\left(1+.09\left(\frac{9}{12}\right)\right) \approx 14{,}411.25$

In six months, the bank will receive $14,411.25. Since the bank wants a 10% return, compute the present value of this amount at 10% for six months.

$P = \dfrac{A}{1+rt} = \dfrac{14{,}411.25}{1+.1\left(\frac{1}{2}\right)} \approx 13{,}725$

The bank pays the plumber $13,725 and receives $14,411.25. This is enough for the plumber to pay a bill for $13,650.

55. y_1 is the future value after t years of $100 invested at 8% simple interest. y_2 is the future value after t years of $200 invested at 3%.

a. $y_1 = 100(.08)t + 100 = 8t + 100$

$y_2 = 200(.03)t + 200 = 6t + 200$

b. The graph of y_1 is a line with slope 8 and y-intercept 100. The graph of y_2 is a line with slope 6 and y-intercept 200.

c.

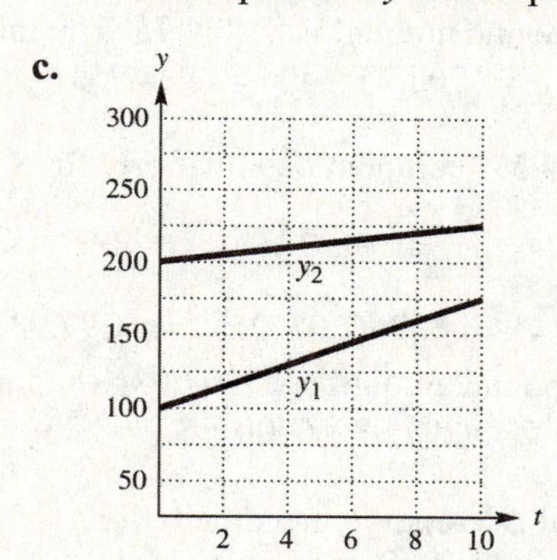

d. The y-intercept of each graph indicates the amount invested. The slope of the graph is the annual amount of interest paid.

Section 5.2 Compound Interest

1. r is the interest rate per year, while i is the interest rate per compounding period. t is the number of years, while n is the number of compounding periods.

3. The interest rate and number of compounding periods determine the amount of interest earned on a fixed principal.

5. The present value is the amount of money to be invested to earn a certain amount of return.

7. \$1000 at 4% compounded annually for 6 years
$P = 1000,\ i = \dfrac{4\%}{1} = .04,$ and $n = 6(1) = 6$
$A = P(1+i)^n = 1000(1.04)^6 \approx 1265.32$
The compound amount is \$1265.32. The interest earned is $I = 1265.32 - 1000 = \$265.32$.

9. \$470 at 8% compounded semiannually for 12 years
$P = 470,\ i = \dfrac{8\%}{2} = .04,\ n = 12(2) = 24$
$A = P(1+i)^n = 470(1.04)^{24} \approx 1204.75$
The compound amount is \$1204.75. The interest earned is $I = 1204.75 - 470 = \$734.75$.

11. \$6500 at 4.5% compounded quarterly for 8 years
$P = 6500,\ i = \dfrac{4.5\%}{4} = .01125,\ n = 8(4) = 32$
$A = P(1+i)^n = 6500(1.01125)^{32} \approx 9297.93$
The compound amount is \$9297.93. The interest earned is $I = 9297.93 - 6500 = \$2797.93$.

13. \$10,000 at .9% compounded daily for 1 yr
$P = 10,000,\ i = .009,\ n = 365$
Find the compound amount and then amount of interest.
$A = P(1+\frac{i}{n})^{nt}$
$= 10,000(1+\frac{.009}{365})^{(365)(1)}$
$\approx 10,090.41$
The compound amount is \$10,090.41. The amount of interest earned is \$10,090.41 – \$10,000 = \$90.41.

15. \$5000 at .81% compounded monthly for 2 years
Find the compound amount and then amount of interest.
$P = 5000,\ i = .0081,\ n = 12$
$A = P(1+\frac{i}{n})^{nt}$
$= 5000(1+\frac{.0081}{12})^{(12)(2)}$
≈ 5081.63
The amount of interest earned is \$5081.63 – \$5000 = \$81.63.

17. \$100,000 at 1.52% compounded daily for 5 years
$P = 100,000,\ i = .0152,\ n = 365$
$A = P(1+\frac{i}{n})^{nt}$
$= 100,000(1+\frac{.0152}{365})^{(365)(5)}$
$\approx 107,896.09$
The amount of interest earned is \$107,896.09 – \$100,000 = \$7896.09.

19. $P = 3000,\ A = 3606,\ n = 5$. Solve for i.
$P(1+i)^n = A \Rightarrow 3000(1+i)^5 = 3606 \Rightarrow$
$(1+i)^5 = \dfrac{3606}{3000} \Rightarrow 1+i = \sqrt[5]{\dfrac{3606}{3000}} \Rightarrow$
$i = \sqrt[5]{\dfrac{3606}{3000}} - 1 \approx .0375$
The interest rate is about 3.75%

21. $P = 8500,\ A = 12,161,\ n = 7$. Solve for i.
$P(1+i)^n = A \Rightarrow 8500(1+i)^7 = 12,161 \Rightarrow$
$(1+i)^7 = \dfrac{12,161}{8500} \Rightarrow \sqrt[7]{(1+i)^7} = \sqrt[7]{\dfrac{12,161}{8500}} \Rightarrow$
$1+i = \sqrt[7]{\dfrac{12,161}{8500}} \Rightarrow i = \sqrt[7]{\dfrac{12,161}{8500}} - 1 \approx 0.0525$
The interest rate is about 5.25%.

23. \$20,000 at 3.5% compounded continuously for 5 years
$P = 20,000,\ r = .035$
$A = Pe^{rt}$
$= 20,000e^{(.035)(5)}$
$\approx 23,824.92$
The amount of interest earned is \$23,824.92 – \$20,000 = \$3824.92.

25. \$30,000 at 1.8% compounded continuously for 3 years
$P = 30{,}000$, $r = .018$

$$A = Pe^{rt}$$
$$= 30{,}000e^{(.018)(3)}$$
$$\approx 31{,}664.54$$

The amount of interest earned is \$31,664.54 – 30,000 = \$1664.54.

For exercises 27–29, we use the compound interest formula. Interest is paid twice per year.

27. $P = 4630$, $r = .052$, $n = 2(15) = 30$

$$A = P\left(1+\frac{i}{n}\right)^{nt} = 4630\left(1+\frac{.052}{2}\right)^{30} \approx 10{,}000$$

The face value is \$10,000.

29. $P = 9992$, $r = .035$, $t = 2(20) = 40$

$$A = P\left(1+\frac{i}{n}\right)^{n} = 9992\left(1+\frac{.035}{2}\right)^{40} \approx 20{,}000$$

The face value is \$20,000.

31. 4% compounded semiannually
Use the formula for effective rate with $r = .04$ and $m = 2$.

$$r_E = \left(1+\frac{r}{m}\right)^{m} - 1 = \left(1+\frac{.04}{2}\right)^{2} - 1$$
$$= (1.02)^2 - 1 = 1.0404 - 1 = .0404$$

The effective rate is 4.04%.

33. 5% compounded quarterly
$r = .05$, $m = 4$

$$r_E = \left(1+\frac{r}{m}\right)^{m} - 1 = \left(1+\frac{.05}{4}\right)^{4} - 1 \approx .05095$$

The effective rate is 5.095%.

35. \$12,000 at 5% compounded annually for 6 yr
Use the present value formula for compound interest with
$A = 12{,}000$, $i = .05$, and $n = 6$.

$$P = A(1+i)^{-n} = 12{,}000(1.05)^{-6} \approx 8954.58$$

The present value is \$8954.58.

37. \$17,230 at 4% compounded quarterly for 10 yr

$$A = 17{,}230,\ i = \frac{.04}{4} = .01,\ n = 10(4) = 40$$

$$P = A(1+i)^{-n} = 17{,}230(1.01)^{-40} \approx 11{,}572.58$$

The present value is \$11,572,58.

39. $A = 5000$, $i = \dfrac{.035}{2}$, $n = 5(2) = 10$

$$P = \frac{A}{(1+i)^n} = \frac{5000}{\left(1+\frac{.035}{2}\right)^{10}} \approx 4203.64$$

A fair price would be \$4203.64.

41. $A = 20{,}000$, $i = \dfrac{.047}{2}$, $n = 15(2) = 30$

$$P = \frac{A}{(1+i)^n} = \frac{20{,}000}{\left(1+\frac{.047}{2}\right)^{30}} \approx 9963.10$$

A fair price would be \$9963.10.

43. $A = 5000$, $i = .0207$, $n = 4$

$$P = \frac{A}{(1+i)^n} = \frac{5000}{(1+.0207)^4} \approx 4604.57$$

The \$5000 item costs about \$4604.57 4 years prior.

45. $A = 500$, $i = .036$, $n = 2$

$$P = \frac{A}{(1+i)^n} = \frac{500}{(1+.036)^2} \approx 465.85$$

The \$500 item costs about \$465.85 2 years prior.

47. $A = 1210$, $i = \dfrac{.08}{4} = .02$, $n = 5(4) = 20$

$$P = A(1+i)^{-n} = 1210(1.02)^{-20} \approx 814.30$$

Since this amount is less than \$1000, "\$1000 now" is greater.

49. $P = 50{,}000$, $i = \dfrac{.09}{12} = .0075$, $n = 4(12) = 48$

Find the compound amount and then the amount of interest.

$$A = P(1+i)^n = 50{,}000(1.0075)^{48} \approx \$71{,}570.27$$

The business will pay interest in the amount of \$71,570.27 – \$50,000 = \$21,570.27.

51. $P = 10{,}000$
Money market fund:
$i = \frac{.058}{365}, n = 365(2) = 730$
$A = P(1+i)^n = 10{,}000\left(1+\frac{.058}{365}\right)^{730}$
$\approx 11{,}229.86$
$\text{Interest} = 11{,}229.86 - 10{,}000 = 1229.86$
Treasury note (recall that treasury notes pay simple interest–see section 5.1):
$A = P(1+rt)$
$A = 10{,}000(1+.06(2)) \approx 11{,}200$
Interest = 11,200 – 10,000 = 1200
The treasury note pays the most interest.

53. $P = 1000, i = .06, n = 5$
$A = P(1+i)^n = 1000(1.06)^5 \approx 1338.23$
Since this amount is greater than $1210, "$1000 now" is larger.

55. $A = 11{,}115, P = 10{,}000, n = 3$
$A = P(1+i)^n \Rightarrow 11{,}115 = 10{,}000(1+i)^3 \Rightarrow$
$1.1115 = (1+i)^3 \Rightarrow i = \sqrt[3]{1.1115} - 1 \approx .0359$
The interest rate is about 3.59%.

57. For Flagstar Bank, $r = 4.38\%, m = 4$.
$r_E = \left(1+\frac{r}{m}\right)^m - 1 = \left(1+\frac{.0438}{4}\right)^4 - 1 \approx .0445$
The effective rate for Flagstar Bank is about 4.45%.
For Principal Bank, $r = 4.37\%, m = 12$.
$r_E = \left(1+\frac{r}{m}\right)^m - 1 = \left(1+\frac{.0437}{12}\right)^{12} - 1 \approx .04459$
The effective rate for Principal Bank is about 4.46%.
Principal Bank pays a higher APY.

59. $A = 2.9, i = \frac{.05}{12}, n = 5(12) = 60$
Use the formula for present value with compound interest.
$P = A(1+i)^{-n} \approx 2.9\left(1+\frac{.05}{12}\right)^{-60} \approx 2.259696$
The company should invest about $2,259,696 now.

61. $A = 23{,}500, i = .0375, n = 3$
Use the formula for present value with compound interest.
$P = A(1+i)^{-n}$
$= 23{,}500(1+.0375)^{-3} \approx 21{,}042.80$
The car would have cost about $21,043 three years ago.

63. To find the number of years it will take $1 to inflate to $2, use the formula for compound amount with $A = 2, P = 1$, and $i = .03$.
$A = P(1+i)^n \Rightarrow 2 = 1(1.03)^n \Rightarrow 2 = (1.03)^n \Rightarrow$
$\ln 2 = n \ln 1.03 \Rightarrow n = \frac{\ln 2}{\ln 1.03} \approx 23.4$
Prices will double in about 23.4 yr.

65. Find the number of years, it will take for $1 to inflate to $2 using the formula for compound amount with $A = 2, P = 1, i = .05$
$A = P(1+i)^n \Rightarrow 2 = 1(1.05)^n \Rightarrow 2 = (1.05)^n \Rightarrow$
$\ln 2 = n \ln 1.05 \Rightarrow n = \frac{\ln 2}{\ln 1.05} \approx 14.21$
Prices will double in about 14.2 yr.

67. To find the number of years it will take for a demand of 1 unit of electricity to increase to a demand of 2 units, use the formula for compound amount with $A = 2, P = 1$, and $i = .06$.
$A = P(1+i)^n \Rightarrow 2 = 1(1.06)^n \Rightarrow 2 = (1.06)^n \Rightarrow$
$\ln 2 = n \ln 1.06 \Rightarrow n = \frac{\ln 2}{\ln 1.06} \approx 11.90$
The electric utilities will need to double their generating capacity in about 11.9 yr.

69. a. $P = 16{,}000 - 30 = 15{,}970, i = \frac{.055}{12}, n = 12$
Use the formula for compound amount.
$A = P(1+i)^n = 15{,}970\left(1+\frac{.055}{12}\right)^{12}$
$\approx 16{,}870.83$
After the annual charges,
$16{,}870.83 - .0125(16{,}870.83) \approx \$16{,}659.95$

b. Note that the amount in part (a) can be written in terms of P.

$$A = P\left(1+\frac{.055}{12}\right)^{12}(.9875) \approx \$16,659.95$$

Therefore, the amount in the account after 7 years can be found by repeatedly multiplying P by the factors $\left(1+\frac{.055}{12}\right)^{12}(.9875)$. So, by commutatively of multiplication this amount is

$$A = P\left(1+\frac{.055}{12}\right)^{12(7)}(.9875)^7 \approx \$21,472.67.$$

71. First consider the case of earning interest at a rate of k per annum compounded quarterly for all eight years and earning \$2203.76 interest on the \$1000 investment.

$$2203.76 = 1000\left(1+\frac{k}{4}\right)^{8(4)} \Rightarrow$$

$$2.20376 = \left(1+\frac{k}{4}\right)^{32}$$

Use a calculator to raise both sides to the power of $\frac{1}{32}$.

$$1.025 = 1+\frac{k}{4} \Rightarrow .025 = \frac{k}{4} \Rightarrow .1 = k$$

Next consider the actual investments. The \$1000 was invested for the first five years at a rate of j per annum compounded semiannually.

$$A = 1000\left(1+\frac{j}{2}\right)^{5(2)} = 1000\left(1+\frac{j}{2}\right)^{10}$$

This amount was then invested for the remaining three years at $k = .1$ per annum compounded quarterly for a final compound amount of \$1990.76.

$$1990.76 = A\left(1+\frac{.1}{4}\right)^{3(4)}$$

$$1990.76 = A(1.025)^{12}$$

$$1480.24 \approx A$$

Recall that $A = 1000\left(1+\frac{j}{2}\right)^{10}$ and substitute this value into the above equation.

$$1480.24 = 1000\left(1+\frac{j}{2}\right)^{10}$$

$$1.48024 = \left(1+\frac{j}{2}\right)^{10}$$

Use a calculator to raise both sides to the power of $\frac{1}{10}$.

$$1.04 \approx 1+\frac{j}{2} \Rightarrow .04 = \frac{j}{2} \Rightarrow .08 = j$$

The ratio of k to j is $\frac{k}{j} = \frac{.1}{.08} = 1.25$, which is choice (a).

Section 5.3 Annuities, Future Value, and Sinking Funds

1. $1+1.05+1.05^2+1.05^3+\cdots+1.05^{14}$

$$= \frac{1.05^{15}-1}{1.05-1} \approx 21.5786$$

In Exercises 3–8, use the formula

$$S = R \cdot s_{\overline{n}|i} \text{ or } S = R\left[\frac{(1+i)^n - 1}{i}\right].$$

3. $R = 12{,}000$, $i = .062$, $n = 8$

$$S = 12{,}000\left[\frac{(1.062)^8 - 1}{.062}\right] \approx 119{,}625.61$$

The future value is \$119,625.61.

5. $R = 865$, $i = \frac{.06}{2} = .03$, $n = 10(2) = 20$

$$S = 865\left[\frac{(1.03)^{20} - 1}{.03}\right] \approx 23{,}242.87$$

The future value is \$23,242.87.

7. $R = 1200$, $i = \frac{.08}{4} = .02$, $n = 10(4) = 40$

$$S = 1200\left[\frac{(1.02)^{40} - 1}{.02}\right] \approx 72{,}482.38$$

The future value is \$72,482.38.

9. $R = 400,\ i = \frac{.04}{12},\ n = 12(10) = 120$

$$S = 400\left[\frac{\left(1+\frac{.04}{12}\right)^{120}-1}{\frac{.04}{12}}\right] \approx 58,899.92$$

At the end of 10 years, there will be about \$58,899.92 in the account. Using the compound amount formula, the future value of this money is $58,899.92\left(1+\frac{.06}{12}\right)^{120} \approx 107,162.32$, for $i = \frac{.06}{12}$, and $n = 12(10) = 120$. Now compute the amount in the account for $R = 600$, $i = \frac{.06}{12}$, and
$n = 12(10) = 120$.

$$S = 600\left[\frac{\left(1+\frac{.06}{12}\right)^{120}-1}{\frac{.06}{12}}\right] \approx 98,327.61.$$

$98,327.61 + 107,162.32 = 205,489.93$
There will be about \$205,490 in the account at the end of 20 years.

11. $R = 1000,\ i = \frac{.042}{4},\ n = 4(10) = 40$

$$S = 1000\left[\frac{\left(1+\frac{.042}{4}\right)^{40}-1}{\frac{.042}{4}}\right] \approx 49,393.58$$

At the end of 10 years, there will be about \$49,393.58 in the account. Using the compound amount formula, the future value of this money is $49,393.58\left(1+\frac{.074}{4}\right)^{60} \approx 148,365.36$, for $i = \frac{.074}{4}$, and $n = 4(15) = 60$. Now compute the amount in the account for $R = 1500$, $i = \frac{.074}{4}$, and $n = 4(15) = 60$.

$$S = 1500\left[\frac{\left(1+\frac{.074}{4}\right)^{60}-1}{\frac{.074}{4}}\right] \approx 162,465.22.$$

$148,365.36 + 162,465.22 = 310,830.58$
There will be about \$310,831 in the account at the end of 25 years.

13. $S = 11,000,\ i = \frac{.05}{2} = .025,\ n = 6(2) = 12$

$$11,000 = R\left[\frac{(1.025)^{12}-1}{.025}\right]$$
$$11,000 \approx R(13.79555)$$
$$R \approx 797.36$$

The periodic payment is \$797.36.

15. $S = 50,000,\ i = \frac{.08}{4} = .02,\ n = \left(2\frac{1}{2}\right)(4) = 10$

$$50,000 = R\left[\frac{(1.02)^{10}-1}{.02}\right]$$
$$50,000 \approx R(10.94972)$$
$$R \approx 4566.33$$

The periodic payment is \$4566.33.

17. $S = 6000,\ i = \frac{.06}{12}, = .005,\ n = 3(12) = 36$

$$6000 = R\left[\frac{(1.005)^{36}-1}{.005}\right]$$
$$6000 \approx R(39.33610)$$
$$R \approx 152.53$$

The periodic payment is \$152.53.

For exercises 19–21, we use the future value formula,
$S = R\left[\frac{(1+i)^n - 1}{i}\right]$.

19. $R = 3940,\ n = 10,\ S = 50,000,\ i = r$

$$50,000 = 3940\left[\frac{(1+i)^{10}-1}{i}\right]$$

Graphing each side of the equation, we find that the intersection is $x \approx .0519$

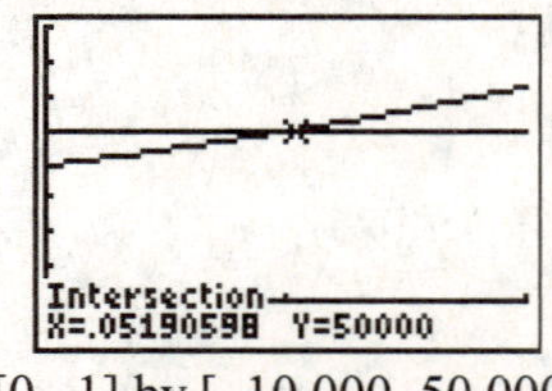

[0, .1] by [–10,000, 50,000]

An interest rate of about 5.19% is needed.

21. $R = 1675, n = 5(4) = 20, S = 38{,}000, i = \frac{r}{4}$

$$38{,}000 = 1675\left[\frac{\left(1+\frac{r}{4}\right)^{20} - 1}{\frac{r}{4}}\right]$$

Graphing each side of the equation, we find that the intersection is $x \approx .05223$

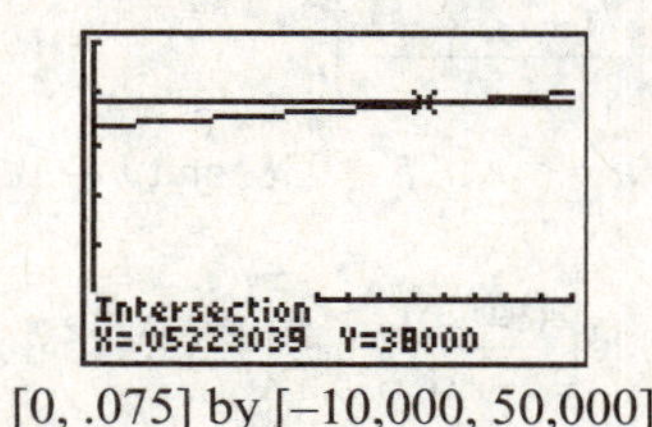

[0, .075] by [–10,000, 50,000]

An interest rate of about 5.223% is needed.

23. A sinking fund is a fund that is set up to receive periodic payments. Businesses and corporations use sinking funds to repay bond issues or to provide for replacement of fixed assets.

25. $R = 500, i = .05, n = 10$

$$S = 500\left[\frac{(1.05)^{11} - 1}{.05}\right] - 500 \approx 6603.39$$

The future value is \$6603.39.

27. $R = 16{,}000, i = .047, n = 11$

$$S = 16{,}000\left[\frac{(1.047)^{12} - 1}{.047}\right] - 16{,}000$$

$$\approx 234{,}295.32$$

The future value is \$234,295.32.

29. $R = 1000, i = \frac{.08}{2} = .04, n = 9(2) = 18$

$$S = 1000\left[\frac{(1.04)^{19} - 1}{.04}\right] - 1000$$

$$\approx 26{,}671.23$$

The future value is \$26,671.23.

31. $R = 100, i = \frac{.09}{4} = .0225, n = 7(4) = 28$

$$S = 100\left[\frac{(1.0225)^{29} - 1}{.0225}\right] - 100$$

$$\approx 3928.88$$

The future value is \$3928.88.

For exercises 33–35, we use the future value of an annuity due formula, $S = R\left[\frac{(1+i)^{n+1} - 1}{i}\right] - R.$

33. $S = 8000, n = 3(4) = 12, i = \frac{.044}{4}$

$$8000 = R\left[\frac{\left(1+\frac{.044}{4}\right)^{13} - 1}{\frac{.044}{4}}\right] - R \Rightarrow$$

$$8000 = R\left(\frac{\left(1+\frac{.044}{4}\right)^{13} - 1}{\frac{.044}{4}} - 1\right) \Rightarrow$$

$$R = \frac{8000}{\frac{\left(1+\frac{.044}{4}\right)^{13} - 1}{\frac{.044}{4}} - 1} \approx 620.46$$

The payment should be \$620.46.

35. $S = 55{,}000, n = 12(12) = 144, i = \frac{.057}{12}$

$$55{,}000 = R\left[\frac{\left(1+\frac{.057}{12}\right)^{145} - 1}{\frac{.057}{12}}\right] - R \Rightarrow$$

$$55{,}000 = R\left(\frac{\left(1+\frac{.057}{12}\right)^{145} - 1}{\frac{.057}{12}} - 1\right) \Rightarrow$$

$$R = \frac{55{,}000}{\frac{\left(1+\frac{.057}{12}\right)^{145} - 1}{\frac{.057}{12}} - 1} \approx 265.71$$

The payment should be \$265.71.

37. $R = 170, i = \frac{.053}{12}, n = 40(12) = 480$

Use the formula for future value of an ordinary annuity.

$$S = R \cdot s_{\overline{n}|i}$$

$$= 170\left[\frac{\left(1+\frac{.053}{12}\right)^{480} - 1}{\frac{.053}{12}}\right] \approx 280{,}686.25$$

After 40 years, the amount in the account will be about \$280,686.25.

39. $R = 800,\ i = \dfrac{.1099}{12},\ n = (10)(12) = 120$

a. Use the formula for future value of an ordinary annuity.

$$S = R \cdot s_{\overline{n}|i}$$
$$= 800\left[\frac{(1+\frac{.1099}{12})^{120} - 1}{\frac{.1099}{12}}\right] \approx 173,497.86$$

She will have \$173,497.86 on deposit after 10 years.

b. $i = \dfrac{.0777}{12}$

$$S = 800\left[\frac{(1+\frac{.0777}{12})^{120} - 1}{\frac{.0777}{12}}\right] \approx 144,493.82$$

She will have \$144,493.82 on deposit after 10 years.

c. 173,497.86 – 144,493.82 = 29,004.04
She would lose \$29,004.04.

41. $R = 200,\ i = \dfrac{.0875}{12},\ n = 35(12) = 420$

a. Use the formula for future value of an ordinary annuity.

$$S = R \cdot s_{\overline{n}|i}$$
$$= 200\left[\frac{\left(1+\frac{.0875}{12}\right)^{420} - 1}{\frac{.0875}{12}}\right] \approx 552,539.96$$

He will have \$552,539.96 in the account when he is 60.

b. $$A = P(1+i)^n = 552,539.96\left(1+\frac{.0875}{12}\right)^{12(5)}$$
$$\approx 854,433.28$$

He will have \$854,433.28 for retirement at age 65.

43. This may be considered an annuity due, since payments are made at the beginning of each year, starting with the day the son is born. However, a payment should not be subtracted at the end, since a nineteenth payment is made on his eighteenth birthday. Thus, the future value is given by

$$S = R\left[\frac{(1+i)^{n+1} - 1}{i}\right],$$

where $R = 1000$, $i = .056$, and $n = 18$.
Therefore,

$$S = 1000\left[\frac{(1.056)^{19} - 1}{.056}\right] = 32,426.46$$

There will be \$32,426.46 in the account at the end of the day on the son's eighteenth birthday.

45. For the first 12 years, we have an annuity due. To find the amount in this account after 12 years, use the formula for the future value of an annuity due with $R = 10,000$, $i = .05$, and $n = 12$.

$$S = R\left[\frac{(1+i)^{n+1} - 1}{i}\right] - R$$
$$= 10,000\left[\frac{(1.05)^{13} - 1}{.05}\right] - 10,000$$
$$\approx 167,129.83$$

This amount, \$167,129.83, now earns 6% interest compounded semiannually for another 9 yr, but no new deposits are made. Use the formula for compound amount with

$P = 167,129.83,\ i = \dfrac{.06}{2} = .03$, and $n = 9(2) = 18$.

$$A = P(1+i)^n = 167,129.83(1.03)^{18}$$
$$= 284,527.35.$$

The final amount on deposit after 21 yr is \$284,527.35.

47. $S = 10{,}000$, $n = 8(4) = 32$

a. $i = \frac{.05}{4} = .0125$

$$S = R \cdot s_{\overline{n}|i}$$

$$10{,}000 = R\left[\frac{(1.0125)^{32} - 1}{.0125}\right] \Rightarrow$$

$$10{,}000 \approx R(39.05044) \Rightarrow R = \$256.08$$

He should deposit \$256.08 at the end of each quarter.

b. $i = \frac{.058}{4} = .0145$

$$S = R \cdot s_{\overline{n}|i}$$

$$10{,}000 = R\left[\frac{(1.0145)^{32} - 1}{.0145}\right]$$

$$10{,}000 \approx R(40.35398) \Rightarrow R = \$247.81$$

He should deposit \$247.81 quarterly.

49. $S = 24{,}000$, $i = \frac{.05}{4} = .0125$, $n = 6(4) = 24$

$$S = R \cdot s_{\overline{n}|i}$$

$$24{,}000 = R\left[\frac{(1.0125)^{24} - 1}{.0125}\right] \Rightarrow$$

$$R = \frac{24{,}000}{\frac{(1.0125)^{24} - 1}{.0125}} \approx 863.68$$

She should deposit \$863.68 at the end of each quarter.

51. **a.** $P = 60{,}000$, $r = .08$, $t = \frac{1}{4}$

$$I = Prt = 60{,}000(.08)\left(\frac{1}{4}\right) = \$1200$$

Each quarterly interest payment is \$1200.

b. $S = 60{,}000$, $i = \frac{.06}{2} = .03$,

$n = 7(2) = 14$

$$S = R \cdot s_{\overline{n}|i}$$

$$60{,}000 = R\left[\frac{(1.03)^{14} - 1}{.03}\right]$$

$$60{,}000 \approx R(17.08632)$$

$$R = \$3511.58$$

The amount of each payment is \$3511.58.

53. $S = 147{,}126$, $R = 300$, $n = 12(20)$, $i = \frac{r}{12}$

$$S = R\left[\frac{(1+i)^n - 1}{i}\right]$$

$$147{,}126 = 300\left[\frac{\left(1 + \frac{r}{12}\right)^{12(20)} - 1}{\frac{r}{12}}\right]$$

Graph each side of this equation. The intersection is (.065, 147,126).

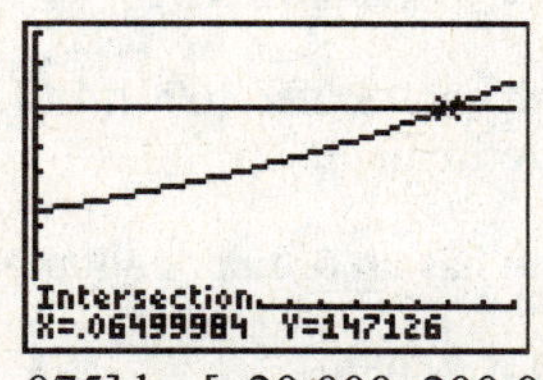

[0, .075] by [–20,000, 200,000]

She received an interest rate of 6.5%.

55. **a.** Answers will vary.

b. Joe contributed (600)(40) = \$24,000 to his retirement.

c. For the first ten years, Joe has an ordinary annuity with $R = 600$, $i = \frac{.081}{4}$, and $n = 10(4) = 40$. Use the formula for the future value of an ordinary annuity.

$$S = \left[\frac{(1+i)^n - 1}{i}\right] = 600\left[\frac{\left(1+\frac{.081}{4}\right)^{40} - 1}{\frac{.081}{4}}\right]$$
$$= 36,438.25$$

He makes no further deposits, but leaves this amount in the account at the same interest rate. Use the formula for compound amount with $P = 36{,}438.25$, $i = \frac{.081}{4}$, and $n = 35(4) = 140$ to find the value of this amount after 35 years.

$$A = P(1+i)^n$$
$$= 36,438.25\left(1+\tfrac{.081}{4}\right)^{140} \approx 603,229$$

After 45 years, there will be about $603,229 in Joe's account.

d. Sarah contributed (600)(4)(35) = $84,000 to her retirement.

e. Sarah has an ordinary annuity for 35 years with $R = 600$, $i = \frac{.081}{4}$, and $n = 35(4) = 140$.

$$S = \left[\frac{(1+i)^n - 1}{i}\right] = 600\left[\frac{\left(1+\frac{.081}{4}\right)^{140} - 1}{\frac{.081}{4}}\right]$$
$$\approx 460,884$$

Sarah will have about $460,884 in her account after 45 years.

Section 5.4 Annuities, Present Value, and Amortization

1. The present value of an annuity is the amount that must be deposited today to provide all the payments for the term of the annuity. The future value of an annuity is the final sum in an account receiving annuity payments after all payments are made.

In Exercises 3–5, use the formula

$$P = R\left[\frac{1-(1+i)^{-n}}{i}\right].$$

3. $R = 1400$, $i = .06$, $n = 8$

$$P = 1400\left[\frac{1-(1.06)^{-8}}{.06}\right] \approx 8693.71$$

The present value is $8693.71.

5. $R = 50{,}000$, $i = \frac{.05}{4} = .0125$, $n = 10(4) = 40$

$$P = 50,000\left[\frac{1-(1.0125)^{-40}}{.0125}\right] \approx 1,566,346.66$$

The present value is $1,566,346.66.

7. $R = 650$, $i = \frac{.049}{4}$, $n = 5(4) = 20$

$$P = 650\left[\frac{1-\left(1+\frac{.049}{4}\right)^{-20}}{\frac{.049}{4}}\right] \approx 11,468.10$$

About $11,468.10 is needed to fund the withdrawals.

9. $R = 425$, $i = \frac{.061}{12}$, $n = 10(12) = 120$

$$P = 425\left[\frac{1-\left(1+\frac{.061}{12}\right)^{-120}}{\frac{.061}{12}}\right] \approx 38,108.61$$

About $38,108.61 is needed to fund the withdrawals.

11. $P = 90{,}000$, $i = \frac{.049}{12}$, $n = 22(12) = 264$

$$90,000 = R\left[\frac{1-\left(1+\frac{.049}{12}\right)^{-264}}{\frac{.049}{12}}\right] \Rightarrow$$

$$R = \frac{90,000}{\frac{1-\left(1+\frac{.049}{12}\right)^{-264}}{\frac{.049}{12}}} \approx 557.68$$

The monthly payment will be $557.68.

13. $P = 275{,}000,\ i = \frac{.06}{4} = .015,\ n = 18(4) = 72$

$$275{,}000 = R\left[\frac{1-(1+.015)^{-72}}{.015}\right] \Rightarrow$$

$$R = \frac{275{,}000}{\frac{1-(1+.015)^{-72}}{.015}} \approx 6272.14$$

The monthly payment will be $6272.14.

In Exercises 15–17, use the formula

$$P = R \cdot a_{\overline{n}|i} \text{ or } P = R\left[\frac{1-(1+i)^{-n}}{i}\right].$$

15. $R = 10{,}000,\ i = .03,\ n = 15$

The lump sum is the same as the present value of the annuity.

$$P = 10{,}000\left[\frac{1-(1.03)^{-15}}{.03}\right] \approx 119{,}379.35$$

The required lump sum is $119,379.35.

17. $R = 10{,}000,\ i = .06,\ n = 15$

$$P = 10{,}000\left[\frac{1-(1.06)^{-15}}{.06}\right] \approx 97{,}122.49$$

The required lump sum is $97,122.49.

19. Find the future value of an annuity with $R = 4000,\ i = \frac{.06}{2} = .03$, and $n = 10(2) = 20$.

$$S = R\left[\frac{(1+i)^n - 1}{i}\right] = 4000\left[\frac{(1.03)^{20} - 1}{.03}\right] \approx 107{,}481.50$$

Now find the present value of $107,481.50 at 8% compounded quarterly for 10 years.

$$A = 107{,}481.50,\ i = \frac{.08}{4} = .02,\ n = 10(4) = 40$$

$$A = P(1+i)^n \Rightarrow 107{,}481.50 = P(1.02)^{40} \Rightarrow$$

$$P = \frac{107{,}481.50}{(1.02)^{40}} \approx 48{,}677.34$$

The required lump sum is $48,677.34.

21. The interest that is paid each half-year is $I = Prt = 15{,}000 \cdot .06 \cdot \frac{1}{2} = \$450$. So, we have a two-part investment, an annuity that pays $450 every six months for 4 years and the $15,000 face value of the bond which will be paid when the bond matures 4 years from now. The purchaser should be willing to pay the present value of each part of the investment, assuming 5% interest compounded semiannually.

Present value of annuity

$$R = 450,\ i = \frac{.05}{2} = .025,\ n = 4(2) = 8$$

$$P = 450\left[\frac{1-(1+.025)^{-8}}{.025}\right] \approx 3226.56$$

Present value of $15,000 in 4 years

$$P = A(1+i)^{-n}$$

$$P = 15{,}000(1+.025)^{-8} \approx 12{,}311.20$$

The purchaser should be willing to pay $3226.56 + $12,311.20 = $15,537.76.

23. The interest that is paid each half-year is $I = Prt = 10{,}000 \cdot .054 \cdot \frac{1}{2} = \$270$. So, we have a two-part investment, an annuity that pays $270 every six months for 12 years and the $10,000 face value of the bond which will be paid when the bond matures 12 years from now. The purchaser should be willing to pay the present value of each part of the investment, assuming 6.5% interest compounded semiannually.

Present value of annuity

$$R = 270,\ i = \frac{.065}{2},\ n = 12(2) = 24$$

$$P = 270\left[\frac{1-\left(1+\frac{.065}{2}\right)^{-24}}{\frac{.065}{2}}\right] \approx 4451.85$$

Present value of $10,000 in 12 years

$$P = A(1+i)^{-n} = 10{,}000\left(1+\tfrac{.065}{2}\right)^{-24} \approx 4641.29$$

The purchaser should be willing to pay $4451.85 + $4641.29 = $9093.14.

25. $P = 2500,\ i = \dfrac{.08}{4} = .02,\ n = 6$

$$R = 2500\left[\frac{.02}{1-(1.02)^{-6}}\right] \approx 446.31$$

Quarterly payments of $446.31 are required to amortize this loan.

27. $P = 90{,}000,\ i = \dfrac{.07}{1} = .07,\ n = 12$

$$R = 90{,}000\left[\frac{.07}{1-(1.07)^{-12}}\right] \approx 11{,}331.18$$

Annual payments of $11,331.18 are required to amortize this loan.

29. $P = 7400,\ i = \dfrac{.082}{2} = .041,\ n = 18$

$$R = 7400\left[\frac{.041}{1-(1.041)^{-18}}\right] \approx 589.31$$

Semiannual payments of $589.31 are required to amortize this loan.

For exercises 31–37, we use the amortization payment formula $R = \dfrac{Pi}{1-(1+i)^{-n}}$.

31. $P = 225{,}000,\ i = \dfrac{.0325}{12},\ n = 30(12) = 360$

$$R = 225{,}000\left[\frac{\frac{.0325}{12}}{1-\left(1+\frac{.0325}{12}\right)^{-360}}\right] \approx 979.21$$

Monthly payments of $979.21 are required to amortize this loan.

33. $P = 140{,}000,\ i = \dfrac{.02375}{12},\ n = 15(12) = 180$

$$R = 140{,}000\left[\frac{\frac{.02375}{12}}{1-\left(1+\frac{.02375}{12}\right)^{-180}}\right] \approx 925.29$$

Monthly payments of $925.29 are required to amortize this loan.

35. $P = 26{,}799,\ i = \dfrac{.0313}{12},\ n = 4(12) = 48$

$$R = 26{,}799\left[\frac{\frac{.0313}{12}}{1-\left(1+\frac{.0313}{12}\right)^{-48}}\right] \approx 594.72$$

After 2 years (24 payments), the remaining 24 payments can be thought of as annuity. The present value of this annuity is the remaining balance.

So, we use the present value formula with $R = 594.72,\ i = \dfrac{.0313}{12},\ n = 24$

$$P = 594.72\left[\frac{1-\left(1+\frac{.0313}{12}\right)^{-24}}{\frac{.0313}{12}}\right] \approx 13{,}818.25$$

The monthly payment is $594.72 and the balance remaining after two years is about $13,818.25.

37. $P = 210{,}000,\ i = \dfrac{.0354}{12},\ n = 30(12) = 360$

$$R = 210{,}000\left[\frac{\frac{.0354}{12}}{1-\left(1+\frac{.0354}{12}\right)^{-360}}\right] \approx 947.69$$

After 12 years (144 payments), the remaining 216 payments can be thought of as annuity. The present value of this annuity is the remaining balance. So, we use the present value formula with $R = 947.69,\ i = \dfrac{.0354}{12},\ n = 216$

$$P = 947.69\left[\frac{1-\left(1+\frac{.0354}{12}\right)^{-216}}{\frac{.0354}{12}}\right] \approx 151{,}223.33$$

The monthly payment is $947.69 and the balance remaining after 12 years is about $151,223.33.

39. Locate the table entry that is in the row labeled Payment Number 5 and in the column labeled Interest for Period. Note that $6.80 of the fifth payment is interest.

41. To find how much interest is paid in the first 5 months of the loan, add the first five nonzero entries in the column labeled Interest for Period.

$$\begin{array}{r} \$10.00 \\ 9.21 \\ 8.42 \\ 7.61 \\ +6.80 \\ \hline \$42.04 \end{array}$$

In the first 5 months of the loan, $42.04 is paid in interest.

For exercises 43–45, we use the formula for the present value of an annuity due, $S = R + R\left[\frac{1-(1+i)^{-(n-1)}}{i}\right]$

43. The yearly payment for \$57.6 million is $\frac{57,600,000}{30} = 1,920,000.$

$R = 1,920,000$, $i = .051$, $n = 30$

$$S = 1,920,000 + 1,920,000\left[\frac{1-(1.051)^{-(30-1)}}{.051}\right] \approx 30,669,881$$

The cash value is about \$30,669,881.
The figure shows the solution using the TVM on a TI-84 Plus calculator.

45. The yearly payment for \$41.6 million is $\frac{41,600,000}{26} = 1,600,000.$

$R = 1,600,000$, $i = .04735$, $n = 26$

$$S = 1,600,000 + 1,600,000\left[\frac{1-(1.04735)^{-25}}{.04735}\right] \approx 24,761,633$$

The cash value is about \$24,761,633.
The figure shows the solution using the TVM on a TI-84 Plus calculator.

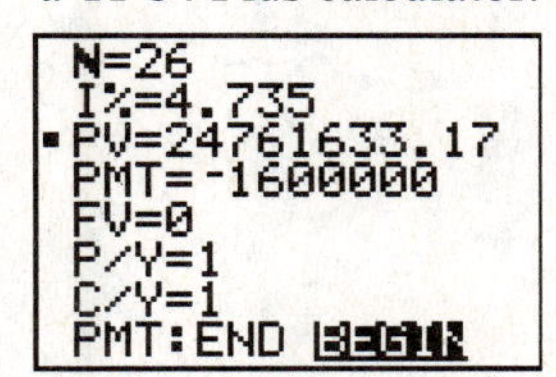

47. The monthly payments form an annuity with $R = 30$, $i = .0125$, and $n = 36$.

a. The price of the stereo system is \$600 plus the present value of the annuity.

$$P = 600 + 30\left[\frac{1-(1.0125)^{-36}}{.0125}\right] = 600 + 865.42 = 1465.42$$

The stereo costs \$1465.42.

b. The total amount paid is

$$600 + 30(36) = 600 + 1080 = \$1680$$

$$\text{Interest} = 1680.00 - 1465.42 = 214.58$$

The total amount of interest paid is \$214.58.

49. $P = 15,000$, $i = \frac{.10}{2} = .05$, $n = 4(2) = 8$

$$R = \frac{P}{a_{\overline{n}|i}} = \frac{15,000}{\left[\frac{1-(1.05)^{-8}}{.05}\right]} \approx \frac{15,000}{6.46321} = \$2320.83$$

Each payment is \$2320.83.

51. $$R = \frac{35,000}{\left[\frac{1-(1+\frac{.0743}{12})^{-120}}{\frac{.0743}{12}}\right]} \approx 414.18$$

Total payments = 10(12)(414.18) = 49,701.60
Interest = 49,701.60 – 35,000 = 14,701.60
The monthly payments are \$414.18 and the total interest paid is \$14,701.60.

53. a. $P = 14,000 + 7200 - 1200 = 20,000$

$i = \frac{.12}{2} = .06$, $n = 5(2) = 10$

$$R = \frac{20,000}{\left[\frac{1-(1.06)^{-10}}{.06}\right]} \approx 2717.36$$

Each payment is \$2717.36.

b. There were 2 payments left at the time she decided to pay off the loan.

55. a. $P = 257,000 - .2(257,000) = 205,600$

$i = \frac{.039}{12} = .00325, n = 30(12) = 360$

$$R = \frac{205,600}{\left[\frac{1-(1.00325)^{-360}}{.00325}\right]} \approx 969.75$$

Their monthly payments are \$969.75.

b. There are 360 – 60 = 300 payments left to be made. This can be thought of as an annuity consisting of 300 payments of \$969.75 at 3.9% interest

$\frac{.039}{12} = .00325$ per period.

$$969.75\left[\frac{1-(1.00325)^{-300}}{.00325}\right] = 185,658.15$$

Their loan balance is \$185,658.15.

c. Total payments = (360)(969.75) = 349,110
Interest = 349,110 – 205,600 = 143,510
The total interest paid is $143,510.

57. We can think of these as two ordinary annuities. First, we determine the future value of her retirement account with $R = 400$, $i = \frac{.07}{12}$, and $n = 45(12) = 540$.

$$S = R\left[\frac{(1+i)^n - 1}{i}\right] = 400\left[\frac{\left(1+\frac{.07}{12}\right)^{540} - 1}{\frac{.07}{12}}\right]$$

$$\approx 1,517,038$$

Now we compute the amount of each withdrawal, using $P = 1,517,038$, $i = \frac{.05}{12}$, and $n = 30(12) = 360$.

$$P = R\left[\frac{1-(1+i)^{-n}}{i}\right] \Rightarrow$$

$$1,517,038 = R\left[\frac{1-\left(1+\frac{.05}{12}\right)^{-360}}{\frac{.05}{12}}\right] \Rightarrow$$

$$R = \frac{1,517,038}{\frac{1-\left(1+\frac{.05}{12}\right)^{-360}}{\frac{.05}{12}}} \approx 8143.79$$

Her maximum monthly withdrawal is $8143.79.

59. Amount needed for retirement:

$$2500 = \frac{P}{\left[\frac{1-(1+\frac{.071}{12})^{-120}}{\frac{.071}{12}}\right]} \Rightarrow P = 214,363.15$$

Monthly contribution to acquire $214,363.15:

$$214,363.15 = R\left[\frac{(1+\frac{.071}{12})^{240} - 1}{\frac{.071}{12}}\right]$$

$$R = 406.53$$

She must make monthly payments of $406.53.

61. Amount needed for retirement:

$$3400 = \frac{P}{\left[\frac{1-(1+\frac{.082}{12})^{-120}}{\frac{.082}{12}}\right]} \Rightarrow P = 277,807.34$$

Amount needed after 120 equal contributions (i.e., how much money is needed to earn interest for 10 years with no contributions to amount to $277,807.34):

$$277,807.34 = P\left[1+\frac{.082}{12}\right]^{10(12)}$$

$$P = 122,696.87$$

Monthly contributions to acquire $122,696.87:

$$122,696.87 = R\left[\frac{(1+\frac{.082}{12})^{120} - 1}{\frac{.082}{12}}\right]$$

$$R = 663.22$$

He needs to make monthly contributions of $663.22.

In Exercises 63–65, the computer program "Explorations in Finite Mathematics" has been used to prepare the amortization schedules. Answers found using other computer programs or calculators may differ slightly.

63. $P = 4000,\ i = .08,\ n = 4$

Each annual payment is

$$R = \frac{P}{a_{\overline{n}|i}} = \frac{4000}{\left[\frac{1-(1.08)^{-4}}{.08}\right]} = \$1207.68$$

(although the last payment may differ slightly).

The interest for each period is 8% of the principal at the end of the previous period.

The portion to principal for each period is the difference between the amount of the payment and the interest for the period.

The principal at the end of each new period is obtained by subtracting the new portion to principal from the principal at the end of the previous period.

Repeat these steps four times to obtain the following amortization schedule.

Payment Number	Amount of Payment	Interest for Period	Portion to Principal	Principal at End of Period
0	------	------	------	$4000.00
1	$1207.68	$320.00	$887.68	3112.32
2	1207.68	248.99	958.69	2153.63
3	1207.68	172.29	1035.39	1118.24
4	1207.70	89.46	1118.24	0

65. $P = 8(1048) - 1200 = 7184,\ i = \frac{.12}{12} = .01,\ n = 4(12) = 48$

Each monthly payment is

$$R = \frac{P}{a_{\overline{n}|i}} = \frac{7184}{\left[\frac{1-(1.01)^{-48}}{.01}\right]} = \$189.18.$$

The interest for the first period is .01(7184) = \$71.84.

The portion to principal for the first period is \$189.18 − 71.84 = \$117.34.

The principal at the end of the first period is \$7184 − 117.34 = \$7066.66.

Repeat these steps to construct three more rows of the table, which will look as follows.

Payment Number	Amount of Payment	Interest for Period	Portion to Principal	Principal at End of Period
0	------	------	------	$7184.00
1	$189.18	$71.84	$117.34	7066.66
2	189.18	70.67	118.51	6948.15
3	189.18	69.48	119.70	6828.45
4	189.18	68.28	120.90	6707.55

Chapter 5 Review Exercises

1. $P = 4902,\ r = .065,\ t = \frac{11}{12}$

$I = Prt = 4902(.065)\left(\frac{11}{12}\right) \approx 292.08$

The simple interest is \$292.08.

2. $P = 42{,}368,\ r = .0922,\ t = \frac{5}{12}$

$I = Prt = (42{,}368)(.0922)\left(\frac{5}{12}\right) \approx 1627.64$

The simple interest is \$1627.64.

3. $P = 3478,\ r = .074,\ t = \frac{88}{365}$

$I = Prt = 3478(.074)\left(\frac{88}{365}\right) \approx 62.05$

The simple interest is \$62.05.

4. $P = 2390,\ r = .087,\ t = \frac{[(31-3)+30+28]}{365} = \frac{86}{365}$

$I = Prt = (2390)(.087)\left(\frac{86}{365}\right) \approx 48.99$

The simple interest is \$48.99.

5. $I = Prt = 12{,}000 \cdot .0475 \cdot \frac{1}{2} = 285$

The semiannual simple interest payment is \$285. The total interest earned over the life of the loan is (\$285)(12) = \$3420.

6. $I = Prt = 20{,}000 \cdot .0525 \cdot \frac{1}{2} = 525$

The semiannual simple interest payment is \$525. The total interest earned over the life of the loan is (\$525)(18) = \$9450.

7. $A = P(1+rt) = 7750\left(1 + .068 \cdot \frac{4}{12}\right) = \$7925.67$

The maturity value is \$7925.67.

8. $A = P(1+rt) = 15{,}600\left(1 + .082 \cdot \frac{9}{12}\right)$

$= \$16{,}559.40$

The maturity value is \$16,559.40.

9. The present value of an amount A is the amount that must be invested now to obtain a final balance of A.

10. $A = 459.57,\ r = .055,\ t = \frac{7}{12}$

$P = \frac{A}{1+rt} = \frac{459.57}{1 + .055\left(\frac{7}{12}\right)} \approx 445.28$

The present value is \$445.28.

11. $A = 80{,}612,\ r = .0677,\ t = \frac{128}{365}$

$P = \frac{A}{1+rt} = \frac{80{,}612}{1 + (.0677)\left(\frac{128}{365}\right)} \approx 78{,}742.54$

The present value is \$78,742.54.

12. Discount $= Prt = 7000 \cdot .035 \cdot \frac{9}{12} = \$183.75$

The price of the T-bill is the face value – the discount. $7000 - 183.75 = \$6816.25$

13. Discount $= Prt = 10{,}000 \cdot .04 \cdot \frac{6}{12} = \$200$

The price of the T-bill is the face value – the discount. $10{,}000 - 200 = \$9800$

Now compute the actual interest rate:

$I = Prt \Rightarrow 200 = 980(r)\left(\frac{1}{2}\right) \Rightarrow$

$r \approx .040816$

The actual interest rate is about 4.082%.

14. Answers vary. Possible answer: Compound interest produces more interest because interest is paid on the previously earned interest as well as on the original principal.

15. $P = 2800,\ i = .06,\ n = 12$

$A = P(1+i)^n = 2800(1.06)^{12} \approx 5634.15$

The compound amount is \$5634.15. The amount of interest earned is \$5634.15 – \$2800 = \$2834.15.

16. $P = 57{,}809.34,\ i = \frac{.04}{4} = .01,\ n = 6(4) = 24$

$A = P(1+i)^n = 57{,}809.34(1.01)^{24} \approx 73{,}402.52$

The compound amount is \$73,402.52. The amount of interest earned is \$73,402.52 – \$57,809.34 = \$15,593.18.

17. $P = 12{,}903.45,\ i = \frac{.0637}{4} = .015925,\ n = 29$

$A = P(1+i)^n$

$= 12{,}903.45(1.015925)^{29} \approx \$20{,}402.98$

The compound amount is \$20,402.98. The amount of interest earned is \$20,402.98 – \$12,903.45 = \$7499.53.

18. $P = 4677.23,\ i = \dfrac{.0457}{12},\ n = 32$

$$A = P(1+i)^n$$
$$= 4677.23\left(1+\frac{.0457}{12}\right)^{32} \approx 5282.19$$

The compound amount is \$5282.19. The amount of interest earned is \$5282.19 – \$4677.23 = \$604.96.

19. $P = 22{,}000,\ i = \dfrac{.055}{4},\ n = 6(4) = 24$

$$A = P(1+i)^n$$
$$= 22{,}000\left(1+\frac{.055}{4}\right)^{24} \approx 30{,}532.58$$

The compound amount is \$30,532.52. The amount of interest earned is \$30,532.52 – \$22,000 = \$8532.58.

20. $P = 2975,\ i = \dfrac{.047}{12},\ n = 4(12) = 48$

$$A = P(1+i)^n$$
$$= 2975\left(1+\frac{.047}{12}\right)^{48} \approx 3589.01$$

The compound amount is \$3580.01. The amount of interest earned is \$3589.01 – \$2975 = \$614.01.

21. $P = 12{,}366,\ i = \dfrac{.039}{2},\ n = 5(2) = 10$

$$A = 12{,}366\left(1+\frac{.039}{2}\right)^{10} \approx 15{,}000$$

The face value is \$15,000.

22. $P = 11{,}575,\ i = \dfrac{.052}{2},\ n = 15(2) = 30$

$$A = 11{,}575\left(1+\frac{.052}{2}\right)^{30} \approx 25{,}000$$

The face value is \$25,000.

23. Use the formula for effective rate with $r = .05$ and $m = 2$.

$$r_E = \left(1+\frac{r}{m}\right)^m - 1 = \left(1+\frac{.05}{2}\right)^2 - 1$$
$$= .050625$$

The effective rate is 5.0625%.

24. Use the formula for effective rate with $r = .065$ and $m = 365$.

$$r_E = \left(1+\frac{r}{m}\right)^m - 1 = \left(1+\frac{.065}{365}\right)^{365} - 1$$
$$= .067153$$

The effective rate is 6.7153%.

25. $A = 42{,}000,\ i = \dfrac{.12}{12} = .01,\ n = 7(12) = 84$

$$P = \frac{A}{(1+i)^n} = \frac{42{,}000}{(1.01)^{84}} \approx 18{,}207.65$$

The present value is \$18,207.65.

26. $A = 17{,}650,\ i = \dfrac{.08}{4} = .02,\ n = 4(4) = 16$

$$P = \frac{A}{(1+i)^n} = \frac{17{,}650}{(1.02)^{16}} \approx 12{,}857.07$$

The present value is \$12,857.07.

27. $A = 1347.89,\ i = \dfrac{.062}{2} = .031,$
$n = (3.5)(2) = 7$

$$P = \frac{A}{(1+i)^n} = \frac{1347.89}{(1.031)^7} \approx 1088.54$$

The present value is \$1088.54.

28. $A = 2388.90,\ i = \dfrac{.0575}{12},\ n = 44$

$$P = \frac{A}{(1+i)^n} = \frac{2388.90}{\left(1+\frac{.0575}{12}\right)^{44}} \approx 1935.77$$

The present value is \$1935.77.

29. $P = \dfrac{A}{(1+i)^n} = \dfrac{15{,}000}{\left(1+\frac{.044}{2}\right)^{2(10)}} \approx 9706.74$

A purchaser should be willing to pay about \$9706.74.

30. $P = \dfrac{A}{(1+i)^n} = \dfrac{30{,}000}{\left(1+\frac{.062}{2}\right)^{2(25)}} \approx 6519.11$

A purchaser should be willing to pay about \$6519.11.

31. The future value of an annuity is the balance in the account at the end of the term.

32. $R = 1288,\ i = .07,\ n = 14$

$$S = R \cdot s_{\overline{n}|i} = 1288\left[\frac{(1.07)^{14} - 1}{.07}\right] \approx 29,045.03$$

The future value of this ordinary annuity is \$29,045.03.

33. $R = 4000,\ i = \frac{.06}{4} = .015,\ n = 8(4) = 32$

$$S = R \cdot s_{\overline{n}|i} = 4000\left[\frac{(1.015)^{32} - 1}{.015}\right] \approx 162,753.15$$

The future value of this ordinary annuity is \$162,753.15.

34. $R = 233,\ i = \frac{.06}{12} = .005,\ n = 4(12) = 48$

$$S = R \cdot s_{\overline{n}|i} = 233\left[\frac{(1.005)^{48} - 1}{.005}\right] \approx 12,604.79$$

The future value of this ordinary annuity is \$12,604.79.

35. $R = 672,\ i = \frac{.05}{4} = .0125,\ n = 7(4) = 28$

Because deposits are made at the beginning of each time period, this is an annuity due.

$$S = R \cdot s_{\overline{n+1}|i} - R = 672\left[\frac{(1.0125)^{29} - 1}{.0125}\right] - 672 \approx 22,643.29$$

The future value of this annuity due is \$22,643.29.

36. $R = 11,900,\ i = \frac{.07}{12} = .005833,\ n = 13$

$$S = R \cdot s_{\overline{n+1}|i} - R = 11,900\left[\frac{1.0058333^{14} - 1}{.0058333}\right] - 11,900 \approx 161,166.70$$

The future value of this annuity due is \$161,166.70.

37. Answers vary. Possible answer: The purpose of a sinking fund is to receive funds deposited regularly for some goal.

38. $S = 6500,\ i = .05,\ n = 6$

$$S = R \cdot s_{\overline{n}|i} \Rightarrow R = \frac{S}{s_{\overline{n}|i}} = \frac{6500}{\left[\frac{(1.05)^6 - 1}{.05}\right]} \approx 955.61$$

The amount of each payment into this sinking fund is \$955.61.

39. $S = 57,000,\ i = \frac{.06}{2} = .03,\ n = \left(8\frac{1}{2}\right)(2) = 17$

$$R = \frac{S}{s_{\overline{n}|i}} = \frac{57,000}{\left[\frac{(1.03)^{17} - 1}{.03}\right]} \approx 2619.29$$

The amount of each payment is \$2619.29.

40. $S = 233,188,\ i = \frac{.057}{4} = .01425,$

$n = \left(7\frac{3}{4}\right)(4) = 31$

$$R = \frac{S}{s_{\overline{n}|i}} = \frac{233,188}{\left[\frac{(1.01425)^{31} - 1}{.01425}\right]} \approx 6035.27$$

The amount of each payment is \$6035.27.

41. $S = 56,788,\ i = \frac{.0612}{12} = .0051,$

$n = \left(4\frac{1}{2}\right)(12) = 54$

$$R = \frac{S}{s_{\overline{n}|i}} = \frac{56,788}{\left[\frac{(1.0051)^{54} - 1}{.0051}\right]} \approx 916.12$$

The amount of each payment is \$916.12.

42. $R = 850,\ i = .05,\ n = 4$

$$P = R \cdot a_{\overline{n}|i} = 850\left[\frac{1 - (1.05)^{-4}}{.05}\right] \approx 3014.06$$

The present value of this ordinary annuity is \$3014.06.

43. $R = 1500,\ i = \frac{.08}{4} = .02,\ n = 7(4) = 28$

$$P = R \cdot a_{\overline{n}|i} = 1500\left[\frac{1 - (1.02)^{-28}}{.02}\right] \approx 31,921.91$$

The present value is \$31,921.91.

44. $R = 4210,\ i = \dfrac{.056}{2} = .028,\ n = 8(2) = 16$

$$P = R \cdot a_{\overline{n}|i} = 4210\left[\frac{1-(1.028)^{-16}}{.028}\right] \approx \$53,699.94$$

The present value is $53,699.94.

45. $R = 877.34,\ i = \dfrac{.064}{12} \approx .0053,\ n = 17$

$$P = R \cdot a_{\overline{n}|i} = 877.34\left[\frac{1-\left(1+\frac{.064}{12}\right)^{-17}}{\frac{.064}{12}}\right] \approx 14,222.42$$

The present value is $14,222.42.

For exercises 46–48, recall that an annuity in which the payments are made at the end of each period is an ordinary annuity. We are seeking the present value.

46. $R = 800,\ i = \dfrac{.046}{4},\ n = 4(4) = 16$

$$P = 800\left[\frac{1-\left(1+\frac{.046}{4}\right)^{-16}}{\frac{.046}{4}}\right] \approx 11,630.63$$

About $11,630.63 is needed to fund the withdrawals.

47. $R = 1500,\ i = \dfrac{.058}{12},\ n = 10(12) = 120$

$$P = 1500\left[\frac{1-\left(1+\frac{.058}{12}\right)^{-120}}{\frac{.058}{12}}\right] \approx 136,340.32$$

About $136,340.32 is needed to fund the withdrawals.

48. $R = 3000,\ i = .062,\ n = 15$

$$P = 3000\left[\frac{1-(1.062)^{-15}}{.062}\right] \approx 28,759.74$$

About $28,759.74 is needed to fund the withdrawals.

49. $P = 150,000,\ i = \dfrac{.051}{12},\ n = 15(12) = 180$

$$150,000 = R\left[\frac{1-\left(1+\frac{.051}{12}\right)^{-180}}{\frac{.051}{12}}\right] \Rightarrow$$

$$R = \frac{150,000}{\frac{1-\left(1+\frac{.051}{12}\right)^{-180}}{\frac{.051}{12}}} \approx 1194.02$$

The monthly payment will be $1194.02.

50. $P = 25,000,\ i = \dfrac{.049}{4},\ n = 8(4) = 32$

$$25,000 = R\left[\frac{1-\left(1+\frac{.049}{4}\right)^{-32}}{\frac{.049}{4}}\right] \Rightarrow$$

$$R = \frac{25,000}{\frac{1-\left(1+\frac{.049}{4}\right)^{-32}}{\frac{.049}{4}}} \approx 949.07$$

The monthly payment will be $949.07.

51. Find the future value of an annuity with $R = 4200,\ i = .045,$ and $n = 12.$

$$S = R\left[\frac{(1+i)^n - 1}{i}\right] = 4200\left[\frac{(1.045)^{12} - 1}{.045}\right] \approx 64,948.93$$

Now find the present value of $64,948.93 at 4.5% compounded annually for 12 years.

$A = 64,948.93,\ i = .045,\ n = 12$

$$A = P(1+i)^n \Rightarrow 64,948.93 = P(1.045)^{12} \Rightarrow$$

$$P = \frac{64,948.93}{(1.045)^{12}} \approx 38,298.04$$

The required lump sum is $38,298.04.

52. The interest that is paid each half-year is $I = Prt = 24{,}000 \cdot .05 \cdot \frac{1}{2} = \$600$. So, we have a two-part investment, an annuity that pays \$600 every six months for 6 years and the \$24,000 face value of the bond which will be paid when the bond matures 6 years from now. The purchaser should be willing to pay the present value of each part of the investment, assuming 6.5% interest compounded semiannually.

Present value of annuity

$R = 600,\ i = \frac{.065}{2} = .0325,\ n = 6(2) = 12$

$$P = 600\left[\frac{1-(1.0325)^{-12}}{.0325}\right] \approx 5884.25$$

Present value of \$24,000 in 6 years

$P = A(1+i)^{-n}$

$P = 24{,}000(1.0325)^{-12} \approx 16{,}350.48$

The purchaser should be willing to pay \$5884.25 + \$16,350.48 = \$22,234.73, or about \$22,235.

53. $P = 32{,}000,\ i = \frac{.084}{4} = .021,\ n = 10$

$$P = R \cdot a_{\overline{n}|i} \Rightarrow R = \frac{P}{a_{\overline{n}|i}} = \frac{32{,}000}{\left[\frac{1-(1.021)^{-10}}{.021}\right]}$$

≈ 3581.11

Quarterly payments of \$3581.11 are necessary to amortize this loan.

54. $P = 5607,\ i = \frac{.076}{12},\ n = 32$

$$R = \frac{P}{a_{\overline{n}|i}} = \frac{5607}{\left[\frac{1-\left(1+\frac{.076}{12}\right)^{-32}}{\frac{.076}{12}}\right]} \approx 194.13$$

Monthly payments of \$194.13 are needed to amortize this loan.

55. $P = 95{,}000,\ i = \frac{.0367}{12},\ n = 30(12) = 360$

$$R = \frac{P}{a_{\overline{n}|i}} = \frac{95{,}000}{\left[\frac{1-\left(1+\frac{.0367}{12}\right)^{-360}}{\frac{.0367}{12}}\right]} \approx 435.66$$

The monthly payment for this mortgage is \$435.66.

56. $P = 167{,}000,\ i = \frac{.0291}{12},\ n = 15(12) = 180$

$$R = \frac{P}{a_{\overline{n}|i}} = \frac{167{,}000}{\left[\frac{1-\left(1+\frac{.0291}{12}\right)^{-180}}{\frac{.0291}{12}}\right]} \approx 1146.06$$

The monthly payment for this mortgage is \$1146.06.

57. After 5 years of payments, the remaining 25 years of payments can be thought of as annuity. The present value of this annuity is the remaining balance. So, we use the present value formula with $R = 435.66,\ i = \frac{.0367}{12},\ n = 300$

$$P = 435.66\left[\frac{1-\left(1+\frac{.0367}{12}\right)^{-300}}{\frac{.0367}{12}}\right] \approx 85{,}459.20$$

The balance remaining after five years is about \$85,459.20.

58. After 7.5 years (90 payments), the remaining 90 payments can be thought of as annuity. The present value of this annuity is the remaining balance. So, we use the present value formula with $R = 1146.06,\ i = \frac{.0291}{12},\ n = 90$

$$P = 1146.06\left[\frac{1-\left(1+\frac{.0291}{12}\right)^{-90}}{\frac{.0291}{12}}\right] \approx 92{,}565.32$$

The balance remaining after 7.5 years of payments is about \$92,565.32.

59. Locate the entry of the table that is in the row labeled "Payment Number 7" and in the column labeled "Interest for Period," to observe that \$81.98 of the seventh payment is interest.

60. Locate the entry of the table that is in the row labeled "Payment Number 4" and in the column labeled "Portion to Principal," to observe that \$89.12 of the forth payment is used to reduce the debt.

61. To find out how much interest is paid in the first 6 months of the loan, add the first six nonzero entries in the column labeled "Interest for Period."

$$\begin{array}{r} \$85.00 \\ 84.50 \\ 84.00 \\ 83.50 \\ 83.00 \\ +\ 82.49 \\ \hline \$502.49 \end{array}$$

62. To find out how much the debt has been reduced by the end of the first 8 months, subtract the balance at the end of the eighth payment from the original value.
$\$15,000 - 14,284.97 = \$715.03$

63. a. $\frac{217,000,000}{30} = 7,233,333.33$
Each yearly payment would have been \$7,233,333.33.

b. We use the formula for the present value of an annuity due, $S = R + R\left[\frac{1-(1+i)^{-(n-1)}}{i}\right]$.

$R = 7,233,333.33,\ i = .0358,\ n = 30$

$$S = 7,233,333.33 + 7,233,333.33\left[\frac{1-(1.0358)^{-29}}{.0358}\right] \approx 136,427,623.40$$

The cash value is about \$136,427,623.33.

64. P = 15,000, $i = \frac{.06}{2} = .03$, n = 7.5(2) = 15

$$I = P(1+i)^n - P = 15,000(1+.03)^{15} - 15,000 \approx 8369.51$$

65. Starting at age 23:

$P = 500,\ i = \frac{.05}{4} = .0125,$
$n = (65-23)4 = 42(4) = 168$
$A = P(1+i)^n = 500(1.0125)^{168} \approx 4030.28$
Starting at age 40:
$t = (65-40)4 = 25(4) = 100$
$A = 500(1.0125)^{100} \approx 1731.70$
$4030.28 - 1731.70 = 2298.58$
He will have \$2298.58 more if he invests now.

66. $r_F = \left(1+\frac{r}{m}\right)^m - 1;\ r_E = \left(1+\frac{r}{m}\right)^m - 1$

Frontenac Bank:

$$r_F = \left(1+\frac{.0394}{4}\right)^4 - 1 \approx .04$$

The effective rate is about 4.00%.
E*TRADE Bank:

$$r_E = \left(1+\frac{.0393}{365}\right)^{365} - 1 \approx .0401$$

The effective rate is about 4.01%.
E*TRADE Bank paid a higher effective rate.

67. $R = 3200,\ i = \frac{.068}{2} = .034,\ n = (3.5)(2) = 7$
Use the formula for the future value of an ordinary annuity.

$$S = R \cdot s_{\overline{n}|i} = 3200\left[\frac{(1.034)^7 - 1}{.034}\right] \approx 24,818.76$$

The final amount in the account will be \$24,818.76.
The interest earned will be \$24,818.76 – 7(\$3200) = \$2418.76.

68. $S = 52,000,\ i = \frac{.075}{12} = .00625,\ n = 20(12) = 240$
Use the formula for future value of an ordinary annuity to find the periodic payment.
$S = R \cdot s_{\overline{n}|i} \Rightarrow$

$$R = \frac{S}{s_{\overline{n}|i}} = \frac{52,000}{\left[\frac{(1.00625)^{240} - 1}{.00625}\right]} \approx 93.91$$

The firm should invest \$93.91 monthly.

69. $S = 2,000,000,\ i = \frac{.055}{12},\ n = 10(12) = 120$
Use the formula for future value of an annuity.

$$R = S\left[\frac{i}{(1+i)^n - 1}\right] \approx 2,000,000\left[\frac{\frac{.055}{12}}{(1+\frac{.055}{12})^{120} - 1}\right] \approx 12,538.59$$

She would have to put \$12,538.59 into her savings account every month.

70. Age 55: $P = 6000,\ i = \frac{.08}{1} = .08,\ n = 20(1) = 20$

$$S = 6000\left[\frac{(1.08)^{20} - 1}{.08}\right] \approx 274{,}571.79$$

Age 65:

$P = 12{,}000,\ i = \frac{.08}{1} = .08,\ n = 10(1) = 10$

$$S = 12{,}000\left[\frac{(1.08)^{10} - 1}{.08}\right] \approx 173{,}838.75$$

If the pension starts at age 55, the total is \$274,571.79, whereas, if the pension starts at age 65, the total is \$173,838.75. Thus, taking the option at age 55 produces the larger amount.

71. $A = 7500,\ i = \frac{.10}{2} = .05,\ n = 3(2) = 6$

$$A = P(1+i)^n$$

$$P = \frac{A}{(1+i)^n} = \frac{7500}{(1.05)^6} \approx 5596.62$$

The required lump sum is \$5596.62.

72. $P = 15{,}000,\ i = \frac{.072}{12} = .006,\ n = 36$

This is an ordinary annuity.

$P = R \cdot a_{\overline{n}|i} \Rightarrow$

$$R = \frac{P}{a_{\overline{n}|i}} = \frac{15{,}000}{\left[\frac{1-(1.006)^{-36}}{.006}\right]} \approx 464.53$$

The amount of each payment will be \$464.53.

73. $P = 40{,}000,\ i = \frac{.09}{2} = .045,\ n = 8(2) = 16$

This is an ordinary annuity.

$$R = \frac{P}{a_{\overline{n}|i}} = \frac{40{,}000}{\left[\frac{1-(1.045)^{-16}}{.045}\right]} \approx \$3560.61$$

The amount of each payment will be \$3560.61.

74. Amount of loan
= \$210,000 – \$42,000 = \$168,000

a. Use the formula for amortization payments with $P = 168{,}000,\ i = \frac{.0375}{12}$, and $n = 30(12) = 360$.

$$R = \frac{Pi}{1-(1+i)^{-n}} = \frac{168{,}000\left(\frac{.0375}{12}\right)}{1-\left(1+\frac{.0375}{12}\right)^{-360}} \approx 778.03$$

The monthly payment for this mortgage is \$778.03.

b. To find the amount of the first payment that goes to interest, use $I = Prt$ with $P = 168{,}000,\ i = .003125$, and $t = 1$.

$$I = (168{,}000)(.003125)(1) = 525.00$$

Of the first payment, \$525.00 is interest.

c. Using method 1, since 180 payments were made, there are 180 remaining payments. The present value is

$$778.03\left[\frac{1-(1.003125)^{-180}}{.003125}\right] \approx 106{,}986.52,$$

so the remaining balance is \$106,986.52. Using method 2, since 180 payments were already made, we have

$$778.03\left[\frac{1-(1.003125)^{-180}}{.003125}\right] \approx 106{,}986.52.$$

They still owe \$106,986.52 after 15 years of payments.

d. Amount of money received
= Selling price–Current mortgage balance
Using method 1, the amount received is
\$255,000 – \$106,986.52
= \$148,013.48.
Using method 2, the amount received is
\$255,000 – \$106,986.52
= \$148,013.48.

75. This is an ordinary annuity. We are seeking the future value.

$R = 250,\ i = \frac{.112}{12},\ n = 20(12) = 240$

$$S = 250\left[\frac{\left(1+\frac{.112}{12}\right)^{240} - 1}{\frac{.112}{12}}\right] \approx 222{,}221.02$$

There will be about \$222,221.02 in the account at the end of 20 years.

76. Total amount after 60 deposits:

$R = 400,\ i = \frac{.031}{12},\ n = 60$

$$S = 400\left[\frac{(1+\frac{.031}{12})^{60} - 1}{\frac{.031}{12}}\right] = 25,923.81$$

Total amount after next 96 months:

$A = 25,923.81(1+\frac{.031}{12})^{96} = 33,209.70$

Total of 60 \$525 deposits:

$R = 525,\ i = \frac{.031}{12},\ n = 60$

$$S = 525\left[\frac{(1+\frac{.031}{12})^{60} - 1}{\frac{.031}{12}}\right] \approx 34,025.00$$

Gene's account balance is \$33,209.70 + \$34,025.00 = \$67,234.70.

77. $P = 10,000,\ i = .05,\ n = 7$

$A = P(1+i)^n = 10,000(1.05)^7 \approx \$14,071.00$

At the end of 7 yr, the value of the death benefit will be \$14,071.00. This balance of \$14,071.00 is to be paid out in 120 equal monthly payments, with $i = \frac{.03}{12} = .0025$. The full balance will be paid out, so we use the amortization formula. Let X be the amount of each payment.

$$X = \frac{14,071.00}{a_{\overline{120}|.0025}} = \frac{14,071.00}{\left[\frac{1-(1.0025)^{-120}}{.0025}\right]} \approx \$135.87$$

This corresponds to choice (d).

78. Amount needed for retirement:

$$75,000 = \frac{x}{\left[\frac{1-(1.101)^{-20}}{.101}\right]}$$

$x = 634,183.19$

Annual contribution to acquire \$634,183.19:

$$634,183.19 = R\left[\frac{(1.101)^{30}-1}{.101}\right]$$

$R \approx 3783.01$

She will have to make annual deposits of \$3783.01.

Case 5 Continuous Compounding

1. a. $A = 20,000e^{.06\cdot 2} \approx 22,549.94$

b. $A = 20,000e^{.06\cdot 10} \approx 36,442.38$

c. $A = 20,000e^{.06\cdot 20} \approx 66,402.34$

2. a. $A = 2500\left(1+\frac{.055}{12}\right)^{12\cdot 2} \approx 2789.99$

b. $A = 2500e^{.055\cdot 2} \approx 2790.70$

3. a. $A = 25,000\left(1+\frac{.05}{1}\right)^{1\cdot 10} \approx \$40,722.37$

b. $A = 25,000\left(1+\frac{.05}{4}\right)^{4\cdot 10} \approx \$41,090.49$

c. $A = 25,000\left(1+\frac{.05}{12}\right)^{12\cdot 10} \approx \$41,175.24$

d. $A = 25,000\left(1+\frac{.05}{365}\right)^{365\cdot 10} \approx \$41,216.62$

e. $A = 25,000e^{.05\cdot 10} \approx \$41,218.03$

4. a. $A = 250,000\left(1+\frac{.05}{1}\right)^{1\cdot 10} \approx \$407,223.66$

b. $A = 250,000\left(1+\frac{.05}{4}\right)^{4\cdot 10} \approx \$410,904.87$

c. $A = 250,000\left(1+\frac{.05}{12}\right)^{12\cdot 10} \approx \$411,752.37$

d. $A = 250,000\left(1+\frac{.05}{365}\right)^{365\cdot 10} \approx \$412,166.20$

e. $A = 250,000e^{.05\cdot 10} \approx \$412,180.32$

5. a. $r_E = e^{.045} - 1 \approx .04603$

The effective rate is about 4.603%.

b. $r_E = e^{.057} - 1 \approx .05866$

The effective rate is about 5.866%.

c. $r_E = e^{.074} - 1 \approx .07681$

The effective rate is about 7.681%.

6. a. $5000 = Pe^{.0375 \cdot 8} \Rightarrow P = \frac{5000}{e^{.0375 \cdot 8}} \approx 3704.09$

The present value of the \$5000 is about \$3704.09.

b. Yes, because \$4000 is more than the present value.

Chapter 6 Systems of Linear Equations and Matrices

Section 6.1 Systems of Two Linear Equations in Two Variables

1. Is (–1, 3) a solution of

$2x+y=1 \quad (1)$

$-3x+2y=9 \quad (2)$

Check (–1, 3) in equation (1).

$2(-1)+3=1\ ?$

$-2+3=1\ ?$ True

Check (–1, 3) in equation (2).

$-3(-1)+2(3)=9\ ?$

$3+6=9\ ?$ True

The ordered pair (–1, 3) is a solution of this system of equations.

3. $3x-y=1 \quad (1)$

$x+2y=-9 \quad (2)$

Solve for y from equation (1) to get $y=3x-1$.

Then substitute $y=3x-1$

into (2) $x+2(3x-1)=-9$. Solve for x.

$x+2(3x-1)=-9 \Rightarrow x+6x-2=-9 \Rightarrow$

$7x-2=-9 \Rightarrow 7x=-7 \Rightarrow x=-1$

Substitute $x=-1$ into $y=3x-1$ to solve for y.

$y=3(-1)-1 \Rightarrow y=-4.$

The solution set is (–1,–4).

5. $3x-2y=4 \quad (1)$

$2x+y=-1 \quad (2)$

Solve for y from equation 2 to get

$y=-1-2x$.

Substitute $y=-1-2x$ into equation (1)

$3x-2(-1-2x)=4 \Rightarrow 3x+2+4x=4 \Rightarrow$

$7x+2=4 \Rightarrow 7x=2 \Rightarrow x=\frac{2}{7}$

Substitute $x=\frac{2}{7}$ into $y=-1-2x$ to solve for y.

$y=-1-2\left(\frac{2}{7}\right)=-\frac{11}{7}$

The solution set is $\left(\frac{2}{7},-\frac{11}{7}\right)$.

7. $x-2y=5 \ (1)$

$2x+y=3 \ (2)$

Multiply equation (2) by 2 and add the result to equation (1).

$$\begin{array}{l} x-2y=5 \quad (1) \\ \underline{4x+2y=6} \\ 5x \qquad =11 \Rightarrow x=\frac{11}{5} \end{array}$$

Substitute $\frac{11}{5}$ for x in equation (1) to get

$\frac{11}{5}-2y=5 \Rightarrow -2y=\frac{14}{5} \Rightarrow y=-\frac{7}{5}.$

The solution of the system is $\left(\frac{11}{5},-\frac{7}{5}\right)$.

9. $2x-2y=12 \ (1)$

$-2x+3y=10 \ (2)$

Add the two equations to get $y = 22$.

Substitute 22 for y in equation (1) to get

$2x-2(22)=12 \Rightarrow 2x-44=12 \Rightarrow$

$2x=56 \Rightarrow x=28$

The solution of the system is (28, 22).

11. $x+3y=-1 \ (1)$

$2x-y=5 \quad (2)$

Multiply equation (2) by 3, then add the result to equation (1)

$$\begin{array}{l} x+3y=-1 \ (1) \\ \underline{6x-3y=15 \ (2)} \\ 7x \qquad =14 \Rightarrow x=2 \end{array}$$

Substitute 2 for x in equation (1) and solve for y.

$2+3y=-1 \Rightarrow 3y=-3 \Rightarrow y=-1$

The solution of the system is (2, –1).

13. $2x+3y=15 \quad (1)$

$8x+12y=40 \ (2)$

Multiply equation (1) by –4 and add the result to equation (1).

$$\begin{array}{l} -8x-12y=-60 \\ \underline{\ 8x+12y=40 \quad (2)} \\ \qquad\quad 0=-20 \ \text{False} \end{array}$$

The system is inconsistent. There is no solution.

15. $2x - 8y = 2$ (1)
$3x - 12y = 3$ (2)
Multiply equation (1) by 3 and equation (2) by –2 to get
$6x - 24y = 6$ (3)
$-6x + 24y = -6.$ (4)
Add the two equations.

$$\begin{array}{rl} 6x - 24y = 6 & (3) \\ -6x + 24y = -6 & (4) \\ \hline 0 = 0 & \end{array}$$

The system is dependent, so there are an infinite number of solutions. To solve the system in terms of one of the parameters, say y, solve equation (1) for x in terms of y.
$2x - 8y = 2 \Rightarrow 2x = 8y + 2 \Rightarrow x = 4y + 1$
The solution to the system is the infinite set of pairs of the form $(4y + 1, y)$ for any real number y.
Note: the system could also be solved in terms of x. The ordered pairs would all have the form $\left(x, \frac{1}{4}x - \frac{1}{4}\right)$.

17. $\frac{x}{5} + 3y = 31$ (1)
$2x - \frac{y}{5} = 8$ (2)
Multiply both equations (1) and (2) by 5 to clear the fractions.
$x + 15y = 155$ (3)
$10x - y = 40$ (4)
Multiply equation (4) by 15, then add the result to equation (3) and solve for x.

$$\begin{array}{rl} 150x - 15y = 600 & (5) \\ x + 15y = 155 & (3) \\ \hline 151x \qquad = 755 \Rightarrow x = 5 & \end{array}$$

Substitute 5 for x in equation (1) to find y.
$\frac{x}{5} + 3y = 31 \Rightarrow \frac{5}{5} + 3y = 31 \Rightarrow$
$3y = 30 \Rightarrow y = 10$
The solution is (5, 10).

19. a. Multiply the second equation by –1 and then add the equations:

$$\begin{array}{l} 43,500x - y = 1,295,000 \\ -27,000x + y = -440,000 \\ \hline 16,500x \qquad = 855,000 \quad \Rightarrow x \approx 51.8 \text{ weeks} \end{array}$$

Substitute $\frac{8550}{165}$ for x in the second equation and solve for y.
$-27,000\left(\frac{8550}{165}\right) + y = -440,000 \Rightarrow$
$y \approx 959,091$
In about 51.8 weeks, the profit will be about \$959,091 for each venue.

b. Answers vary.

21. $-5x + y = 11,400$ (1)
$-140x + y = 8,000$ (2)
We are asked to find the year in which the two states have the same population, which is represented by the x-coordinate of the solution. Multiply equation (1) by –1, then add the result to equation (2) and solve for x.

$$\begin{array}{l} 5x - y = -11,400 \\ -140x + y = 8,000 \\ \hline -135x \qquad = -3400 \Rightarrow x \approx 25.2 \end{array}$$

$2000 + 25.2 = 2025.2$
Thus, the two cities will have the same population in 2025.

23. $2y - 9x = 1248$ (1)
$10y - 13x = 8110$ (2)
We are asked to find when the median income of men and women will be the same, i.e., the year. This is represented by the x-coordinate of the solution. Multiply equation (1) by –5 and then add the result to equation (2) to solve for x.

$$\begin{array}{l} -10y + 45x = -6240 \\ 10y - 13x = 8110 \\ \hline 32x \qquad = 1870 \Rightarrow x \approx 58.4 \end{array}$$

$2000 + 58.4 = 2058.4$
According to the model, the median income of men and women will be the same in 2058.

25. Let x = the number of adults and let y = the number of children.

$$x + y = 200 \quad (1)$$
$$8x + 5y = 1435 \quad (2)$$

Solve equation (1) for y, then substitute that expression into equation (2) and solve for x.

$$x + y = 200 \Rightarrow y = 200 - x$$
$$8x + 5(200 - x) = 1435 \Rightarrow$$
$$8x + 1000 - 5x = 1435 \Rightarrow 3x = 435 \Rightarrow x = 145$$

Substitute x = 145 into equation (1) and solve for y.

$$145 + y = 200 \Rightarrow y = 55$$

There were 145 adults and 55 children.

27. We use the formula rate × time = distance. Let x = the speed of the plane and let y = the speed of the wind.

$$5(x + y) = 3000 \Rightarrow x + y = 600$$
$$6(x - y) = 3000 \Rightarrow x - y = 500$$

Add the equations and solve for x.

$$2x = 1100 \Rightarrow x = 550$$

Substitute x = 550 into the first equation and solve for y.

$$550 + y = 600 \Rightarrow y = 50$$

The plane's speed is 550 mph and the wind speed is 50 mph.

29. Let x = the number of shares of Boeing stock and y = the number of shares of GE stock.

$$30x + 70y = 16,000 \quad (1)$$
$$1.50(30x) + 3(70y) = 345,000 \quad (2)$$

Simplify equation (2)

$$30x + 70y = 16,000 \quad (1)$$
$$45x + 210y = 34,500 \quad (2)$$

Multiply equation (1) by – 3, add the resulting equation to equation (2) and solve for x.

$$\begin{array}{r} -90x - 210y = -48,000 \\ 45x + 210y = 34,500 \\ \hline -45x = -13,500 \Rightarrow x = 300 \end{array}$$

Substitute x = 300 into (1) and solve for y.

$$30(300) + 70y = 16000$$
$$9000 + 70y = 16000 \Rightarrow 70y = 7000 \Rightarrow y = 100$$

Shirley owns 300 shares of Boeing stock and 100 of GE stock.

31. a. To find the equation of the line through (1, 2) and (3, 4), first find the slope.

$$m = \frac{4-2}{3-1} = \frac{2}{2} = 1$$

Use the point-slope form with $(x_1, y_1) = (1, 2)$.

$$y - y_1 = m(x - x_1) \Rightarrow y - 2 = 1(x - 1) \Rightarrow$$
$$y - 2 = x - 1 \Rightarrow y = x + 1$$

An alternate solution can be written using a system of equations. Using the form $y = mx + b$, the line through (1, 2) and (3, 4) would yield the system of equations.

$$2 = m(1) + b \quad (1)$$
$$4 = m(3) + b. \quad (2)$$

This simplifies to

$$m + b = 2 \quad (1)$$
$$3m + b = 4. \quad (2)$$

This system can be solved by multiplying equation (1) by –1 and adding the result to equation (2).

$$\begin{array}{r} -m - b = -2 \\ 3m + b = 4 \\ \hline 2m = 2 \Rightarrow m = 1 \end{array}$$

Substitute 1 for m in equation (1).

$$m + b = 2 \Rightarrow 1 + b = 2 \Rightarrow b = 1$$

So, the equation is $y = x + 1$.

b. To find the equation of the line with slope $m = 3$ through (–1, 1), use the form $y = mx + b$. Since $m = 3$, the equation becomes $y = 3x + b$.

Substitute –1 for x and 1 for y.

$$1 = 3(-1) + b \Rightarrow 1 = -3 + b \Rightarrow 4 = b$$

The equation is $y = 3x + 4$.

c. To find a point on both lines, solve the system

$$y = x + 1 \quad (1)$$
$$y = 3x + 4 \quad (2)$$

Equate the expressions for y, then solve.

$$x + 1 = 3x + 4 \Rightarrow -2x = 3 \Rightarrow x = -\frac{3}{2}$$

Substitute $-\frac{3}{2}$ for x in equation (1).

$$y = x + 1 \Rightarrow y = -\frac{3}{2} + 1 \Rightarrow y = -\frac{1}{2}$$

The point $\left(-\frac{3}{2}, -\frac{1}{2}\right)$ is on both lines.

Section 6.2 Larger Systems of Linear Equations

1.
$$\begin{aligned} x \quad\quad - 3z &= 2 \\ 2x - 4y + 5z &= 1 \\ 5x - 8y + 7z &= 6 \\ 3x - 4y + 2z &= 3 \end{aligned}$$

3.
$$\begin{aligned} 3x + \quad\quad z + 2w + 18v &= 0 \\ -4x + y - \quad\quad w - 24v &= 0 \\ 7x - y + z + 3w + 42v &= 0 \\ 4x + \quad\quad z + 2w + 24v &= 0 \end{aligned}$$

5.
$$\begin{aligned} x + y + 2z + 3w &= 1 \\ -y - z - 2w &= -1 \\ 3x + y + 4z + 5w &= 2 \end{aligned}$$

7.
$$\begin{aligned} x + 12y - 3z + 4w &= 10 \\ 2y + 3z + w &= 4 \\ -z &= -7 \\ 6y - 2z - 3w &= 0 \end{aligned}$$

9.
$$\begin{aligned} x + 3y - 4z + 2w &= 1 \quad (1) \\ y + z - w &= 4 \quad (2) \\ 2z + 2w &= -6 \quad (3) \\ 3w &= 9 \quad (4) \end{aligned}$$

Divide (4) by 3 to obtain $w = 3$. Substitute $w = 3$ into (3) and solve for z.

$2z + 2(3) = -6 \Rightarrow 2z = -12 \Rightarrow z = -6$

Substitute $w = 3$ and $z = -6$ into (2) to solve for y.

$y + (-6) - 3 = 4 \Rightarrow y = 13$

Substitute $y = 13$, $w = 3$, and $z = -6$ into (1) to solve for x.

$x + 3(13) - 4(-6) + 2(3) = 1 \Rightarrow x = -68$

The solution set is $(-68, 13, -6, 3)$.

11.
$$\begin{aligned} 2x + 2y - 4z + w &= -5 \quad (1) \\ 3y + 4z - w &= 0 \quad (2) \\ 2z - 7w &= -6 \quad (3) \\ 5w &= 15 \quad (4) \end{aligned}$$

Divide (4) by 5 to obtain $w = 3$. Substitute $w = 3$ into (3) to solve for z.

$$2z - 7(3) = -6 \Rightarrow 2z - 21 = -6 \Rightarrow z = \frac{15}{2}$$

Substitute $w = 3$ and $z = \frac{15}{2}$ into (2) to solve for y.

$$3y + 4\left(\frac{15}{2}\right) - 3 = 0 \Rightarrow 3y + 30 = 3 \Rightarrow$$
$$3y = -27 \Rightarrow y = -9$$

Substitute $y = -9$, $z = \frac{15}{2}$, and $w = 3$ into (1) to solve for x.

$$2x + 2(-9) - 4\left(\frac{15}{2}\right) + 3 = -5 \Rightarrow$$
$$2x - 18 - 30 + 3 = -5 \Rightarrow 2x - 45 = -5 \Rightarrow$$
$$2x = 40 \Rightarrow x = 20$$

The solution set is $\left(20, -9, \frac{15}{2}, 3\right)$.

13. The augmented matrix for the system
$$\begin{aligned} 2x + y + z &= 3 \\ 3x - 4y + 2z &= -5 \\ x + y + z &= 2 \end{aligned}$$
is $\left[\begin{array}{ccc|c} 2 & 1 & 1 & 3 \\ 3 & -4 & 2 & -5 \\ 1 & 1 & 1 & 2 \end{array}\right]$.

15. The system of equations for the augmented matrix
$\left[\begin{array}{ccc|c} 2 & 3 & 8 & 20 \\ 1 & 4 & 6 & 12 \\ 0 & 3 & 5 & 10 \end{array}\right]$ is
$$\begin{aligned} 2x + 3y + 8z &= 20 \\ x + 4y + 6z &= 12 \\ 3y + 5z &= 10. \end{aligned}$$

17. $\left[\begin{array}{ccc|c} 1 & 2 & 3 & -1 \\ 6 & 5 & 4 & 6 \\ 2 & 0 & 7 & -4 \end{array}\right]$

Interchange R_2 and R_3 to obtain

$$\left[\begin{array}{ccc|c} 1 & 2 & 3 & -1 \\ 2 & 0 & 7 & -4 \\ 6 & 5 & 4 & 6 \end{array}\right].$$

19. $\left[\begin{array}{cccc|c} -4 & -3 & 1 & -1 & 2 \\ 8 & 2 & 5 & 0 & 6 \\ 0 & -2 & 9 & 4 & 5 \end{array}\right]$

Replace R_2 with $2R_1 + R_2$ to obtain

$$\left[\begin{array}{cccc|c} -4 & -3 & 1 & -1 & 2 \\ 0 & -4 & 7 & -2 & 10 \\ 0 & -2 & 9 & 4 & 5 \end{array}\right].$$

21. $\left[\begin{array}{cccc|c} 1 & 0 & 0 & 0 & \frac{3}{2} \\ 0 & 1 & 0 & 0 & 17 \\ 0 & 0 & 1 & 0 & -5 \\ 0 & 0 & 0 & 1 & 0 \end{array}\right]$

The solution of the system is $\left(\frac{3}{2}, 17, -5, 0\right)$.

23. $\left[\begin{array}{cccc|c} 1 & 0 & 0 & 1 & 12 \\ 0 & 1 & 0 & 2 & -3 \\ 0 & 0 & 1 & 0 & -5 \\ 0 & 0 & 0 & 0 & 0 \end{array}\right]$

The last row, which represents 0 = 0, indicates that the system has no unique solution. The linear system associated with this matrix is

$$\begin{aligned} x + w &= 12 \quad (1) \\ y + 2w &= -3 \quad (2) \\ z &= 5 \quad (3) \end{aligned}$$

To put the solution in terms of w, solve for x in equation (1) and y in equation (2).
The solution is $(12 - w, -3 - 2w, -5, w)$.

25. $\begin{aligned} x + 2y \quad &= 0 \quad (1) \\ y - z &= 2 \quad (2) \\ x + y + z &= -2 \quad (3) \end{aligned}$

Multiply equation (3) by −1, then add the result to equation (1).

$$\begin{array}{r} x + 2y \quad = 0 \\ -x - y - z = 2 \\ \hline y - z = 2 \end{array}$$

The system now becomes

$$\begin{aligned} y - z &= 2 \quad (4) \\ y - z &= 2 \quad (2) \\ x + y + z &= -2 \quad (3) \end{aligned}$$

Multiply equation (4) by −1, then add the result to equation (2).

$$\begin{array}{r} y - z = 2 \\ -y + z = -2 \\ \hline 0 = 0 \quad (5) \end{array}$$

Since equation (5) is true, the system is dependent.

27. $\begin{aligned} x + 2y + 4z &= 6 \quad (1) \\ y + z &= 1 \quad (2) \\ x + 3y + 5z &= 10 \quad (3) \end{aligned}$

Multiply equation (3) by −1 and add the result to equation (1).

$$\begin{array}{r} x + 2y + 4z = 6 \\ -x - 3y - 5z = -10 \\ \hline -y - z = -4 \end{array}$$

The system now becomes

$$\begin{aligned} -y - z &= -4 \quad (1) \\ y + z &= 1 \quad (2) \\ x + 3y + 5z &= 10 \quad (4) \end{aligned}$$

Add equation (2) to equation (4).

$$\begin{array}{r} y + z = 1 \quad (2) \\ -y - z = -4 \quad (4) \\ \hline 0 = -3 \quad (5) \end{array}$$

Since equation (5) is false, the system is inconsistent.

29. $a-3b-2c=-3$ (1)
$3a+2b-c=12$ (2)
$-a-b+4c=3$ (3)

Multiply equation (1) by –3 and add the result to equation (2). Also, add equation (1) to equation (3).

The new system is

$a-3b-2c=-3$ (1)
$11b+5c=21$ (4)
$-4b+2c=0.$ (5)

Multiply equation (4) by $\frac{1}{11}$. This gives

$a-3b-2c=-3$ (1)
$b+\frac{5}{11}c=\frac{21}{11}$ (6)
$-4b+2c=0.$ (5)

Multiply equation (6) by 4 and add the result to equation (5). This gives

$a-3b-2c=-3$ (1)
$b+\frac{5}{11}c=\frac{21}{11}$ (6)
$\frac{42}{11}c=\frac{84}{11}.$ (7)

Multiply equation (7) by $\frac{11}{42}$ to get

$a-3b-2c=-3$ (1)
$b+\frac{5}{11}c=\frac{21}{11}$ (6)
$c=2.$ (8)

Substitute 2 for c in equation (6) to get

$b+\frac{10}{11}=\frac{21}{11}$, or $b=\frac{11}{11}=1.$

Substitute 2 for c and 1 for b in equation (1) to get $a=4$. The system is independent.

31. The augmented matrix is

$$\left[\begin{array}{ccc|c} -1 & 3 & 2 & 0 \\ 2 & -1 & -1 & 3 \\ 1 & 2 & 3 & 0 \end{array}\right].$$

Now use the matrix method to solve the system.

$$\left[\begin{array}{ccc|c} 1 & -3 & -2 & 0 \\ 2 & -1 & -1 & 3 \\ 1 & 2 & 3 & 0 \end{array}\right] \begin{array}{l} -R_1 \\ \\ \\ \end{array}$$

$$\left[\begin{array}{ccc|c} 1 & -3 & -2 & 0 \\ 0 & 5 & 3 & 3 \\ 1 & 2 & 3 & 0 \end{array}\right] \begin{array}{l} \\ -2R_1+R_2 \\ \\ \end{array}$$

$$\left[\begin{array}{ccc|c} 1 & -3 & -2 & 0 \\ 0 & 5 & 3 & 3 \\ 0 & -5 & -5 & 0 \end{array}\right] \begin{array}{l} \\ \\ -R_3+R_1 \end{array}$$

$$\left[\begin{array}{ccc|c} 1 & -3 & -2 & 0 \\ 0 & 5 & 3 & 3 \\ 0 & 0 & -2 & 3 \end{array}\right] \begin{array}{l} \\ \\ R_2+R_3 \end{array}$$

$$\left[\begin{array}{ccc|c} 1 & -3 & -2 & 0 \\ 0 & 1 & \frac{3}{5} & \frac{3}{5} \\ 0 & 0 & 1 & -\frac{3}{2} \end{array}\right] \begin{array}{l} \\ \frac{1}{5}R_2 \\ -\frac{1}{2}R_3 \end{array}$$

The system is now in row echelon form, and $z=-\frac{3}{2}$.

Using back substitution, we find that

$$y+\frac{3}{5}\left(-\frac{3}{2}\right)=\frac{3}{5}\Rightarrow y=\frac{3}{2}$$

Now solve for x.

$$-x-3\left(-\frac{3}{2}\right)-2\left(-\frac{3}{2}\right)=0\Rightarrow x=\frac{3}{2}$$

The solution set is $\left(\frac{3}{2},\frac{3}{2},-\frac{3}{2}\right)$.

33. The augmented matrix is

$$\left[\begin{array}{ccc|c} 1 & -2 & 4 & 6 \\ 1 & 2 & 13 & 6 \\ -2 & 6 & -1 & -10 \end{array}\right].$$

Now use the matrix method to solve the system.

$$\left[\begin{array}{ccc|c} 1 & -2 & 4 & 6 \\ 0 & -4 & -9 & 0 \\ 0 & 2 & 7 & 2 \end{array}\right]\begin{array}{l} \\ R_1 - R_2 \\ 2R_1 + R_3 \end{array}$$

$$\left[\begin{array}{ccc|c} 1 & -2 & 4 & 6 \\ 0 & -4 & -9 & 0 \\ 0 & 0 & 5 & 4 \end{array}\right]\begin{array}{l} \\ \\ R_2 + 2R_3 \end{array}$$

$$\left[\begin{array}{ccc|c} 1 & -2 & 4 & 6 \\ 0 & 1 & \frac{9}{4} & 0 \\ 0 & 0 & 1 & \frac{4}{5} \end{array}\right]\begin{array}{l} \\ -\frac{1}{4}R_2 \\ \frac{1}{5}R_3 \end{array}$$

The system is now in row echelon form, and $z = \frac{4}{5}$. Using back substitution, we find that

$$y + \frac{9}{4}\left(\frac{4}{5}\right) = 0 \Rightarrow y = -\frac{9}{5}$$

$$x - 2\left(-\frac{9}{5}\right) + 4\left(\frac{4}{5}\right) = 6 \Rightarrow x = -\frac{4}{5}$$

The solution set for the system is $(-.8, -1.8, .8)$.

35. Note that we have four equations in three unknowns. The augmented matrix is

$$\left[\begin{array}{ccc|c} 1 & 1 & 1 & 200 \\ 1 & -2 & 0 & 0 \\ 2 & 3 & 5 & 600 \\ 2 & -1 & 1 & 200 \end{array}\right]$$

Use the matrix method to solve the system.

$$\left[\begin{array}{ccc|c} 1 & 1 & 1 & 200 \\ 0 & 3 & 1 & 200 \\ 0 & 4 & 4 & 400 \\ 2 & -1 & 1 & 200 \end{array}\right]\begin{array}{l} \\ R_1 - R_2 \\ R_3 - R_4 \\ \\ \end{array}$$

$$\left[\begin{array}{ccc|c} 1 & 1 & 1 & 200 \\ 0 & 3 & 1 & 200 \\ 0 & 4 & 4 & 400 \\ 0 & 3 & 1 & 200 \end{array}\right]\begin{array}{l} \\ \\ \\ 2R_1 - R_4 \end{array}$$

$$\left[\begin{array}{ccc|c} 1 & 1 & 1 & 200 \\ 0 & 3 & 1 & 200 \\ 0 & 1 & 1 & 100 \\ 0 & 0 & 0 & 0 \end{array}\right]\begin{array}{l} \\ \\ \frac{1}{4}R_3 \\ R_2 - R_4 \end{array}$$

$$\left[\begin{array}{ccc|c} 1 & 1 & 1 & 200 \\ 0 & 3 & 1 & 200 \\ 0 & 0 & 2 & 100 \\ 0 & 0 & 0 & 0 \end{array}\right]\begin{array}{l} \\ \\ 3R_3 - R_2 \\ \\ \end{array}$$

$$\left[\begin{array}{ccc|c} 1 & 1 & 1 & 200 \\ 0 & 1 & \frac{1}{3} & \frac{200}{3} \\ 0 & 0 & 1 & 50 \\ 0 & 0 & 0 & 0 \end{array}\right]\begin{array}{l} \\ \frac{1}{3}R_2 \\ \frac{1}{2}R_3 \\ \\ \end{array}$$

The system is now in row echelon form, and $z = 50$. Use back-substitution to solve for x and y.

$$y + \frac{1}{3}(50) = \frac{200}{3} \Rightarrow y = 50$$

$$x + 1(50) + 1(50) = 200 \Rightarrow x = 100$$

The solution to the system is $(100, 50, 50)$.

37.

$$x + y + z = 5$$
$$2x + y - z = 2$$
$$x - y + z = -2$$

The augmented matrix is

$$\left[\begin{array}{ccc|c} 1 & 1 & 1 & 5 \\ 2 & 1 & -1 & 2 \\ 1 & -1 & 1 & -2 \end{array}\right]$$

Now use the matrix method to solve the system:

$$\left[\begin{array}{ccc|c} 1 & 1 & 1 & 5 \\ 0 & -1 & -3 & -8 \\ 0 & -2 & 0 & -7 \end{array}\right]\begin{array}{l} \\ -2R_1 + R_2 \\ -R_1 + R_3 \end{array}$$

$$\left[\begin{array}{ccc|c} 1 & 1 & 1 & 5 \\ 0 & -1 & -3 & -8 \\ 0 & 0 & 6 & 9 \end{array}\right]\begin{array}{l} \\ \\ -2R_2 + R_3 \end{array}$$

$$\left[\begin{array}{ccc|c} 1 & 1 & 1 & 5 \\ 0 & 1 & 3 & 8 \\ 0 & 0 & 1 & \frac{3}{2} \end{array}\right]\begin{array}{l} \\ -R_2 \\ \frac{1}{6}R_3 \end{array}$$

The system is now in row echelon form and $z = \frac{3}{2}$. Use back-substitution to solve for x and y.

$$y + 3\left(\frac{3}{2}\right) = 8 \Rightarrow y = \frac{7}{2}$$

$$x + \frac{7}{2} + \frac{3}{2} = 5 \Rightarrow x = 0$$

The solution set is $(0, 3.5, 1.5)$.

39. $x+2y+z=5$
$2x+y-3z=-2$
$3x+y+4z=-5$

The matrix is

$$\left[\begin{array}{ccc|c} 1 & 2 & 1 & 5 \\ 2 & 1 & -3 & -2 \\ 3 & 1 & 4 & -5 \end{array}\right]$$

$$\left[\begin{array}{ccc|c} 1 & 2 & 1 & 5 \\ 0 & -3 & -5 & -12 \\ 0 & -5 & 1 & -20 \end{array}\right]\begin{array}{l} \\ -2R_1+R_2 \\ -3R_1+R_3 \end{array}$$

$$\left[\begin{array}{ccc|c} 1 & 2 & 1 & 5 \\ 0 & 1 & \frac{5}{3} & 4 \\ 0 & -5 & 1 & -20 \end{array}\right]\begin{array}{l} \\ -\frac{1}{3}R_2 \\ \\ \end{array}$$

$$\left[\begin{array}{ccc|c} 1 & 0 & -\frac{7}{3} & -3 \\ 0 & 1 & \frac{5}{3} & 4 \\ 0 & 0 & \frac{28}{3} & 0 \end{array}\right]\begin{array}{l} -2R_2+R_1 \\ \\ 5R_2+R_3 \end{array}$$

$$\left[\begin{array}{ccc|c} 1 & 0 & -\frac{7}{3} & -3 \\ 0 & 1 & \frac{5}{3} & 4 \\ 0 & 0 & 1 & 0 \end{array}\right]\begin{array}{l} \\ \\ \frac{3}{28}R_3 \end{array}$$

$$\left[\begin{array}{ccc|c} 1 & 0 & 0 & -3 \\ 0 & 1 & 0 & 4 \\ 0 & 0 & 1 & 0 \end{array}\right]\begin{array}{l} \frac{7}{3}R_3+R_1 \\ -\frac{5}{3}R_3+R_2 \\ \\ \end{array}$$

The solution is (–3, 4, 0).

41. $x+3y-6z=7$
$2x-y+2z=0$
$x+y+2z=-1$

The matrix is

$$\left[\begin{array}{ccc|c} 1 & 3 & -6 & 7 \\ 2 & -1 & 2 & 0 \\ 1 & 1 & 2 & -1 \end{array}\right]$$

$$\left[\begin{array}{ccc|c} 1 & 3 & -6 & 7 \\ 0 & -7 & 14 & -14 \\ 0 & -2 & 8 & -8 \end{array}\right]\begin{array}{l} \\ -2R_1+R_2 \\ -1R_1+R_3 \end{array}$$

$$\left[\begin{array}{ccc|c} 1 & 3 & -6 & 7 \\ 0 & 1 & -2 & 2 \\ 0 & -2 & 8 & -8 \end{array}\right]\begin{array}{l} \\ -\frac{1}{7}R_2 \\ \\ \end{array}$$

$$\left[\begin{array}{ccc|c} 1 & 0 & 0 & 1 \\ 0 & 1 & -2 & 2 \\ 0 & 0 & 4 & -4 \end{array}\right]\begin{array}{l} -3R_2+R_1 \\ \\ 2R_2+R_3 \end{array}$$

$$\left[\begin{array}{ccc|c} 1 & 0 & 0 & 1 \\ 0 & 1 & -2 & 2 \\ 0 & 0 & 1 & -1 \end{array}\right]\begin{array}{l} \\ \\ \frac{1}{4}R_3 \end{array}$$

$$\left[\begin{array}{ccc|c} 1 & 0 & 0 & 1 \\ 0 & 1 & 0 & 0 \\ 0 & 0 & 1 & -1 \end{array}\right]\begin{array}{l} \\ 2R_3+R_2 \\ \\ \end{array}$$

The solution is (1, 0, –1).

43. $x-2y+4z=9$
$x+y+13z=6$
$-2x+6y-z=-10$

The augmented matrix is

$$\left[\begin{array}{ccc|c} 1 & -2 & 4 & 9 \\ 1 & 1 & 13 & 6 \\ -2 & 6 & -1 & -10 \end{array}\right]$$

$$\left[\begin{array}{ccc|c} 1 & -2 & 4 & 9 \\ 0 & 3 & 9 & -3 \\ 0 & 2 & 7 & 8 \end{array}\right]\begin{array}{l} \\ R_2-R_1 \\ 2R_1+R_3 \end{array}$$

$$\left[\begin{array}{ccc|c} 1 & -2 & 4 & 9 \\ 0 & 1 & 3 & -1 \\ 0 & 2 & 7 & 8 \end{array}\right]\begin{array}{l} \\ \frac{1}{3}R_2 \\ \\ \end{array}$$

$$\left[\begin{array}{ccc|c} 1 & -2 & 4 & 9 \\ 0 & 1 & 3 & -1 \\ 0 & 0 & 1 & 10 \end{array}\right]\begin{array}{l} \\ \\ R_3-2R_2 \end{array}$$

$$\left[\begin{array}{ccc|c} 1 & -2 & 0 & -31 \\ 0 & 1 & 0 & -31 \\ 0 & 0 & 1 & 10 \end{array}\right]\begin{array}{l} R_1-4R_3 \\ R_2-3R_3 \\ \\ \end{array}$$

$$\left[\begin{array}{ccc|c} 1 & 0 & 0 & -93 \\ 0 & 1 & 0 & -31 \\ 0 & 0 & 1 & 10 \end{array}\right]\begin{array}{l} R_1+2R_2 \\ \\ \\ \end{array}$$

The solution is (–93, –31, 10).

45. $x+3y+4z=14$
$2x-3y+2z=10$
$3x-\ \ y+\ z=9$
$4x+2y+5z=9$

The augmented matrix is

$$\left[\begin{array}{ccc|c} 1 & 3 & 4 & 14 \\ 2 & -3 & 2 & 10 \\ 3 & -1 & 1 & 9 \\ 4 & 2 & 5 & 9 \end{array}\right]$$

Now use the matrix method to solve the system.

$$\left[\begin{array}{ccc|c} 1 & 3 & 4 & 14 \\ 0 & -9 & -6 & -18 \\ 0 & -10 & -11 & -33 \\ 0 & -10 & -11 & -47 \end{array}\right] \begin{array}{l} \\ -2R_1+R_2 \\ -3R_1+R_3 \\ -4R_1+R_4 \end{array}$$

$$\left[\begin{array}{ccc|c} 1 & 3 & 4 & 14 \\ 0 & -9 & -6 & -18 \\ 0 & -10 & -11 & -33 \\ 0 & 0 & 0 & -14 \end{array}\right] \begin{array}{l} \\ \\ \\ -R_3+R_4 \end{array}$$

The last row is equivalent to the equation $0=-14$, which is false. Therefore the system is inconsistent and has no solution.

47. $x+8y+8z=8$
$3x-\ \ y+3z=5$
$-2x-4y-6z=5$

The augmented matrix is

$$\left[\begin{array}{ccc|c} 1 & 8 & 8 & 8 \\ 3 & -1 & 3 & 5 \\ -2 & -4 & -6 & 5 \end{array}\right]$$

Now use the matrix method to solve the system:

$$\left[\begin{array}{ccc|c} 1 & 8 & 8 & 8 \\ 0 & 25 & 21 & 19 \\ 0 & 12 & 10 & 21 \end{array}\right] \begin{array}{l} \\ 3R_1-R_2 \\ 2R_1+R_3 \end{array}$$

$$\left[\begin{array}{ccc|c} 1 & 8 & 8 & 8 \\ 0 & 1 & 0.84 & 0.76 \\ 0 & 12 & 10 & 21 \end{array}\right] \begin{array}{l} \\ 0.04R_2 \\ \\ \end{array}$$

$$\left[\begin{array}{ccc|c} 1 & 0 & 1.28 & 1.92 \\ 0 & 1 & 0.84 & 0.76 \\ 0 & 0 & 0.08 & -11.88 \end{array}\right] \begin{array}{l} R_1-8R_2 \\ \\ 12R_2-R_3 \end{array}$$

$$\left[\begin{array}{ccc|c} 1 & 0 & 1.28 & 1.92 \\ 0 & 1 & 0.84 & 0.76 \\ 0 & 0 & 1 & -148.5 \end{array}\right] \begin{array}{l} \\ \\ \frac{R_3}{.08} \end{array}$$

$$\left[\begin{array}{ccc|c} 1 & 0 & 0 & 192 \\ 0 & 1 & 0 & 125.5 \\ 0 & 0 & 1 & -148.5 \end{array}\right] \begin{array}{l} R_1-1.28R_3 \\ R_2-0.84R_3 \\ \\ \end{array}$$

The solution is (192, 125.5, –148.5).

49. $5x+3y+4z=19$
$3x-\ \ y+\ z=-4$

The augmented matrix is

$$\left[\begin{array}{ccc|c} 5 & 3 & 4 & 19 \\ 3 & -1 & 1 & -4 \end{array}\right]$$

Now use the matrix method to solve the system:

$$\left[\begin{array}{ccc|c} 14 & 0 & 7 & 7 \\ 3 & -1 & 1 & -4 \end{array}\right] \begin{array}{l} 3R_2+R_1 \\ \\ \end{array}$$

$$\left[\begin{array}{ccc|c} 1 & 0 & \frac{1}{2} & \frac{1}{2} \\ 3 & -1 & 1 & -4 \end{array}\right] \begin{array}{l} \frac{1}{14}R_1 \\ \\ \end{array}$$

Since there are only two equations, the system has an infinite number of solutions. Solve the equation represented by the first row of the matrix for x in terms of the parameter z.

$$x+\frac{1}{2}z=\frac{1}{2} \Rightarrow x=\frac{1}{2}-\frac{1}{2}z=\frac{1-z}{2}$$

Now substitute $\frac{1-z}{2}$ for x in the second equation and solve for y in terms of the parameter z.

(*continued next page*)

$$3\left(\frac{1-z}{2}\right)-y+z=-4$$
$$3\left(\frac{1-z}{2}\right)+z+4=y$$
$$\frac{3(1-z)+2z+8}{2}=y$$
$$\frac{3-3z+2z+8}{2}=y \Rightarrow \frac{11-z}{2}=y$$

The solution is $\left(\frac{1-z}{2}, \frac{11-z}{2}, z\right)$ for any real number z.

51. $x-2y+z=5$
$2x+y-z=2$
$-2x+4y-2z=2$

The matrix is

$$\left[\begin{array}{ccc|c} 1 & -2 & 1 & 5 \\ 2 & 1 & -1 & 2 \\ -2 & 4 & -2 & 2 \end{array}\right]$$

$$\left[\begin{array}{ccc|c} 1 & -2 & 1 & 5 \\ 0 & 5 & -3 & -8 \\ 0 & 0 & 0 & 12 \end{array}\right]\begin{array}{l} \\ -2R_1+R_2 \\ 2R_1+R_3 \end{array}$$

The third row represents the equation $0 = 12$, which is false. The system is inconsistent and has no solution.

53. $-8x-9y=11$
$24x+34y=2$
$16x+11y=-57$

The matrix is

$$\left[\begin{array}{cc|c} -8 & -9 & 11 \\ 24 & 34 & 2 \\ 16 & 11 & -57 \end{array}\right]$$

$$\left[\begin{array}{cc|c} -8 & -9 & 11 \\ 0 & 7 & 35 \\ 0 & -7 & -35 \end{array}\right]\begin{array}{l} \\ 3R_1+R_2 \\ 2R_1+R_3 \end{array}$$

$$\left[\begin{array}{cc|c} 1 & \frac{9}{8} & -\frac{11}{8} \\ 0 & 1 & 5 \\ 0 & 1 & 5 \end{array}\right]\begin{array}{l} -\frac{1}{8}R_1 \\ \frac{1}{7}R_2 \\ -\frac{1}{7}R_3 \end{array}$$

$$\left[\begin{array}{cc|c} 1 & 0 & -7 \\ 0 & 1 & 5 \\ 0 & 0 & 0 \end{array}\right]\begin{array}{l} -\frac{9}{8}R_2+R_1 \\ \\ -1R_2+R_3 \end{array}$$

The solution is $(-7, 5)$.

55. $x+2y=3$
$2x+3y=4$
$3x+4y=5$
$4x+5y=6$

The matrix is

$$\left[\begin{array}{cc|c} 1 & 2 & 3 \\ 2 & 3 & 4 \\ 3 & 4 & 5 \\ 4 & 5 & 6 \end{array}\right]$$

$$\left[\begin{array}{cc|c} 1 & 2 & 3 \\ 0 & -1 & -2 \\ 0 & -2 & -4 \\ 0 & -3 & -6 \end{array}\right]\begin{array}{l} \\ -2R_1+R_2 \\ -3R_1+R_3 \\ -4R_1+R_4 \end{array}$$

$$\left[\begin{array}{cc|c} 1 & 2 & 3 \\ 0 & 1 & 2 \\ 0 & -2 & -4 \\ 0 & -3 & -6 \end{array}\right]\begin{array}{l} \\ -R_2 \\ \\ \\ \end{array}$$

$$\left[\begin{array}{cc|c} 1 & 0 & -1 \\ 0 & 1 & 2 \\ 0 & 0 & 0 \\ 0 & 0 & 0 \end{array}\right]\begin{array}{l} R_1-2R_2 \\ \\ 2R_2+R_3 \\ 3R_2+R_4 \end{array}$$

The solution is $(-1, 2)$.

57. $x+y-z=-20$
$2x-y+z=11$

The matrix is

$$\left[\begin{array}{ccc|c} 1 & 1 & -1 & -20 \\ 2 & -1 & 1 & 11 \end{array}\right]$$

$$\left[\begin{array}{ccc|c} 1 & 1 & -1 & -20 \\ 0 & -3 & 3 & 51 \end{array}\right]\begin{array}{l} \\ -2R_1+R_2 \end{array}$$

$$\left[\begin{array}{ccc|c} 1 & 1 & -1 & -20 \\ 0 & 1 & -1 & -17 \end{array}\right]\begin{array}{l} \\ -\frac{1}{3}R_2 \end{array}$$

$$\left[\begin{array}{ccc|c} 1 & 0 & 0 & -3 \\ 0 & 1 & -1 & -17 \end{array}\right]\begin{array}{l} -R_2+R_1 \\ \\ \end{array}$$

This last matrix is the augmented matrix for the system

$x=-3$ (1)
$y-z=-17$ (2)

Solve equation (2) for y in terms of z. $y = z - 17$. The solution is all ordered triples of the form $(-3, z - 17, z)$ for any real number z.

59.
$$\begin{aligned} 2x + y + 3z - 2w &= -6 \\ 4x + 3y + z - w &= -2 \\ x + y + z + w &= -5 \\ -2x - 2y - 2z + 2w &= -10 \end{aligned}$$

The matrix is

$$\left[\begin{array}{cccc|c} 2 & 1 & 3 & -2 & -6 \\ 4 & 3 & 1 & -1 & -2 \\ 1 & 1 & 1 & 1 & -5 \\ -2 & -2 & -2 & 2 & -10 \end{array}\right]$$

Interchange R_1 and R_3.

$$\left[\begin{array}{cccc|c} 1 & 1 & 1 & 1 & -5 \\ 4 & 3 & 1 & -1 & -2 \\ 2 & 1 & 3 & -2 & -6 \\ -2 & -2 & -2 & 2 & -10 \end{array}\right]$$

$$\left[\begin{array}{cccc|c} 1 & 1 & 1 & 1 & -5 \\ 0 & -1 & -3 & -5 & 18 \\ 0 & -1 & 1 & -4 & 4 \\ 0 & 0 & 0 & 4 & -20 \end{array}\right] \begin{array}{l} \\ -4R_1 + R_2 \\ -2R_1 + R_3 \\ 2R_1 + R_4 \end{array}$$

$$\left[\begin{array}{cccc|c} 1 & 1 & 1 & 1 & -5 \\ 0 & 1 & 3 & 5 & -18 \\ 0 & -1 & 1 & -4 & 4 \\ 0 & 0 & 0 & 1 & -5 \end{array}\right] \begin{array}{l} \\ -R_2 \\ \\ \frac{1}{4}R_4 \end{array}$$

$$\left[\begin{array}{cccc|c} 1 & 0 & -2 & -4 & 13 \\ 0 & 1 & 3 & 5 & -18 \\ 0 & 0 & 4 & 1 & -14 \\ 0 & 0 & 0 & 1 & -5 \end{array}\right] \begin{array}{l} -R_2 + R_1 \\ \\ R_2 + R_3 \\ \\ \end{array}$$

$$\left[\begin{array}{cccc|c} 1 & 0 & -2 & -4 & 13 \\ 0 & 1 & 3 & 5 & -18 \\ 0 & 0 & 1 & \frac{1}{4} & -\frac{7}{2} \\ 0 & 0 & 0 & 1 & -5 \end{array}\right] \begin{array}{l} \\ \\ \frac{1}{4}R_3 \\ \\ \end{array}$$

$$\left[\begin{array}{cccc|c} 1 & 0 & -2 & -4 & 13 \\ 0 & 1 & 3 & 5 & -18 \\ 0 & 0 & 1 & 0 & -\frac{9}{4} \\ 0 & 0 & 0 & 1 & -5 \end{array}\right] \begin{array}{l} \\ \\ -\frac{1}{4}R_4 + R_3 \\ \\ \end{array}$$

$$\left[\begin{array}{cccc|c} 1 & 0 & -2 & 0 & -7 \\ 0 & 1 & 3 & 0 & 7 \\ 0 & 0 & 1 & 0 & -\frac{9}{4} \\ 0 & 0 & 0 & 1 & -5 \end{array}\right] \begin{array}{l} 4R_4 + R_1 \\ -5R_4 + R_2 \\ \\ \\ \end{array}$$

$$\left[\begin{array}{cccc|c} 1 & 0 & 0 & 0 & -\frac{23}{2} \\ 0 & 1 & 0 & 0 & \frac{55}{4} \\ 0 & 0 & 1 & 0 & -\frac{9}{4} \\ 0 & 0 & 0 & 1 & -5 \end{array}\right] \begin{array}{l} 2R_3 + R_1 \\ -3R_3 + R_2 \\ \\ \\ \end{array}$$

The solution is $(-11.5, 13.75, -2.25, -5)$.

61.
$$\begin{aligned} x + 2y - z \quad &= 3 \\ 3x + y + \quad w &= 4 \\ 2x - y + z + w &= 2 \end{aligned}$$

The matrix is

$$\left[\begin{array}{cccc|c} 1 & 2 & -1 & 0 & 3 \\ 3 & 1 & 0 & 1 & 4 \\ 2 & -1 & 1 & 1 & 2 \end{array}\right]$$

$$\left[\begin{array}{cccc|c} 1 & 2 & -1 & 0 & 3 \\ 0 & -5 & 3 & 1 & -5 \\ 0 & -5 & 3 & 1 & -4 \end{array}\right] \begin{array}{l} \\ -3R_1 + R_2 \\ -2R_1 + R_3 \end{array}$$

$$\left[\begin{array}{cccc|c} 1 & 2 & -1 & 0 & 3 \\ 0 & -5 & 3 & 1 & -5 \\ 0 & 0 & 0 & 0 & 1 \end{array}\right] \begin{array}{l} \\ \\ -1R_2 + R_3 \end{array}$$

Since the last row yields a false statement, there is no solution.

63. $\frac{3}{x}-\frac{1}{y}+\frac{4}{z}=-13$
$\frac{1}{x}+\frac{2}{y}-\frac{1}{z}=12$
$\frac{4}{x}-\frac{1}{y}+\frac{3}{z}=-7$

Let $u=\frac{1}{x}, v=\frac{1}{y}, w=\frac{1}{z}$. Then the system becomes

$$\begin{aligned}3u - v + 4w &= -13\\ u + 2v - w &= 12\\ 4u - v + 3w &= -7\end{aligned}$$

The matrix is

$$\left[\begin{array}{ccc|c}3 & -1 & 4 & -13\\ 1 & 2 & -1 & 12\\ 4 & -1 & 3 & -7\end{array}\right]$$

Interchange rows 1 and 2:

$$\left[\begin{array}{ccc|c}1 & 2 & -1 & 12\\ 3 & -1 & 4 & -13\\ 4 & -1 & 3 & -7\end{array}\right]$$

$$\left[\begin{array}{ccc|c}1 & 2 & -1 & 12\\ 0 & -7 & 7 & -49\\ 0 & -9 & 7 & -55\end{array}\right]\begin{matrix}\\ -3R_1+R_2\\ -4R_1+R_3\end{matrix}$$

$$\left[\begin{array}{ccc|c}1 & 2 & -1 & 12\\ 0 & 1 & -1 & 7\\ 0 & -9 & 7 & -55\end{array}\right]\begin{matrix}\\ -\frac{1}{7}R_2\\ \\ \end{matrix}$$

$$\left[\begin{array}{ccc|c}1 & 2 & -1 & 12\\ 0 & 1 & -1 & 7\\ 0 & 0 & -2 & 8\end{array}\right]\begin{matrix}\\ \\ 9R_2+R_3\end{matrix}$$

$$\left[\begin{array}{ccc|c}1 & 2 & -1 & 12\\ 0 & 1 & -1 & 7\\ 0 & 0 & 1 & -4\end{array}\right]\begin{matrix}\\ \\ -\frac{1}{2}R_3\end{matrix}$$

$$\left[\begin{array}{ccc|c}1 & 2 & 0 & 8\\ 0 & 1 & 0 & 3\\ 0 & 0 & 1 & -4\end{array}\right]\begin{matrix}R_1+R_3\\ R_2+R_3\\ \\ \end{matrix}$$

$$\left[\begin{array}{ccc|c}1 & 0 & 0 & 2\\ 0 & 1 & 0 & 3\\ 0 & 0 & 1 & -4\end{array}\right]\begin{matrix}-2R_2+R_1\\ \\ \\ \end{matrix}$$

Thus, $u = 2$, $v = 3$, and $w = -4$. Substituting, we have $2=\frac{1}{x}\Rightarrow x=\frac{1}{2}$, $3=\frac{1}{y}\Rightarrow y=\frac{1}{3}$, $-4=\frac{1}{z}\Rightarrow z=-\frac{1}{4}$. The solution is $\left(\frac{1}{2}, \frac{1}{3}, -\frac{1}{4}\right)$.

65. We are given the system

$$\begin{aligned}x+20y &= 102\\ -x+10y &= 48\\ y &= 5\end{aligned}.$$

The associated matrix is

$$\left[\begin{array}{cc|c}1 & 20 & 102\\ -1 & 10 & 48\\ 0 & 1 & 5\end{array}\right].$$

$$\left[\begin{array}{cc|c}1 & 0 & 2\\ -1 & 0 & -2\\ 0 & 1 & 5\end{array}\right]\begin{matrix}-20R_3+R_1\\ -10R_3+R_2\\ \\ \end{matrix}$$

$$\left[\begin{array}{cc|c}1 & 0 & 2\\ 1 & 0 & 2\\ 0 & 1 & 5\end{array}\right]\begin{matrix}\\ -R_2\\ \\ \end{matrix}$$

Thus, $x = 2$ and $y = 5$.
All three counties had the same population in $2010 + 2 = 2012$. The population was 5000.

67.
$$\begin{aligned} x + 1.25y + .25z &= 457.5 \\ x + \ \ .6y + \ .4z &= 390 \\ 3.16x + 3.48y + \ .4z &= 1297.2 \end{aligned}$$

The associated matrix is

$$\left[\begin{array}{ccc|c} 1 & 1.25 & .25 & 457.5 \\ 1 & .6 & .4 & 390 \\ 3.16 & 3.48 & .4 & 1297.2 \end{array}\right].$$

$$\left[\begin{array}{ccc|c} 1 & 1.25 & .25 & 457.5 \\ 0 & .65 & -.15 & 67.5 \\ 0 & -.47 & -.39 & -148.5 \end{array}\right] \begin{array}{l} \\ -R_2 + R_1 \\ -3.16R_1 + R_3 \end{array}$$

$$\left[\begin{array}{ccc|c} 1 & 1.25 & .25 & 457.5 \\ 0 & 1 & -\frac{3}{13} & \frac{1350}{13} \\ 0 & -.47 & -.39 & -148.5 \end{array}\right] \begin{array}{l} \\ \frac{1}{.65}R_2 \\ \\ \end{array}$$

At this point, it is probably easier to convert to fractions.

$$\left[\begin{array}{ccc|c} 1 & \frac{5}{4} & \frac{1}{4} & \frac{4575}{10} \\ 0 & 1 & -\frac{3}{13} & \frac{1350}{13} \\ 0 & -\frac{47}{100} & -\frac{39}{100} & -\frac{1485}{10} \end{array}\right]$$

$$\left[\begin{array}{ccc|c} 1 & \frac{5}{4} & \frac{1}{4} & \frac{4575}{10} \\ 0 & 1 & -\frac{3}{13} & \frac{1350}{13} \\ 0 & 0 & -\frac{162}{325} & -\frac{1296}{13} \end{array}\right] \begin{array}{l} \\ \\ \frac{47}{100}R_2 + R_3 \end{array}$$

$$\left[\begin{array}{ccc|c} 1 & 0 & \frac{7}{13} & \frac{4260}{13} \\ 0 & 1 & -\frac{3}{13} & \frac{1350}{13} \\ 0 & 0 & 1 & 200 \end{array}\right] \begin{array}{l} -\frac{5}{4}R_2 + R_1 \\ \\ -\frac{325}{162}R_3 \end{array}$$

$$\left[\begin{array}{ccc|c} 1 & 0 & 0 & 220 \\ 0 & 1 & 0 & 150 \\ 0 & 0 & 1 & 200 \end{array}\right] \begin{array}{l} -\frac{7}{13}R_3 + R_1 \\ \frac{3}{13}R_3 + R_2 \\ \\ \end{array}$$

The TI-84 Plus calculator solution is shown below.

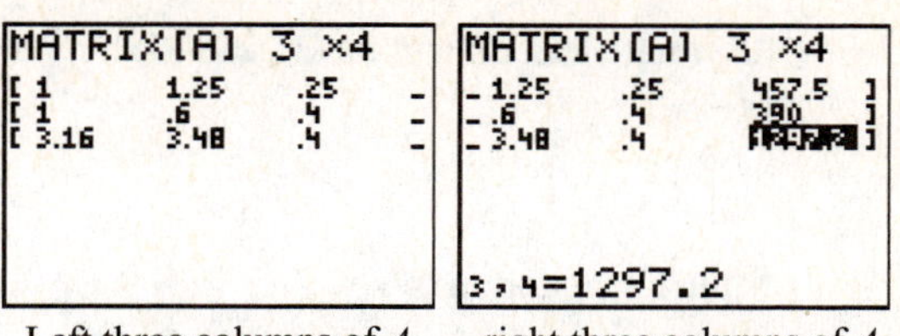

Left three columns of A right three columns of A

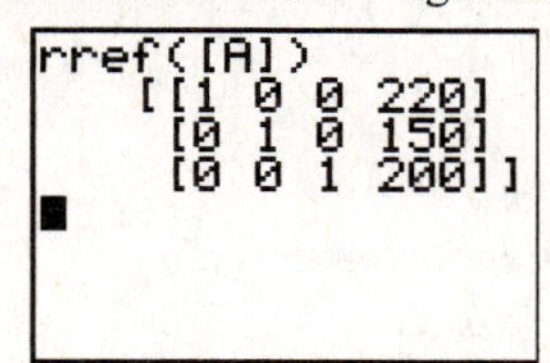

There were 220 adults, 150 teenagers, and 200 preteen children present.

69.

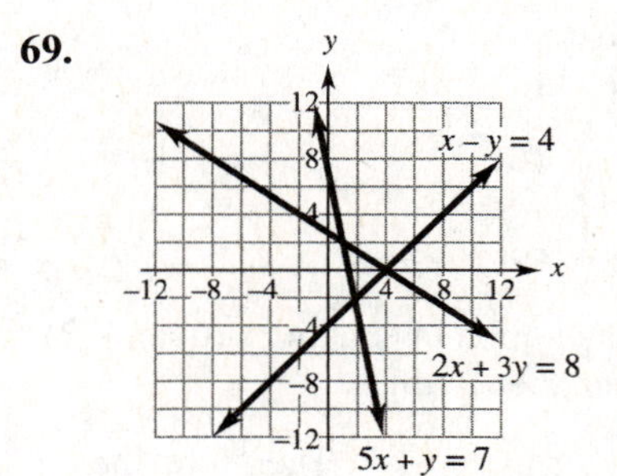

Answers vary. Sample answer: The three lines do not have one common intersection point.

71. $y = ax^2 + bx + c$

Since (2, 3), (–1, 0), and (–2, 2) satisfy the above equation, the following system of equations can be obtained.

$$\begin{aligned} 3 &= a(2)^2 + b(2) + c & (1) \\ 0 &= a(-1)^2 + b(-1) + c & (2) \\ 2 &= a(-2)^2 + b(-2) + c & (3) \end{aligned}$$

The system simplifies to

$$\begin{aligned} 4a + 2b + c &= 3 & (1) \\ a - b + c &= 0 & (2) \\ 4a - 2b + c &= 2. & (3) \end{aligned}$$

Interchange equations (2) and (1).

$$\begin{aligned} a - b + c &= 0 & (2) \\ 4a + 2b + c &= 3 & (1) \\ 4a - 2b + c &= 2 & (3) \end{aligned}$$

(*continued next page*)

Eliminate a in equations (1) and (3).

$$\begin{aligned} a-b+c&=0 \quad (2)\\ 6b-3c&=3 \quad (4)\\ 2b-3c&=2 \quad (5) \end{aligned}$$

Multiply equation (4) by $\frac{1}{6}$.

$$\begin{aligned} a-b+c&=0 \quad (2)\\ b-\frac{1}{2}c&=\frac{1}{2} \quad (6)\\ 2b-3c&=2 \quad (5) \end{aligned}$$

Eliminate b in equation (5).

$$\begin{aligned} a-b+c&=0 \quad (2)\\ b-\frac{1}{2}c&=\frac{1}{2} \quad (6)\\ -2c&=1 \quad (7) \end{aligned}$$

Solve for c in equation (7).

$$\begin{aligned} a-b+c&=0 \quad (2)\\ b-\frac{1}{2}c&=\frac{1}{2} \quad (6)\\ c&=-\frac{1}{2} \quad (8) \end{aligned}$$

Use back-substitution in equations (6) and (2) to solve for b and then a. thus,

$a=\frac{3}{4}$, $b=\frac{1}{4}$, and $c=-\frac{1}{2}$. Therefore, the equation is $y=.75x^2+.25x-.5$.

Section 6.3 Applications of Systems of Linear Equations

For these exercises, we show just one of the various solution methods that could be used.

1. From example 1, let x = the number of vans, let y = the number of small trucks, and let z = the number of large trucks. We have the system

$$\begin{aligned} x+\quad y+\quad z&=175 \quad (1)\\ 25{,}000x+30{,}000y+40{,}000z&=5{,}000{,}000 \quad (2)\\ x-\quad 2y\quad &=0 \quad (3) \end{aligned}$$

We divide equation (2) by 5000, so the system becomes

$$\begin{aligned} x+\ y+\ z&=175\\ 5x+6y+8z&=1000\\ x-\ 2y\quad&=0 \end{aligned}$$

Form the augmented matrix and transform it into reduced row echelon form.

$$\left[\begin{array}{ccc|c} 1 & 1 & 1 & 175\\ 5 & 6 & 8 & 1000\\ 1 & -2 & 0 & 0 \end{array}\right]$$

$$\left[\begin{array}{ccc|c} 1 & 1 & 1 & 175\\ 0 & 1 & 3 & 125\\ 0 & 3 & 1 & 175 \end{array}\right] \begin{array}{l} \\ -5R_1+R_2\\ -R_3+R_1 \end{array}$$

$$\left[\begin{array}{ccc|c} 1 & 1 & 1 & 175\\ 0 & 1 & 3 & 125\\ 0 & 0 & -8 & -200 \end{array}\right] \begin{array}{l} \\ \\ -3R_2+R_3 \end{array}$$

$$\left[\begin{array}{ccc|c} 1 & 1 & 1 & 175\\ 0 & 1 & 3 & 125\\ 0 & 0 & 1 & 25 \end{array}\right] \begin{array}{l} \\ \\ -\frac{5}{31}R_3 \end{array}$$

$$\left[\begin{array}{ccc|c} 1 & 1 & 0 & 150\\ 0 & 1 & 0 & 50\\ 0 & 0 & 1 & 25 \end{array}\right] \begin{array}{l} -R_3+R_1\\ -3R_3+R_2\\ \\ \end{array}$$

$$\left[\begin{array}{ccc|c} 1 & 0 & 0 & 100\\ 0 & 1 & 0 & 50\\ 0 & 0 & 1 & 25 \end{array}\right] \begin{array}{l} -R_3+R_1\\ \\ \\ \end{array}$$

The company should buy 100 vans, 50 small trucks, and 25 large trucks.

3. From example 3, let x = the number of units of corn, let y = the number of units of soybeans, and let z = the number of units of cottonseed. We have the system

$$\begin{aligned} 10x+20y+30z&=1800\\ 30x+20y+40z&=2400\\ 20x+40y+25z&=2200 \end{aligned}$$

Form the augmented matrix and transform it into reduced row echelon form.

$$\left[\begin{array}{ccc|c} 10 & 20 & 30 & 1800\\ 30 & 20 & 40 & 2400\\ 20 & 40 & 25 & 2200 \end{array}\right]$$

$$\left[\begin{array}{ccc|c} 10 & 20 & 30 & 1800\\ 0 & 40 & 50 & 3000\\ 0 & 0 & 35 & 1400 \end{array}\right] \begin{array}{l} \\ 3R_1-R_2\\ 2R_1-R_3 \end{array}$$

(*continued next page*)

$$\left[\begin{array}{ccc|c} 1 & 2 & 3 & 180 \\ 0 & 40 & 50 & 3000 \\ 0 & 0 & 1 & 40 \end{array}\right] \begin{array}{l} \frac{1}{10}R_1 \\ \\ \frac{1}{35}R_3 \end{array}$$

$$\left[\begin{array}{ccc|c} 1 & 2 & 0 & 60 \\ 0 & 40 & 0 & 1000 \\ 0 & 0 & 1 & 40 \end{array}\right] \begin{array}{l} -3R_3 + R_1 \\ -50R_3 + R_2 \\ \\ \end{array}$$

$$\left[\begin{array}{ccc|c} 1 & 2 & 0 & 60 \\ 0 & 1 & 0 & 25 \\ 0 & 0 & 1 & 40 \end{array}\right] \begin{array}{l} \\ \frac{1}{40}R_2 \\ \\ \end{array}$$

$$\left[\begin{array}{ccc|c} 1 & 0 & 0 & 10 \\ 0 & 1 & 0 & 25 \\ 0 & 0 & 1 & 40 \end{array}\right] \begin{array}{l} -2R_2 + R_1 \\ \\ \\ \end{array}$$

The feed should contain 10 units of corn, 25 units of soybeans, and 40 units of cottonseed.

5. Let x = the amount charged for a shirt, let y = the amount charged for a pair of slack, and let z = the amount charged for a sport coat. We have the system

$$\begin{aligned} 3x + y \quad &= 10.96 \\ 7x + 2y + z &= 30.40 \\ 4x + \quad z &= 14.45 \end{aligned}$$

which has the augmented matrix

$$\left[\begin{array}{ccc|c} 3 & 1 & 0 & 10.96 \\ 7 & 2 & 1 & 30.40 \\ 4 & 0 & 1 & 14.45 \end{array}\right].$$

$$\left[\begin{array}{ccc|c} 4 & 0 & 1 & 14.45 \\ 7 & 2 & 1 & 30.40 \\ 3 & 1 & 0 & 10.96 \end{array}\right] \text{ interchange } R_1 \text{ and } R_3$$

$$\left[\begin{array}{ccc|c} 1 & 0 & \frac{1}{4} & 3.6125 \\ 7 & 2 & 1 & 30.40 \\ 3 & 1 & 0 & 10.96 \end{array}\right] \begin{array}{l} \frac{1}{4}R_1 \\ \\ \\ \end{array}$$

$$\left[\begin{array}{ccc|c} 1 & 0 & \frac{1}{4} & 3.6125 \\ 0 & 2 & -\frac{3}{4} & 5.1125 \\ 0 & 1 & -\frac{3}{4} & .1225 \end{array}\right] \begin{array}{l} \\ -7R_1 + R_2 \\ -3R_1 + R_3 \end{array}$$

$$\left[\begin{array}{ccc|c} 1 & 0 & \frac{1}{4} & 3.6125 \\ 0 & 1 & 0 & 4.99 \\ 0 & 1 & -\frac{3}{4} & .1225 \end{array}\right] \begin{array}{l} \\ R_2 - R_3 \\ \\ \end{array}$$

$$\left[\begin{array}{ccc|c} 1 & 0 & \frac{1}{4} & 3.6125 \\ 0 & 1 & 0 & 4.99 \\ 0 & 0 & 1 & 6.49 \end{array}\right] \begin{array}{l} \\ \\ \frac{4}{3}(R_2 - R_3) \end{array}$$

$$\left[\begin{array}{ccc|c} 1 & 0 & 0 & 1.99 \\ 0 & 1 & 0 & 4.99 \\ 0 & 0 & 1 & 6.49 \end{array}\right] \begin{array}{l} R_1 - \frac{1}{4}R_2 \\ \\ \\ \end{array}$$

Shirts cost \$1.99 each, slacks cost \$4.99 each, and sports coats cost \$6.49 each.

7. Let x = the number of adult tickets, let y = the number of teenager tickets, let z = the number of preteen tickets. Then we have

$$\begin{aligned} x + y + z &= 570 \quad (1) \\ 5x + 3y + 2z &= 1950 \quad (2). \\ \frac{3}{4}z &= y \quad (3) \end{aligned}$$

Using substitution, we have

$$x + \frac{3}{4}z + z = 570 \quad (4)$$

$$5x + 3\left(\frac{3}{4}z\right) + 2z = 1950 \quad (5)$$

Clear the fractions and combine like terms.

$$\begin{aligned} 4x + 7z &= 2280 \quad (6) \\ 20x + 17z &= 7800 \quad (7) \end{aligned}$$

Multiply (6) by –5, then add the result to (7) and solve for z.

$$\begin{aligned} -20x - 35z &= -11400 \\ 20x + 17z &= \quad 7800 \\ \hline -18z &= -3600 \Rightarrow z = 200 \end{aligned}$$

Substitute $z = 200$ into equation (6) and solve for x, and into equation (3) to solve for y.

$$4x + 7(200) = 2280 \Rightarrow 4x = 880 \Rightarrow x = 220$$

$$y = \frac{3}{4}(200) = 150$$

There were 220 adult tickets, 150 teenager tickets, and 200 preteen tickets sold.

9. Let x = investment in B bonds, and let $2x$ = investment in AAA bonds. Then, $30{,}000 - 3x$ = the investment in A bonds. The sum of the interests of the three is \$2000. Thus, we have

$$0.10(x) + 0.06(30{,}000 - 3x) + 0.05(2x) = 2000.00$$
$$0.10x + 1800 - 0.18x + 0.10x = 2000.00$$
$$0.02x + 1800 = 2000 \Rightarrow 0.02x = 200 \Rightarrow$$
$$x = 10{,}000; \quad 2x = 20{,}000; \quad 30{,}000 - 3x = 0$$

\$10,000 was invested in B bonds, \$20,000 was invested in the AAA bonds, and no money was invested in A bonds.

11. Let x = the number of pounds of pretzels, y = the number of pounds of dried fruit, and z = the number of pounds of nuts.
The system of equations is

$$\begin{aligned} x+y+z&=140\\ 3x+4y+8z&=6(140)\\ x&=2y. \end{aligned}$$

The system simplifies to

$$\begin{aligned} x+y+z&=140\\ 3x+4y+8z&=840\\ x-2y&=0. \end{aligned}$$

The matrix is

$$\left[\begin{array}{ccc|c} 1 & 1 & 1 & 140\\ 3 & 4 & 8 & 840\\ 1 & -2 & 0 & 0 \end{array}\right]$$

$$\left[\begin{array}{ccc|c} 1 & 1 & 1 & 140\\ 0 & 1 & 5 & 420\\ 0 & -3 & -1 & -140 \end{array}\right]\begin{array}{l} \\ -3R_1+R_2\\ -R_1+R_3 \end{array}$$

$$\left[\begin{array}{ccc|c} 1 & 0 & -4 & -280\\ 0 & 1 & 5 & 420\\ 0 & 0 & 14 & 1120 \end{array}\right]\begin{array}{l} -R_2+R_1\\ \\ 3R_2+R_3 \end{array}$$

$$\left[\begin{array}{ccc|c} 1 & 0 & -4 & -280\\ 0 & 1 & 5 & 420\\ 0 & 0 & 1 & 80 \end{array}\right]\begin{array}{l} \\ \\ \frac{1}{14}R_3 \end{array}$$

$$\left[\begin{array}{ccc|c} 1 & 0 & 0 & 40\\ 0 & 1 & 0 & 20\\ 0 & 0 & 1 & 80 \end{array}\right]\begin{array}{l} 4R_3+R_1\\ -5R_3+R_2\\ \\ \end{array}$$

Use 40 lb of pretzels, 20 lb of dried fruit, and 80 lb of nuts.

13. Let x = the number of cases of Brand A, let y = the number of cases of Brand B, let z = the number of cases of Brand C, and let w = the number of cases of Brand D.
The system of equations is

$$\begin{aligned} 25x+50y+75z+100w&=1200\\ 30x+30y+30z+60w&=600\\ 30x+20y+20z+30w&=400. \end{aligned}$$

This system can be simplified to

$$\begin{aligned} x+2x+3z+4w&=48\\ x+y+z+2w&=20\\ 3x+2y+2z+3w&=40. \end{aligned}$$

The matrix is

$$\left[\begin{array}{cccc|c} 1 & 2 & 3 & 4 & 48\\ 1 & 1 & 1 & 2 & 20\\ 3 & 2 & 2 & 3 & 40 \end{array}\right]$$

$$\left[\begin{array}{cccc|c} 1 & 2 & 3 & 4 & 48\\ 0 & -1 & -2 & -2 & -28\\ 0 & -4 & -7 & -9 & -104 \end{array}\right]\begin{array}{l} \\ -R_1+R_2\\ -3R_1+R_3 \end{array}$$

$$\left[\begin{array}{cccc|c} 1 & 2 & 3 & 4 & 48\\ 0 & 1 & 2 & 2 & 28\\ 0 & -4 & -7 & -9 & -104 \end{array}\right]\begin{array}{l} \\ -R_2\\ \\ \end{array}$$

$$\left[\begin{array}{cccc|c} 1 & 0 & -1 & 0 & -8\\ 0 & 1 & 2 & 2 & 28\\ 0 & 0 & 1 & -1 & 8 \end{array}\right]\begin{array}{l} -2R_2+R_1\\ \\ 4R_2+R_3 \end{array}$$

$$\left[\begin{array}{cccc|c} 1 & 0 & 0 & -1 & 0\\ 0 & 1 & 0 & 4 & 12\\ 0 & 0 & 1 & -1 & 8 \end{array}\right]\begin{array}{l} R_3+R_1\\ -2R_3+R_2\\ \\ \end{array}$$

This last matrix is the augmented matrix for the dependent system

$$\begin{aligned} x-w&=0 \quad (1)\\ y+4w&=12 \quad (2)\\ z-w&=8 \quad (3) \end{aligned}$$

Solve equations (1), (2), and (3) in terms of w.

$$\begin{aligned} x&=w \quad &(1)\\ y&=-4w+12 \quad &(2)\\ z&=w+8 \quad &(3) \end{aligned}$$

The solution is all points of the form $(w, -4w+12, w+8, w)$. But in order for y to be positive, w has to be 0, 1, 2, or 3. So, there are four possibilities, as follows:
0 cases of A, 12 cases of B, 8 cases of C, 0 cases of D;
1 case of A, 8 cases of B, 9 cases of C, 1 case of D;
2 cases of A, 4 cases of B, 10 cases of C, 2 cases of D;
3 cases of A, 0 cases of B, 11 cases of C, 3 cases of D.

15. Let x = amount in mutual fund
let y = amount in corporate bonds
let z = amount in fast food franchise
$x+y+z=x+2x+z=3x+z=70{,}000$
$.02x+.06(2x)+.1z=.14x+.1z=4800$
The matrix is

$$\left[\begin{array}{cc|c} 3 & 1 & 70{,}000 \\ .14 & .1 & 4800 \end{array}\right]$$

$$\left[\begin{array}{cc|c} 3 & 1 & 70{,}000 \\ 14 & 10 & 480{,}000 \end{array}\right]$$

$$\left[\begin{array}{cc|c} 1 & \frac{1}{3} & \frac{70{,}000}{3} \\ 14 & 10 & 480{,}000 \end{array}\right] \frac{1}{3}R_1$$

$$\left[\begin{array}{cc|c} 1 & \frac{1}{3} & \frac{70{,}000}{3} \\ 0 & \frac{16}{3} & \frac{460{,}000}{3} \end{array}\right] -14R_1+R_2$$

$$\left[\begin{array}{cc|c} 1 & \frac{1}{3} & \frac{70{,}000}{3} \\ 0 & 1 & 28{,}750 \end{array}\right] \frac{3}{16}R_2$$

$$\left[\begin{array}{cc|c} 1 & 0 & 13{,}750 \\ 0 & 1 & 28{,}750 \end{array}\right] -\frac{1}{3}R_2+R_1$$

Therefore, she should invest \$13,750 in the mutual fund, \$27,500 in bonds, and \$28,750 in the food franchise.

17. a. The attenuation value for beam 3 is $b+c$.

b. The system of equations is
$a+b=.8$
$a+c=.55$
$b+c=.65$

The augmented matrix is

$$\left[\begin{array}{ccc|c} 1 & 1 & 0 & .8 \\ 1 & 0 & 1 & .55 \\ 0 & 1 & 1 & .65 \end{array}\right]$$

$$\left[\begin{array}{ccc|c} 1 & 1 & 0 & .8 \\ 0 & 1 & -1 & .25 \\ 0 & 1 & 1 & .65 \end{array}\right] -R_2+R_1$$

$$\left[\begin{array}{ccc|c} 1 & 1 & 0 & .8 \\ 0 & 1 & 0 & .45 \\ 0 & 1 & 1 & .65 \end{array}\right] \frac{1}{2}(R_2+R_3)$$

$$\left[\begin{array}{ccc|c} 1 & 0 & 0 & .35 \\ 0 & 1 & 0 & .45 \\ 0 & 0 & 1 & .2 \end{array}\right] \begin{array}{l} R_1-R_2 \\ \\ R_3-R_2 \end{array}$$

So $a=.35$, $b=.45$, and $c=.2$. Therefore, A is tumorous, B is bone, and C is healthy tissue.

19. a. For intersection C, x_2 cars leave on 11th street and x_3 leave on N street. The number of cars entering C must equal the number leaving, so that
$x_2+x_3=300+400$
$x_2+x_3=700$

For intersection D, x_3 cars enter on N street and x_4 cars enter on 10th street. The figure shows that 200 cars leave D on N street and 400 on 10th street.
$x_3+x_4=200+400$
$x_3+x_4=600$

b. The system of equations is
$x_1+x_4=1000$
$x_1+x_2=1100$
$x_2+x_3=700$
$x_3+x_4=600$
Solve each equation in terms of x_4.
$x_1=1000-x_4$
$x_3=600-x_4$
$x_2=1100-(1000-x_4)=100+x_4$
The solution is
$(1000-x_4, 100+x_4, 600-x_4, x_4)$.

c. The largest possible value is 600 and the smallest possible value is 0 for x_4, the number of cars leaving intersection A on 10th street.

d. For x_1 : the largest possible value: 1000
smallest possible value: 400

x_2 : largest possible value: 700
smallest possible value: 100

x_3 : largest possible value: 600
smallest possible value: 0

e. If you know the number of cars entering or leaving three of the intersections, then the number entering or leaving the fourth is automatically determined, because the number leaving must equal the number entering.

21. Let x = number of cups of Hearty Chicken Rotini let y = number of cups of Hearty Chicken, and let z = number of cups of Chunky Chicken Noodle

The system of equations is

$$\begin{aligned} 100x+130y+110z &= 1710 \\ 960x+480y+890z &= 11,580 \\ 7x+9y+8z &= 121 \end{aligned}$$

After simplifying the equations, the matrix is

$$\left[\begin{array}{ccc|c} 10 & 13 & 11 & 171 \\ 96 & 48 & 89 & 1158 \\ 7 & 9 & 8 & 121 \end{array}\right]$$

$$\left[\begin{array}{ccc|c} 1 & \frac{13}{10} & \frac{11}{10} & \frac{171}{10} \\ 96 & 48 & 89 & 1158 \\ 7 & 9 & 8 & 121 \end{array}\right] \begin{array}{l} \frac{1}{10}R_1 \\ \\ \\ \end{array}$$

$$\left[\begin{array}{ccc|c} 1 & \frac{13}{10} & \frac{11}{10} & \frac{171}{10} \\ 0 & -\frac{384}{5} & -\frac{83}{5} & -\frac{2418}{5} \\ 0 & -\frac{1}{10} & \frac{3}{10} & \frac{13}{10} \end{array}\right] \begin{array}{l} \\ -96R_1+R_3 \\ -7R_1+R_3 \end{array}$$

$$\left[\begin{array}{ccc|c} 1 & \frac{13}{10} & \frac{11}{10} & \frac{171}{10} \\ 0 & -\frac{384}{5} & -\frac{83}{5} & -\frac{2418}{5} \\ 0 & 1 & -3 & -13 \end{array}\right] \begin{array}{l} \\ \\ -10R_3 \end{array}$$

Interchange R_3 and R_2.

$$\left[\begin{array}{ccc|c} 1 & \frac{13}{10} & \frac{11}{10} & \frac{171}{10} \\ 0 & 1 & -3 & -13 \\ 0 & -\frac{384}{5} & -\frac{83}{5} & -\frac{2418}{5} \end{array}\right]$$

$$\left[\begin{array}{ccc|c} 1 & 0 & 5 & 34 \\ 0 & 1 & -3 & -13 \\ 0 & 0 & -247 & -1482 \end{array}\right] \begin{array}{l} -\frac{13}{10}R_2+R_1 \\ \\ \frac{384}{5}R_2+R_3 \end{array}$$

$$\left[\begin{array}{ccc|c} 1 & 0 & 5 & 34 \\ 0 & 1 & -3 & -13 \\ 0 & 0 & 1 & 6 \end{array}\right] \begin{array}{l} \\ \\ -\frac{1}{247}R_3 \end{array}$$

$$\left[\begin{array}{ccc|c} 1 & 0 & 0 & 4 \\ 0 & 1 & 0 & 5 \\ 0 & 0 & 1 & 6 \end{array}\right] \begin{array}{l} R_1-5R_3 \\ R_2+3R_3 \\ \\ \end{array}$$

Therefore, use 4 cups of Hearty Chicken Rotini, 5 cups of Hearty Chicken, and 6 cups of Chunky Chicken Noodle.

$$\text{Serving size} = \frac{4+5+6}{10} = 1.5 \text{ cups}$$

23. Let x = the amount invested in AAA bonds; let y = the amount invested in A bonds; and let z = the amount invested in B bonds.

a. Solve the system

$$\begin{aligned} x+y+z &= 25,000 && (1) \\ .06x+.07y+.1z &= 1810 && (2) \\ x &= 2z. && (3) \end{aligned}$$

Equation (3) should be rewritten so that the system becomes

$$\begin{aligned} x+y+z &= 25,000 && (1) \\ .06x+.07y+.1z &= 1810 && (2) \\ x-2z &= 0. && (3) \end{aligned}$$

The matrix is

$$\left[\begin{array}{ccc|c} 1 & 1 & 1 & 25,000 \\ .06 & .07 & .01 & 1810 \\ 1 & 0 & -2 & 0 \end{array}\right]$$

$$\left[\begin{array}{ccc|c} 1 & 1 & 1 & 25,000 \\ 6 & 7 & 10 & 181,000 \\ 1 & 0 & -2 & 0 \end{array}\right] \begin{array}{l} \\ 100R_2 \\ \\ \end{array}$$

$$\left[\begin{array}{ccc|c} 1 & 1 & 1 & 25,000 \\ 0 & 1 & 4 & 31,000 \\ 0 & -1 & -3 & -25,000 \end{array}\right] \begin{array}{l} \\ -6R_1+R_2 \\ -R_1+R_3 \end{array}$$

(continued next page)

$$\left[\begin{array}{ccc|c} 1 & 0 & -3 & -6000 \\ 0 & 1 & 4 & 31,000 \\ 0 & 0 & 1 & 6000 \end{array}\right] \begin{array}{l} -R_2 + R_1 \\ \\ R_2 + R_3 \end{array}$$

$$\left[\begin{array}{ccc|c} 1 & 0 & 0 & 12,000 \\ 0 & 1 & 0 & 7000 \\ 0 & 0 & 1 & 6000 \end{array}\right] \begin{array}{l} 3R_3 + R_1 \\ -4R_3 + R_2 \\ \\ \end{array}$$

The client should invest $12,000 in AAA bonds at 6%, $7000 in A bonds at 7%, and $6000 in B bonds at 10%.

b. The new system is

$$\begin{aligned} x + y + z &= 30,000 \\ .06x + .07y + .1z &= 2150 \\ x - 2z &= 0. \end{aligned}$$

The matrix is

$$\left[\begin{array}{ccc|c} 1 & 1 & 1 & 30,000 \\ .06 & .07 & .1 & 2150 \\ 1 & 0 & -2 & 0 \end{array}\right]$$

$$\left[\begin{array}{ccc|c} 1 & 1 & 1 & 30,000 \\ 6 & 7 & 10 & 215,000 \\ 1 & 0 & -2 & 0 \end{array}\right] \begin{array}{l} \\ 100R_2 \\ \\ \end{array}$$

$$\left[\begin{array}{ccc|c} 1 & 1 & 1 & 30,000 \\ 0 & 1 & 4 & 35,000 \\ 0 & -1 & -3 & -30,000 \end{array}\right] \begin{array}{l} \\ -6R_1 + R_2 \\ -R_1 + R_3 \end{array}$$

$$\left[\begin{array}{ccc|c} 1 & 0 & -3 & -5000 \\ 0 & 1 & 4 & 35,000 \\ 0 & 0 & 1 & 5000 \end{array}\right] \begin{array}{l} -R_2 + R_1 \\ \\ R_2 + R_3 \end{array}$$

$$\left[\begin{array}{ccc|c} 1 & 0 & 0 & 10,000 \\ 0 & 1 & 0 & 15,000 \\ 0 & 0 & 1 & 5000 \end{array}\right] \begin{array}{l} 3R_3 + R_1 \\ -4R_3 + R_2 \\ \\ \end{array}$$

The client should invest $10,000 in AAA bonds at 6%, $15,000 in A bonds at 7%, and $5000 in B bonds at 10%.

c. The new system is

$$\begin{aligned} x + y + z &= 40,000 \\ .06x + .07y + .1z &= 2900 \\ x - 2z &= 0. \end{aligned}$$

The matrix is

$$\left[\begin{array}{ccc|c} 1 & 1 & 1 & 40,000 \\ .06 & .07 & .1 & 2900 \\ 1 & 0 & -2 & 0 \end{array}\right]$$

$$\left[\begin{array}{ccc|c} 1 & 1 & 1 & 40,000 \\ 6 & 7 & 10 & 290,000 \\ 1 & 0 & -2 & 0 \end{array}\right] \begin{array}{l} \\ 100R_2 \\ \\ \end{array}$$

$$\left[\begin{array}{ccc|c} 1 & 1 & 1 & 40,000 \\ 0 & 1 & 4 & 50,000 \\ 0 & -1 & -3 & -40,000 \end{array}\right] \begin{array}{l} \\ -6R_1 + R_2 \\ -R_1 + R_3 \end{array}$$

$$\left[\begin{array}{ccc|c} 1 & 0 & -3 & -10,000 \\ 0 & 1 & 4 & 50,000 \\ 0 & 0 & 1 & 10,000 \end{array}\right] \begin{array}{l} -R_2 + R_1 \\ \\ R_2 + R_3 \end{array}$$

$$\left[\begin{array}{ccc|c} 1 & 0 & 0 & 20,000 \\ 0 & 1 & 0 & 10,000 \\ 0 & 0 & 1 & 10,000 \end{array}\right] \begin{array}{l} 3R_3 + R_1 \\ -4R_3 + R_2 \\ \\ \end{array}$$

The client should invest $20,000 in AAA bonds at 6%, $10,000 in A bonds at 7%, and $10,000 in B bonds at 10%.

25. **a.** $f(x) = ax^2 + bx + c$

We have the system

$$163 = a(15)^2 + b(15) + c$$
$$187 = a(35)^2 + b(35) + c\ .$$
$$192 = a(40)^2 + b(40) + c$$

This becomes

$$225a + 15b + c = 163$$
$$1225a + 35b + c = 187\ .$$
$$1600a + 40b + c = 192$$

The augmented matrix is

$$\left[\begin{array}{ccc|c} 225 & 15 & 1 & 163 \\ 1225 & 35 & 1 & 187 \\ 1600 & 40 & 1 & 192 \end{array}\right].$$

$$\left[\begin{array}{ccc|c} 1375 & 25 & 0 & 29 \\ 375 & 5 & 0 & 5 \\ 1600 & 40 & 1 & 192 \end{array}\right] \begin{array}{l} -R_1 + R_3 \\ -R_2 + R_3 \\ \\ \end{array}$$

$$\left[\begin{array}{ccc|c} 1375 & 25 & 0 & 29 \\ 75 & 1 & 0 & 1 \\ 1600 & 40 & 1 & 192 \end{array}\right] \begin{array}{l} \\ \frac{1}{5}R_2 \\ \\ \end{array}$$

$$\left[\begin{array}{ccc|c} -500 & 0 & 0 & 4 \\ 75 & 1 & 0 & 1 \\ -1400 & 0 & 1 & 152 \end{array}\right] \begin{array}{l} -25R_2 + R_1 \\ \\ R_3 - 40R_2 \end{array}$$

$$\left[\begin{array}{ccc|c} 1 & 0 & 0 & -0.008 \\ 75 & 1 & 0 & 1 \\ -1400 & 0 & 1 & 152 \end{array}\right] \begin{array}{l} -\frac{1}{500}R_1 \\ \\ \\ \end{array}$$

$$\left[\begin{array}{ccc|c} 1 & 0 & 0 & -0.008 \\ 0 & 1 & 0 & 1.6 \\ 0 & 0 & 1 & 140.8 \end{array}\right] \begin{array}{l} \\ -75R_1 + R_2 \\ 1400R_1 + R_3 \end{array}$$

Thus, $a = -0.008$, $b = 1.6$, and $c = 140.8$.
The equation is
$f(x) = -0.008x^2 + 1.6x + 140.8$.

b. The year 2020 corresponds to $x = 20$.

$$f(20) = -0.008(20)^2 + 1.6(20) + 140.8$$
$$= 169.6$$

The U.S. female population will be about 169.6 million in 2020.

27. **a.** An equation $y = ax^2 + bx + c$ is sought.
When $x = 6$, $y = 2.80$. Therefore
$36a + 6b + c = 2.80$.
When $x = 8$, $y = 2.48$. Therefore,
$64a + 8b + c = 2.48$.
When $x = 10$, $y = 2.24$. Therefore,
$100a + 10b + c = 2.24$.
Thus, the system of equations is

$$36a + 6b + c = 2.80 \quad (1)$$
$$64a + 8b + c = 2.48 \quad (2)$$
$$100a + 10b + c = 2.24. \quad (3)$$

Solve the system by elimination. Subtract equation (1) from equation (2) to obtain
$28a + 2b = -.32$,
and subtract equation (2) from equation (3) to obtain
$36a + 2b = -.24$.
Now the system is

$$28a + 2b = -.32 \quad (4)$$
$$36a + 2b = -.24 \quad (5)$$

Subtract equation (5) from equation (4).
$-8a = -.08$
Thus, $a = .01$. Substituting into
$28a + 2b = -.32$ gives

$$28(.01) + 2b = -.32$$
$$2b = -.60$$
$$b = -.30.$$

Substituting into equation (1) gives

$$36(.01) + 6(-.30) + c = 2.80$$
$$.36 - 1.80 + c = 2.80$$
$$-1.44 + c = 2.80$$
$$c = 4.24.$$

Thus, the equation is
$y = .01x^2 - .3x + 4.24$.

b. Write $y = .01x^2 - .3x + 4.24$ in the form
$y = a(x - h)^2 + k$.

$$y = .01(x^2 - 30x) + 4.24$$
$$y = .01(x^2 - 30x + 225) + 4.24 - 2.25$$
$$y = .01(x - 15)^2 + 1.99$$

The minimum value of y is 1.99, occurring when $x = 15$. Thus, 15 platters should be fired at one time to minimize the fuel cost. The minimum fuel cost is \$1.99.

29. a. $C = aS^2 + bS + c$

$33 = a(320)^2 + b(320) + c$

$40 = a(600)^2 + b(600) + c$

$50 = a(1283)^2 + b(1283) + c$

Simplify the equations

$c + 320b + 102,400a = 33$

$c + 600b + 360,000a = 40$

$c + 1283b + 1,646,089a = 50$

The matrix is

$$\left[\begin{array}{ccc|c} 1 & 320 & 102,400 & 33 \\ 1 & 600 & 360,000 & 40 \\ 1 & 1283 & 1,646,089 & 50 \end{array}\right]$$

$$\left[\begin{array}{ccc|c} 1 & 320 & 102,400 & 33 \\ 0 & 280 & 257,600 & 7 \\ 0 & 963 & 1,543,689 & 17 \end{array}\right] \begin{array}{l} \\ -R_1 + R_2 \\ -R_1 + R_3 \end{array}$$

$$\left[\begin{array}{ccc|c} 1 & 320 & 102,400 & 33 \\ 0 & 280 & 257,600 & 7 \\ 0 & 0 & 657,729 & -\frac{283}{40} \end{array}\right] \begin{array}{l} \\ \\ -\frac{963}{280}R_2 + R_3 \end{array}$$

$$\left[\begin{array}{ccc|c} 1 & 320 & 102,400 & 33 \\ 0 & 1 & 920 & \frac{1}{40} \\ 0 & 0 & 1 & -\frac{283}{26,309,160} \end{array}\right] \begin{array}{l} \\ \frac{1}{280}R_2 \\ \frac{1}{657,729}R_3 \end{array}$$

$$\left[\begin{array}{ccc|c} 1 & 0 & -192,000 & 25 \\ 0 & 1 & 0 & \frac{918,089}{26,309,160} \\ 0 & 0 & 1 & -\frac{283}{26,309,160} \end{array}\right] \begin{array}{l} -320R_2 + R_1 \\ -920R_3 + R_2 \\ \end{array}$$

$$\left[\begin{array}{ccc|c} 1 & 0 & 0 & \frac{5,028,275}{219,243} \\ 0 & 1 & 0 & \frac{918,089}{26,309,160} \\ 0 & 0 & 1 & -\frac{283}{26,309,160} \end{array}\right] \begin{array}{l} 192,000R_3 + R_1 \\ \\ \end{array}$$

The relationship is expressed as

$C = aS^2 + bS + c$

$C = -.0000108S^2 + .034896S + 22.9$

b. $45 = -.0000108S^2 + .034896S + 22.9$

Plot $y_1 = 45$ and

$y_2 = -.0000108x^2 + .034896x + 22.9$

They intersect at (864.7, 45). The top speed is approximately 864.7 knots.

Section 6.4 Basic Matrix Operations

1. $\begin{bmatrix} 7 & -8 & 4 \\ 0 & 13 & 9 \end{bmatrix}$ is a 2×3 matrix.

Its additive inverse is $\begin{bmatrix} -7 & 8 & -4 \\ 0 & -13 & -9 \end{bmatrix}$.

3. $\begin{bmatrix} -3 & 0 & 11 \\ 1 & \frac{1}{4} & -7 \\ 5 & -3 & 9 \end{bmatrix}$ is a 3×3 square matrix.

Its additive inverse is $\begin{bmatrix} 3 & 0 & -11 \\ -1 & -\frac{1}{4} & 7 \\ -5 & 3 & -9 \end{bmatrix}$.

5. $\begin{bmatrix} 7 \\ 11 \end{bmatrix}$ is a 2×1 column matrix.

Its additive inverse is $\begin{bmatrix} -7 \\ -11 \end{bmatrix}$.

7. If $A + B = A$, then B must be a zero matrix. Because A is a 5×3 matrix and only matrices of the same size can be added, B must also be 5×3. Therefore, B is a 5×3 zero matrix.

9. $\begin{bmatrix} 1 & 2 & 7 & -1 \\ 8 & 0 & 2 & -4 \end{bmatrix} + \begin{bmatrix} -8 & 12 & -5 & 5 \\ -2 & -3 & 0 & 0 \end{bmatrix}$

$= \begin{bmatrix} 1+(-8) & 2+12 & 7+(-5) & -1+5 \\ 8+(-2) & 0+(-3) & 2+0 & -4+0 \end{bmatrix}$

$= \begin{bmatrix} -7 & 14 & 2 & 4 \\ 6 & -3 & 2 & -4 \end{bmatrix}$

11. $\begin{bmatrix} -1 & -5 & 9 \\ 2 & 2 & 3 \end{bmatrix} + \begin{bmatrix} 4 & 4 & -7 \\ 1 & -1 & 2 \end{bmatrix}$

$= \begin{bmatrix} -1+4 & -5+4 & 9+(-7) \\ 2+1 & 2+(-1) & 3+2 \end{bmatrix}$

$= \begin{bmatrix} 3 & -1 & 2 \\ 3 & 1 & 5 \end{bmatrix}$

13. $\begin{bmatrix} -3 & -2 & 5 \\ 3 & 9 & 0 \end{bmatrix} - \begin{bmatrix} 1 & 5 & -2 \\ -3 & 6 & 8 \end{bmatrix}$

$= \begin{bmatrix} -3-1 & -2-5 & 5-(-2) \\ 3-(-3) & 9-6 & 0-8 \end{bmatrix}$

$= \begin{bmatrix} -4 & -7 & 7 \\ 6 & 3 & -8 \end{bmatrix}$

15. $\begin{bmatrix} 9 & 1 \\ 0 & -3 \\ 4 & 10 \end{bmatrix} - \begin{bmatrix} 1 & 9 & -4 \\ -1 & 1 & 0 \end{bmatrix}$

This subtraction cannot be performed since the matrices have different sizes.

17. $2A = 2\begin{bmatrix} -2 & 0 \\ 5 & 3 \end{bmatrix} = \begin{bmatrix} -4 & 0 \\ 10 & 6 \end{bmatrix}$

19. $-4B = -4\begin{bmatrix} 0 & 2 \\ 4 & -6 \end{bmatrix} = \begin{bmatrix} 0 & -8 \\ -16 & 24 \end{bmatrix}$

21. $-4A + 5B = -4\begin{bmatrix} -2 & 0 \\ 5 & 3 \end{bmatrix} + 5\begin{bmatrix} 0 & 2 \\ 4 & -6 \end{bmatrix}$

$= \begin{bmatrix} 8 & 0 \\ -20 & -12 \end{bmatrix} + \begin{bmatrix} 0 & 10 \\ 20 & -30 \end{bmatrix}$

$= \begin{bmatrix} 8 & 10 \\ 0 & -42 \end{bmatrix}$

23. $A = \begin{bmatrix} 1 & -2 \\ 4 & 3 \end{bmatrix}$, $B = \begin{bmatrix} 2 & -1 \\ 0 & 5 \end{bmatrix}$

$2A + 3B = 2\begin{bmatrix} 1 & -2 \\ 4 & 3 \end{bmatrix} + 3\begin{bmatrix} 2 & -1 \\ 0 & 5 \end{bmatrix}$

$= \begin{bmatrix} 2 & -4 \\ 8 & 6 \end{bmatrix} + \begin{bmatrix} 6 & -3 \\ 0 & 15 \end{bmatrix} = \begin{bmatrix} 8 & -7 \\ 8 & 21 \end{bmatrix}$

If $2X = 2A + 3B$, then

$2X = \begin{bmatrix} 8 & -7 \\ 8 & 21 \end{bmatrix}$ and $X = \begin{bmatrix} 4 & -\frac{7}{2} \\ 4 & \frac{21}{2} \end{bmatrix}$.

25. $X + T = \begin{bmatrix} x & y \\ z & w \end{bmatrix} + \begin{bmatrix} r & s \\ t & u \end{bmatrix} = \begin{bmatrix} x+r & y+s \\ z+t & w+u \end{bmatrix}$,

which is another 2×2 matrix.

27. Show that $X + (T + P) = (X + T) + P$. On the left-hand side, the sum of $T + P$ is obtained first, and then $X + (T + P)$. This gives the matrix

$\begin{bmatrix} x+(r+m) & y+(s+n) \\ z+(t+p) & w+(u+q) \end{bmatrix}$.

For the right-hand side, first the sum $X + T$ is obtained, and then $(X + T) + P$. This gives the matrix

$\begin{bmatrix} (x+r)+m & (y+s)+n \\ (z+t)+p & (w+u)+q \end{bmatrix}$.

Comparing corresponding elements shows that they are equal by the associative property of addition of real numbers. Thus, $X + (T + P) = (X + T) + P$.

29. Show that $P + O = P$.

$P + O = \begin{bmatrix} m & n \\ p & q \end{bmatrix} + \begin{bmatrix} 0 & 0 \\ 0 & 0 \end{bmatrix} = \begin{bmatrix} m+0 & n+0 \\ p+0 & q+0 \end{bmatrix}$

$= \begin{bmatrix} m & n \\ p & q \end{bmatrix} = P$

31. Several possible answers, including:

$$\begin{array}{r} \\ \text{percent of no shows} \\ \text{lost revenue per fan (\$)} \\ \text{lost annual revenue (millions \$)} \end{array} \begin{array}{c} \begin{array}{cccc} \text{basketball} & \text{hockey} & \text{football} & \text{baseball} \end{array} \\ \begin{bmatrix} 16 & 16 & 20 & 18 \\ 18.20 & 18.25 & 19 & 15.40 \\ 22.7 & 35.8 & 51.9 & 96.3 \end{bmatrix} \end{array}$$

33.

$$\begin{array}{r} \\ \text{heart} \\ \text{lung} \\ \text{liver} \\ \text{kidney} \end{array} \begin{array}{c} \begin{array}{ccc} 2009 & 2010 & 2011 \end{array} \\ \begin{bmatrix} 2674 & 2874 & 2813 \\ 1799 & 1759 & 1630 \\ 15{,}094 & 15{,}394 & 15{,}330 \\ 79{,}397 & 83{,}919 & 86{,}547 \end{bmatrix} \end{array}$$

35. a.

$$\begin{array}{r} \\ \text{I} \\ \text{II} \\ \text{III} \end{array} \begin{array}{c} \begin{array}{cccc} \text{Bread} & \text{Milk} & \text{Peanut butter} & \text{Cold cuts} \end{array} \\ \begin{bmatrix} 88 & 48 & 16 & 112 \\ 105 & 72 & 21 & 147 \\ 60 & 40 & 0 & 50 \end{bmatrix} \end{array}$$

b. For store I,
$1.25(88) = 110$, $1.25(48) = 60$, $1.25(16) = 20$, $1.25(112) = 140$.

For store II,
$\frac{4}{3}(105) = 140$, $\frac{4}{3}(72) = 96$, $\frac{4}{3}(21) = 28$, $\frac{4}{3}(147) = 196$.

For store III,
$1.10(60) = 66$, $1.10(40) = 44$, $1.10(0) = 0$, $1.10(50) = 55$.

The new matrix is

$$\begin{bmatrix} 110 & 60 & 20 & 140 \\ 140 & 96 & 28 & 196 \\ 66 & 44 & 0 & 55 \end{bmatrix}.$$

c. To find the total sales, add the matrices from parts (a) and (b).

$$\begin{bmatrix} 88 & 48 & 16 & 112 \\ 105 & 72 & 21 & 147 \\ 60 & 40 & 0 & 50 \end{bmatrix} + \begin{bmatrix} 110 & 60 & 20 & 140 \\ 140 & 96 & 28 & 196 \\ 66 & 44 & 0 & 55 \end{bmatrix} = \begin{bmatrix} 198 & 108 & 36 & 252 \\ 245 & 168 & 49 & 343 \\ 126 & 84 & 0 & 105 \end{bmatrix}$$

37. a. A matrix for the death rate of male drivers is

$$A = \begin{bmatrix} 2.61 & 4.39 & 6.29 & 9.08 \\ 1.63 & 2.77 & 4.61 & 6.92 \\ .92 & .75 & .62 & .54 \end{bmatrix}$$

b. A matrix for the death rate of female drivers is

$$B = \begin{bmatrix} 1.38 & 1.72 & 1.94 & 3.31 \\ 1.26 & 1.48 & 2.82 & 2.28 \\ .41 & .33 & .27 & .40 \end{bmatrix}$$

c. Subtract matrix B from matrix A to see the difference between the death rate of males and females.

$$\begin{bmatrix} 2.61 & 4.39 & 6.29 & 9.08 \\ 1.63 & 2.77 & 4.61 & 6.92 \\ .92 & .75 & .62 & .54 \end{bmatrix} - \begin{bmatrix} 1.38 & 1.72 & 1.94 & 3.31 \\ 1.26 & 1.48 & 2.82 & 2.28 \\ .41 & .33 & .27 & .40 \end{bmatrix} = \begin{bmatrix} 1.23 & 2.67 & 4.35 & 5.77 \\ .37 & 1.29 & 1.79 & 4.64 \\ .51 & .42 & .35 & .14 \end{bmatrix}$$

Section 6.5 Matrix Products and Inverses

1. AB is 2×2. A has 2 rows, B has 2 columns.
   BA is 2×2. B has 2 rows, A has 2 columns.

3. AB is 3×3. A has 3 rows, B has 3 columns.
   BA is 5×5. B has 5 rows, A has 5 columns.

5. AB does not exist because the number of columns of A is not the same as the number of rows of B.
   BA is 3×2. B has 3 rows, A has 2 columns.

7. Columns; rows

9. $\begin{bmatrix} 1 & 2 \\ 3 & 4 \end{bmatrix}\begin{bmatrix} -1 \\ 3 \end{bmatrix} = \begin{bmatrix} 1(-1)+2(3) \\ 3(-1)+4(3) \end{bmatrix} = \begin{bmatrix} 5 \\ 9 \end{bmatrix}$

11. $\begin{bmatrix} 2 & 2 & -1 \\ 5 & 0 & 1 \end{bmatrix}\begin{bmatrix} 0 & -2 \\ -1 & 5 \\ 0 & 2 \end{bmatrix} = \begin{bmatrix} (2)(0)+(2)(-1)+(-1)(0) & (2)(-2)+(2)(5)+(-1)(2) \\ (5)(0)+(0)(-1)+(1)(0) & (5)(-2)+(0)(5)+(1)(2) \end{bmatrix} = \begin{bmatrix} -2 & 4 \\ 0 & -8 \end{bmatrix}$

13. $\begin{bmatrix} -4 & 1 \\ 2 & -3 \end{bmatrix}\begin{bmatrix} 1 & 0 \\ 0 & 1 \end{bmatrix} = \begin{bmatrix} (-4)(1)+(1)(0) & (-4)(0)+(1)(1) \\ (2)(1)+(-3)(0) & (2)(0)+(-3)(1) \end{bmatrix} = \begin{bmatrix} -4 & 1 \\ 2 & -3 \end{bmatrix}$

15. $\begin{bmatrix} 1 & 0 & 0 \\ 0 & 1 & 0 \\ 0 & 0 & 1 \end{bmatrix}\begin{bmatrix} 3 & -5 & 7 \\ -2 & 1 & 6 \\ 0 & -3 & 4 \end{bmatrix} = \begin{bmatrix} (1)(3)+(0)(-2)+(0)(0) & (1)(-5)+(0)(1)+(0)(-3) & (1)(7)+(0)(6)+(0)(4) \\ (0)(3)+(1)(-2)+(0)(0) & (0)(-5)+(1)(1)+(0)(-3) & (0)(7)+(1)(6)+(0)(4) \\ (0)(3)+(0)(-2)+(1)(0) & (0)(-5)+(0)(1)+(1)(-3) & (0)(7)+(0)(6)+(1)(4) \end{bmatrix}$

$= \begin{bmatrix} 3 & -5 & 7 \\ -2 & 1 & 6 \\ 0 & -3 & 4 \end{bmatrix}$

17. $\begin{bmatrix} 1 & 2 & 3 \\ 4 & 0 & 6 \\ 7 & 8 & 9 \end{bmatrix}\begin{bmatrix} -1 & 4 \\ 7 & 0 \\ 1 & 2 \end{bmatrix} = \begin{bmatrix} (1)(-1)+(2)(7)+(3)(1) & (1)(4)+(2)(0)+(3)(2) \\ (4)(-1)+(0)(7)+(6)(1) & (4)(4)+(0)(0)+(6)(2) \\ (7)(-1)+(8)(7)+(9)(1) & (7)(4)+(8)(0)+(9)(2) \end{bmatrix} = \begin{bmatrix} 16 & 10 \\ 2 & 28 \\ 58 & 46 \end{bmatrix}$

19. $AB = \begin{bmatrix} -3 & -9 \\ 2 & 6 \end{bmatrix}\begin{bmatrix} 4 & 6 \\ 2 & 3 \end{bmatrix} = \begin{bmatrix} -30 & -45 \\ 20 & 30 \end{bmatrix}$, $BA = \begin{bmatrix} 4 & 6 \\ 2 & 3 \end{bmatrix}\begin{bmatrix} -3 & -9 \\ 2 & 6 \end{bmatrix} = \begin{bmatrix} 0 & 0 \\ 0 & 0 \end{bmatrix}$

Since $\begin{bmatrix} -30 & -45 \\ 20 & 30 \end{bmatrix} \neq \begin{bmatrix} 0 & 0 \\ 0 & 0 \end{bmatrix}$, $AB \neq BA$.
Therefore, matrix multiplication is not commutative.

21.

$$A+B=\begin{bmatrix}-3 & -9\\ 2 & 6\end{bmatrix}+\begin{bmatrix}4 & 6\\ 2 & 3\end{bmatrix}=\begin{bmatrix}1 & -3\\ 4 & 9\end{bmatrix}$$

$$A-B=\begin{bmatrix}-3 & -9\\ 2 & 6\end{bmatrix}-\begin{bmatrix}4 & 6\\ 2 & 3\end{bmatrix}=\begin{bmatrix}-7 & -15\\ 0 & 3\end{bmatrix}$$

$$(A+B)(A-B)=\begin{bmatrix}1 & -3\\ 4 & 9\end{bmatrix}\begin{bmatrix}-7 & -15\\ 0 & 3\end{bmatrix}=\begin{bmatrix}-7 & -24\\ -28 & -33\end{bmatrix}$$

$$A^2=\begin{bmatrix}-3 & -9\\ 2 & 6\end{bmatrix}\begin{bmatrix}-3 & -9\\ 2 & 6\end{bmatrix}=\begin{bmatrix}-9 & -27\\ 6 & 18\end{bmatrix}$$

$$B^2=\begin{bmatrix}4 & 6\\ 2 & 3\end{bmatrix}\begin{bmatrix}4 & 6\\ 2 & 3\end{bmatrix}=\begin{bmatrix}28 & 42\\ 14 & 21\end{bmatrix}$$

$$A^2-B^2=\begin{bmatrix}-9 & -27\\ 6 & 18\end{bmatrix}-\begin{bmatrix}28 & 42\\ 14 & 21\end{bmatrix}=\begin{bmatrix}-37 & -69\\ -8 & -3\end{bmatrix}$$

Since $\begin{bmatrix}-7 & -24\\ -28 & -33\end{bmatrix}\neq\begin{bmatrix}-37 & -69\\ -8 & -3\end{bmatrix}$, $(A+B)(A-B)\neq A^2-B^2$.

23. Verify that $(PX)T = P(XT)$.

$$(PX)T=\left(\begin{bmatrix}m & n\\ p & q\end{bmatrix}\begin{bmatrix}x & y\\ z & w\end{bmatrix}\right)\begin{bmatrix}r & s\\ t & u\end{bmatrix}=\begin{bmatrix}mx+nz & my+nw\\ px+qz & py+qw\end{bmatrix}\begin{bmatrix}r & s\\ t & u\end{bmatrix}$$

$$=\begin{bmatrix}(mx+nz)r+(my+nw)t & (mx+nz)s+(my+nw)u\\ (px+qz)r+(py+qw)t & (px+qz)s+(py+qw)u\end{bmatrix}$$

$$=\begin{bmatrix}mxr+nzr+myt+nwt & mxs+nzs+myu+nwu\\ pxr+qzr+pyt+qwt & pxs+qzs+pyu+qwu\end{bmatrix}$$ Distributive property for real numbers

$$P(XT)=\begin{bmatrix}m & n\\ p & q\end{bmatrix}\left(\begin{bmatrix}x & y\\ z & w\end{bmatrix}\begin{bmatrix}r & s\\ t & u\end{bmatrix}\right)=\begin{bmatrix}m & n\\ p & q\end{bmatrix}\begin{bmatrix}xr+yt & xs+yu\\ zr+wt & zs+wu\end{bmatrix}$$

$$=\begin{bmatrix}m(xr+yt)+n(zr+wt) & m(xs+yu)+n(zs+wu)\\ p(xr+yt)+q(zr+wt) & p(xs+yu)+q(zs+wu)\end{bmatrix}$$

$$=\begin{bmatrix}mxr+myt+nzr+nwt & mxs+myu+nzs+nwu\\ pxr+pyt+qzr+qwt & pxs+pyu+qzs+qwu\end{bmatrix}$$ Distributive property for real numbers

$$=\begin{bmatrix}mxr+nzr+myt+nwt & mxs+nzs+myu+nwu\\ pxr+qzr+pyt+qwt & pxs+qzs+pyu+qwu\end{bmatrix}$$ Commutative property for real numbers

Thus, $(PX)T = P(XT)$.

25. Verify $k(X+T) = kX + kT$ for any real number k.

$$k(X+T)=k\left(\begin{bmatrix}x & y\\ z & w\end{bmatrix}+\begin{bmatrix}r & s\\ t & u\end{bmatrix}\right)=k\begin{bmatrix}x+r & y+s\\ z+t & w+u\end{bmatrix}=\begin{bmatrix}k(x+r) & k(y+s)\\ k(z+t) & k(w+u)\end{bmatrix}=\begin{bmatrix}kx+kr & ky+ks\\ kz+kt & kw+ku\end{bmatrix}$$

$$=\begin{bmatrix}kx & ky\\ kz & kw\end{bmatrix}+\begin{bmatrix}kr & ks\\ kt & ku\end{bmatrix}=k\begin{bmatrix}x & y\\ z & w\end{bmatrix}+k\begin{bmatrix}r & s\\ t & u\end{bmatrix}=kX+kT$$

27. $\begin{bmatrix} 5 & 2 \\ 3 & -1 \end{bmatrix}\begin{bmatrix} -1 & 2 \\ 3 & -4 \end{bmatrix} = \begin{bmatrix} 1 & 2 \\ -6 & 10 \end{bmatrix} \neq I$,

so the given matrices are not inverses of each other.

29. $\begin{bmatrix} 3 & -1 \\ -4 & 2 \end{bmatrix}\begin{bmatrix} 1 & \frac{1}{2} \\ 2 & \frac{3}{2} \end{bmatrix} = \begin{bmatrix} 1 & 0 \\ 0 & 1 \end{bmatrix} = I$

$\begin{bmatrix} 1 & \frac{1}{2} \\ 2 & \frac{3}{2} \end{bmatrix}\begin{bmatrix} 3 & -1 \\ -4 & 2 \end{bmatrix} = \begin{bmatrix} 1 & 0 \\ 0 & 1 \end{bmatrix} = I$

Therefore, the given matrices are inverses of each other.

31. $\begin{bmatrix} 1 & 1 & 1 \\ 2 & 3 & 0 \\ 1 & 2 & 1 \end{bmatrix}\begin{bmatrix} 1.5 & .5 & -1.5 \\ -1 & 0 & 1 \\ .5 & -.5 & .5 \end{bmatrix} = \begin{bmatrix} 1 & 0 & 0 \\ 0 & 1 & 0 \\ 0 & 0 & 1 \end{bmatrix} = I$

$\begin{bmatrix} 1.5 & .5 & -1.5 \\ -1 & 0 & 1 \\ .5 & -.5 & .5 \end{bmatrix}\begin{bmatrix} 1 & 1 & 1 \\ 2 & 3 & 0 \\ 1 & 2 & 1 \end{bmatrix} = \begin{bmatrix} 1 & 0 & 0 \\ 0 & 1 & 0 \\ 0 & 0 & 1 \end{bmatrix} = I$

The given matrices are inverses of each other.

33. To find the inverse of $\begin{bmatrix} 2 & -3 \\ -1 & 2 \end{bmatrix}$, write the augmented matrix $[A \mid I]$.

$\left[\begin{array}{cc|cc} 2 & -3 & 1 & 0 \\ -1 & 2 & 0 & 1 \end{array}\right].$

$\left[\begin{array}{cc|cc} 1 & -1 & 1 & 1 \\ -1 & 2 & 0 & 1 \end{array}\right] R_1 + R_2$

$\left[\begin{array}{cc|cc} 1 & -1 & 1 & 1 \\ 0 & 1 & 1 & 2 \end{array}\right] R_1 + R_2$

$\left[\begin{array}{cc|cc} 1 & 0 & 2 & 3 \\ 0 & 1 & 1 & 2 \end{array}\right] R_1 + R_2$

The inverse is $\begin{bmatrix} 2 & 3 \\ 1 & 2 \end{bmatrix}$.

35. To find the inverse of $\begin{bmatrix} -1 & 2 \\ 1 & -1 \end{bmatrix}$, write the augmented matrix $[A \mid I]$.

$\left[\begin{array}{cc|cc} -1 & 2 & 1 & 0 \\ 1 & -1 & 0 & 1 \end{array}\right]$

$\left[\begin{array}{cc|cc} -1 & 2 & 1 & 0 \\ 0 & 1 & 1 & 1 \end{array}\right] R_1 + R_2$

$\left[\begin{array}{cc|cc} -1 & 0 & -1 & -2 \\ 0 & 1 & 1 & 1 \end{array}\right] -2R_2 + R_1$

$\left[\begin{array}{cc|cc} 1 & 0 & 1 & 2 \\ 0 & 1 & 1 & 1 \end{array}\right] -R_1$

The inverse is $\begin{bmatrix} 1 & 2 \\ 1 & 1 \end{bmatrix}$.

37. To find the inverse of $\begin{bmatrix} 1 & 2 \\ 3 & 6 \end{bmatrix}$, write the augmented matrix $[A \mid I]$.

$\left[\begin{array}{cc|cc} 1 & 2 & 1 & 0 \\ 3 & 6 & 0 & 1 \end{array}\right]$

$\left[\begin{array}{cc|cc} 1 & 2 & 1 & 0 \\ 0 & 0 & -3 & 1 \end{array}\right] -3R_1 + R_2$

Since there is no way to continue the transformation, the given matrix has no inverse.

39. To find the inverse of $\begin{bmatrix} 1 & -1 & 0 \\ -1 & 2 & 3 \\ 1 & 0 & 2 \end{bmatrix}$, write the augmented matrix $[A \mid I]$.

$$\left[\begin{array}{ccc|ccc} 1 & -1 & 0 & 1 & 0 & 0 \\ -1 & 2 & 3 & 0 & 1 & 0 \\ 1 & 0 & 2 & 0 & 0 & 1 \end{array}\right]$$

$$\left[\begin{array}{ccc|ccc} 1 & -1 & 0 & 1 & 0 & 0 \\ 0 & 1 & 3 & 1 & 1 & 0 \\ 0 & -1 & -2 & 1 & 0 & -1 \end{array}\right] \begin{array}{l} \\ R_1 + R_2 \\ R_1 - R_3 \end{array}$$

$$\left[\begin{array}{ccc|ccc} 1 & 0 & 3 & 2 & 1 & 0 \\ 0 & 1 & 3 & 1 & 1 & 0 \\ 0 & 0 & 1 & 2 & 1 & -1 \end{array}\right] \begin{array}{l} R_1 + R_2 \\ \\ R_2 + R_3 \end{array}$$

$$\left[\begin{array}{ccc|ccc} 1 & 0 & 0 & -4 & -2 & 3 \\ 0 & 1 & 0 & -5 & -2 & 3 \\ 0 & 0 & 1 & 2 & 1 & -1 \end{array}\right] \begin{array}{l} R_1 - 3R_3 \\ R_2 - 3R_3 \\ \\ \end{array}$$

The inverse of the given matrix is

$$\begin{bmatrix} -4 & -2 & 3 \\ -5 & -2 & 3 \\ 2 & 1 & -1 \end{bmatrix}.$$

41. To find the inverse of $\begin{bmatrix} 1 & 4 & 3 \\ 1 & -3 & -2 \\ 2 & 5 & 4 \end{bmatrix}$, write the augmented matrix $[A \mid I]$.

$$\left[\begin{array}{ccc|ccc} 1 & 4 & 3 & 1 & 0 & 0 \\ 1 & -3 & -2 & 0 & 1 & 0 \\ 2 & 5 & 4 & 0 & 0 & 1 \end{array}\right]$$

$$\left[\begin{array}{ccc|ccc} 1 & 4 & 3 & 1 & 0 & 0 \\ 0 & -7 & -5 & -1 & 1 & 0 \\ 0 & -3 & -2 & -2 & 0 & 1 \end{array}\right] \begin{array}{l} \\ -1R_1 + R_2 \\ -2R_1 + R_3 \end{array}$$

$$\left[\begin{array}{ccc|ccc} 1 & 4 & 3 & 1 & 0 & 0 \\ 0 & 1 & \frac{5}{7} & \frac{1}{7} & -\frac{1}{7} & 0 \\ 0 & -3 & -2 & -2 & 0 & 1 \end{array}\right] \begin{array}{l} \\ -\frac{1}{7}R_2 \\ \\ \end{array}$$

$$\left[\begin{array}{ccc|ccc} 1 & 0 & \frac{1}{7} & \frac{3}{7} & \frac{4}{7} & 0 \\ 0 & 1 & \frac{5}{7} & \frac{1}{7} & -\frac{1}{7} & 0 \\ 0 & 0 & \frac{1}{7} & -\frac{11}{7} & -\frac{3}{7} & 1 \end{array}\right] \begin{array}{l} -4R_2 + R_1 \\ \\ 3R_2 + R_3 \end{array}$$

$$\left[\begin{array}{ccc|ccc} 1 & 0 & \frac{1}{7} & \frac{3}{7} & \frac{4}{7} & 0 \\ 0 & 1 & \frac{5}{7} & \frac{1}{7} & -\frac{1}{7} & 0 \\ 0 & 0 & 1 & -11 & -3 & 7 \end{array}\right] \begin{array}{l} \\ \\ 7R_3 \end{array}$$

$$\left[\begin{array}{ccc|ccc} 1 & 0 & 0 & 2 & 1 & -1 \\ 0 & 1 & 0 & 8 & 2 & -5 \\ 0 & 0 & 1 & -11 & -3 & 7 \end{array}\right] \begin{array}{l} -\frac{1}{7}R_3 + R_1 \\ -\frac{5}{7}R_3 + R_2 \\ \\ \end{array}$$

The inverse of the given matrix is

$$\begin{bmatrix} 2 & 1 & -1 \\ 8 & 2 & -5 \\ -11 & -3 & 7 \end{bmatrix}.$$

43. To find the inverse of $\begin{bmatrix} 1 & 2 & 0 \\ 3 & -1 & 2 \\ -2 & 3 & -2 \end{bmatrix}$,

write the augmented matrix $[A \mid I]$.

$$\left[\begin{array}{rrr|rrr} 1 & 2 & 0 & 1 & 0 & 0 \\ 3 & -1 & 2 & 0 & 1 & 0 \\ -2 & 3 & -2 & 0 & 0 & 1 \end{array}\right]$$

$$\left[\begin{array}{rrr|rrr} 1 & 2 & 0 & 1 & 0 & 0 \\ 0 & -7 & 2 & -3 & 1 & 0 \\ 0 & 7 & -2 & 2 & 0 & 1 \end{array}\right] \begin{array}{l} \\ -3R_1 + R_2 \\ 2R_1 + R_3 \end{array}$$

$$\left[\begin{array}{rrr|rrr} 1 & 2 & 0 & 1 & 0 & 0 \\ 0 & -7 & 2 & -3 & 1 & 0 \\ 0 & 0 & 0 & -1 & 1 & 1 \end{array}\right] \begin{array}{l} \\ \\ R_2 + R_3 \end{array}$$

Since there is no way to continue the desired transformation, the given matrix has no inverse.

45. $A = \begin{bmatrix} 1 & 2 & 3 \\ 1 & 4 & 2 \\ 0 & 1 & -1 \end{bmatrix}$

```
[A]-1
   [[6   -5 8 ]
    [-1 1   -1]
    [-1 1   -2]]
```

47. Use a graphing calculator to find the inverse of

$$\begin{bmatrix} 1 & 0 & -2 & 0 \\ -2 & 1 & 2 & 2 \\ 3 & -1 & -2 & -3 \\ 0 & 1 & 4 & 1 \end{bmatrix}.$$

Using a graphing calculator, we find that the inverse of the given matrix is

$$\begin{bmatrix} \frac{1}{2} & -1 & -\frac{1}{2} & \frac{1}{2} \\ \frac{1}{2} & 4 & \frac{5}{2} & -\frac{1}{2} \\ -\frac{1}{4} & -\frac{1}{2} & -\frac{1}{4} & \frac{1}{4} \\ \frac{1}{2} & -2 & -\frac{3}{2} & \frac{1}{2} \end{bmatrix}.$$

49. a. $A = \begin{bmatrix} 3719 & 727 & 521 & 313 \\ 4164 & 738 & 590 & 345 \\ 4566 & 744 & 652 & 374 \end{bmatrix}$

b. $B = \begin{bmatrix} .019 & .007 \\ .011 & .011 \\ .019 & .006 \\ .024 & .008 \end{bmatrix}$

c. $AB = \begin{bmatrix} 3719 & 727 & 521 & 313 \\ 4164 & 738 & 590 & 345 \\ 4566 & 744 & 652 & 374 \end{bmatrix} \begin{bmatrix} .019 & .007 \\ .011 & .011 \\ .019 & .006 \\ .024 & .008 \end{bmatrix}$

$= \begin{bmatrix} 96 & 40 \\ 107 & 44 \\ 116 & 47 \end{bmatrix}$

d. The rows represent the years 2000, 2010, 2020. Column 1 gives the total births (in millions) in those years, column 2 the total deaths (in millions).

e. The total number of births in 2010 was about 107 million.
The total number of deaths projected for 2020 is about 47 million.

51. a. $A = \begin{bmatrix} 195 & 143 & 1225 & 1341 \\ 210 & 141 & 1387 & 1388 \end{bmatrix}$

b. $B = \begin{bmatrix} .016 & .006 \\ .011 & .014 \\ .023 & .008 \\ .013 & .007 \end{bmatrix}$

c. $AB = \begin{bmatrix} 195 & 143 & 1225 & 1341 \\ 210 & 141 & 1387 & 1388 \end{bmatrix} \begin{bmatrix} .016 & .006 \\ .011 & .014 \\ .023 & .008 \\ .013 & .007 \end{bmatrix} = \begin{bmatrix} 50 & 22 \\ 55 & 24 \end{bmatrix}$

d. The rows correspond to years 2010 and 2020. The entries in column 1 give the total number of births (in millions) in those years and column 2 the total deaths (in millions).

e. The total number of deaths in these four countries combined in 2010 was about 22 million. The projected number of births in these four countries in 2020 will be about 55 million.

53. a. Let matrix P contain the amount of products needed.

$P =$	Paper	Tape	Print Rib.	Memo Pads	Pens
Dept. 1	10	4	3	5	6
Dept. 2	7	2	2	3	8
Dept. 3	4	5	1	0	10
Dept. 4	0	3	4	5	5

Let matrix C contain the cost from each supplier.

$C =$	A	B
Paper	9	12
Tape	6	6
Ink Cartridges	24	18
Memo Pads	4	4
Pens	8	12

To find the total departmental cost from each supplier, multiply P times C.

$$PC = \begin{bmatrix} 10 & 4 & 3 & 5 & 6 \\ 7 & 2 & 2 & 3 & 8 \\ 4 & 5 & 1 & 0 & 10 \\ 0 & 3 & 4 & 5 & 5 \end{bmatrix} \begin{bmatrix} 9 & 12 \\ 6 & 6 \\ 24 & 18 \\ 4 & 4 \\ 8 & 12 \end{bmatrix} = \begin{matrix} \text{Dept. 1} \\ \text{Dept. 2} \\ \text{Dept. 3} \\ \text{Dept. 4} \end{matrix} \overset{\begin{matrix} A & B \end{matrix}}{\begin{bmatrix} 254 & 290 \\ 199 & 240 \\ 170 & 216 \\ 174 & 170 \end{bmatrix}}$$

b. The total cost from each supplier would be found by adding each column. The total cost from Supplier A is \$797; the total cost from Supplier B is \$916. The company should buy from Supplier A.

55. a.

$$S = \begin{array}{r} \text{Burgers} \\ \text{Fries} \\ \text{Drinks} \end{array} \overset{\begin{array}{ccc} \text{Barn I} & \text{Barn II} & \text{Barn III} \end{array}}{\begin{bmatrix} 900 & 1500 & 1150 \\ 600 & 950 & 800 \\ 750 & 900 & 825 \end{bmatrix}}$$

b.

$$P = \overset{\begin{array}{ccc} \text{Burgers} & \text{Fries} & \text{Drinks} \end{array}}{\begin{bmatrix} 3.00 & 1.80 & 1.20 \end{bmatrix}}$$

c. The product of P and S gives the daily revenue at each location.

$$PS = \begin{bmatrix} 3.00 & 1.80 & 1.20 \end{bmatrix} \begin{bmatrix} 900 & 1500 & 1150 \\ 600 & 950 & 800 \\ 750 & 900 & 825 \end{bmatrix} = \begin{bmatrix} 4680 & 7290 & 5880 \end{bmatrix}$$

d.

$$PS = \overset{\begin{array}{ccc} \text{Barn I} & \text{Barn II} & \text{Barn III} \end{array}}{\begin{bmatrix} 4680 & 7290 & 5880 \end{bmatrix}}$$

Adding the revenues from the three locations gives the total revenue as $17,850.

Section 6.6 Applications of Matrices

1. The solution to the matrix equation $AX = B$ is $X = A^{-1}B$.

$$A = \begin{bmatrix} 1 & -1 \\ 5 & 6 \end{bmatrix}, \; B = \begin{bmatrix} 4 \\ -2 \end{bmatrix}$$

Use a graphing calculator or row operations on $[A \mid I]$ to find A^{-1}.

$$\left[\begin{array}{rr|rr} 1 & -1 & 1 & 0 \\ 5 & 6 & 0 & 1 \end{array}\right]$$

$$\left[\begin{array}{rr|rr} 1 & -1 & 1 & 0 \\ 0 & 11 & -5 & 1 \end{array}\right] \; -5R_1 + R_2$$

$$\left[\begin{array}{rr|rr} 1 & -1 & 1 & 0 \\ 0 & 1 & -\frac{5}{11} & \frac{1}{11} \end{array}\right] \; \tfrac{1}{11}R_2$$

$$\left[\begin{array}{rr|rr} 1 & 0 & \frac{6}{11} & \frac{1}{11} \\ 0 & 1 & -\frac{5}{11} & \frac{1}{11} \end{array}\right] \; R_1 + R_2$$

$$A^{-1} = \begin{bmatrix} \frac{6}{11} & \frac{1}{11} \\ -\frac{5}{11} & \frac{1}{11} \end{bmatrix}$$

$$X = A^{-1}B = \begin{bmatrix} \frac{6}{11} & \frac{1}{11} \\ -\frac{5}{11} & \frac{1}{11} \end{bmatrix} \begin{bmatrix} 4 \\ -2 \end{bmatrix} = \begin{bmatrix} 2 \\ -2 \end{bmatrix}.$$

3. The solution to the matrix equation $AX = B$ is $X = A^{-1}B$.

$$A = \begin{bmatrix} 3 & 1 \\ 4 & 2 \end{bmatrix}, \; B = \begin{bmatrix} 3 & 4 \\ 5 & 6 \end{bmatrix}.$$

Use a graphing calculator or row operations on $[A \mid I]$ to find A^{-1}.

$$\left[\begin{array}{rr|rr} 3 & 1 & 1 & 0 \\ 4 & 2 & 0 & 1 \end{array}\right]$$

$$\left[\begin{array}{rr|rr} 1 & \frac{1}{3} & \frac{1}{3} & 0 \\ 4 & 2 & 0 & 1 \end{array}\right] \; \tfrac{1}{3}R_1$$

$$\left[\begin{array}{rr|rr} 1 & \frac{1}{3} & \frac{1}{3} & 0 \\ 0 & \frac{2}{3} & -\frac{4}{3} & 1 \end{array}\right] \; R_2 - 4R_1$$

$$\left[\begin{array}{rr|rr} 1 & \frac{1}{3} & \frac{1}{3} & 0 \\ 0 & 1 & -2 & \frac{3}{2} \end{array}\right] \; \tfrac{3}{2}R_2$$

$$\left[\begin{array}{rr|rr} 1 & 0 & 1 & -\frac{1}{2} \\ 0 & 1 & -2 & \frac{3}{2} \end{array}\right] \; R_1 - \tfrac{1}{3}R_2$$

$$A^{-1} = \begin{bmatrix} 1 & -\frac{1}{2} \\ -2 & \frac{3}{2} \end{bmatrix}$$

$$X = A^{-1}B = \begin{bmatrix} 1 & -\frac{1}{2} \\ -2 & \frac{3}{2} \end{bmatrix} \begin{bmatrix} 3 & 4 \\ 5 & 6 \end{bmatrix} = \begin{bmatrix} \frac{1}{2} & 1 \\ \frac{3}{2} & 1 \end{bmatrix}$$

5. The solution to the matrix equation $AX = B$ is $X = A^{-1}B$.

$$A = \begin{bmatrix} 2 & 1 & 0 \\ -4 & -1 & 3 \\ 3 & 1 & -2 \end{bmatrix}, B = \begin{bmatrix} 1 \\ 4 \\ 0 \end{bmatrix}.$$

Use a graphing calculator or row operations on $[A \mid I]$ to find that

$$A^{-1} = \begin{bmatrix} 1 & -2 & -3 \\ -1 & 4 & 6 \\ 1 & -1 & -2 \end{bmatrix}.$$

$$X = A^{-1}B = \begin{bmatrix} 1 & -2 & -3 \\ -1 & 4 & 6 \\ 1 & -1 & -2 \end{bmatrix} \begin{bmatrix} 1 \\ 4 \\ 0 \end{bmatrix} = \begin{bmatrix} -7 \\ 15 \\ -3 \end{bmatrix}.$$

7.
$$\begin{aligned} x + 2y + 3z &= 10 \\ 2x + 3y + 2z &= 6 \\ -x - 2y - 4z &= -1 \end{aligned}$$

has coefficient matrix

$$A = \begin{bmatrix} 1 & 2 & 3 \\ 2 & 3 & 2 \\ -1 & -2 & -4 \end{bmatrix}.$$

Find A^{-1}:

$$[A \mid I] = \left[\begin{array}{rrr|rrr} 1 & 2 & 3 & 1 & 0 & 0 \\ 2 & 3 & 2 & 0 & 1 & 0 \\ -1 & -2 & -4 & 0 & 0 & 1 \end{array}\right]$$

$$\left[\begin{array}{rrr|rrr} 1 & 2 & 3 & 1 & 0 & 0 \\ 0 & -1 & -4 & -2 & 1 & 0 \\ 0 & 0 & -1 & 1 & 0 & 1 \end{array}\right] \begin{array}{l} \\ -2R_1 + R_2 \\ R_1 + R_3 \end{array}$$

$$\left[\begin{array}{rrr|rrr} 1 & 0 & -5 & -3 & 2 & 0 \\ 0 & 1 & 4 & 2 & -1 & 0 \\ 0 & 0 & 1 & -1 & 0 & -1 \end{array}\right] \begin{array}{l} 2R_2 + R_1 \\ -1R_2 \\ -1R_3 \end{array}$$

$$\left[\begin{array}{rrr|rrr} 1 & 0 & 0 & -8 & 2 & -5 \\ 0 & 1 & 0 & 6 & -1 & 4 \\ 0 & 0 & 1 & -1 & 0 & -1 \end{array}\right] \begin{array}{l} 5R_3 + R_1 \\ -4R_3 + R_2 \\ \end{array}$$

$$A^{-1} = \begin{bmatrix} -8 & 2 & -5 \\ 6 & -1 & 4 \\ -1 & 0 & -1 \end{bmatrix}$$

$$X = A^{-1}B = \begin{bmatrix} -8 & 2 & -5 \\ 6 & -1 & 4 \\ -1 & 0 & -1 \end{bmatrix} \begin{bmatrix} 10 \\ 6 \\ -1 \end{bmatrix} = \begin{bmatrix} -63 \\ 50 \\ -9 \end{bmatrix}$$

The solution is (–63, 50, –9).

9.
$$\begin{aligned} x + 4y + 3z &= -12 \\ x - 3y - 2z &= 0 \\ 2x + 5y + 4z &= 7 \end{aligned}$$

has the coefficient matrix

$$A = \begin{bmatrix} 1 & 4 & 3 \\ 1 & -3 & -2 \\ 2 & 5 & 4 \end{bmatrix}.$$

From Exercise 41 of Section 6.5,

$$A^{-1} = \begin{bmatrix} 2 & 1 & -1 \\ 8 & 2 & -5 \\ -11 & -3 & 7 \end{bmatrix}.$$

$$X = A^{-1}B = \begin{bmatrix} 2 & 1 & -1 \\ 8 & 2 & -5 \\ -11 & -3 & 7 \end{bmatrix} \begin{bmatrix} -12 \\ 0 \\ 7 \end{bmatrix} = \begin{bmatrix} -31 \\ -131 \\ 181 \end{bmatrix}.$$

The solution is (–31, –131, 181).

11.
$$\begin{aligned} x + 2y + 3z &= 4 \\ x + 4y + 2z &= 8 \\ y - z &= -4 \end{aligned}$$

has coefficient matrix

$$A = \begin{bmatrix} 1 & 2 & 3 \\ 1 & 4 & 2 \\ 0 & 1 & -1 \end{bmatrix}.$$

From Exercise 45 of Section 6.5,

$$A^{-1} = \begin{bmatrix} 6 & -5 & 8 \\ -1 & 1 & -1 \\ -1 & 1 & -2 \end{bmatrix}.$$

$$X = A^{-1}B = \begin{bmatrix} 6 & -5 & 8 \\ -1 & 1 & -1 \\ -1 & 1 & -2 \end{bmatrix} \begin{bmatrix} 4 \\ 8 \\ -4 \end{bmatrix} = \begin{bmatrix} -48 \\ 8 \\ 12 \end{bmatrix}$$

The solution is (–48, 8, 12).

13. $x + y + 2w = 3$
$2x - y + z - w = 3$
$3x + 3y + 2z - 2w = 5$
$x + 2y + z = 3$

The coefficient matrix is

$$\begin{bmatrix} 1 & 1 & 0 & 2 \\ 2 & -1 & 1 & -1 \\ 3 & 3 & 2 & -2 \\ 1 & 2 & 1 & 0 \end{bmatrix}.$$

The inverse of this coefficient matrix was calculated in Exercise 48 of Section 6.5. Use that result to obtain

$$\begin{bmatrix} x \\ y \\ z \\ w \end{bmatrix} = \begin{bmatrix} \frac{1}{2} & 0 & \frac{1}{2} & -1 \\ \frac{1}{10} & -\frac{2}{5} & \frac{3}{10} & -\frac{1}{5} \\ -\frac{7}{10} & \frac{4}{5} & -\frac{11}{10} & \frac{12}{5} \\ \frac{1}{5} & \frac{1}{5} & -\frac{2}{5} & \frac{3}{5} \end{bmatrix} \begin{bmatrix} 3 \\ 3 \\ 5 \\ 3 \end{bmatrix} = \begin{bmatrix} 1 \\ 0 \\ 2 \\ 1 \end{bmatrix}.$$

The solution is (1, 0, 2, 1).

15. Since $N = X - MX$, $N = IX - MX \Rightarrow$
$N = (I - M)X \Rightarrow$
$(I - M)^{-1}N = (I - M)^{-1}(I - M)X \Rightarrow$
$(I - M)^{-1}N = IX.$

Thus, $X = (I - M)^{-1}N$.

$$I - M = \begin{bmatrix} 1 & 0 \\ 0 & 1 \end{bmatrix} - \begin{bmatrix} 0 & 1 \\ -2 & 1 \end{bmatrix} = \begin{bmatrix} 1 & -1 \\ 2 & 0 \end{bmatrix}$$

If $I - M = \begin{bmatrix} 1 & -1 \\ 2 & 0 \end{bmatrix}$,

$$(I - M)^{-1} = \begin{bmatrix} 0 & \frac{1}{2} \\ -1 & \frac{1}{2} \end{bmatrix}.$$

Since $X = (I - M)^{-1}N$,

$$X = \begin{bmatrix} 0 & \frac{1}{2} \\ -1 & \frac{1}{2} \end{bmatrix} \begin{bmatrix} 8 \\ -12 \end{bmatrix} = \begin{bmatrix} -6 \\ -14 \end{bmatrix}.$$

17. Let x = the number of buffets, let y = the number of chairs, and let z = the number of tables. The information can be summarized in a table.

	Cutting hours	Assembly hours	Finishing hours
Buffets	$15x$	$20x$	$5x$
Chairs	$5y$	$8y$	$5y$
Tables	$10z$	$6z$	$6z$
Total	4900	6600	3900

The system of equations is
$15x + 5y + 10z = 4900$
$20x + 8y + 6z = 6600$
$5x + 5y + 6z = 3900$

Then, we have

$$A = \begin{bmatrix} 15 & 5 & 10 \\ 20 & 8 & 6 \\ 5 & 5 & 6 \end{bmatrix} \text{ and } B = \begin{bmatrix} 4900 \\ 6600 \\ 3900 \end{bmatrix}.$$

Using a calculator, we find that the inverse is

$$\begin{bmatrix} \frac{3}{70} & \frac{1}{21} & -\frac{5}{42} \\ -\frac{3}{14} & \frac{2}{21} & \frac{11}{42} \\ \frac{1}{7} & -\frac{5}{42} & \frac{1}{21} \end{bmatrix}.$$

$$X = A^{-1}B = \begin{bmatrix} \frac{3}{70} & \frac{1}{21} & -\frac{5}{42} \\ -\frac{3}{14} & \frac{2}{21} & \frac{11}{42} \\ \frac{1}{7} & -\frac{5}{42} & \frac{1}{21} \end{bmatrix} \begin{bmatrix} 4900 \\ 6600 \\ 3900 \end{bmatrix} = \begin{bmatrix} 60 \\ 600 \\ 100 \end{bmatrix}$$

Therefore, 60 buffets, 600 chairs, and 100 tables should be produced each week.

19. The following solution presupposes the use of a TI-82 graphing calculator. Similar results can be obtained from other graphing calculators.
Let x = number of bacterium 1, let y = number of bacterium 2, and let z = number of bacterium 3.

	Food		
	I	II	III
Bacterium 1	1.3	1.3	2.3
Bacterium 2	1.1	2.4	3.7
Bacterium 3	8.1	2.9	5.1
Totals	16,000	28,000	44,000

The data in the table produce the system of equations

$$1.3x + 1.1y + 8.1z = 16,000$$
$$1.3x + 2.4y + 2.9z = 28,000$$
$$2.3x + 3.7y + 5.1z = 44,000.$$

We store the following matrix as matrix A:

$$\left[\begin{array}{ccc|c} 1.3 & 1.1 & 8.1 & 16,000 \\ 1.3 & 2.4 & 2.9 & 28,000 \\ 2.3 & 3.7 & 5.1 & 44,000 \end{array}\right].$$

Using row operations, we transform this to obtain

$$\left[\begin{array}{ccc|c} 1 & 0 & 0 & 2339.74359 \\ 0 & 1 & 0 & 10,128.20513 \\ 0 & 0 & 1 & 224.3589744 \end{array}\right]$$

Thus, 2340 of the first species, 10,128 of the second species, and 224 of the third species can be maintained.

21. Let x = the wholesale price of jeans, let y = the wholesale price of jackets, let z = the wholesale price of sweaters, and let w = the wholesale price of shirts. The system is

$$3000x + 3000y + 2200z + 4200w = 507,650$$
$$2700x + 2500y + 2100z + 4300w = 459,075$$
$$5000x + 2000y + 1400z + 7500w = 541,225$$
$$7000x + 1800y + 600z + 8000w = 571,500$$

Store the following matrix as matrix A.

$$\left[\begin{array}{cccc|c} 3000 & 3000 & 2200 & 4200 & 507,650 \\ 2700 & 2500 & 2100 & 4300 & 459,075 \\ 5000 & 2000 & 1400 & 7500 & 541,225 \\ 7000 & 1800 & 600 & 8000 & 571,500 \end{array}\right]$$

Using row operations, transform this matrix to obtain

$$\left[\begin{array}{cccc|c} 1 & 0 & 0 & 0 & 34.5 \\ 0 & 1 & 0 & 0 & 72 \\ 0 & 0 & 1 & 0 & 44 \\ 0 & 0 & 0 & 1 & 21.75 \end{array}\right]$$

Therefore, the wholesale price for jeans is \$34.50, jacket is \$72, sweater is \$44, and a shirt is \$21.75.

23. $A = \begin{bmatrix} .1 & .03 \\ .07 & .6 \end{bmatrix}, D = \begin{bmatrix} 5 \\ 10 \end{bmatrix}$

First calculate $I - A$.

$$I - A = \begin{bmatrix} .9 & -.03 \\ -.07 & .4 \end{bmatrix}$$

Use row operations to find the inverse of $I - A$.

$$(I - A)^{-1} = \begin{bmatrix} 1.118 & .084 \\ .196 & 2.515 \end{bmatrix}$$

Since $X = (I - A)^{-1}D$,

$$X = \begin{bmatrix} 1.118 & .084 \\ .196 & 2.515 \end{bmatrix}\begin{bmatrix} 5 \\ 10 \end{bmatrix} = \begin{bmatrix} 6.43 \\ 26.12 \end{bmatrix}.$$

25. $A = \begin{bmatrix} .25 & .08 \\ .33 & .11 \end{bmatrix}, D = \begin{bmatrix} 690 \\ 920 \end{bmatrix}$

$$(I - A)^{-1} = \begin{bmatrix} 1.388 & 0.1248 \\ 0.5147 & 1.1698 \end{bmatrix}$$

$$X = (I - A)^{-1}D = \begin{bmatrix} 1.388 & 0.1248 \\ 0.5147 & 1.1699 \end{bmatrix}\begin{bmatrix} 690 \\ 920 \end{bmatrix}$$

$$= \begin{bmatrix} 1073 \\ 1431 \end{bmatrix}$$

Produce 1073 metric tons of wheat, and 1431 metric tons of oil.

27. First, write the input-output matrix A.

$$A = \begin{bmatrix} .4 & .6 \\ .5 & .25 \end{bmatrix} = \begin{bmatrix} \frac{2}{5} & \frac{3}{5} \\ \frac{1}{2} & \frac{1}{4} \end{bmatrix}$$

$$I - A = \begin{bmatrix} 1 & 0 \\ 0 & 1 \end{bmatrix} - \begin{bmatrix} \frac{2}{5} & \frac{3}{5} \\ \frac{1}{2} & \frac{1}{4} \end{bmatrix} = \begin{bmatrix} \frac{3}{5} & -\frac{3}{5} \\ -\frac{1}{2} & \frac{3}{4} \end{bmatrix}$$

Form $[I - A \mid I]$.

$$\left[\begin{array}{cc|cc} \frac{3}{5} & -\frac{3}{5} & 1 & 0 \\ -\frac{1}{2} & \frac{3}{4} & 0 & 1 \end{array}\right]$$

$$\left[\begin{array}{cc|cc} 1 & -1 & \frac{5}{3} & 0 \\ -\frac{1}{2} & \frac{3}{4} & 0 & 1 \end{array}\right] \frac{5}{3}R_1$$

$$\left[\begin{array}{cc|cc} 1 & -1 & \frac{5}{3} & 0 \\ 0 & \frac{1}{4} & \frac{5}{6} & 1 \end{array}\right] \frac{1}{2}R_1 + R_2$$

$$\left[\begin{array}{cc|cc} 1 & -1 & \frac{5}{3} & 0 \\ 0 & 1 & \frac{10}{3} & 4 \end{array}\right] 4R_2$$

$$\left[\begin{array}{cc|cc} 1 & 0 & 5 & 4 \\ 0 & 1 & \frac{10}{3} & 4 \end{array}\right] R_2 + R_1$$

Thus, $(I - A)^{-1} = \begin{bmatrix} 5 & 4 \\ \frac{10}{3} & 4 \end{bmatrix}$

Since $D = \begin{bmatrix} 15 \text{ million} \\ 12 \text{ million} \end{bmatrix}$ and $X = (I - A)^{-1} D$,

$$X = \begin{bmatrix} 5 & 4 \\ \frac{10}{3} & 4 \end{bmatrix} \begin{bmatrix} 15 \text{ million} \\ 12 \text{ million} \end{bmatrix} = \begin{bmatrix} 123 \text{ million} \\ 98 \text{ million} \end{bmatrix}.$$

The output should be \$123 million of electricity and \$98 million of gas.

29. $A = \begin{bmatrix} \frac{1}{4} & \frac{1}{6} \\ \frac{1}{2} & 0 \end{bmatrix}, I - A = \begin{bmatrix} \frac{3}{4} & -\frac{1}{6} \\ -\frac{1}{2} & 1 \end{bmatrix}$

$$(I - A)^{-1} = \begin{bmatrix} \frac{3}{2} & \frac{1}{4} \\ \frac{3}{4} & \frac{9}{8} \end{bmatrix}$$

a. $X = \begin{bmatrix} \frac{3}{2} & \frac{1}{4} \\ \frac{3}{4} & \frac{9}{8} \end{bmatrix} \begin{bmatrix} 1 \\ 1 \end{bmatrix} = \begin{bmatrix} \frac{7}{4} \\ \frac{15}{8} \end{bmatrix}$

Thus, $\frac{7}{4}$ bushes of yams and $\frac{15}{8} \approx 2$ pigs should be produced.

b. $X = \begin{bmatrix} \frac{3}{2} & \frac{1}{4} \\ \frac{3}{4} & \frac{9}{8} \end{bmatrix} \begin{bmatrix} 100 \\ 70 \end{bmatrix} = \begin{bmatrix} 167.5 \\ 153.75 \end{bmatrix}$

Thus, 167.5 bushels of yams and $153.75 \approx 154$ pigs should be produced.

31. a. From the input-output matrix, we see that .40 unit of agriculture, .12 unit of manufacturing, and 3.60 units of households are required for the manufacturing sector to produce 1 unit.

b. $A = \begin{bmatrix} .25 & .40 & .133 \\ .14 & .12 & .100 \\ .80 & 3.60 & .133 \end{bmatrix}$

$$I - A = \begin{bmatrix} .75 & -.40 & -.133 \\ -.14 & .88 & -.100 \\ -.80 & -3.60 & .867 \end{bmatrix}$$

Next, calculate $(I - A)^{-1}$.

$$(I - A)^{-1} \approx \begin{bmatrix} 6.61 & 13.53 & 2.57 \\ 3.3 & 8.91 & 1.53 \\ 19.8 & 49.5 & 9.9 \end{bmatrix}$$

$$X = (I - A)^{-1} D$$

$$\approx \begin{bmatrix} 6.61 & 13.53 & 2.57 \\ 3.3 & 8.91 & 1.53 \\ 19.8 & 49.5 & 9.9 \end{bmatrix} \begin{bmatrix} 35 \\ 38 \\ 40 \end{bmatrix} \approx \begin{bmatrix} 848 \\ 516 \\ 2970 \end{bmatrix}$$

Therefore, 848 units of agriculture, 516 units of manufacturing, and 2970 units of households need to be produced.

c. Since .25 units of agriculture are used in producing each unit of agriculture, $(848)(.25) = 212$ units are used. Since .4 units of agriculture are used in producing each unit of manufacturing, $(.4)(516) = 206.4$ units are used. Since .133 units of agriculture are used in producing each unit of households, $(.133)(2970) = 395.01$. Thus, $212 + 206 + 395 = 813$ units of agriculture are used in the production process.

33. a. The energy sector requires .017 unit of manufacturing and .216 unit of energy to produce one unit.

b. We are given the input-output matrix A and the production matrix X.

$$A = \begin{bmatrix} .293 & 0 & 0 \\ .014 & .207 & .017 \\ .044 & .010 & .216 \end{bmatrix},\ X = \begin{bmatrix} 175,000 \\ 22,000 \\ 12,000 \end{bmatrix}$$

$$D = (I - A)X$$

$$= \begin{bmatrix} .707 & 0 & 0 \\ -.014 & .793 & -.017 \\ .044 & -.010 & .784 \end{bmatrix}\begin{bmatrix} 175,000 \\ 22,000 \\ 12,000 \end{bmatrix}$$

$$= \begin{bmatrix} 123,725 \\ 14,792 \\ 1488 \end{bmatrix}$$

Therefore, about 123,725,000 pounds of agriculture, 14,792,000 pounds of manufacturing, and 1,488,000 pounds of energy is available to satisfy demand.

c. $X = (I - A)^{-1}D$

$$\approx \begin{bmatrix} 1.414 & 0 & 0 \\ .027 & 1.261 & .027 \\ .080 & .016 & 1.276 \end{bmatrix}\begin{bmatrix} 138,213 \\ 17,597 \\ 1786 \end{bmatrix}$$

$$\approx \begin{bmatrix} 195,492 \\ 25,933 \\ 13,580 \end{bmatrix}$$

Therefore, about 195,492,000 pounds of agriculture, 25,933,000 pounds of manufacturing, and 13,580,000 pounds of energy should be produced.

35. $A = \begin{bmatrix} .1045 & .0428 & .0029 & .0031 \\ .0826 & .1087 & .0584 & .0321 \\ .0867 & .1019 & .2032 & .3555 \\ .6253 & .3448 & .6106 & .0798 \end{bmatrix},\ D = \begin{bmatrix} 450 \\ 300 \\ 125 \\ 100 \end{bmatrix}$

$$(I - A)^{-1} \approx \begin{bmatrix} 1.133 & .062 & .019 & .013 \\ .202 & 1.19 & .171 & .108 \\ .748 & .536 & 1.87 & .742 \\ 1.343 & .845 & 1.315 & 1.63 \end{bmatrix}$$

$$X = (I - A)^{-1}D$$

$$X \approx \begin{bmatrix} 1.133 & .062 & .019 & .013 \\ .202 & 1.19 & .171 & .108 \\ .748 & .536 & 1.87 & .742 \\ 1.343 & .845 & 1.315 & 1.63 \end{bmatrix}\begin{bmatrix} 450 \\ 300 \\ 125 \\ 100 \end{bmatrix} \approx \begin{bmatrix} 532 \\ 481 \\ 805 \\ 1185 \end{bmatrix}$$

This means \$532 million of natural resources, \$481 million of manufacturing, \$805 million of trade and services, and \$1185 million of personal consumption.

37. a. The input-output matrix A, and the matrix $I - A$, are

$$A = \begin{bmatrix} .2 & .1 & .1 \\ .1 & .1 & 0 \\ .5 & .6 & .7 \end{bmatrix} \text{ and}$$

$$I - A = \begin{bmatrix} .8 & -.1 & -.1 \\ -.1 & .9 & 0 \\ -.5 & -.6 & .3 \end{bmatrix}.$$

Next, calculate $(I - A)^{-1}$.

$$(I - A)^{-1} \approx \begin{bmatrix} 1.67 & .56 & .56 \\ .19 & 1.17 & .06 \\ 3.15 & 3.27 & 4.38 \end{bmatrix}$$

b. These multipliers imply that if the demand for one community's output increases by \$1 then the output in the other community will increase by the amount in the row and column of that matrix. For example, if the demand for Hermitage's output increases by \$1, then output from Sharon will increase by \$.56, Farrell by \$.06, and Hermitage by \$4.38.

39. The message *Head for the hills* broken into groups of 2 letters would have the following matrices.

$$\begin{bmatrix} 8 \\ 5 \end{bmatrix}, \begin{bmatrix} 1 \\ 4 \end{bmatrix}, \begin{bmatrix} 27 \\ 6 \end{bmatrix}, \begin{bmatrix} 15 \\ 18 \end{bmatrix}, \begin{bmatrix} 27 \\ 20 \end{bmatrix},$$

$$\begin{bmatrix} 8 \\ 5 \end{bmatrix}, \begin{bmatrix} 27 \\ 8 \end{bmatrix}, \begin{bmatrix} 9 \\ 12 \end{bmatrix}, \begin{bmatrix} 12 \\ 19 \end{bmatrix}.$$

Multiply $M = \begin{bmatrix} 1 & 3 \\ 2 & 7 \end{bmatrix}$ by each matrix in the code to obtain

$$\begin{bmatrix} 23 \\ 51 \end{bmatrix}, \begin{bmatrix} 13 \\ 30 \end{bmatrix}, \begin{bmatrix} 45 \\ 96 \end{bmatrix}, \begin{bmatrix} 69 \\ 156 \end{bmatrix}, \begin{bmatrix} 87 \\ 194 \end{bmatrix},$$

$$\begin{bmatrix} 23 \\ 51 \end{bmatrix}, \begin{bmatrix} 51 \\ 110 \end{bmatrix}, \begin{bmatrix} 45 \\ 102 \end{bmatrix}, \begin{bmatrix} 69 \\ 157 \end{bmatrix}.$$

41. $A = \begin{bmatrix} 0 & 1 & 2 & 2 \\ 1 & 0 & 1 & 0 \\ 2 & 1 & 0 & 1 \\ 2 & 0 & 1 & 0 \end{bmatrix}$

$A^2 = \begin{bmatrix} 9 & 2 & 3 & 2 \\ 2 & 2 & 2 & 3 \\ 3 & 2 & 6 & 4 \\ 2 & 3 & 4 & 5 \end{bmatrix}$

a. The number of ways to travel from city 1 to city 3 by passing through exactly one city is the entry in row 1, column 3 of A^2, which is 3.

b. The number of ways to travel from city 2 to city 4 by passing through exactly one city is the entry in row 2, column 4 of A^2, which is 3.

c. The number of ways to travel from city 1 to city 3 by passing through at most one city is the sum of the entries in row 1, column 3 of A and A^2, which is 2 + 3 or 5.

d. The number of ways to travel from city 2 to city 4 by passing through at most one city is the sum of the entries in row 2, column 4 of A and A^2, which is 0 + 3 or 3.

43. a. $B = \begin{array}{c} \\ 1 \\ 2 \\ 3 \end{array}\!\!\begin{array}{c} \begin{matrix} 1 & 2 & 3 \end{matrix} \\ \begin{bmatrix} 0 & 2 & 3 \\ 2 & 0 & 4 \\ 3 & 4 & 0 \end{bmatrix} \end{array}$

b. $B^2 = \begin{bmatrix} 0 & 2 & 3 \\ 2 & 0 & 4 \\ 3 & 4 & 0 \end{bmatrix}\begin{bmatrix} 0 & 2 & 3 \\ 2 & 0 & 4 \\ 3 & 4 & 0 \end{bmatrix} = \begin{bmatrix} 13 & 12 & 8 \\ 12 & 20 & 6 \\ 8 & 6 & 25 \end{bmatrix}$

c. Use the entry in row 1, column 2 of B^2, which is 12.

d. Use the sum of the entries in row 1, column 2 of B and B^2, which is 2 + 12 or 14.

45. a. $C = \begin{array}{c} \\ d \\ r \\ c \\ m \end{array}\!\!\begin{array}{c} \begin{matrix} d & r & c & m \end{matrix} \\ \begin{bmatrix} 0 & 1 & 1 & 1 \\ 0 & 0 & 0 & 1 \\ 0 & 1 & 0 & 1 \\ 0 & 0 & 0 & 0 \end{bmatrix} \end{array}$

b. $C^2 = \begin{bmatrix} 0 & 1 & 1 & 1 \\ 0 & 0 & 0 & 1 \\ 0 & 1 & 0 & 1 \\ 0 & 0 & 0 & 0 \end{bmatrix}\begin{bmatrix} 0 & 1 & 1 & 1 \\ 0 & 0 & 0 & 1 \\ 0 & 1 & 0 & 1 \\ 0 & 0 & 0 & 0 \end{bmatrix} = \begin{bmatrix} 0 & 1 & 0 & 2 \\ 0 & 0 & 0 & 0 \\ 0 & 0 & 0 & 1 \\ 0 & 0 & 0 & 0 \end{bmatrix}$

C^2 gives the number of food sources once removed from the feeder. Thus, since dogs eat rats and rats eat mice, mice are an indirect as well as a direct food source of dogs

Chapter 6 Review Exercises

1. $-5x - 3y = -3 \quad (1)$
$2x + y = 4 \quad (2)$

Multiply equation (2) by 3 and add to equation (1).

$$\begin{aligned} -5x - 3y &= -3 \\ 6x + 3y &= 12 \\ \hline x \quad\quad &= 9 \end{aligned}$$

Substitute 9 for x in equation (2) to solve for y.

$2(9) + y = 4 \Rightarrow y = 14$

The solution is (9, –14).

2. $3x - y = 8 \quad (1)$
$2x + 3y = 6 \quad (2)$
Multiply equation (1) by 3 and add to equation (2).

$$\begin{array}{l} 9x - 3y = 24 \\ \underline{2x + 3y = \ \ 6} \\ 11x \qquad = 30 \Rightarrow x = \frac{30}{11} \end{array}$$

Substitute $\frac{30}{11}$ for x in equation (1) to solve for y.

$$3\left(\frac{30}{11}\right) - y = 8 \Rightarrow y = \frac{2}{11}$$

The solution is $\left(\frac{30}{11}, \frac{2}{11}\right)$.

3. $3x - 5y = 16 \quad (1)$
$2x + 3y = -2 \quad (2)$
Multiply equation (1) by 2 and equation (2) by –3 and add the results.

$$\begin{array}{r} 6x - 10y = 32 \\ \underline{-6x - \ 9y = 6} \\ -19y = 38 \Rightarrow y = -2 \end{array}$$

Substitute –2 for y in equation (1) to solve for x.
$3x - 5(-2) = 16 \Rightarrow 3x = 6 \Rightarrow x = 2$
The solution is (2, –2).

4. $\frac{1}{4}x - \frac{1}{3}y = -\frac{1}{4} \quad (1)$
$\frac{1}{10}x + \frac{2}{5}y = \frac{2}{5} \quad (2)$
Multiply equation (1) by 12 and equation (2) by 10 and add the results.

$$\begin{array}{l} 3x - 4y = -3 \quad (3) \\ \underline{x + 4y = \ \ 4} \quad (4) \\ 4x \qquad = \ 1 \Rightarrow x = \frac{1}{4} \end{array}$$

Substitute $\frac{1}{4}$ for x in equation (4) to solve for y.

$$\frac{1}{4} + 4y = 4 \Rightarrow 4y = \frac{15}{4} \Rightarrow y = \frac{15}{16}$$

The solution is $\left(\frac{1}{4}, \frac{15}{16}\right)$.

5. Let x = the number of shares of the first stock, and let y = the number of shares of the second stock.
We will solve the following system.
$32x + 23y = 10{,}100 \quad (1)$
$1.2x + 1.4y = 540 \quad (2)$
Multiply equation (1) by 1.2 and equation (2) by –32 and add the results.

$$\begin{array}{r} 38.4x + 27.6y = 12{,}120 \\ \underline{-38.4x - 44.8y = -17{,}280} \\ -17.2y = -5160 \Rightarrow y = 300 \end{array}$$

Substitute 300 for y in equation (2) and solve for x.
$1.2x + 1.4(300) = 540 \Rightarrow 1.2x = 120 \Rightarrow x = 100$
She should buy 100 shares of the first stock and 300 shares of the second stock.

6. Let x = the amount invested in the first fund, and let y = the amount invested in the second fund. We will solve the system

$$.08x + .02y = 780 \quad (1)$$
$$.1x + .01y = 810 \quad (2)$$

Multiply equation (2) by –2 and add to equation (1).

$$\begin{array}{l} .08x + .02y = 780 \\ \underline{-.2x - .02y = -1620} \\ -.12x = -840 \Rightarrow x = 7000 \end{array}$$

Substitute 7000 for x in equation (1) and solve for y.

$.08(7000) + .02y = 780 \Rightarrow .02y = 220 \Rightarrow$
$y = 11{,}000$

Emma has \$7000 invested in the first fund and \$11,000 in the second.

7.
$$x - 2y = 1 \quad (1)$$
$$4x + 4y = 2 \quad (2)$$
$$10x + 8y = 4 \quad (3)$$

Multiply equation (1) by 2 and add to equation (2).

$$\begin{array}{l} 2x - 4y = 2 \\ \underline{4x + 4y = 2} \\ 6x = 4 \Rightarrow x = \frac{2}{3} \end{array}$$

Substitute $\frac{2}{3}$ for x in equation (1) to solve for y.

$$\frac{2}{3} - 2y = 1 \Rightarrow -2y = \frac{1}{3} \Rightarrow y = -\frac{1}{6}$$

The ordered pair $\left(\frac{2}{3}, -\frac{1}{6}\right)$ satisfies equations (1) and (2), but not equation (3). The system is inconsistent; there is no solution.

8.
$$4x - y - 2z = 4 \quad (1)$$
$$x - y - \frac{1}{2}z = 1 \quad (2)$$
$$2x - y - z = 8 \quad (3)$$

Multiply equation (2) by –4 and add to equation (1).

$$\begin{array}{ll} 4x - y - 2z = 4 & (1) \\ \underline{-4x + 4y + 2z = -4} & (4) \\ 3y = 0 \Rightarrow y = 0 & \end{array}$$

Substitute 0 for y in equations (1) and (3).

$$4x - 2z = 4 \quad (5)$$
$$2x - z = 8 \quad (6)$$

Multiply equation (6) by –2 and add to equation (5).

$$\begin{array}{ll} 4x - 2z = 4 & (5) \\ \underline{-4x + 2z = -16} & (7) \\ 0 = -12 & (8) \end{array}$$

Since equation (8) is false, this system is inconsistent; there is no solution.

9.
$$3x + y - z = 3 \quad (1)$$
$$x + 2z = 6 \quad (2)$$
$$-3x - y + 2z = 9 \quad (3)$$

Add equations (1) and (3).

$$\begin{array}{l} 3x + y - z = 3 \\ \underline{-3x - y + 2z = 9} \\ z = 12 \end{array}$$

Substitute 12 for z in equation (2) and solve for x.

$x + 2(12) = 6 \Rightarrow x = -18$

Substitute –18 for x and 12 for z in equation (1) and solve for y.

$3(-18) + y - 12 = 3 \Rightarrow y = 69$

The solution is (–18, 69, 12).

10. $x + y - 4z = 0$ (1)

$2x + y - 3z = 2$ (2)

Multiply equation (1) by –1 and add to equation (2).

$$\begin{array}{rl} -x - y + 4z = 0 & (3) \\ 2x + y - 3z = 2 & (2) \\ \hline x \quad + z = 2 & (4) \end{array}$$

Solve equation (4) for x in terms of z.

$x + z = 2 \Rightarrow x = 2 - z$

Substitute $2 - z$ for x in equation (1) and solve for y in terms of z.

$$(2 - z) + y - 4z = 0$$
$$2 + y - 5z = 0$$
$$y = 5z - 2$$

The system is dependent, and the solution is all ordered triples of the form $(2 - z, 5z - 2, z)$ for any real number z.

We solve exercises 11–16 using matrix methods.

11. $x + z = -3$

$y - z = 6$

$2x + 3z = 5$

The augmented matrix is

$$\left[\begin{array}{ccc|c} 1 & 0 & 1 & -3 \\ 0 & 1 & -1 & 6 \\ 2 & 0 & 3 & 5 \end{array}\right]$$

$$\left[\begin{array}{ccc|c} 1 & 0 & 1 & -3 \\ 0 & 1 & -1 & 6 \\ 0 & 0 & 1 & 11 \end{array}\right] \begin{array}{l} \\ \\ -2R_1 + R_3 \end{array}$$

$$\left[\begin{array}{ccc|c} 1 & 0 & 0 & -14 \\ 0 & 1 & 0 & 17 \\ 0 & 0 & 1 & 11 \end{array}\right] \begin{array}{l} -1R_3 + R_1 \\ R_3 + R_2 \\ \\ \end{array}$$

The solution is (–14, 17, 11).

12. $2x + 3y + 4z = 8$

$-x + y - 2z = -9$

$2x + 2y + 6z = 16$

The augmented matrix is

$$\left[\begin{array}{ccc|c} 2 & 3 & 4 & 8 \\ -1 & 1 & -2 & -9 \\ 2 & 2 & 6 & 16 \end{array}\right]$$

$$\left[\begin{array}{ccc|c} 2 & 3 & 4 & 8 \\ -1 & 1 & -2 & -9 \\ 1 & 1 & 3 & 8 \end{array}\right] \begin{array}{l} \\ \\ \frac{1}{2}R_3 \end{array}$$

$$\left[\begin{array}{ccc|c} 1 & 1 & 3 & 8 \\ -1 & 1 & -2 & -9 \\ 2 & 3 & 4 & 8 \end{array}\right] \begin{array}{l} \text{Interchange } R_1 \text{ and } R_3 \\ \\ \\ \end{array}$$

$$\left[\begin{array}{ccc|c} 1 & 1 & 3 & 8 \\ 0 & 2 & 1 & -1 \\ 0 & 1 & -2 & -8 \end{array}\right] \begin{array}{l} \\ R_1 + R_2 \\ -2R_1 + R_3 \end{array}$$

$$\left[\begin{array}{ccc|c} 1 & 0 & 5 & 16 \\ 0 & 5 & 0 & -10 \\ 0 & 1 & -2 & -8 \end{array}\right] \begin{array}{l} R_1 - R_3 \\ 2R_2 + R_3 \\ \\ \end{array}$$

$$\left[\begin{array}{ccc|c} 1 & 0 & 5 & 16 \\ 0 & 1 & 0 & -2 \\ 0 & 1 & -2 & -8 \end{array}\right] \begin{array}{l} \\ \frac{1}{5}R_2 \\ \\ \end{array}$$

$$\left[\begin{array}{ccc|c} 1 & 0 & 5 & 16 \\ 0 & 1 & 0 & -2 \\ 0 & 0 & -2 & -6 \end{array}\right] \begin{array}{l} \\ \\ -R_2 + R_3 \end{array}$$

$$\left[\begin{array}{ccc|c} 1 & 0 & 5 & 16 \\ 0 & 1 & 0 & -2 \\ 0 & 0 & 1 & 3 \end{array}\right] \begin{array}{l} \\ \\ -\frac{1}{2}R_3 \end{array}$$

$$\left[\begin{array}{ccc|c} 1 & 0 & 0 & 1 \\ 0 & 1 & 0 & -2 \\ 0 & 0 & 1 & 3 \end{array}\right] \begin{array}{l} -5R_3 + R_1 \\ \\ \\ \end{array}$$

The solution is (1, –2, 3).

13. $x-2y+5z=3$
$4x+3y-4z=1$
$3x+5y-9z=7$
The augmented matrix is

$$\left[\begin{array}{ccc|c} 1 & -2 & 5 & 3 \\ 4 & 3 & -4 & 1 \\ 3 & 5 & -9 & 7 \end{array}\right]$$

$$\left[\begin{array}{ccc|c} 1 & -2 & 5 & 3 \\ 0 & 11 & -24 & -11 \\ 0 & 11 & -24 & -2 \end{array}\right]\begin{array}{l} \\ -4R_1+R_2 \\ -3R_1+R_2 \end{array}$$

$$\left[\begin{array}{ccc|c} 1 & -2 & 5 & 3 \\ 0 & 11 & -24 & -11 \\ 0 & 0 & 0 & 9 \end{array}\right]\begin{array}{l} \\ \\ -R_2+R_3 \end{array}$$

Since row 3 has all zeros except for the last entry, there is no solution.

14. $5x-8y+z=1$
$3x-2y+4z=3$
$10x-16y+2z=3$
The augmented matrix is

$$\left[\begin{array}{ccc|c} 5 & -8 & 1 & 1 \\ 3 & -2 & 4 & 3 \\ 10 & -16 & 2 & 3 \end{array}\right]$$

$$\left[\begin{array}{ccc|c} 1 & -\frac{8}{5} & \frac{1}{5} & \frac{1}{5} \\ 3 & -2 & 4 & 3 \\ 10 & -16 & 2 & 3 \end{array}\right]\begin{array}{l} \frac{1}{5}R_1 \\ \\ \\ \end{array}$$

$$\left[\begin{array}{ccc|c} 1 & -\frac{8}{5} & \frac{1}{5} & \frac{1}{5} \\ 0 & \frac{14}{5} & \frac{17}{5} & \frac{12}{5} \\ 0 & 0 & 0 & 1 \end{array}\right]\begin{array}{l} \\ -3R_1+R_2 \\ -10R_1+R_3 \end{array}$$

Since row 3 has all zeros except for the last entry, there is no solution.

15. $x-2y+3z=4$
$2x+y-4z=3$
$-3z+4y-z=-2$
The augmented matrix is

$$\left[\begin{array}{ccc|c} 1 & -2 & 3 & 4 \\ 2 & 1 & -4 & 3 \\ -3 & 4 & -1 & -2 \end{array}\right]$$

$$\left[\begin{array}{ccc|c} 1 & -2 & 3 & 4 \\ 0 & 5 & -10 & -5 \\ 0 & -2 & 8 & 10 \end{array}\right]\begin{array}{l} \\ -2R_1+R_2 \\ 3R_1+R_3 \end{array}$$

$$\left[\begin{array}{ccc|c} 1 & -2 & 3 & 4 \\ 0 & 1 & -2 & -1 \\ 0 & -2 & 8 & 10 \end{array}\right]\begin{array}{l} \\ \frac{1}{5}R_2 \\ \\ \end{array}$$

$$\left[\begin{array}{ccc|c} 1 & 0 & -1 & 2 \\ 0 & 1 & -2 & -1 \\ 0 & 0 & 4 & 8 \end{array}\right]\begin{array}{l} 2R_2+R_1 \\ \\ 2R_2+R_3 \end{array}$$

$$\left[\begin{array}{ccc|c} 1 & 0 & -1 & 2 \\ 0 & 1 & -2 & -1 \\ 0 & 0 & 1 & 2 \end{array}\right]\begin{array}{l} \\ \\ \frac{1}{4}R_3 \end{array}$$

$$\left[\begin{array}{ccc|c} 1 & 0 & 0 & 4 \\ 0 & 1 & 0 & 3 \\ 0 & 0 & 1 & 2 \end{array}\right]\begin{array}{l} R_3+R_1 \\ 2R_3+R_2 \\ \\ \end{array}$$

The solution is (4, 3, 2).

16. $3x+2y-6z=3$
$x+y+2z=2$
$2x+2y+5z=0$
The augmented matrix is
$$\left[\begin{array}{ccc|c} 3 & 2 & -6 & 3 \\ 1 & 1 & 2 & 2 \\ 2 & 2 & 5 & 0 \end{array}\right]$$
Interchange R_1 and R_2.
$$\left[\begin{array}{ccc|c} 1 & 1 & 2 & 2 \\ 3 & 2 & -6 & 3 \\ 2 & 2 & 5 & 0 \end{array}\right]$$
$$\left[\begin{array}{ccc|c} 1 & 1 & 2 & 2 \\ 0 & -1 & -12 & -3 \\ 0 & 0 & 1 & -4 \end{array}\right]\begin{array}{l} \\ -3R_1+R_2 \\ -2R_1+R_3 \end{array}$$
$$\left[\begin{array}{ccc|c} 1 & 1 & 2 & 2 \\ 0 & 1 & 12 & 3 \\ 0 & 0 & 1 & -4 \end{array}\right]\begin{array}{l} \\ -1R_2 \\ \\ \end{array}$$
$$\left[\begin{array}{ccc|c} 1 & 0 & -10 & -1 \\ 0 & 1 & 12 & 3 \\ 0 & 0 & 1 & -4 \end{array}\right]\begin{array}{l} -1R_2+R_1 \\ \\ \\ \end{array}$$
$$\left[\begin{array}{ccc|c} 1 & 0 & 0 & -41 \\ 0 & 1 & 0 & 51 \\ 0 & 0 & 1 & -4 \end{array}\right]\begin{array}{l} 10R_3+R_1 \\ -12R_3+R_2 \\ \\ \end{array}$$
The solution is (–41, 51, –4).

Exercises 17–20 can be solved using either the elimination method or matrix methods.

17. Let x = the number of one dollar bills, let y = the number of five dollar bills, and let z = the number of ten dollar bills.
From the given information, we have the system
$$\begin{aligned} x+y+z&=35 \quad (1)\\ x+5y+10z&=144 \quad (2)\\ z&=y+2 \quad (3)\end{aligned}$$

Substitute $y + 2$ for z in equation (1) to get
$$\begin{aligned} x+y+y+2&=35\\ x+2y+2&=35\\ x+2y&=33 \quad (4)\end{aligned}$$
Substitute $y + 2$ for z in equation (2) to get
$$\begin{aligned} x+5y+10(y+2)&=144\\ x+5y+10y+20&=144\\ x+15y&=124 \quad (5)\end{aligned}$$
Multiply equation (4) by –1 and add to equation (5).
$$\begin{aligned} -x-2y&=-33\\ x+15y&=124\\ \hline 13y&=91 \Rightarrow y=7\end{aligned}$$
Substitute 7 for y in equation (4) and solve for x.
$x+2(7)=33 \Rightarrow x=19$
Substitute 7 for y in equation (3) and solve for z.
$z=7+2=9$
The solution is 19 ones, 7 fives, and 9 tens.

18. Organize the information into a table.

	I	II	III
Food	100	200	150
Shelter	250	0	200
Counseling	0	100	100

Let x = the number of clients from source I, let y = the number of clients from source II, and let z = the number of clients from source III.
The system is
$$\begin{aligned} 100x+200y+150z&=50,000\\ 250x\qquad\quad +200z&=32,500\\ 100y+100z&=25,000\end{aligned}$$
Using matrix methods, we have
$$\left[\begin{array}{ccc|c} 100 & 200 & 150 & 50,000 \\ 250 & 0 & 200 & 32,500 \\ 0 & 100 & 100 & 25,000 \end{array}\right]$$
$$\left[\begin{array}{ccc|c} 1 & 2 & \frac{3}{2} & 500 \\ 1 & 0 & \frac{4}{5} & 130 \\ 0 & 1 & 1 & 250 \end{array}\right]\begin{array}{l} \frac{1}{100}R_1 \\ \frac{1}{250}R_2 \\ \frac{1}{100}R_3 \end{array}$$
$$\left[\begin{array}{ccc|c} 1 & 0 & -\frac{1}{2} & 0 \\ 0 & 2 & \frac{7}{10} & 370 \\ 0 & 1 & 1 & 250 \end{array}\right]\begin{array}{l} R_1-2R_3 \\ R_1-R_2 \\ \\ \end{array}$$
$$\left[\begin{array}{ccc|c} 1 & 0 & -\frac{1}{2} & 0 \\ 0 & 2 & \frac{7}{10} & 370 \\ 0 & 0 & \frac{13}{10} & 130 \end{array}\right]\begin{array}{l} \\ \\ 2R_3-R_2 \end{array}$$
$$\left[\begin{array}{ccc|c} 1 & 0 & -\frac{1}{2} & 0 \\ 0 & 2 & \frac{7}{10} & 370 \\ 0 & 0 & 1 & 100 \end{array}\right]\begin{array}{l} \\ \\ \frac{10}{13}R_3 \end{array}$$

(continued next page)

$$\left[\begin{array}{ccc|c} 1 & 0 & 0 & 50 \\ 0 & 2 & 0 & 300 \\ 0 & 0 & 1 & 100 \end{array}\right] \begin{array}{l} R_1 + \frac{1}{2}R_3 \\ R_2 - \frac{7}{10}R_3 \\ \end{array}$$

$$\left[\begin{array}{ccc|c} 1 & 0 & 0 & 50 \\ 0 & 1 & 0 & 150 \\ 0 & 0 & 1 & 100 \end{array}\right] \begin{array}{l} \\ \frac{1}{2}R_2 \\ \end{array}$$

Thus, the agency can serve 50 clients from source I, 150 from source II, and 100 from source III.

19. Let x = the number of blankets, let y = the number of rugs, and let z = the number of skirts. Solve the following system.

$$\begin{aligned} 24x + 30y + 12z &= 306 \quad (1) \\ 4x + 5y + 3z &= 59 \quad (2) \\ 15x + 18y + 9z &= 201 \quad (3) \end{aligned}$$

Simplify the system by dividing equation (1) by 6 and equation (3) by 3.

$$\begin{aligned} 4x + 5y + 2z &= 51 \quad (4) \\ 4x + 5y + 3z &= 59 \quad (2) \\ 5x + 6y + 3z &= 67 \quad (5) \end{aligned}$$

Multiply equation (4) by –1 and add to equation (2).

$$\begin{aligned} -4x - 5y - 2z &= -51 \\ 4x + 5y + 3z &= 59 \\ \hline z &= 8 \end{aligned}$$

Substitute 8 for z in equation (4).

$$\begin{aligned} 4x + 5y + 2(8) &= 51 \\ 4x + 5y &= 35 \quad (5) \end{aligned}$$

Substitute 8 for z in equation (5).

$$\begin{aligned} 5x + 6y + 3(8) &= 67 \\ 5x + 6y &= 43 \quad (6) \end{aligned}$$

Multiply equation (5) by 5 and equation (6) by –4 and add the results.

$$\begin{aligned} 20x + 25y &= 175 \\ -20x - 24y &= -172 \\ \hline y &= 3 \end{aligned}$$

Substitute 3 for y in equation (5) and solve for x.

$4x + 5(3) = 35 \Rightarrow 4x = 20 \Rightarrow x = 5$

They can make 5 blankets, 3 rugs, and 8 skirts.

20. Let x = the number of chairs, and let y = number tables, and let z = number of chests.

	Construction	Painting	Packing
Chair	2	1	2
Table	4	3	3
Chest	8	6	4
Totals	2000	1400	1300

$$\begin{aligned} 2x + 4y + 8z &= 2000 \\ x + 3y + 6z &= 1400 \\ 2x + 3y + 4z &= 1300 \end{aligned}$$

Divide the first equation by 2 and write the augmented matrix.

$$\left[\begin{array}{ccc|c} 1 & 2 & 4 & 1000 \\ 1 & 3 & 6 & 1400 \\ 2 & 3 & 4 & 1300 \end{array}\right]$$

$$\left[\begin{array}{ccc|c} 1 & 2 & 4 & 1000 \\ 0 & 1 & 2 & 400 \\ 0 & -1 & -4 & -700 \end{array}\right] \begin{array}{l} \\ -1R_1 + R_2 \\ -2R_1 + R_3 \end{array}$$

$$\left[\begin{array}{ccc|c} 1 & 0 & 0 & 200 \\ 0 & 1 & 2 & 400 \\ 0 & 0 & -2 & -300 \end{array}\right] \begin{array}{l} -2R_2 + R_1 \\ \\ R_2 + R_3 \end{array}$$

$$\left[\begin{array}{ccc|c} 1 & 0 & 0 & 200 \\ 0 & 1 & 0 & 100 \\ 0 & 0 & -2 & -300 \end{array}\right] \begin{array}{l} \\ R_3 + R_2 \\ \end{array}$$

$$\left[\begin{array}{ccc|c} 1 & 0 & 0 & 200 \\ 0 & 1 & 0 & 100 \\ 0 & 0 & 1 & 150 \end{array}\right] \begin{array}{l} \\ \\ -\frac{1}{2}R_3 \end{array}$$

The factory can produce 200 chairs, 100 tables, and 150 chests.

21. $\begin{bmatrix} 2 & 3 \\ 5 & 9 \end{bmatrix}$

The matrix is 2×2. Since the matrix is 2×2, it is square.

22. $\begin{bmatrix} 2 & -1 \\ 4 & 6 \\ 5 & 7 \end{bmatrix}$

The size of this matrix is 3×2.

23. $\begin{bmatrix} 12 & 4 & -8 & -1 \end{bmatrix}$

The matrix is 1×4. The matrix is a row matrix.

24. $\begin{bmatrix} -7 & 5 & 6 & 4 \\ 3 & 2 & -1 & 2 \\ -1 & 12 & 8 & -1 \end{bmatrix}$

This is a 3×4 matrix.

25. $\begin{bmatrix} 6 & 8 & 10 \\ 5 & 3 & -2 \end{bmatrix}$

This matrix is 2×3.

26. $\begin{bmatrix} -9 \\ 15 \\ 4 \end{bmatrix}$

This matrix is 3×1. It is a column matrix.

27. As a 3 x 3 matrix, the data is

$$\begin{bmatrix} 542.10 & -6.93 & 12{,}605{,}900 \\ 52.37 & .53 & 3{,}616{,}100 \\ 44.06 & -.21 & 11{,}227{,}700 \end{bmatrix}$$

28. $\begin{bmatrix} 8 & 8 & 8 \\ 10 & 5 & 9 \\ 7 & 10 & 7 \\ 8 & 9 & 7 \end{bmatrix}$

29. $B = \begin{bmatrix} 1 & 2 & -3 \\ 2 & 3 & 0 \\ 0 & 1 & 4 \end{bmatrix}$; $-B = \begin{bmatrix} -1 & -2 & 3 \\ -2 & -3 & 0 \\ 0 & -1 & -4 \end{bmatrix}$

30. $D = \begin{bmatrix} 6 \\ 1 \\ 0 \end{bmatrix}$; $-D = \begin{bmatrix} -6 \\ -1 \\ 0 \end{bmatrix}$

31. $A = \begin{bmatrix} 4 & 6 \\ -2 & -2 \\ 5 & 9 \end{bmatrix}, C = \begin{bmatrix} 5 & 0 \\ -1 & 3 \\ 4 & 7 \end{bmatrix}$

$A + 2C$

$$A + 2C = \begin{bmatrix} 4 & 6 \\ -2 & -2 \\ 5 & 9 \end{bmatrix} + \begin{bmatrix} 10 & 0 \\ -2 & 6 \\ 8 & 14 \end{bmatrix} = \begin{bmatrix} 14 & 6 \\ -4 & 4 \\ 13 & 23 \end{bmatrix}$$

32. $F = \begin{bmatrix} -1 & 2 \\ 6 & 7 \end{bmatrix}, G = \begin{bmatrix} 2 & 5 \\ 1 & 6 \end{bmatrix}$

$$F + 3G = \begin{bmatrix} -1 & 2 \\ 6 & 7 \end{bmatrix} + \begin{bmatrix} 6 & 15 \\ 3 & 18 \end{bmatrix} = \begin{bmatrix} 5 & 17 \\ 9 & 25 \end{bmatrix}$$

33. $B = \begin{bmatrix} 1 & 2 & -3 \\ 2 & 3 & 0 \\ 0 & 1 & 4 \end{bmatrix}, C = \begin{bmatrix} 5 & 0 \\ -1 & 3 \\ 4 & 7 \end{bmatrix}$

$2B - 5C$ is not defined, since the matrices have different sizes.

34. $D = \begin{bmatrix} 6 \\ 1 \\ 0 \end{bmatrix}, E = \begin{bmatrix} 1 & 3 & -4 \end{bmatrix}$

$D + E$ is not defined, since the matrices have different sizes.

35. $A = \begin{bmatrix} 4 & 6 \\ -2 & -2 \\ 5 & 9 \end{bmatrix}, C = \begin{bmatrix} 5 & 0 \\ -1 & 3 \\ 4 & 7 \end{bmatrix}$

$3A - 2C$

$$3A - 2C = \begin{bmatrix} 12 & 18 \\ -6 & -6 \\ 15 & 27 \end{bmatrix} - \begin{bmatrix} 10 & 0 \\ -2 & 6 \\ 8 & 14 \end{bmatrix} = \begin{bmatrix} 2 & 18 \\ -4 & -12 \\ 7 & 13 \end{bmatrix}$$

36. $G = \begin{bmatrix} 2 & 5 \\ 1 & 6 \end{bmatrix}, F = \begin{bmatrix} -1 & 2 \\ 6 & 7 \end{bmatrix}$

$$G - 2F = \begin{bmatrix} 2 & 5 \\ 1 & 6 \end{bmatrix} - \begin{bmatrix} -2 & 4 \\ 12 & 14 \end{bmatrix} = \begin{bmatrix} 4 & 1 \\ -11 & -8 \end{bmatrix}$$

37. The 3×2 matrix that represents the first day's data is:

$$\begin{bmatrix} -6.93 & 12{,}605{,}900 \\ .53 & 3{,}616{,}100 \\ -.21 & 11{,}227{,}700 \end{bmatrix}$$

The 3×2 matrix that represents the next day's data is:

$$\begin{bmatrix} -15.1 & 21{,}226{,}200 \\ .51 & 3{,}397{,}000 \\ .24 & 14{,}930{,}400 \end{bmatrix}$$

The sum of these matrices is:

$$\begin{bmatrix} -22.03 & 33{,}832{,}100 \\ 1.04 & 7{,}013{,}100 \\ .03 & 26{,}158{,}100 \end{bmatrix}$$

38. a. First shipment:

$$\begin{array}{l} \\ \text{Chicago} \\ \text{Dallas} \\ \text{Atlanta} \end{array} \begin{array}{c} \text{Tulsa} \quad \text{New Orleans} \\ \begin{bmatrix} 110{,}000 & 85{,}000 \\ 73{,}000 & 108{,}000 \\ 95{,}000 & 69{,}000 \end{bmatrix} \end{array}$$

Second shipment:

$$\begin{bmatrix} 58{,}000 & 40{,}000 \\ 33{,}000 & 52{,}000 \\ 80{,}000 & 30{,}000 \end{bmatrix}$$

b. Add the 2 matrices in part (a).

$$\begin{bmatrix} 168{,}000 & 125{,}000 \\ 106{,}000 & 160{,}000 \\ 175{,}000 & 99{,}000 \end{bmatrix}$$

39. $A = \begin{bmatrix} 4 & 6 \\ -2 & -2 \\ 5 & 9 \end{bmatrix}, G = \begin{bmatrix} 2 & 5 \\ 1 & 6 \end{bmatrix}$

$$AG = \begin{bmatrix} (4)(2)+(6)(1) & (4)(5)+(6)(6) \\ (-2)(2)+(-2)(1) & (-2)(5)+(-2)(6) \\ (5)(2)+(9)(1) & (5)(5)+(9)(6) \end{bmatrix} = \begin{bmatrix} 14 & 56 \\ -6 & -22 \\ 19 & 79 \end{bmatrix}$$

40. $E = \begin{bmatrix} 1 & 3 & -4 \end{bmatrix}, B = \begin{bmatrix} 1 & 2 & -3 \\ 2 & 3 & 0 \\ 0 & 1 & 4 \end{bmatrix}$

$EB = \begin{bmatrix} 7 & 7 & -19 \end{bmatrix}$

41. $G = \begin{bmatrix} 2 & 5 \\ 1 & 6 \end{bmatrix}, A = \begin{bmatrix} 4 & 6 \\ -2 & -2 \\ 5 & 9 \end{bmatrix}$

GA is not defined since the number of columns in G is not equal to the number of rows in A

42. $C = \begin{bmatrix} 5 & 0 \\ -1 & 3 \\ 4 & 7 \end{bmatrix}, A = \begin{bmatrix} 4 & 6 \\ -2 & -2 \\ 5 & 9 \end{bmatrix}$

CA is not defined since the number of columns in C is not equal to the number of rows in A.

43. $A = \begin{bmatrix} 4 & 6 \\ -2 & -2 \\ 5 & 9 \end{bmatrix}, G = \begin{bmatrix} 2 & 5 \\ 1 & 6 \end{bmatrix}, F = \begin{bmatrix} -1 & 2 \\ 6 & 7 \end{bmatrix}$

$GF = \begin{bmatrix} 2 & 5 \\ 1 & 6 \end{bmatrix}\begin{bmatrix} -1 & 2 \\ 6 & 7 \end{bmatrix} = \begin{bmatrix} 28 & 39 \\ 35 & 44 \end{bmatrix}$.

Then, $AGF = A(GF)$

$$AGF = A(GF) = \begin{bmatrix} 4 & 6 \\ -2 & -2 \\ 5 & 9 \end{bmatrix}\begin{bmatrix} 28 & 39 \\ 35 & 44 \end{bmatrix} = \begin{bmatrix} 322 & 420 \\ -126 & -166 \\ 455 & 591 \end{bmatrix}$$

44. $E = \begin{bmatrix} 1 & 3 & -4 \end{bmatrix}, B = \begin{bmatrix} 1 & 2 & -3 \\ 2 & 3 & 0 \\ 0 & 1 & 4 \end{bmatrix}, D = \begin{bmatrix} 6 \\ 1 \\ 0 \end{bmatrix}$

From Exercise 40, $EB = \begin{bmatrix} 7 & 7 & -19 \end{bmatrix}$.

Then, $EBD = (EB)D$

$$EBD = (EB)D = \begin{bmatrix} 7 & 7 & -19 \end{bmatrix}\begin{bmatrix} 6 \\ 1 \\ 0 \end{bmatrix} = \begin{bmatrix} 49 \end{bmatrix}$$

45. There were 8000 athlete exposures for the league. The column matrix used will have values of 4 because there are an equal number and they are measured in thousands. Therefore, the matrix is

$$\begin{bmatrix} 3.54 & 1.41 \\ 1.53 & 1.57 \\ .34 & .29 \\ 7.53 & 6.21 \end{bmatrix} \cdot \begin{bmatrix} 4 \\ 4 \end{bmatrix} = \begin{bmatrix} 3.54(4)+1.41(4) \\ 1.53(4)+1.57(4) \\ .34(4)+.29(4) \\ 7.53(4)+6.21(4) \end{bmatrix} = \begin{bmatrix} 19.8 \\ 12.4 \\ 2.52 \\ 54.96 \end{bmatrix}$$

Therefore, there are about 20 head and face injuries, about 12 concussions, about 3 neck injuries, and about 55 other injuries.

46. a.

$$\begin{array}{l} \\ \text{Standard} \\ \text{Extra Large} \end{array} \begin{array}{c} \text{Cutting} \quad \text{Shaping} \\ \begin{bmatrix} \frac{1}{4} & \frac{1}{2} \\ \frac{1}{3} & \frac{1}{3} \end{bmatrix} \end{array}$$

b. $\begin{bmatrix} 48 & 66 \end{bmatrix} \begin{bmatrix} \frac{1}{4} & \frac{1}{2} \\ \frac{1}{3} & \frac{1}{3} \end{bmatrix} = \begin{bmatrix} 34 & 46 \end{bmatrix}$

The cutting machine will operate for 34 hr and the shaping machine for 46 hr.

47. a.

	Cost Per Share	Earnings Per Share
Stock 1	61.17	1
Stock 2	60.50	.84
Stock 3	96.21	1.88

b.

	Stock 1	Stock 2	Stock 3
Number of shares	100	500	200

c. $\begin{bmatrix} 100 & 500 & 200 \end{bmatrix} \begin{bmatrix} 61.17 & 1 \\ 60.50 & .84 \\ 96.21 & 1.88 \end{bmatrix}$

$= \begin{bmatrix} 55{,}609 & 896 \end{bmatrix}$

Total cost = \$55,609.
Total dividend = \$896.

48. a.

	Cost Per Share	Earnings Per Share
Stock 1	9.45	0
Stock 2	6.59	.24
Stock 3	18.21	0
Stock 4	60.70	.88

b.

	Stock 1	Stock 2	Stock 3	Stock 4
shares	4000	1000	5000	2000

c. $\begin{bmatrix} 4000 & 1000 & 5000 & 2000 \end{bmatrix} \begin{bmatrix} 9.45 & 0 \\ 6.59 & .24 \\ 18.21 & 0 \\ 60.70 & .88 \end{bmatrix}$

$= \begin{bmatrix} 256{,}840 & 2000 \end{bmatrix}$

Total cost = \$256,840.
Total dividend = \$2000.

49. There are many correct answers. Here is one example.

$A = \begin{bmatrix} 3 & 0 \\ 2 & 1 \end{bmatrix}$; let $B = \begin{bmatrix} 1 & 2 \\ 3 & 4 \end{bmatrix}$.

$AB = \begin{bmatrix} 3 & 6 \\ 5 & 8 \end{bmatrix}$; $BA = \begin{bmatrix} 7 & 2 \\ 17 & 4 \end{bmatrix}$

Thus, AB and BA are both defined, and $AB \neq BA$.

50. No. $A = 4I$, so $AB = BA = 4B$.

51. Let $A = \begin{bmatrix} -2 & 2 \\ 0 & 5 \end{bmatrix}$.

$$[A \mid I] = \left[\begin{array}{cc|cc} -2 & 2 & 1 & 0 \\ 0 & 5 & 0 & 1 \end{array}\right]$$

$$= \left[\begin{array}{cc|cc} 1 & -1 & -\frac{1}{2} & 0 \\ 0 & 1 & 0 & \frac{1}{5} \end{array}\right] \begin{array}{l} -\frac{1}{2}R_1 \\ \frac{1}{5}R_2 \end{array}$$

$$= \left[\begin{array}{cc|cc} 1 & 0 & -\frac{1}{2} & \frac{1}{5} \\ 0 & 1 & 0 & \frac{1}{5} \end{array}\right] R_2 + R_1$$

$$A^{-1} = \begin{bmatrix} -.5 & .2 \\ 0 & .2 \end{bmatrix}$$

52. Let $A = \begin{bmatrix} 3 & -1 \\ -5 & 2 \end{bmatrix}$.

$$[A \mid I] = \left[\begin{array}{cc|cc} 3 & -1 & 1 & 0 \\ -5 & 2 & 0 & 1 \end{array}\right]$$

$$\left[\begin{array}{cc|cc} 1 & -\frac{1}{3} & \frac{1}{3} & 0 \\ -5 & 2 & 0 & 1 \end{array}\right] \begin{matrix} \frac{1}{3}R_1 \\ \end{matrix}$$

$$\left[\begin{array}{cc|cc} 1 & -\frac{1}{3} & \frac{1}{3} & 0 \\ 0 & \frac{1}{3} & \frac{5}{3} & 1 \end{array}\right] \begin{matrix} \\ 5R_1 + R_2 \end{matrix}$$

$$\left[\begin{array}{cc|cc} 1 & 0 & 2 & 1 \\ 0 & 1 & 5 & 3 \end{array}\right] \begin{matrix} R_1 + R_2 \\ 3R_2 \end{matrix}$$

$$A^{-1} = \begin{bmatrix} 2 & 1 \\ 5 & 3 \end{bmatrix}$$

53. Let $A = \begin{bmatrix} 6 & 4 \\ 3 & 2 \end{bmatrix}$.

$$[A \mid I] = \left[\begin{array}{cc|cc} 6 & 4 & 1 & 0 \\ 3 & 2 & 0 & 1 \end{array}\right]$$

$$\left[\begin{array}{cc|cc} 1 & \frac{2}{3} & \frac{1}{6} & 0 \\ 3 & 2 & 0 & 1 \end{array}\right] \begin{matrix} \frac{1}{6}R_1 \\ \end{matrix}$$

$$\left[\begin{array}{cc|cc} 1 & \frac{2}{3} & \frac{1}{6} & 0 \\ 0 & 0 & -\frac{1}{2} & 1 \end{array}\right] \begin{matrix} \\ -3R_1 + R_2 \end{matrix}$$

The second row can never become $[0 \quad 1]$, so A has no inverse.

54. Let $A = \begin{bmatrix} 15 & -12 \\ 5 & -4 \end{bmatrix}$.

$$[A \mid I] = \left[\begin{array}{cc|cc} 15 & -12 & 1 & 0 \\ 5 & -4 & 0 & 1 \end{array}\right]$$

$$\left[\begin{array}{cc|cc} 1 & -\frac{4}{5} & \frac{1}{15} & 0 \\ 5 & -4 & 0 & 1 \end{array}\right] \begin{matrix} \frac{1}{15}R_1 \\ \end{matrix}$$

$$\left[\begin{array}{cc|cc} 1 & -\frac{4}{5} & \frac{1}{15} & 0 \\ 0 & 0 & -\frac{1}{3} & 1 \end{array}\right] \begin{matrix} \\ -5R_1 + R_2 \end{matrix}$$

The second row can never become $[0 \quad 1]$, so A has no inverse.

55. Let $A = \begin{bmatrix} 2 & 0 & 6 \\ 1 & -1 & 0 \\ 0 & 1 & -3 \end{bmatrix}$

$$[A \mid I] = \left[\begin{array}{ccc|ccc} 2 & 0 & 6 & 1 & 0 & 0 \\ 1 & -1 & 0 & 0 & 1 & 0 \\ 0 & 1 & -3 & 0 & 0 & 1 \end{array}\right]$$

Interchange rows.

$$\left[\begin{array}{ccc|ccc} 1 & -1 & 0 & 0 & 1 & 0 \\ 0 & 1 & -3 & 0 & 0 & 1 \\ 2 & 0 & 6 & 1 & 0 & 0 \end{array}\right]$$

$$\left[\begin{array}{ccc|ccc} 1 & -1 & 0 & 0 & 1 & 0 \\ 0 & 1 & -3 & 0 & 0 & 1 \\ 0 & 2 & 6 & 1 & -2 & 0 \end{array}\right] \begin{matrix} \\ \\ -2R_1 + R_3 \end{matrix}$$

$$\left[\begin{array}{ccc|ccc} 1 & 0 & -3 & 0 & 1 & 1 \\ 0 & 1 & -3 & 0 & 0 & 1 \\ 0 & 0 & 12 & 1 & -2 & -2 \end{array}\right] \begin{matrix} R_2 + R_1 \\ \\ -2R_2 + R_3 \end{matrix}$$

$$\left[\begin{array}{ccc|ccc} 1 & 0 & -3 & 0 & 1 & 1 \\ 0 & 1 & -3 & 0 & 0 & 1 \\ 0 & 0 & 1 & \frac{1}{12} & -\frac{1}{6} & -\frac{1}{6} \end{array}\right] \begin{matrix} \\ \\ \frac{1}{12}R_3 \end{matrix}$$

$$[I \mid B] = \left[\begin{array}{ccc|ccc} 1 & 0 & 0 & \frac{1}{4} & \frac{1}{2} & \frac{1}{2} \\ 0 & 1 & 0 & \frac{1}{4} & -\frac{1}{2} & \frac{1}{2} \\ 0 & 0 & 1 & \frac{1}{12} & -\frac{1}{6} & -\frac{1}{6} \end{array}\right] \begin{matrix} 3R_3 + R_1 \\ 3R_3 + R_2 \\ \end{matrix}$$

$$A^{-1} = \begin{bmatrix} \frac{1}{4} & \frac{1}{2} & \frac{1}{2} \\ \frac{1}{4} & -\frac{1}{2} & \frac{1}{2} \\ \frac{1}{12} & -\frac{1}{6} & -\frac{1}{6} \end{bmatrix}$$

56. Let $A = \begin{bmatrix} 2 & -1 & 0 \\ 1 & 0 & 2 \\ 1 & -4 & 0 \end{bmatrix}$

$$\left[\begin{array}{ccc|ccc} 2 & -1 & 0 & 1 & 0 & 0 \\ 1 & 0 & 2 & 0 & 1 & 0 \\ 1 & -4 & 0 & 0 & 0 & 1 \end{array}\right]$$

Interchange rows 1 and 2.

$$\left[\begin{array}{ccc|ccc} 1 & 0 & 2 & 0 & 1 & 0 \\ 2 & -1 & 0 & 1 & 0 & 0 \\ 1 & -4 & 0 & 0 & 0 & 1 \end{array}\right]$$

$$\left[\begin{array}{ccc|ccc} 1 & 0 & 2 & 0 & 1 & 0 \\ 0 & -1 & -4 & 1 & -2 & 0 \\ 0 & -4 & -2 & 0 & -1 & 1 \end{array}\right]\begin{array}{l} \\ -2R_1 + R_2 \\ -1R_1 + R_3 \end{array}$$

Multiply row 2 by −1 and continue.

$$\left[\begin{array}{ccc|ccc} 1 & 0 & 2 & 0 & 1 & 0 \\ 0 & 1 & 4 & -1 & 2 & 0 \\ 0 & 0 & 14 & -4 & 7 & 1 \end{array}\right]\begin{array}{l} \\ \\ -4R_2 + R_3 \end{array}$$

$$\left[\begin{array}{ccc|ccc} 1 & 0 & 2 & 0 & 1 & 0 \\ 0 & 1 & 4 & -1 & 2 & 0 \\ 0 & 0 & 1 & -\frac{2}{7} & \frac{1}{2} & \frac{1}{14} \end{array}\right]\begin{array}{l} \\ \\ \frac{1}{14}R_3 \end{array}$$

$$\left[\begin{array}{ccc|ccc} 1 & 0 & 0 & \frac{4}{7} & 0 & -\frac{1}{7} \\ 0 & 1 & 0 & \frac{1}{7} & 0 & -\frac{2}{7} \\ 0 & 0 & 1 & -\frac{2}{7} & \frac{1}{2} & \frac{1}{14} \end{array}\right]\begin{array}{l} -2R_3 + R_1 \\ -4R_3 + R_2 \\ \\ \end{array}$$

$$A^{-1} = \begin{bmatrix} \frac{4}{7} & 0 & -\frac{1}{7} \\ \frac{1}{7} & 0 & -\frac{2}{7} \\ -\frac{2}{7} & \frac{1}{2} & \frac{1}{14} \end{bmatrix}$$

57. Let $A = \begin{bmatrix} 2 & 3 & 5 \\ -2 & -3 & -5 \\ 1 & 4 & 2 \end{bmatrix}$.

$$[A \mid I] = \left[\begin{array}{ccc|ccc} 2 & 3 & 5 & 1 & 0 & 0 \\ -2 & -3 & -5 & 0 & 1 & 0 \\ 1 & 4 & 2 & 0 & 0 & 1 \end{array}\right]$$

$$\left[\begin{array}{ccc|ccc} 1 & 4 & 2 & 0 & 0 & 1 \\ -2 & -3 & -5 & 0 & 1 & 0 \\ 2 & 3 & 5 & 1 & 0 & 0 \end{array}\right]\text{Interchange } R_1 \text{ and } R_3$$

$$\left[\begin{array}{ccc|ccc} 1 & 4 & 2 & 0 & 0 & 1 \\ -2 & -3 & -5 & 0 & 1 & 0 \\ 0 & 0 & 0 & 1 & 1 & 0 \end{array}\right]\begin{array}{l} \\ \\ R_2 + R_3 \end{array}$$

The third row can never become $[0 \quad 0 \quad 1]$, so A does not have an inverse.

58. Let $A = \begin{bmatrix} 1 & 3 & 6 \\ 4 & 0 & 9 \\ 5 & 15 & 30 \end{bmatrix}$

$$\left[\begin{array}{ccc|ccc} 1 & 3 & 6 & 1 & 0 & 0 \\ 4 & 0 & 9 & 0 & 1 & 0 \\ 5 & 15 & 30 & 0 & 0 & 1 \end{array}\right]$$

$$\left[\begin{array}{ccc|ccc} 1 & 3 & 6 & 1 & 0 & 0 \\ 0 & -12 & -15 & -4 & 1 & 0 \\ 0 & 0 & 0 & -5 & 0 & 1 \end{array}\right]\begin{array}{l} \\ -4R_1 + R_2 \\ -5R_1 + R_3 \end{array}$$

The last row is all zeros, so no inverse exists.

59. Let $A = \begin{bmatrix} 1 & 3 & -2 & -1 \\ 0 & 1 & 1 & 2 \\ -1 & -1 & 1 & -1 \\ 1 & -1 & -3 & -2 \end{bmatrix}$.

Use a graphing calculator to find A^{-1}.

$$A^{-1} = \begin{bmatrix} -\frac{2}{3} & -\frac{17}{3} & -\frac{14}{3} & -3 \\ \frac{1}{3} & \frac{1}{3} & \frac{1}{3} & 0 \\ -\frac{1}{3} & -\frac{10}{3} & -\frac{7}{3} & -2 \\ 0 & 2 & 1 & 1 \end{bmatrix}$$

60. Let $A = \begin{bmatrix} 3 & 2 & 0 & -1 \\ 2 & 0 & 1 & 2 \\ 1 & 2 & -1 & 0 \\ 2 & -1 & 1 & 1 \end{bmatrix}$.

Use a graphing calculator to find A^{-1}.

$$A^{-1} = \begin{bmatrix} 0 & -\frac{1}{3} & \frac{1}{3} & \frac{2}{3} \\ \frac{1}{3} & \frac{2}{3} & -\frac{1}{3} & -1 \\ \frac{2}{3} & 1 & -\frac{4}{3} & -\frac{4}{3} \\ -\frac{1}{3} & \frac{1}{3} & \frac{1}{3} & 0 \end{bmatrix}$$

61. $F = \begin{bmatrix} -1 & 2 \\ 6 & 7 \end{bmatrix}$

Form the augmented matrix $[F \mid I]$.

$$\left[\begin{array}{cc|cc} -1 & 2 & 1 & 0 \\ 6 & 7 & 0 & 1 \end{array}\right]$$

$$\left[\begin{array}{cc|cc} 1 & -2 & -1 & 0 \\ 6 & 7 & 0 & 1 \end{array}\right] -1R_1$$

$$\left[\begin{array}{cc|cc} 1 & -2 & -1 & 0 \\ 0 & 19 & 6 & 1 \end{array}\right] -6R_1 + R_2$$

$$\left[\begin{array}{cc|cc} 1 & -2 & -1 & 0 \\ 0 & 1 & \frac{6}{19} & \frac{1}{19} \end{array}\right] \frac{1}{19}R_2$$

$$\left[\begin{array}{cc|cc} 1 & 0 & -\frac{7}{19} & \frac{2}{19} \\ 0 & 1 & \frac{6}{19} & \frac{1}{19} \end{array}\right] R_1 + 2R_2$$

$$F^{-1} = \begin{bmatrix} -\frac{7}{19} & \frac{2}{19} \\ \frac{6}{19} & \frac{1}{19} \end{bmatrix}$$

62. $G = \begin{bmatrix} 2 & 5 \\ 1 & 6 \end{bmatrix}$

Form the augmented matrix $[G \mid I]$.

$$\left[\begin{array}{cc|cc} 2 & 5 & 1 & 0 \\ 1 & 6 & 0 & 1 \end{array}\right]$$

Interchange R_1 and R_2.

$$\left[\begin{array}{cc|cc} 1 & 6 & 0 & 1 \\ 2 & 5 & 1 & 0 \end{array}\right]$$

$$\left[\begin{array}{cc|cc} 1 & 6 & 0 & 1 \\ 0 & -7 & 1 & -2 \end{array}\right] -2R_1 + R_2$$

$$\left[\begin{array}{cc|cc} 1 & 6 & 0 & 1 \\ 0 & 1 & -\frac{1}{7} & \frac{2}{7} \end{array}\right] -\frac{1}{7}R_2$$

$$\left[\begin{array}{cc|cc} 1 & 0 & \frac{6}{7} & -\frac{5}{7} \\ 0 & 1 & -\frac{1}{7} & \frac{2}{7} \end{array}\right] -6R_2 + R_1$$

$$G^{-1} = \begin{bmatrix} \frac{6}{7} & -\frac{5}{7} \\ -\frac{1}{7} & \frac{2}{7} \end{bmatrix}$$

63. $G = \begin{bmatrix} 2 & 5 \\ 1 & 6 \end{bmatrix}, F = \begin{bmatrix} -1 & 2 \\ 6 & 7 \end{bmatrix}$

$$G - F = \begin{bmatrix} 3 & 3 \\ -5 & -1 \end{bmatrix}$$

Form the augmented matrix $[G - F \mid I]$.

$$\left[\begin{array}{cc|cc} 3 & 3 & 1 & 0 \\ -5 & -1 & 0 & 1 \end{array}\right]$$

$$\left[\begin{array}{cc|cc} 1 & 1 & \frac{1}{3} & 0 \\ -5 & -1 & 0 & 1 \end{array}\right] \frac{1}{3}R_1$$

$$\left[\begin{array}{cc|cc} 1 & 1 & \frac{1}{3} & 0 \\ 0 & 4 & \frac{5}{3} & 1 \end{array}\right] 5R_1 + R_2$$

$$\left[\begin{array}{cc|cc} 1 & 1 & \frac{1}{3} & 0 \\ 0 & 1 & \frac{5}{12} & \frac{1}{4} \end{array}\right] \frac{1}{4}R_2$$

$$\left[\begin{array}{cc|cc} 1 & 0 & -\frac{1}{12} & -\frac{1}{4} \\ 0 & 1 & \frac{5}{12} & \frac{1}{4} \end{array}\right] -R_2 + R_1$$

$$(G - F)^{-1} = \begin{bmatrix} -\frac{1}{12} & -\frac{1}{4} \\ \frac{5}{12} & \frac{1}{4} \end{bmatrix}$$

64. $F = \begin{bmatrix} -1 & 2 \\ 6 & 7 \end{bmatrix}, G = \begin{bmatrix} 2 & 5 \\ 1 & 6 \end{bmatrix}$

$$F + G = \begin{bmatrix} 1 & 7 \\ 7 & 13 \end{bmatrix}$$

Form the augmented matrix $[F + G \mid I]$.

$$\left[\begin{array}{cc|cc} 1 & 7 & 1 & 0 \\ 7 & 13 & 0 & 1 \end{array}\right]$$

$$\left[\begin{array}{cc|cc} 1 & 7 & 1 & 0 \\ 0 & -36 & -7 & 1 \end{array}\right] -7R_1 + R_2$$

$$\left[\begin{array}{cc|cc} 1 & 7 & 1 & 0 \\ 0 & 1 & \frac{7}{36} & -\frac{1}{36} \end{array}\right] -\frac{1}{36}R_2$$

$$\left[\begin{array}{cc|cc} 1 & 0 & -\frac{13}{36} & \frac{7}{36} \\ 0 & 1 & \frac{7}{36} & -\frac{1}{36} \end{array}\right] -7R_2 + R_1$$

$$(F + G)^{-1} = \begin{bmatrix} -\frac{13}{36} & \frac{7}{36} \\ \frac{7}{36} & -\frac{1}{36} \end{bmatrix}$$

65. $B=\begin{bmatrix}1&2&-3\\2&3&0\\0&1&4\end{bmatrix}$

Form the augmented matrix $[B\mid I]$.

$$\left[\begin{array}{ccc|ccc}1&2&-3&1&0&0\\2&3&0&0&1&0\\0&1&4&0&0&1\end{array}\right]$$

$$\left[\begin{array}{ccc|ccc}1&0&-11&1&0&-2\\0&1&-6&2&-1&0\\0&1&4&0&0&1\end{array}\right]\begin{array}{l}-2R_3+R_1\\-R_2+2R_1\\ \end{array}$$

$$\left[\begin{array}{ccc|ccc}1&0&-11&1&0&-2\\0&1&-6&2&-1&0\\0&0&1&-\frac{1}{5}&\frac{1}{10}&\frac{1}{10}\end{array}\right]\begin{array}{l}\\ \\ \frac{1}{10}(-R_2+R_3)\end{array}$$

$$\left[\begin{array}{ccc|ccc}1&0&0&-\frac{6}{5}&\frac{11}{10}&-\frac{9}{10}\\0&1&0&\frac{4}{5}&-\frac{2}{5}&\frac{3}{5}\\0&0&1&-\frac{1}{5}&\frac{1}{10}&\frac{1}{10}\end{array}\right]\begin{array}{l}R_1+11R_3\\R_2+6R_3\\ \end{array}$$

$$B^{-1}=\begin{bmatrix}-1.2&1.1&-.9\\.8&-.4&.6\\-.2&.1&.1\end{bmatrix}$$

66. Answers vary. Sample answer: There is no way to transform the given matrix into the identity matrix using row operations

67. $A=\begin{bmatrix}-3&4\\-1&2\end{bmatrix}$, $B=\begin{bmatrix}3\\-1\end{bmatrix}$

If $AX=B$, then $X=A^{-1}B$.

If $A=\begin{bmatrix}-3&4\\-1&2\end{bmatrix}$, then $A^{-1}=\begin{bmatrix}-1&2\\-\frac{1}{2}&\frac{3}{2}\end{bmatrix}$.

Thus, $X=\begin{bmatrix}-1&2\\-\frac{1}{2}&\frac{3}{2}\end{bmatrix}\begin{bmatrix}3\\-1\end{bmatrix}=\begin{bmatrix}-5\\-3\end{bmatrix}$.

68. $A=\begin{bmatrix}1&3\\-2&4\end{bmatrix}$, $B=\begin{bmatrix}9\\6\end{bmatrix}$

$$A^{-1}=\begin{bmatrix}\frac{2}{5}&-\frac{3}{10}\\\frac{1}{5}&\frac{1}{10}\end{bmatrix}$$

Then $X=A^{-1}B$

$$\begin{bmatrix}\frac{2}{5}&-\frac{3}{10}\\\frac{1}{5}&\frac{1}{10}\end{bmatrix}\begin{bmatrix}9\\6\end{bmatrix}=\begin{bmatrix}\frac{9}{5}\\\frac{12}{5}\end{bmatrix}.$$

69. $A=\begin{bmatrix}1&0&2\\-1&1&0\\3&0&4\end{bmatrix}$, $B=\begin{bmatrix}8\\4\\-6\end{bmatrix}$

If $AX=B$, then $X=A^{-1}B$.

If $A=\begin{bmatrix}1&0&2\\-1&1&0\\3&0&4\end{bmatrix}$,

then $A^{-1}=\begin{bmatrix}-2&0&1\\-2&1&1\\\frac{3}{2}&0&-\frac{1}{2}\end{bmatrix}$.

$$X=\begin{bmatrix}-2&0&1\\-2&1&1\\\frac{3}{2}&0&-\frac{1}{2}\end{bmatrix}\begin{bmatrix}8\\4\\-6\end{bmatrix}=\begin{bmatrix}-22\\-18\\15\end{bmatrix}.$$

70. $A=\begin{bmatrix}2&4&0\\1&-2&0\\0&0&3\end{bmatrix}$, $B=\begin{bmatrix}72\\-24\\48\end{bmatrix}$

$$A^{-1}=\begin{bmatrix}\frac{1}{4}&\frac{1}{2}&0\\\frac{1}{8}&-\frac{1}{4}&0\\0&0&\frac{1}{3}\end{bmatrix}$$

$$X=A^{-1}B=\begin{bmatrix}\frac{1}{4}&\frac{1}{2}&0\\\frac{1}{8}&-\frac{1}{4}&0\\0&0&\frac{1}{3}\end{bmatrix}\begin{bmatrix}72\\-24\\48\end{bmatrix}=\begin{bmatrix}6\\15\\16\end{bmatrix}.$$

71. $x+y=-2$
$2x+5y=2$

The system as a matrix equation is

$$\begin{bmatrix}1&1\\2&5\end{bmatrix}\begin{bmatrix}x\\y\end{bmatrix}=\begin{bmatrix}-2\\2\end{bmatrix}.$$

Let $A=\begin{bmatrix}1&1\\2&5\end{bmatrix}$ and $B=\begin{bmatrix}-2\\2\end{bmatrix}$.

$$A^{-1}=\begin{bmatrix}\frac{5}{3}&-\frac{1}{3}\\-\frac{2}{3}&\frac{1}{3}\end{bmatrix}$$

$$\begin{bmatrix}x\\y\end{bmatrix}=\begin{bmatrix}\frac{5}{3}&-\frac{1}{3}\\-\frac{2}{3}&\frac{1}{3}\end{bmatrix}\begin{bmatrix}-2\\2\end{bmatrix}=\begin{bmatrix}-4\\2\end{bmatrix}.$$

The solution is (–4, 2).

72. $5x - 3y = -2$
$2x + 7y = -9$
The system as a matrix equation is
$\begin{bmatrix} 5 & -3 \\ 2 & 7 \end{bmatrix}\begin{bmatrix} x \\ y \end{bmatrix} = \begin{bmatrix} -2 \\ -9 \end{bmatrix}$.
Let $A = \begin{bmatrix} 5 & -3 \\ 2 & 7 \end{bmatrix}$ and $B = \begin{bmatrix} -2 \\ -9 \end{bmatrix}$.
$A^{-1} = \begin{bmatrix} \frac{7}{41} & \frac{3}{41} \\ -\frac{2}{41} & \frac{5}{41} \end{bmatrix}$
$X = A^{-1}B = \begin{bmatrix} \frac{7}{41} & \frac{3}{41} \\ -\frac{2}{41} & \frac{5}{41} \end{bmatrix}\begin{bmatrix} -2 \\ -9 \end{bmatrix} = \begin{bmatrix} -1 \\ -1 \end{bmatrix}$
The solution is (–1, –1).

73. $2x + y = 10$
$3x - 2y = 8$
The system as a matrix equation is
$\begin{bmatrix} 2 & 1 \\ 3 & -2 \end{bmatrix}\begin{bmatrix} x \\ y \end{bmatrix} = \begin{bmatrix} 10 \\ 8 \end{bmatrix}$.
Let $A = \begin{bmatrix} 2 & 1 \\ 3 & -2 \end{bmatrix}$ and $B = \begin{bmatrix} 10 \\ 8 \end{bmatrix}$.
$A^{-1} = \begin{bmatrix} \frac{2}{7} & \frac{1}{7} \\ \frac{3}{7} & -\frac{2}{7} \end{bmatrix}$
$X = \begin{bmatrix} \frac{2}{7} & \frac{1}{7} \\ \frac{3}{7} & -\frac{2}{7} \end{bmatrix}\begin{bmatrix} 10 \\ 8 \end{bmatrix} = \begin{bmatrix} 4 \\ 2 \end{bmatrix}$.
The solution is (4, 2).

74. $x - 2y = 7$
$3x + y = 7$
The system as a matrix equation is
$\begin{bmatrix} 1 & -2 \\ 3 & 1 \end{bmatrix}\begin{bmatrix} x \\ y \end{bmatrix} = \begin{bmatrix} 7 \\ 7 \end{bmatrix}$.
$A = \begin{bmatrix} 1 & -2 \\ 3 & 1 \end{bmatrix}$, $B = \begin{bmatrix} 7 \\ 7 \end{bmatrix}$. $A^{-1} = \begin{bmatrix} \frac{1}{7} & \frac{2}{7} \\ -\frac{3}{7} & \frac{1}{7} \end{bmatrix}$
$X = A^{-1}B = \begin{bmatrix} 3 \\ -2 \end{bmatrix}$
The solution is (3, –2).

75. $x + y + z = 1$
$2x - y = -2$
$3y + z = 2$
The system as a matrix equation is
$\begin{bmatrix} 1 & 1 & 1 \\ 2 & -1 & 0 \\ 0 & 3 & 1 \end{bmatrix}\begin{bmatrix} x \\ y \\ z \end{bmatrix} = \begin{bmatrix} 1 \\ -2 \\ 2 \end{bmatrix}$.
Let $A = \begin{bmatrix} 1 & 1 & 1 \\ 2 & -1 & 0 \\ 0 & 3 & 1 \end{bmatrix}$ and $B = \begin{bmatrix} 1 \\ -2 \\ 2 \end{bmatrix}$.
$A^{-1} = \begin{bmatrix} -\frac{1}{3} & \frac{2}{3} & \frac{1}{3} \\ -\frac{2}{3} & \frac{1}{3} & \frac{2}{3} \\ 2 & -1 & -1 \end{bmatrix}$
$X = \begin{bmatrix} -\frac{1}{3} & \frac{2}{3} & \frac{1}{3} \\ -\frac{2}{3} & \frac{1}{3} & \frac{2}{3} \\ 2 & -1 & -1 \end{bmatrix}\begin{bmatrix} 1 \\ -2 \\ 2 \end{bmatrix} = \begin{bmatrix} -1 \\ 0 \\ 2 \end{bmatrix}$.
The solution is (–1, 0, 2).

76. $x = -3$
$y + z = 6$
$2x - 3z = -9$
The system as a matrix equation is
$\begin{bmatrix} 1 & 0 & 0 \\ 0 & 1 & 1 \\ 2 & 0 & -3 \end{bmatrix}\begin{bmatrix} x \\ y \\ z \end{bmatrix} = \begin{bmatrix} -3 \\ 6 \\ -9 \end{bmatrix}$.
$A = \begin{bmatrix} 1 & 0 & 0 \\ 0 & 1 & 1 \\ 2 & 0 & -3 \end{bmatrix}$, $B = \begin{bmatrix} -3 \\ 6 \\ -9 \end{bmatrix}$
$A^{-1} = \begin{bmatrix} 1 & 0 & 0 \\ -\frac{2}{3} & 1 & \frac{1}{3} \\ \frac{2}{3} & 0 & -\frac{1}{3} \end{bmatrix}$
$X = A^{-1}B = \begin{bmatrix} -3 \\ 5 \\ 1 \end{bmatrix}$
The solution is (–3, 5, 1).

77. $3x - 2y + 4z = 4$
$4x + y - 5z = 2$
$-6x + 4y - 8z = -2$

The system as a matrix equation is
$$\begin{bmatrix} 3 & -2 & 4 \\ 4 & 1 & -5 \\ -6 & 4 & -8 \end{bmatrix}\begin{bmatrix} x \\ y \\ z \end{bmatrix} = \begin{bmatrix} 4 \\ 2 \\ -2 \end{bmatrix}.$$

Let $A = \begin{bmatrix} 3 & -2 & 4 \\ 4 & 1 & -5 \\ -6 & 4 & -8 \end{bmatrix}$.

Since row 3 is –2 times row 1, the matrix will have no inverse, and the system cannot be solved by this method. Another method should be used to complete the solution. Use the elimination method. Multiply equation (1) by 2 and add the result to equation (3).

$6x - 4y + 8z = 8$
$-6x + 4y - 8z = -2$
$0 = 6$

This false result indicates that the system has no solution.

78. $-2x + 3y - z = 1$
$5x - 7y + 8z = 4$
$6x - 9y + 3z = 2$

The system as a matrix equation is
$$\begin{bmatrix} -2 & 3 & -1 \\ 5 & -7 & 8 \\ 6 & -9 & 3 \end{bmatrix}\begin{bmatrix} x \\ y \\ z \end{bmatrix} = \begin{bmatrix} 1 \\ 4 \\ 2 \end{bmatrix}.$$

Let $A = \begin{bmatrix} -2 & 3 & -1 \\ 5 & -7 & 8 \\ 6 & -9 & 3 \end{bmatrix}$.

Since row 3 is –3 times row 1, the matrix will have no inverse, and the system cannot be solved by this method. Another method should be used to complete the solution. Use the elimination method. Multiply equation (1) by 3 and add the result to equation (3).

$-6x + 9y - 3z = 3$
$6x - 9y + 3z = 2$
$0 = 5$

This false result indicates that the system has no solution.

Exercises 79–86 can be solved using either the elimination method or matrix methods.

79. Let x = the amount of the 9% wine and let y = the amount of the 14% wine.
We have the system

$x + y = 40$
$.09x + .14y = .12(40).$

This system can be simplified to

$x + y = 40 \quad (1)$
$.09x + .14y = 4.8. \quad (2)$

Multiply equation (1) by –.09 and add to equation (2).

$-.09x - .09y = -3.6$
$.09x + .14y = 4.8$
$.05y = 1.2 \Rightarrow y = 24$

Substitute 24 for y in equation (1) to get $x = 16$. The wine maker should mix 16 liters of the 9% wine and 24 liters of the 14% wine.

80. Let x = number of grams of 12 carat gold and let y = number of grams of 22 carat gold.
The system is

$x + y = 25 \quad (1)$
$\frac{12}{24}x + \frac{22}{24}y = \frac{15}{24}(25) \quad (2)$

Multiply equation (2) by 24.

$x + y = 25 \quad (1)$
$12x + 22y = 375 \quad (3)$

Multiply equation (1) by –12 and add the result to equation (3).

$-12x - 12y = -300$
$12x + 22y = 375$
$10y = 75 \Rightarrow y = 7.5$

Substitute 7.5 for y in equation (1)
$x + 7.5 = 25 \Rightarrow x = 17.5$

The merchant should mix 17.5 grams of 12 carat gold and 7.5 grams of 22 carat gold.

81. Let x = the number of liters of 40% solution and let y = the number of liters of 60% solution.

$$\begin{aligned} x + \quad y &= 40 \\ .4x + .6y &= .45(40) = 18 \end{aligned}$$

Multiply the second equation by 5.

$$\begin{aligned} x + \quad y &= 40 \\ 2x + 3y &= 90 \end{aligned}$$

Use the Gauss-Jordan method.

$$\left[\begin{array}{cc|c} 1 & 1 & 40 \\ 2 & 3 & 90 \end{array}\right]$$

$$\left[\begin{array}{cc|c} 1 & 1 & 40 \\ 0 & 1 & 10 \end{array}\right] \begin{array}{l} \\ -2R_1 + R_2 \end{array}$$

$$\left[\begin{array}{cc|c} 1 & 0 & 30 \\ 0 & 1 & 10 \end{array}\right] \begin{array}{l} -R_2 + R_1 \\ \\ \end{array}$$

The chemist should use 30 liters of the 40% solution and 10 liters of the 60% solution.

82. Let x = the number of pounds of tea worth \$4.60 a pound and let y = the number of pounds of tea worth \$6.50 a pound.
The system is

$$\begin{aligned} x + \qquad y &= 10 \qquad (1) \\ 4.60x + 6.50y &= 5.74(10). \quad (2) \end{aligned}$$

Multiply equation (1) by –4.60 and add the result to equation (2).

$$\begin{aligned} -4.60x - 4.60y &= -46.0 \\ 4.60x + 6.50y &= 57.4 \\ \hline 1.90y &= 11.4 \Rightarrow y = 6 \end{aligned}$$

Substitute 6 for y in equation (1) and solve for x.

$x + 6 = 10 \Rightarrow x = 4$

Thus, 4 pounds of the tea worth \$4.60 a pound should be used.

83. Let x = the number of bowls and let y = the number of plates.
After converting 8 hours to 480 minutes, we will solve the system

$$\begin{aligned} 3x + \ 2y &= 480 \qquad (1) \\ .25x + .2y &= \ 44 \qquad (2) \end{aligned}$$

Multiply equation (2) by –10 and add to equation (1).

$$\begin{aligned} 3x + 2y &= \ 480 \\ -2.5x - 2y &= -440 \\ \hline .5x \qquad &= \ 40 \Rightarrow x = 80 \end{aligned}$$

Substitute 80 for x in equation (1) and solve for y.

$3(80) + 2y = 480 \Rightarrow y = 120$

The factory can produce 80 bowls and 120 plates.

84. Let x = the speed of the boat and let y = the speed of the current. The system is

$$\begin{aligned} 3(x + y) &= 57 \quad (1) \\ 5(x - y) &= 55. \quad (2) \end{aligned}$$

Dividing equation (1) by 3 and equation (2) by 5 gives

$$\begin{aligned} x + y &= 19 \quad (3) \\ x - y &= 11. \quad (4) \end{aligned}$$

Adding the above two equations gives
$2x = 30 \Rightarrow x = 15.$
Substituting into (3) gives
$15 + y = 19 \Rightarrow y = 4.$
The speed of the boat is 15 km/hr, and the speed of the current is 4 km/hr.

85. Let x = the amount invested at 8%, let y = the amount invested at $8\frac{1}{2}$%, and let z = the amount invested at 11%. The system is

$$\begin{aligned} x + y + z &= 50{,}000 \\ .08x + .085y + .11z &= 4436.25 \\ .11z &= .08x + 80. \end{aligned}$$

Write the system in the proper form.

$$\begin{aligned} x + \qquad y + \qquad z &= 50{,}000 \\ .08x + .085y + .11z &= 4436.25 \\ -.08x \qquad\qquad + .11z &= 80 \end{aligned}$$

(*continued next page*)

Write the augmented matrix of the system.

$$\left[\begin{array}{ccc|c} 1 & 1 & 1 & 50,000 \\ .08 & .085 & .11 & 4436.25 \\ -.08 & 0 & .11 & 80 \end{array}\right]$$

$$\left[\begin{array}{ccc|c} 1 & 1 & 1 & 50,000 \\ 0 & .005 & .03 & 436.25 \\ 0 & .08 & .19 & 4080 \end{array}\right] \begin{array}{l} \\ -.08R_1 + R_2 \\ .08R_1 + R_3 \end{array}$$

$$\left[\begin{array}{ccc|c} 1 & 1 & 1 & 50,000 \\ 0 & 1 & 6 & 87,250 \\ 0 & .08 & .19 & 4080 \end{array}\right] \begin{array}{l} \\ \frac{1}{.005}R_2 \\ \\ \end{array}$$

$$\left[\begin{array}{ccc|c} 1 & 0 & -5 & -37,250 \\ 0 & 1 & 6 & 87,250 \\ 0 & 0 & -.29 & -2900 \end{array}\right] \begin{array}{l} -1R_2 + R_1 \\ \\ -.08R_2 + R_3 \end{array}$$

$$\left[\begin{array}{ccc|c} 1 & 0 & -5 & -37,250 \\ 0 & 1 & 6 & 87,250 \\ 0 & 0 & 1 & 10,000 \end{array}\right] \begin{array}{l} \\ \\ -\frac{1}{.29}R_3 \end{array}$$

$$\left[\begin{array}{ccc|c} 1 & 0 & 0 & 12,750 \\ 0 & 1 & 0 & 27,250 \\ 0 & 0 & 1 & 10,000 \end{array}\right] \begin{array}{l} 5R_3 + R_1 \\ -6R_3 + R_2 \\ \\ \end{array}$$

Thus, $x = 12{,}750$, $y = 27{,}250$, and $z = 10{,}000$. Mr. Miller invested \$12,750 at 8%, \$27,250 at $8\frac{1}{2}$%, and \$10,000 at 11%.

86. Let x = the number of student tickets, let y = the number of alumni tickets, let z = the number of other adult tickets, and let w = the number of children's tickets.

$$\begin{array}{rl} x + y + z + w = 3750 & (1) \\ 4x + 10y + 12z + 6w = 29{,}100 & (2) \\ x = 6w & (3) \\ y = \frac{4}{5}x & (4) \end{array}$$

Solve equation (3) for w in terms of x.

$x = 6w \Rightarrow w = \frac{1}{6}x.$ (5)

Now substitute the values for y and w into equations (1) and (2)

$$\begin{array}{rl} x + \frac{4}{5}x + z + \frac{1}{6}x = 3750 & (6) \\ 4x + 10\left(\frac{4}{5}x\right) + 12z + 6\left(\frac{1}{6}x\right) = 29{,}100 & (7) \end{array}$$

Simplify by multiplying equation (6) by 30 and combining like terms, and combining like terms in (7).

$$\begin{array}{rl} 59x + 30z = 112{,}500 & (8) \\ 13x + 12z = 29{,}100 & (9) \end{array}$$

Multiply (8) by 2 and (9) by –5, then add the resulting equations and solve for x.

$$\begin{array}{rcl} 118x + 60z & = & 225{,}000 \\ -65x - 60z & = & -145{,}500 \\ \hline 53x & = & 79{,}500 \Rightarrow x = 1500 \end{array}$$

Substitute $x = 1500$ into (3), (4), and (8) to solve for the remaining variables.

$1500 = 6w \Rightarrow w = 250$

$y = \frac{4}{5}(1500) = 1200$

$59(1500) + 30z = 112{,}500 \Rightarrow 30z = 24{,}000 \Rightarrow$
$z = 800$

There were 1500 students, 1200 alumni, 800 other adults, and 250 children at the game.

87. $A = \begin{bmatrix} 0 & \frac{1}{4} \\ \frac{1}{2} & 0 \end{bmatrix}, D = \begin{bmatrix} 2100 \\ 1400 \end{bmatrix}$

a. $I - A = \begin{bmatrix} 1 & 0 \\ 0 & 1 \end{bmatrix} - \begin{bmatrix} 0 & \frac{1}{4} \\ \frac{1}{2} & 0 \end{bmatrix} = \begin{bmatrix} 1 & -\frac{1}{4} \\ -\frac{1}{2} & 1 \end{bmatrix}$

b.

$$\left[\begin{array}{cc|cc} 1 & -\frac{1}{4} & 1 & 0 \\ -\frac{1}{2} & 1 & 0 & 1 \end{array}\right]$$

$$\left[\begin{array}{cc|cc} 1 & -\frac{1}{4} & 1 & 0 \\ 0 & \frac{7}{8} & \frac{1}{2} & 1 \end{array}\right] \frac{1}{2}R_1 + R_2$$

$$\left[\begin{array}{cc|cc} 1 & -\frac{1}{4} & 1 & 0 \\ 0 & 1 & \frac{4}{7} & \frac{8}{7} \end{array}\right] \frac{8}{7}R_2$$

$$\left[\begin{array}{cc|cc} 1 & 0 & \frac{8}{7} & \frac{2}{7} \\ 0 & 1 & \frac{4}{7} & \frac{8}{7} \end{array}\right] \frac{1}{4}R_2 + R_1$$

$$(I - A)^{-1} = \begin{bmatrix} \frac{8}{7} & \frac{2}{7} \\ \frac{4}{7} & \frac{8}{7} \end{bmatrix}.$$

c. $X = (I - A)^{-1}D = \begin{bmatrix} \frac{8}{7} & \frac{2}{7} \\ \frac{4}{7} & \frac{8}{7} \end{bmatrix} \begin{bmatrix} 2100 \\ 1400 \end{bmatrix} = \begin{bmatrix} 2800 \\ 2800 \end{bmatrix}$

88. **a.** The input-output matrix is

$$\begin{matrix} & c & g \\ c & 0 & \frac{1}{2} \\ g & \frac{2}{3} & 0 \end{matrix} = A.$$

b. $I - A = \begin{bmatrix} 1 & -\frac{1}{2} \\ -\frac{2}{3} & 1 \end{bmatrix}, D = \begin{bmatrix} 400 \\ 800 \end{bmatrix}$

$$(I - A)^{-1} = \begin{bmatrix} \frac{3}{2} & \frac{3}{4} \\ 1 & \frac{3}{2} \end{bmatrix}$$

$$X = (I - A)^{-1}D = \begin{bmatrix} 1200 \\ 1600 \end{bmatrix}$$

The production required is 1200 units of cheese and 1600 units of goats.

89. Write the input-output matrix.

$$A = \begin{matrix} & A & M \\ A & .10 & .70 \\ M & .40 & .20 \end{matrix} = \begin{bmatrix} \frac{1}{10} & \frac{7}{10} \\ \frac{4}{10} & \frac{2}{10} \end{bmatrix}$$

$$I - A = \begin{bmatrix} 1 & 0 \\ 0 & 1 \end{bmatrix} - \begin{bmatrix} \frac{1}{10} & \frac{7}{10} \\ \frac{4}{10} & \frac{2}{10} \end{bmatrix} = \begin{bmatrix} \frac{9}{10} & -\frac{7}{10} \\ -\frac{4}{10} & \frac{8}{10} \end{bmatrix}$$

Find $(I - A)^{-1}$.

$$\left[\begin{array}{cc|cc} \frac{9}{10} & -\frac{7}{10} & 1 & 0 \\ -\frac{4}{10} & \frac{8}{10} & 0 & 1 \end{array}\right]$$

$$\left[\begin{array}{cc|cc} 1 & -\frac{7}{9} & \frac{10}{9} & 0 \\ -\frac{4}{10} & \frac{8}{10} & 0 & 1 \end{array}\right] \frac{10}{9}R_1$$

$$\left[\begin{array}{cc|cc} 1 & -\frac{7}{9} & \frac{10}{9} & 0 \\ 0 & \frac{44}{90} & \frac{4}{9} & 1 \end{array}\right] \frac{4}{10}R_1 + R_2$$

$$\left[\begin{array}{cc|cc} 1 & -\frac{7}{9} & \frac{10}{9} & 0 \\ 0 & 1 & \frac{10}{11} & \frac{90}{44} \end{array}\right] \frac{90}{44}R_2$$

$$\left[\begin{array}{cc|cc} 1 & 0 & \frac{180}{99} & \frac{70}{44} \\ 0 & 1 & \frac{10}{11} & \frac{90}{44} \end{array}\right] \frac{7}{9}R_2 + R_1$$

$$(I - A)^{-1} = \begin{bmatrix} \frac{180}{99} & \frac{70}{44} \\ \frac{10}{11} & \frac{90}{44} \end{bmatrix}$$

$$X = (I - A)^{-1}D$$

$$X = \begin{bmatrix} \frac{180}{99} & \frac{70}{44} \\ \frac{10}{11} & \frac{90}{44} \end{bmatrix} \begin{bmatrix} 60,000 \\ 20,000 \end{bmatrix}$$

$$X = \begin{bmatrix} 140,909 \\ 95,455 \end{bmatrix} \text{ Rounded}$$

The agriculture industry should produce \$140,909, while the manufacturing industry should produce \$95,455.

90. **a.** From the input-output matrix, we know that .9 unit agriculture; .4 unit services; .02 unit mining; .9 unit manufacturing are required to produce 1 unit.

b. The input-output matrix, A, and the matrix, $I - A$, are

$$A = \begin{bmatrix} .02 & .9 & 0 & .001 \\ 0 & .4 & 0 & .06 \\ .01 & .02 & .06 & .07 \\ .25 & .9 & .9 & .4 \end{bmatrix}$$

$$I - A = \begin{bmatrix} .980 & -.900 & 0 & -.001 \\ 0 & .600 & 0 & -.060 \\ -.010 & -.020 & .940 & -.070 \\ -.250 & -.900 & -.900 & .600 \end{bmatrix}$$

Next, calculate $(I - A)^{-1}$.

$$(I - A)^{-1} \approx \begin{bmatrix} 1.079 & 1.960 & .2132 & .223 \\ .064 & 2.129 & .230 & .240 \\ .060 & .4107 & 1.242 & .186 \\ .635 & 4.627 & 2.296 & 2.398 \end{bmatrix}$$

$$D = \begin{bmatrix} 760 \\ 1600 \\ 1000 \\ 2000 \end{bmatrix}$$

$$X = (I - A)^{-1}D \approx \begin{bmatrix} 4615 \\ 4165 \\ 2317 \\ 14,979 \end{bmatrix}$$

This is about 4615 units of agriculture, 4165 units of services, 2317 units of mining, and 14,979 units of manufacturing.

c. Since .25 units of manufacturing are used in producing each unit of agriculture, $(4615)(.25) = 1153.75$ units are used. Since .9 units of manufacturing are used in producing each unit of service, $(.9)(4165) = 3748.5$ units are used. Since .9 units of manufacturing are used in producing each unit of mining, $(.9)(2317) = 2085.3$. Since .4 units of manufacturing are used in producing each unit of manufacturing, $(.4)(14,979) = 5991.6$.
Thus, $1153.75 + 3748.5 + 2085.3 + 5991.6 \approx 12,979$ units of manufacturing are used in the production process.

91. a. From the input-output matrix, we know that .4 unit agriculture; .09 unit construction, .4 unit energy, .1 unit manufacturing, .9 unit transportation are required to produce 1 unit.

b. The input-output matrix, A and the matrix, $I - A$, are

$$A = \begin{bmatrix} .18 & .017 & .4 & .005 & 0 \\ .14 & .018 & .09 & .001 & 0 \\ .9 & 0 & .4 & .06 & .002 \\ .19 & .16 & .1 & .008 & .5 \\ .14 & .25 & .9 & .4 & .12 \end{bmatrix}$$

$$(I - A) = \begin{bmatrix} .82 & -.017 & -.4 & -.005 & 0 \\ -.14 & .982 & -.09 & -.001 & 0 \\ -.9 & 0 & .6 & -.06 & -.002 \\ -.19 & -.16 & -.1 & .992 & -.5 \\ -.14 & -.25 & -.9 & -.4 & .88 \end{bmatrix}$$

$$X = \begin{bmatrix} 28{,}067 \\ 9383 \\ 51{,}372 \\ 61{,}364 \\ 90{,}403 \end{bmatrix}$$

$$D = (I - A)X = \begin{bmatrix} 1999.809 \\ 599.882 \\ 1700.254 \\ 3700.378 \\ 2499.110 \end{bmatrix}$$

This represents about 2000 units of agriculture, 600 units of construction, 1700 units of energy, 3700 units of manufacturing, and 2500 units of transportation.

c. Multiply matrix $(I - A)^{-1}$ by matrix D where

$$D = \begin{bmatrix} 2400 \\ 850 \\ 1400 \\ 3200 \\ 1800 \end{bmatrix}$$

$$(I - A)^{-1} \approx \begin{bmatrix} 7.744 & .294 & 5.745 & .509 & .302 \\ 2.319 & 1.111 & 1.890 & .167 & .099 \\ 13.100 & .543 & 11.546 & 1.006 & .598 \\ 14.119 & .976 & 12.001 & 2.357 & 1.367 \\ 21.707 & 1.361 & 18.714 & 2.229 & 2.445 \end{bmatrix}$$

$$X = (I - A)^{-1} D \approx \begin{bmatrix} 29{,}049 \\ 9869 \\ 52{,}362 \\ 61{,}520 \\ 90{,}987 \end{bmatrix}$$

This represents 29,049 units of agriculture, 9869 units of construction, 52,362 units of energy, 61,520 units of manufacturing, and 90,987 units of transportation.

92.

$$M = \begin{array}{c} \\ A \\ B \\ C \\ D \end{array} \begin{array}{c} \begin{matrix} A & B & C & D \end{matrix} \\ \begin{bmatrix} 0 & 1 & 0 & 1 \\ 1 & 0 & 0 & 1 \\ 0 & 0 & 0 & 1 \\ 1 & 1 & 1 & 0 \end{bmatrix} \end{array}$$

a. M^2 gives the number of one-step flights between cities.

$$M^2 = \begin{bmatrix} 0 & 1 & 0 & 1 \\ 1 & 0 & 0 & 1 \\ 0 & 0 & 0 & 1 \\ 1 & 1 & 1 & 0 \end{bmatrix} \begin{bmatrix} 0 & 1 & 0 & 1 \\ 1 & 0 & 0 & 1 \\ 0 & 0 & 0 & 1 \\ 1 & 1 & 1 & 0 \end{bmatrix} = \begin{bmatrix} 2 & 1 & 1 & 1 \\ 1 & 2 & 1 & 1 \\ 1 & 1 & 1 & 0 \\ 1 & 1 & 0 & 3 \end{bmatrix}$$

$m_{13}^2 = 1$ so there is 1 one-stop flight between cities A and C.

b. The number of direct or one-stop flights from B to C is $m_{23}^2 + m_{23} = 1 + 0 = 1$.

c. The number of two-stop flights is given by

$$M^3 = M^2 \cdot M = \begin{bmatrix} 2 & 3 & 1 & 4 \\ 3 & 2 & 1 & 4 \\ 1 & 1 & 0 & 3 \\ 4 & 4 & 3 & 2 \end{bmatrix}$$

93. The message "leave now" broken into groups of 2 letters would have the following matrices:

$$\begin{bmatrix} 12 \\ 5 \end{bmatrix}, \begin{bmatrix} 1 \\ 22 \end{bmatrix}, \begin{bmatrix} 5 \\ 27 \end{bmatrix}, \begin{bmatrix} 14 \\ 15 \end{bmatrix}, \begin{bmatrix} 23 \\ 27 \end{bmatrix}$$

Multiply $M = \begin{bmatrix} 2 & 6 \\ 1 & 4 \end{bmatrix}$ by each matrix in the code to obtain $\begin{bmatrix} 54 \\ 32 \end{bmatrix}, \begin{bmatrix} 134 \\ 89 \end{bmatrix}, \begin{bmatrix} 172 \\ 113 \end{bmatrix}, \begin{bmatrix} 118 \\ 74 \end{bmatrix}, \begin{bmatrix} 208 \\ 131 \end{bmatrix}$.

94. To decode the message, use $M^{-1} = \begin{bmatrix} 2 & -3 \\ -\frac{1}{2} & 1 \end{bmatrix}$.

Case 6 Matrix Operations and Airline Route Maps

1. As the matrix A^2 indicates, the only city *not* reachable from San Antonio in a two-flight sequence is Lubbock, since this is the only city whose column has a zero in the row for San Antonio. Thus, the cities reachable by a two-flight sequence from San Antonio are all Stampede cities except Lubbock.
   In the matrix A^3, there are no zero entries in the row for San Antonio, so all Stampede Air cities may be reached by a three-flight sequence.

2. El Paso—Austin—Dallas—San Antonio
   El Paso—Austin—Houston—San Antonio
   El Paso—Dallas—Houston—San Antonio
   El Paso—Houston—Dallas—San Antonio

3. The trips between two different cities for which the matrix entries remain zero until A^3 are trips between Lubbock and Corpus Christi, and Lubbock and San Antonio. These trips each take three flights.

4. There are 5 vertices, so the adjacency matrix is a 5×5 matrix. If the vertices in the Vacation Air graph respectively correspond to Jacksonville (J), Orlando (O), West Palm Beach (W), Miami (M), and Tampa (T), then the adjacency matrix for Vacation Air is

$$\begin{matrix} & J & O & W & M & T \end{matrix}$$
$$B = \begin{bmatrix} 0 & 0 & 0 & 1 & 1 \\ 0 & 0 & 0 & 0 & 1 \\ 0 & 0 & 0 & 0 & 1 \\ 1 & 0 & 0 & 0 & 1 \\ 1 & 1 & 1 & 1 & 0 \end{bmatrix}.$$

5. To find the cities that may be reached by a two-flight sequence, find the matrix that represents B^2.

$$B^2 = \begin{array}{c} \begin{matrix} J & O & W & M & T \end{matrix} \\ \begin{bmatrix} 2 & 1 & 1 & 1 & 1 \\ 1 & 1 & 1 & 1 & 0 \\ 1 & 1 & 1 & 1 & 0 \\ 1 & 1 & 1 & 2 & 1 \\ 1 & 0 & 0 & 1 & 4 \end{bmatrix} \end{array}.$$

This matrix shows the two-flight sequences between cities. An entry is nonzero whenever there is a two-step sequence between the cities.

6. Add the matrices found in Exercises 4 and 5.

$$B + B^2 = \begin{array}{c} \begin{matrix} J & O & W & M & T \end{matrix} \\ \begin{bmatrix} 2 & 1 & 1 & 2 & 2 \\ 1 & 1 & 1 & 1 & 1 \\ 1 & 1 & 1 & 1 & 1 \\ 2 & 1 & 1 & 2 & 2 \\ 2 & 1 & 1 & 2 & 4 \end{bmatrix} \end{array}.$$

From this matrix, we can conclude that all entries are nonzero. This means that one can fly between and two of these cities in at most two steps.

Chapter 7 Linear Programming

Section 7.1 Graphing Linear Inequalities in Two Variables

1. $y \geq -x - 2$

Graph of $y = -x - 2$ is a solid line. The intercepts are (–2, 0) and (0, –2). Use the origin as a test point. Since $0 \geq -0 - 2$ is true, the origin will be included in the region, so shade the half-plane above the line. This matches choice F.

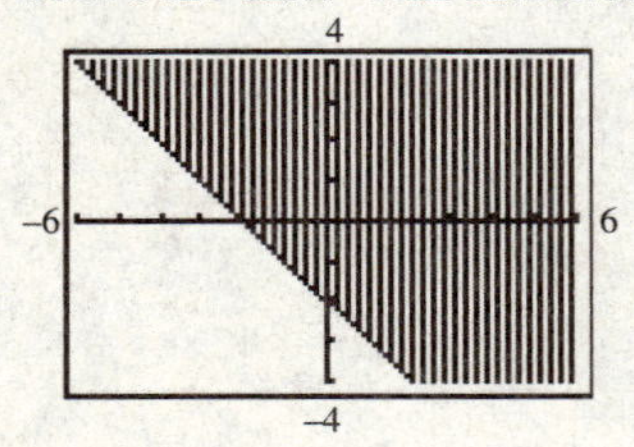

3. $y \leq x + 2$

Graph $y = x + 2$ is a solid line. The intercepts are (–2, 0) and (0, 2). Use the origin as a test point. Since $0 \leq 0 + 2$ is true, the origin will be included in the region, so shade the half-plane below the line. This matches choice A.

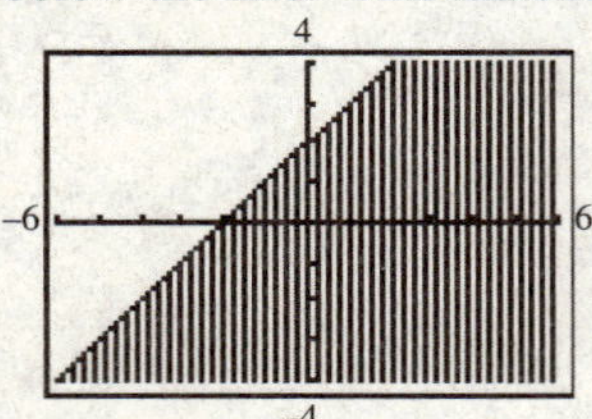

5. $6x + 4y \geq -12$

Graph $6x + 4y = -12$ is a solid line. The intercepts are (–2, 0) and (0, –3). Use the origin as a test point. Since $6(0) + 4(0) \geq -12$ is true, the origin will be included in the region, so shade the half-plane above the line. This matches choice E.

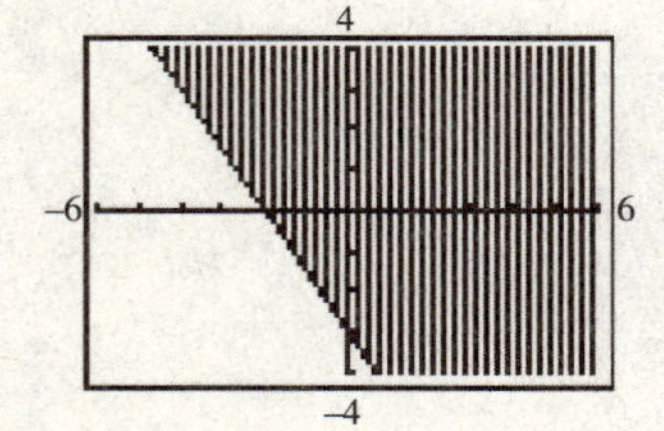

7. $y < 5 - 2x$

Graph $y = 5 - 2x$ as a dashed line. The intercepts are (0, 5) and $\left(\frac{5}{2}, 0\right)$. Test (0, 0) in the original inequality. The result, $0 < 5$, is true, so shade the region that contains (0, 0).

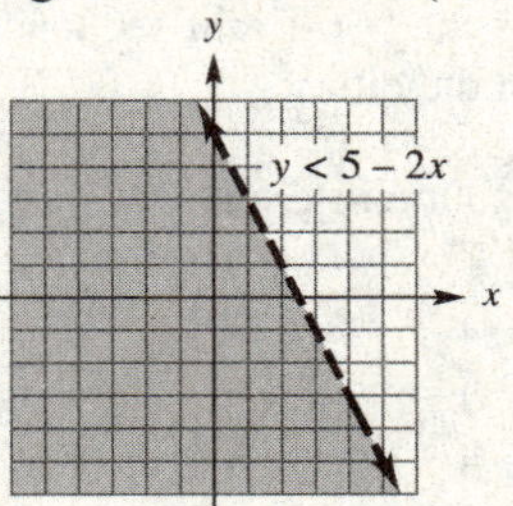

9. $3x - 2y \geq 18$

Graph $3x - 2y = 18$ as a solid line. The intercepts are (0, –9) and (6, 0). Use the origin as a test point. Since $3(0) - 2(0) \geq 18$ is false, the origin will not be included in the region, so shade the half-plane below the line.

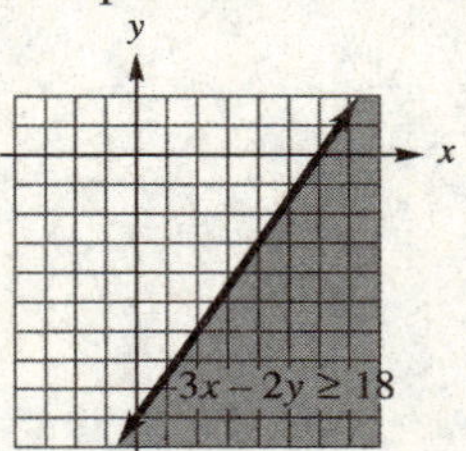

11. $2x - y \leq 4$

Graph $2x - y = 4$ as a solid line. The intercepts are (0, –4) and (2, 0). Use the origin as a test point. Since $2(0) - 0 \leq 4$ is true, the origin will be included in the region, so shade the half-plane above the line.

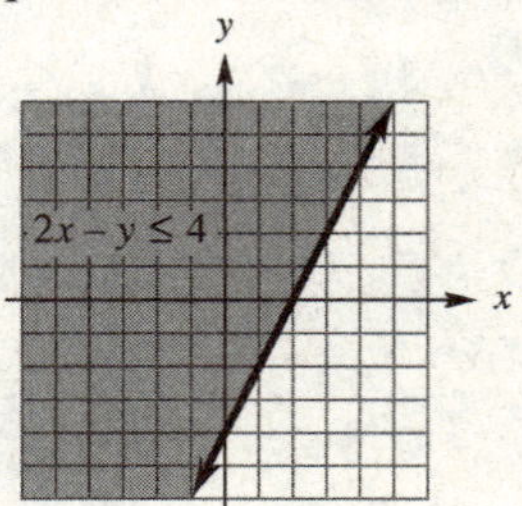

13. $y \le -4$
Graph $y = -4$ as a solid line. $y = -4$ is a horizontal line crossing the y-axis at (0, –4). Since the inequality symbol is ≤, shade the half-plane below the line.

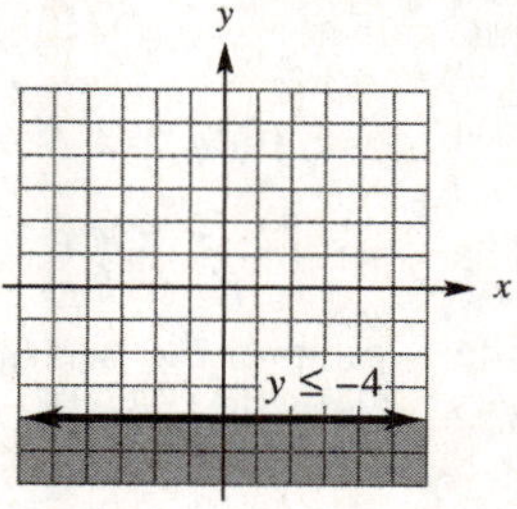

15. $3x - 2y \ge 18$
Graph the line $3x - 2y = 18$ as a solid line. The intercepts are (0, –9) and (6, 0). Use the origin as a test point. Since $3(0) - 2(0) \ge 18$ is false, the origin will not be included in the region, so shade the half-plane below the line.

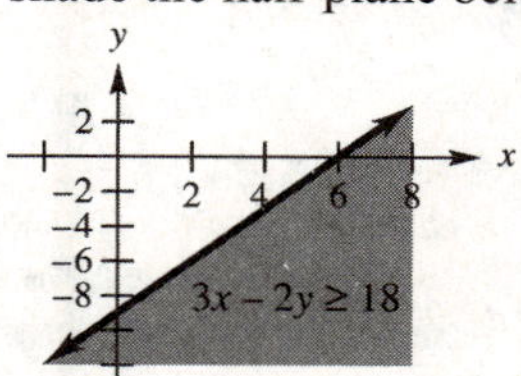

17. $3x + 4y \ge 12$
Graph $3x + 4y = 12$ as a solid line. The intercepts are (0, 3) and (4, 0). Use the origin as a test point. Since $3(0) + 4(0) > 12$ is false, the origin will not be included in the region, so shade the half-plane above the line.

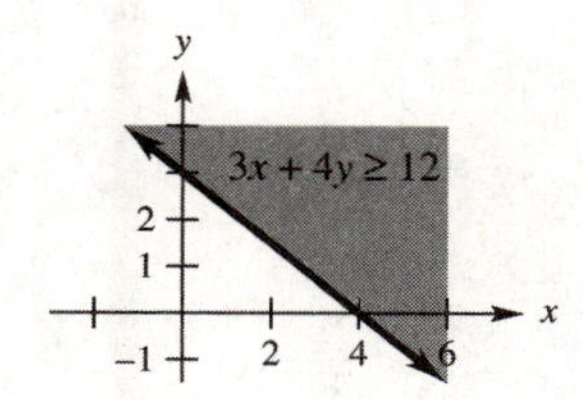

19. $2x - 4y \le 3$
Graph $2x - 4y = 3$ as a solid line. The intercepts are $\left(\frac{3}{2}, 0\right)$ and $\left(0, -\frac{3}{4}\right)$. Use the origin as a test point. $2(0) - 4(0) < 3$ is true, so the region above the line, which includes the origin, is the correct region to shade.

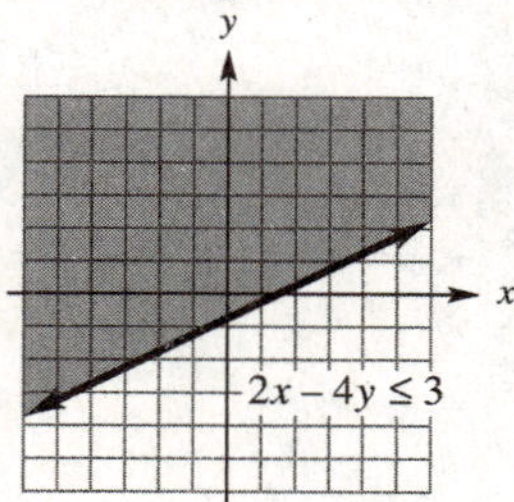

21. $x \le 5y$
Graph $x = 5y$ as a solid line. Since this line contains the origin, some point other than (0, 0) must be used as a test point. The point (1, 2) gives $1 \le 5(2)$ or $1 \le 10$, a true sentence. Shade the side of the line containing (1, 2), that is, the side above the line.

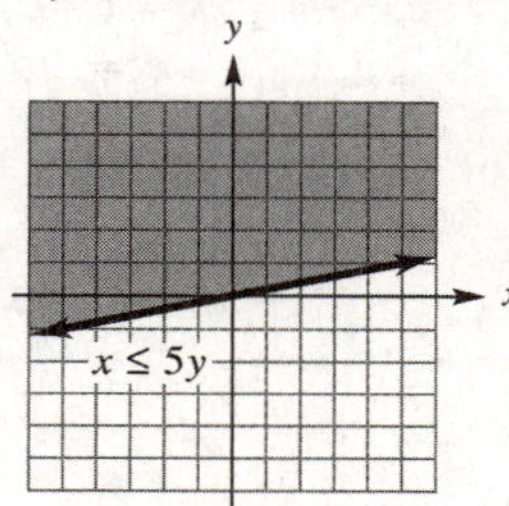

23. $-3x \le y$
Graph $y = -3x$ as a solid line. Since this line contains the origin, use some point other than (0, 0) as a test point. (1, 1) used as a test point gives $-3 < 1$, a true sentence. Shade the region containing (1, 1), which is the region above the line.

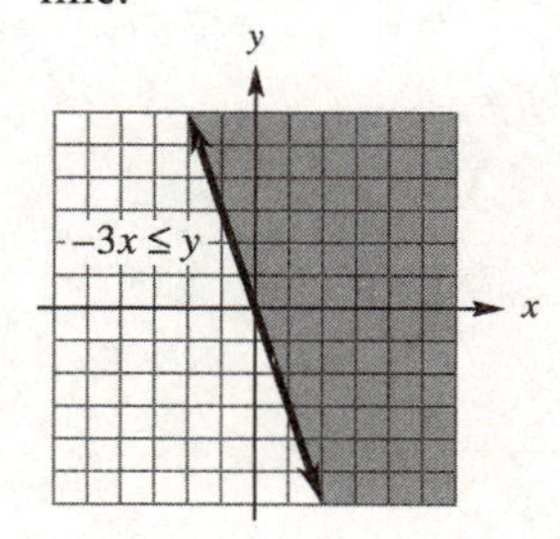

25. $y \le x$

Graph $y = x$ as a dashed line. Since this line contains the origin, choose a point other than (0, 0) as a test point. (2, 3) gives $3 < 2$, which is false. Shade the region that does not contain the test point, that is, the region below the line.

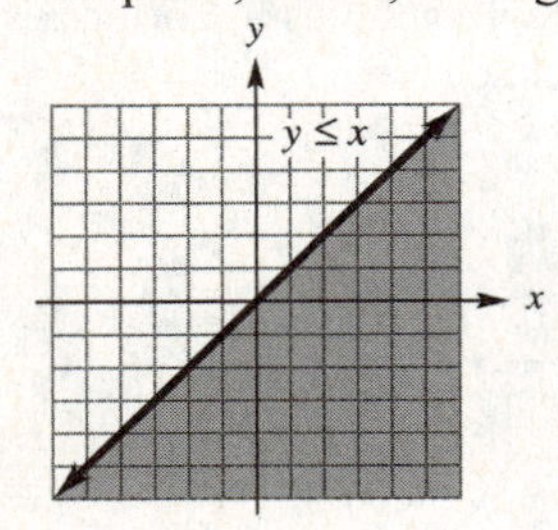

27. Answers may vary.

Possible answer: When the inequality is $<$ or $>$, the line is dashed. When the inequality is $\le$ or $\ge$, the line is solid.

29. $y \ge 3x - 6$
$y \ge -x + 1$

Graph $y \ge 3x - 6$ as the region on or above the solid line $y = 3x - 6$. Graph $y \ge -x + 1$ as the region on or above the solid line $y = -x + 1$. The feasible region is the overlap of the two half-planes.

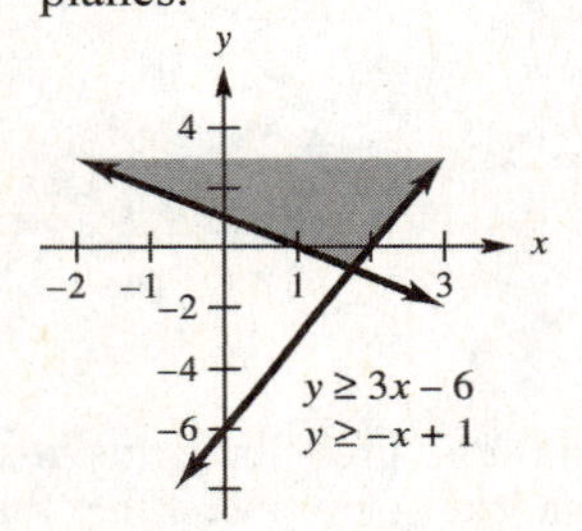

31. $2x + y \le 5$
$x + 2y \le 5$

Graph $2x + y \le 5$ as the region on or below the solid line $2x + y = 5$, which has intercepts (0, 5) and $\left(\frac{5}{2}, 0\right)$. Graph $x + 2y \le 5$ as the region on or below the solid line $x + 2y = 5$, which has intercepts $\left(0, \frac{5}{2}\right)$ and (5, 0). The feasible region is the overlap of the two half-planes.

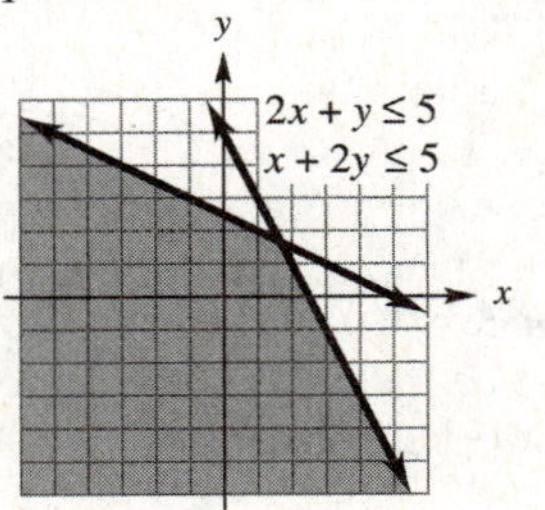

33. $2x + y \ge 8$
$4x - y \le 3$

Graph $2x + y \ge 8$ as the region on or above the dashed line $2x + y = 8$, which has intercepts (0, 8) and (4, 0). Graph $4x - y \le 3$ as the region above the solid line $4x - y = 3$, which has intercepts (0, –3) and $\left(\frac{3}{4}, 0\right)$.

The overlap of these two regions is the feasible region.

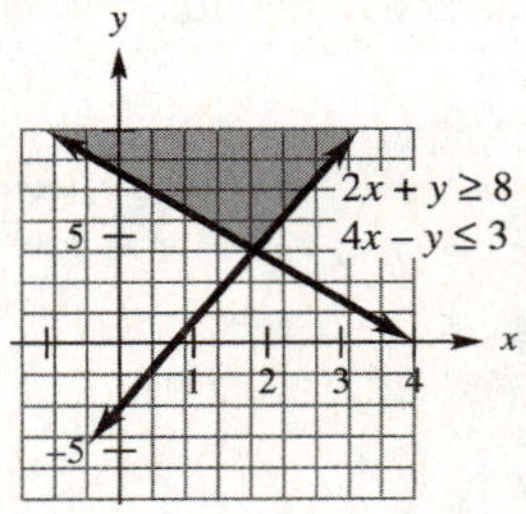

35. $2x - y \le 1$
$3x + y \le 6$

Graph $2x - y \le 1$ as the region on or above the dashed line $2x - y = 1$. Graph $3x + y \le 6$ as the region on or below the solid line $3x + y = 6$. Shade the overlapping part of these two regions to show the feasible region.

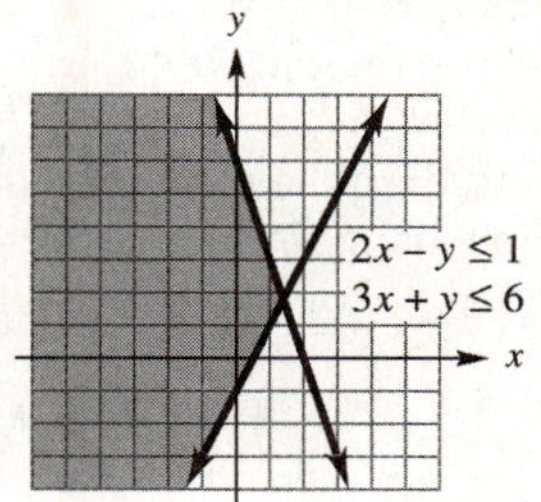

37. $-x - y \le 5$
$2x - y \le 4$

Graph $-x - y \le 5$ as the region on or above the solid line $-x - y = 5$. Graph $2x - y \le 4$ as the region on or above the solid line $2x - y = 4$. Shade the overlapping part of these two regions to show the feasible region.

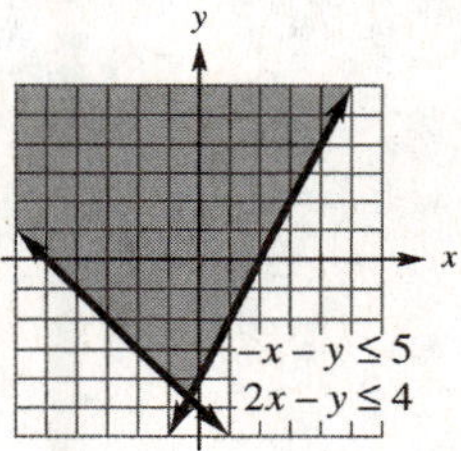

39. $3x + y \ge 6$
$x + 2y \ge 7$
$x \ge 0$
$y \ge 0$

The inequalities $x \ge 0$ and $y \ge 0$ restrict the feasible region to the first quadrant. Graph $3x + y \ge 6$ as the region on or above the solid line $3x + y = 6$. Graph $x + 2y \ge 7$ as the region on or above the solid line $x + 2y = 7$. The feasible region is the overlap of these half-planes.

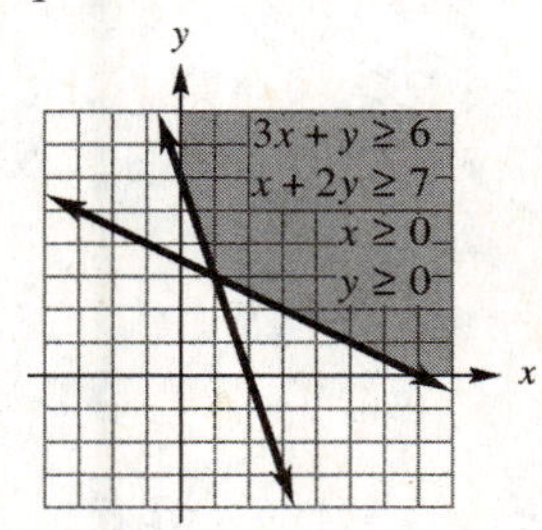

41. $-2 \le x \le 3$
$-1 \le y \le 5$
$2x + y \le 6$

The graph of $-2 \le x \le 3$ is the region between the vertical line $x = -2$ and $x = 3$, including the lines. The graph of $-1 \le y \le 5$ is the region between the horizontal lines $y = -1$ and $y = 5$, including the lines. The graph of $2x + y \le 6$ is the region on or below the solid line $2x + y = 6$. Shade the region common to all three graphs to show the feasible region.

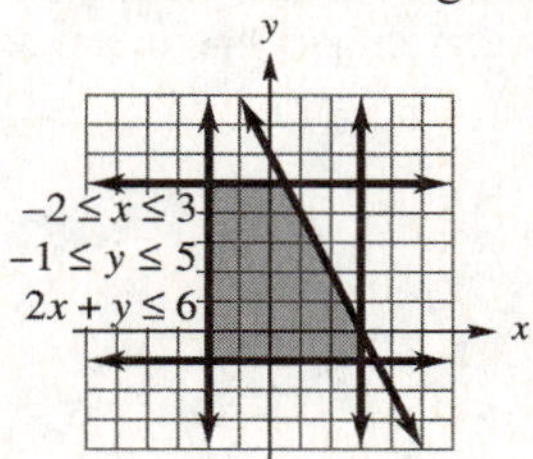

43. $2y - x \ge -5$
$y \le 3 + x$
$x \ge 0$
$y \ge 0$

The graph of $2y - x \ge -5$ consists of the boundary line $2y - x = 5$ and the region above it. The graph of $y \le 3 + x$ consists of the boundary line $y = 3 + x$ and the region below it. The inequalities $x \ge 0$ and $y \ge 0$ restrict the feasible region to the first quadrant. Shade the region common to all of these graphs to show the feasible region.

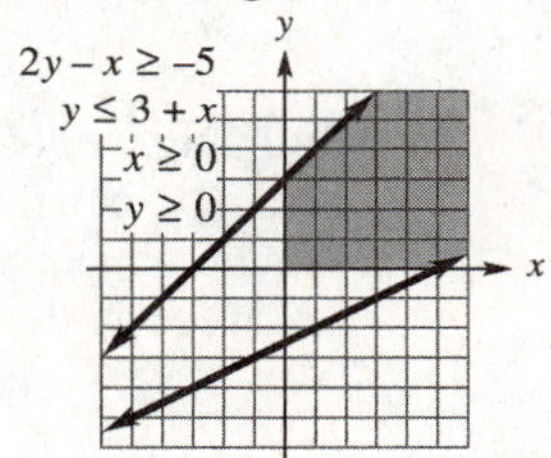

45. $3x + 4y \geq 12$
$2x - 3y \leq 6$
$0 \leq y \leq 2$
$x \geq 0$

$3x + 4y \geq 12$ is the set of points on or above the solid line $3x + 4y = 12$. $2x - 3y \leq 6$ is the set of points on or above the solid line $2x - 3y = 6$. $0 \leq y \leq 2$ is the rectangular strip of points lying on or between the horizontal lines $y = 0$ and $y = 2$. $x \geq 0$ consists of all the points on and to the right of the y-axis. The feasible region is the triangular region satisfying all the inequalities.

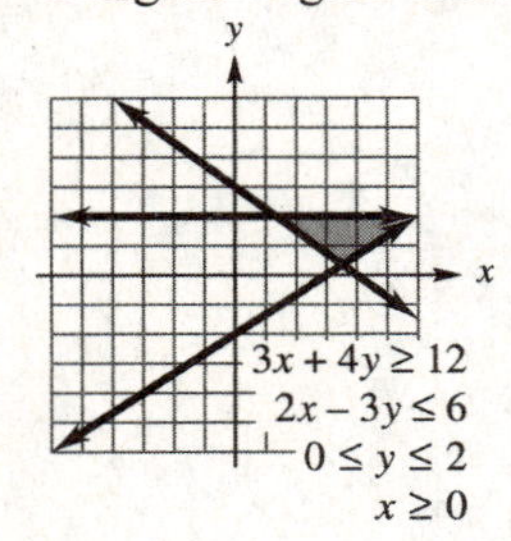

47. The shaded region lies between the horizontal lines $y = 0$ and $y = 4$, so this inequality is $0 \leq y \leq 4$. The shaded region also lies to the right of the vertical line $x = 0$, so this inequality is $x \geq 0$. The shaded region also lies below the line containing the points (3, 4) and (6, 0) which has the equation $4x + 3y = 24$.

Using the origin as a test point, we determine that
$4(0) + 3(0) \leq 24$ is true, so the inequality must be $\leq$. So the third inequality is $4x + 3y \leq 24$. So, the inequalities are
$x \geq 0$
$0 \leq y \leq 4$
$4x + 3y \leq 24$

49. The feasible region is the interior of the rectangle with vertices (2, 3), (2, –1), (7, 3), and (7, –1). The x-values range from 2 to 7, not including 2 or 7. The y-values range from –1 to 3, not including –1 or 3. Thus, the system of inequalities is

$$2 < x < 7$$
$$-1 < y < 3.$$

51.

a.

	Number	Hours Spinning	Hours Dyeing	Hours Weaving
Shawls	x	1	1	1
Afghans	y	2	1	4
Maximum number of hours available		8	6	14

b.

$x + 2y \leq 8$	Spinning inequality
$x + y \leq 6$	Dyeing inequality
$x + 4y \leq 14$	Weaving inequality
$x \geq 0$	Ensures a non-negative number of each
$y \geq 0$	

c. Graph the solid lines $x + 2y = 8$, $x + y = 6$, $x + 4y = 14$, $x = 0$, and $y = 0$, and shade the appropriate half-planes to get the feasible region.

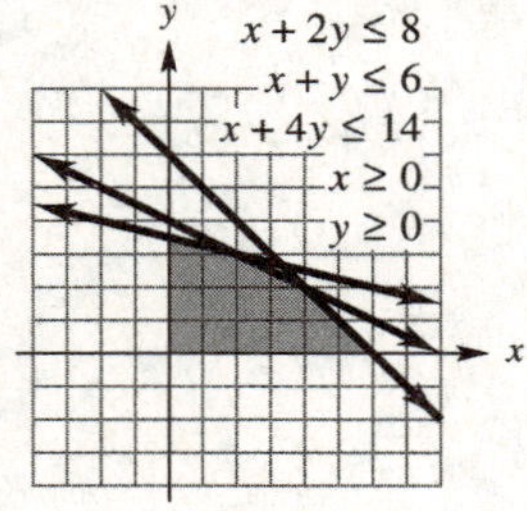

53. The system is
$$x + y \le 50{,}000$$
$$.036x + .05y \ge 2100$$
$$x \ge 0, y \ge 0$$

This system will reduce to:
$$y \le -x + 50{,}000$$
$$y \ge -.72x + 45{,}000$$
$$x \ge 0, y \ge 0$$

The first inequality gives the region below the line that crosses the y axis at 50,000, including the points on the line. The second inequality gives the region above the line that crosses the y axis at 45,000, including the points on the line. The other inequalities are the x and y axes.

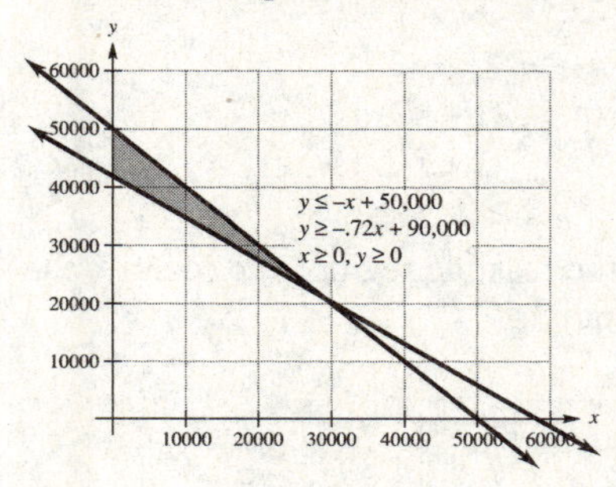

55. The system of inequalities is
$$1.3x + 1.1y \ge 8$$
$$x \ge 2, y \ge 2$$

This system will reduce to:
$$y \ge -1.18x + 7.27$$
$$x \ge 2, y \ge 2$$

The first inequality gives the region above the line that crosses the y axis at 7.27, including the points on the line. The other inequalities are a horizontal and vertical line crossing the the x and y axes at 2.

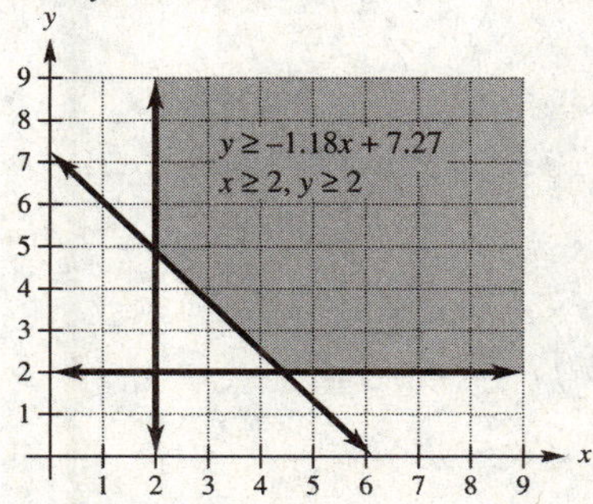

Section 7.2 Linear Programming: The Graphical Method

1. Make a table indicating the value of the objective function $z = 6x + y$ at each corner point.

Corner Point	Value of $z = 6x + y$
(1, 1)	$6(1) + 1 = 7$ Minimum
(2, 7)	$6(2) + 1(7) = 19$
(5, 10)	$6(5) + 10 = 40$ Maximum
(6, 3)	$6(6) + 3 = 39$

The maximum value of 40 occurs at (5, 10). The minimum value of 7 occurs at (1, 1).

3.

Corner Point	Value of $z = .3x + .5y$	
(0, 0)	0	Minimum
(0, 12)	6	Maximum
(4, 8)	5.2	
(7, 3)	3.6	
(9, 0)	2.7	

The maximum is 6 at (0, 12); the minimum is 0 at (0, 0).

5. a.

Corner Point	Value of $z = x + 5y$	
$(0, 8)$	40	
$(3, 4)$	23	
$\left(\frac{13}{2}, 2\right)$	16.5	
$(12, 0)$	12	Minimum

The minimum is 12 at (12, 0). There is no maximum because the feasible region is unbounded.

b.

Corner Point	Value of $z = 2x + 3y$	
$(0, 8)$	24	
$(3, 4)$	18	Minimum
$\left(\frac{13}{2}, 2\right)$	19	
$(12, 0)$	24	

The minimum is 18 at (3, 4). There is no maximum because the feasible region is unbounded.

c.

Corner Point	Value of $z = 2x + 4y$
$(0, 8)$	32
$(3, 4)$	22
$\left(\frac{13}{2}, 2\right)$	21 Minimum
$(12, 0)$	24

The minimum is 21 at $\left(\frac{13}{2}, 2\right)$. There is no maximum.

d.

Corner Point	Value of $z = 4x + y$
$(0, 8)$	8 Minimum
$(3, 4)$	16
$\left(\frac{13}{2}, 2\right)$	28
$(12, 0)$	48

The minimum is 8 at (0, 8). There is no maximum.

7. Maximize $z = 4x + 3y$
subject to: $2x + 3y \le 6$
$4x + y \le 6$
$x \ge 0,\ y \ge 0.$

Graph the feasible region, and identify the corner points.

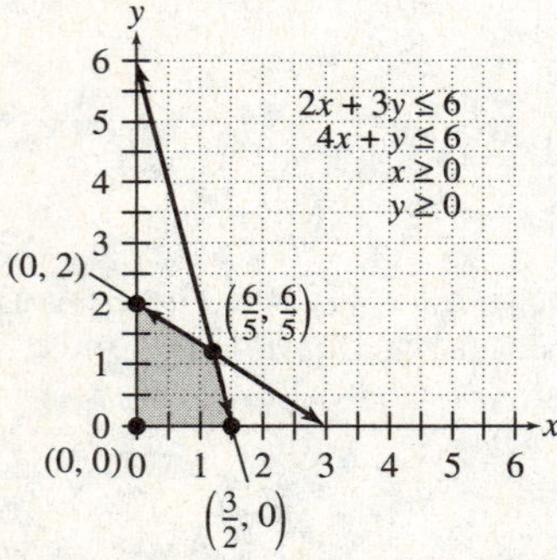

The graph shows the feasible region is bounded. The corner points are

$(0, 0), (0, 2), \left(\frac{3}{2}, 0\right), \left(\frac{6}{5}, \frac{6}{5}\right)$, which is the intersection of $2x + 3y = 6$ and $4x + y = 6$. Use the corner points to find the maximum value of the objective function.

Corner Point	Value of $z = 4x + 3y$
$(0, 0)$	0
$(0, 2)$	6
$\left(\frac{6}{5}, \frac{6}{5}\right)$	$\frac{42}{5} = 8.4$ Maximum
$\left(\frac{3}{2}, 0\right)$	6

The maximum value of $z = 4x + 3y$ is 8.4 at $x = 1.2, y = 1.2$.

9. Minimize $z = 2x + y$
subject to: $3x - y \ge 12$
$x + y \le 15$
$x \ge 2,\ y \ge 3.$

Graph the feasible region, and identify the corner points

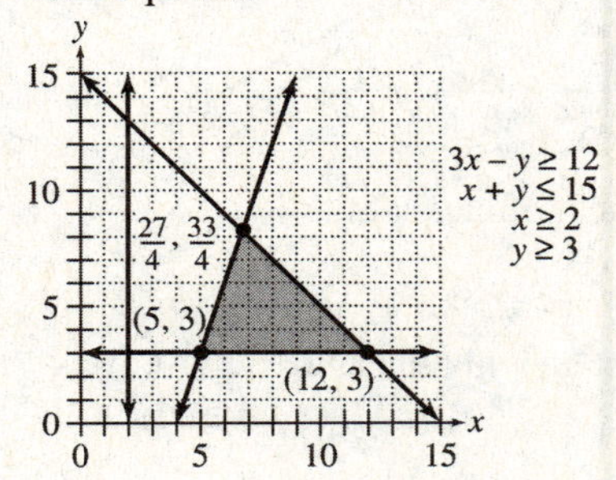

The feasible region is bounded with corner points $(5, 3), \left(\frac{27}{4}, \frac{33}{4}\right), (12, 3)$.

Corner Point	Value of $z = 2x + y$
$(5, 3)$	13 Minimum
$\left(\frac{27}{4}, \frac{33}{4}\right)$	$\frac{87}{4}$
$(12, 3)$	27

The minimum is 13 at (5, 3).

11. Maximize $z = 5x + y$
subject to: $x - y \le 10$
$5x + 3y \le 75$
$x \ge 0,\ y \ge 0.$

Graph the feasible region, and identify the corner points.

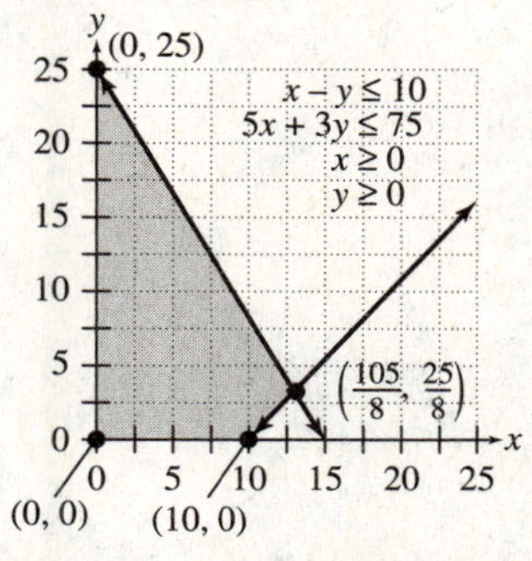

This region is bounded, with corner points $(0, 0)$, $(0, 25)$, $\left(\frac{105}{8}, \frac{25}{8}\right)$, $(10, 0)$.

Corner Point	Value of $z = 5x + y$
$(0, 0)$	0
$(0, 25)$	25
$\left(\frac{105}{8}, \frac{25}{8}\right)$	$\frac{275}{4} = 68.75$ Maximum
$(10, 0)$	50

The maximum is 68.75 at $\left(\frac{105}{8}, \frac{25}{8}\right)$.

13. $3x + 2y \ge 6$
$x + 2y \ge 4$
$x \ge 0,\ y \ge 0$

Graph the feasible region and identify the corner points.

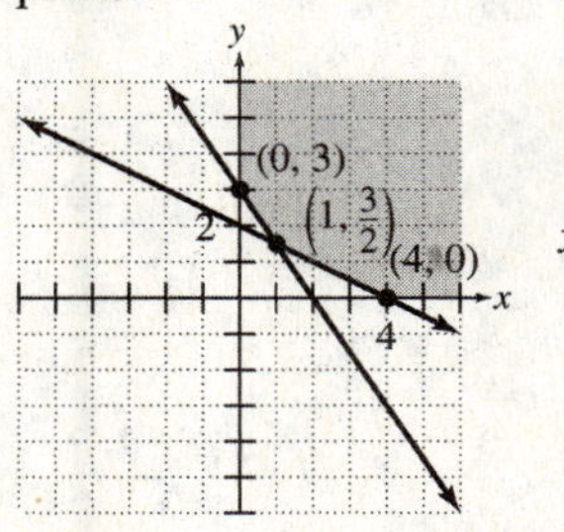

$3x + 2y \ge 6$
$x + 2y \ge 4$
$x \ge 0,\ y \ge 0$

Corner Point	Value of $z = 3x + 4y$
$(0, 3)$	12
$\left(1, \frac{3}{2}\right)$	9 Minimum
$(4, 0)$	12

The minimum value of 9 occurs at $\left(1, \frac{3}{2}\right)$.

There is no maximum value because the feasible region is unbounded.

15. $x + y \le 6$
$-x + y \le 2$
$2x - y \le 8$

Graph the feasible region, and identify the corner points.

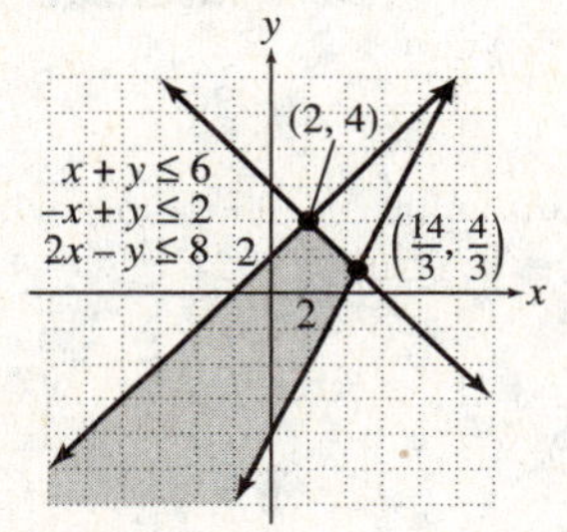

Corner Point	Value of $z = 3x + 4y$
$(2, 4)$	22 Maximum
$\left(\frac{14}{3}, \frac{4}{3}\right)$	$\frac{58}{3}$

The maximum value of z is 22 at (2, 4). There is no minimum because the feasible region is unbounded.

17. a.

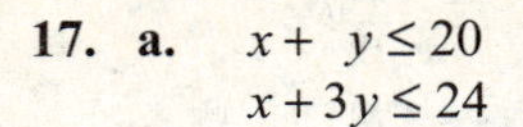

$$x + y \le 20$$
$$x + 3y \le 24$$

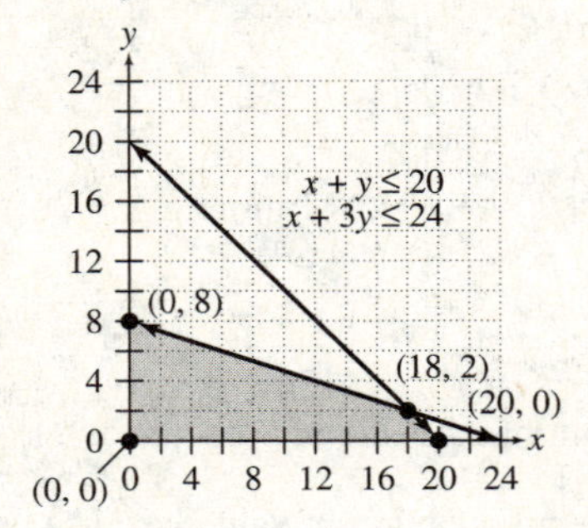

Corner Point	Value of $z = 10x + 12y$
(0, 0)	0
(0, 8)	96
(18, 2)	204 Maximum
(20, 0)	200

The maximum value of 204 occurs when $x = 18$ and $y = 2$, or at (18, 2).

b. $3x + y \le 15$
$x + 2y \le 18$

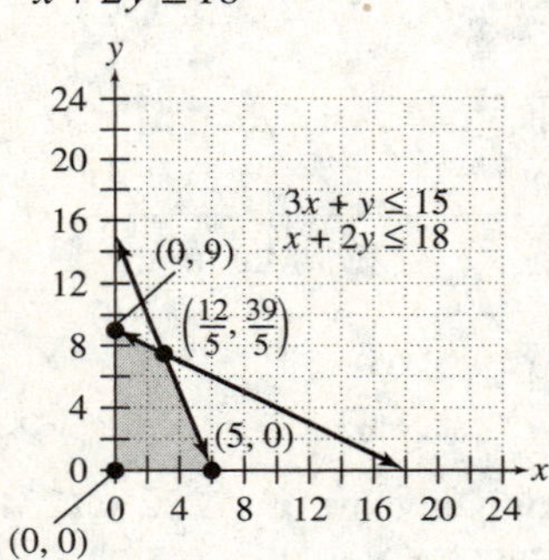

Corner Point	Value of $z = 10x + 12y$
(0, 0)	0
(0, 9)	108
$\left(\frac{12}{5}, \frac{39}{5}\right)$	$\frac{588}{5}$
(5, 0)	50

The maximum value of $\frac{588}{5}$ or $117\frac{3}{5}$ occurs when $x = \frac{12}{5}$ and $y = \frac{39}{5}$, or at $\left(\frac{12}{5}, \frac{39}{5}\right)$.

c. $x + 2y \ge 10$
$2x + y \ge 12$
$x - y \le 8$

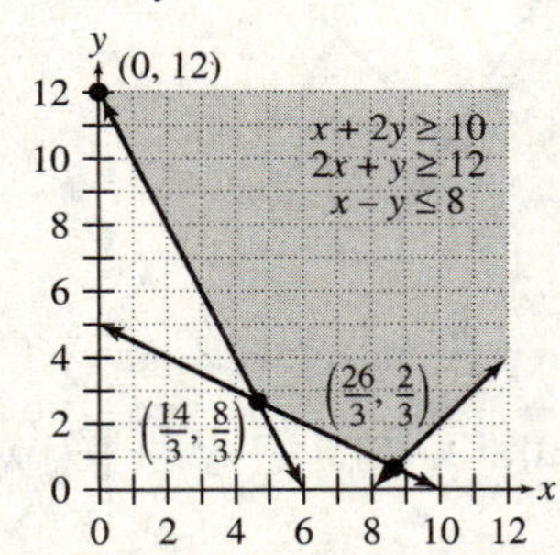

The feasible region is unbounded, so there is no maximum.

19. Answer varies. Sample response: The constraints do not describe a feasible region, i.e. there does not exist a point that satisfies all five constraints.

Section 7.3 Applications of Linear Programming

1. Let x = the number of canoes and let y = the number of rowboats.
The constraints are $8x + 5y \le 110$, $x \ge 0$, $y \ge 0$.

3. Let x = the number of radio spots and let y = the number of TV ads.
The constraints are $250x + 750y \le 9500$, $x \ge 0$ $y \ge 0$.

5. Let x = the number of chain saws and let y = the number of wood chippers.
Assembling x chain saws at 4 hours each takes $4x$ hours while assembling y wood chippers at 6 hours each takes $6y$ hours. There are only 48 available hours, the first constraint is $4x + 6y \leq 48$.
The number of chain saws and wood chippers assembled cannot be negative, so $x \geq 0, y \geq 0$.
Each chain saw produces a profit of \$150 and each wood chipper, \$220. If z represents total profit, then $z = 150x + 220y$ is the objective function to be maximized.
Maximize $z = 150x + 220y$
subject to: $4x + 6y \leq 48$
$x \geq 0,\ y \geq 0$

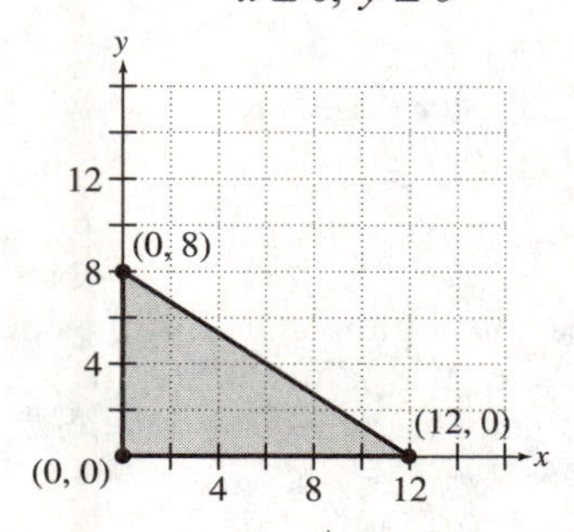

Corner Point	$z = 150x + 220y$
(0, 0)	150(0)+220(0) = 0
(12, 0)	150(12) + 220(0) = 1800 (maximum)
(0, 8)	150(0) + 220(8) = 1760

12 chain saws and no wood chippers should be assembled for maximum profit.

7. Let x = the number of pounds of deluxe coffee and let y = the number of pounds of regular coffee. The mixture of deluxe and regular coffee needs to be at least 50 pounds, so $x + y \geq 50$
The mixture must have at least 10 pounds of deluxe coffee, so $x \geq 10$.
The pounds of coffee cannot be negative, so $x \geq 0, y \geq 0$.
At \$6 per pound of deluxe coffee and \$5 per pound of regular coffee, the total cost, z, is $z = 6x + 5y$, which is the objective function to be minimized.
Minimize: $z = 6x + 5y$
subject to: $x + y \geq 50$
$x \geq 10$
$y \geq 0$

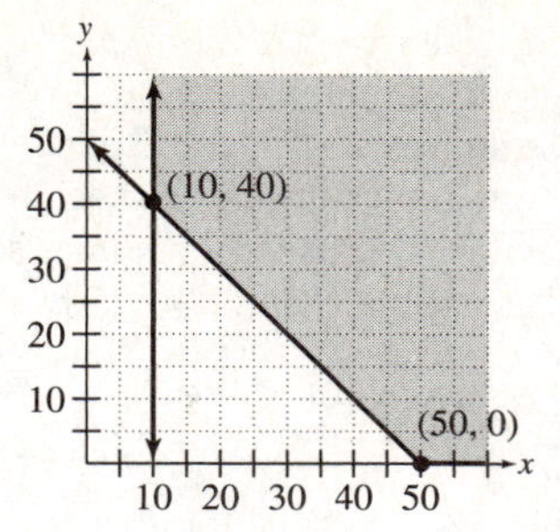

Corner Point	$z = 6x + 5y$
(50, 0)	6(50) + 5(0) = 300
(10, 40)	6(10) + 5(40) = 260 (minimum)

The mixture should contain 10 pounds of deluxe and 40 pounds of regular coffee to minimize cost.

9. Let x = the number of shock replacements and let y = the number of brake replacements.
The constraint imposed by his 48 hour week is $3x + 2y \leq 48$.
He sells at least two shock replacements and at least six brake replacements , so $x \geq 2$ and $y \geq 6$.
Thus, the constraints are
$3x + 2y \leq 48$
$x \geq 2$
$y \geq 6$

Graph the feasibility region

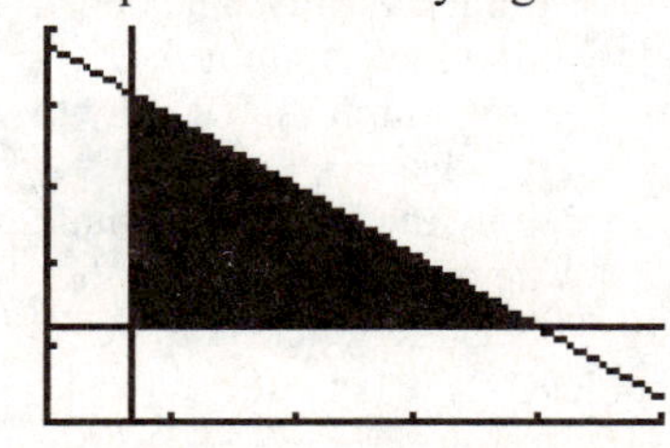

The points of intersection are summarized in the table below.

His labor cost, z, is $z = 500x + 300y$, which is the objective function to be maximized.

Corner Point	$z = 500x + 300y$
(2, 6)	500(2) + 300(6) = 2800
(2, 21)	500(2) + 300(21) = 7300
(12, 6)	500(12) + 300(6) = 7800 (maximum)

He should do 12 shock replacements and 6 brake replacements.

11. From exercise 3, we let x = the number of radio spots and let y = the number of TV ads. Since each radio spot costs \$250 and each TV ad cost \$750, the constraint imposed by the cost is $250x + 750y \le 9500$. The constraints imposed by the number of ads are $x \ge 8$ and $y \ge 3$. Thus, the constraints are

$250x + 750y \le 9500$
$x \ge 8,\ y \ge 3$

The function that describes the number of people reached by the ads is $z = 600x + 2000y$. This is the function to be maximized.

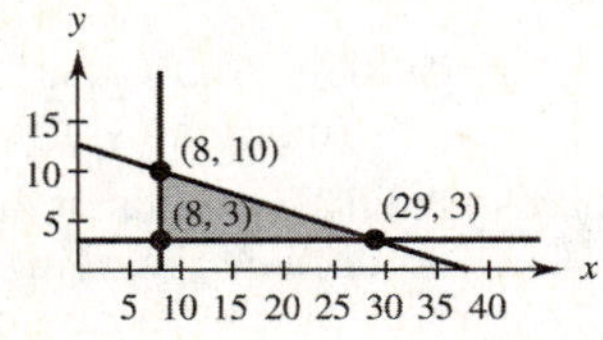

Corner Point	$z = 600x + 2000y$
(8, 3)	$600(8) + 2000(3) = 10{,}800$
(8, 10)	$600(8) + 2000(10) = 24{,}800$ maximum
(29, 3)	$600(29) + 2000(3) = 23{,}400$

The candidate should use 8 radio spots and 10 tv ads to maximize the number of people reached.

13. Let x = the number of brand X pills, let y = the number of brand Z pills. The constraint imposed by the amount of vitamin A is $8x + 2y \ge 16$. The constraint imposed by the amount of vitamin B-1 is $x + y \ge 5$. The constraint imposed by the amount of vitamin C is $2x + 7y \ge 20$. The cost function z is $z = 15x + 30y$. This is the function to be minimized subject to the constraints

$8x + 2y \ge 16$
$x + y \ge 5$
$2x + 7y \ge 20$
$x \ge 0,\ y \ge 0$

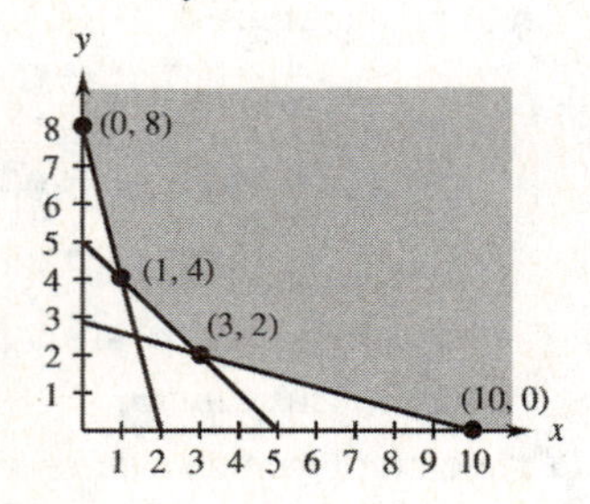

Corner Point	$z = .15x + .30y$
(0, 8)	$.15(0) + .30(8) = 2.40$
(1, 4)	$.15(1) + .30(4) = 1.35$
(3, 2)	$.15(3) + .30(2) = 1.05$ (minimum)
(10, 0)	$.15(10) + .30(0) = 1.50$

He can satisfy the requirements with 3 brand X pills and 2 brand Z pills for a minimum cost of \$1.05.

15. Let x = the number of Type 1 bolts and let y = the number of Type 2 bolts.
Maximize $z = .10x + .12y$
subject to: $.1x + .1y \le 240$
$.1x + .4y \le 720$
$.1x + .02y \le 160$
$x \ge 0$
$y \ge 0$

Graph the feasible region, and label the corner points. Use the corner points to find the maximum value of the objective function.

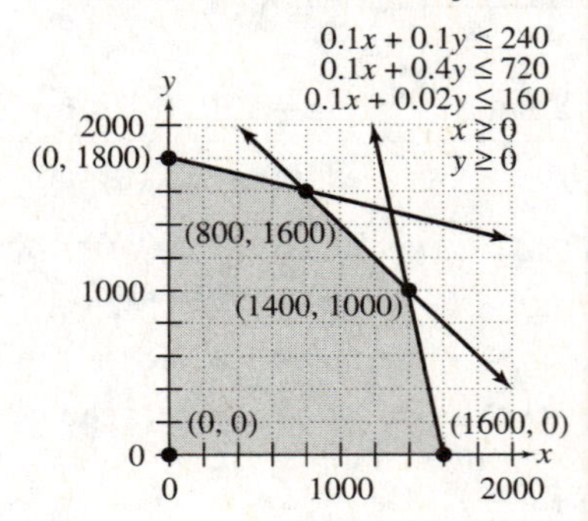

Corner Point	Value of $z = .10x + .12y$
(0, 0)	$.10(0) + .12(0) = 0$
(0, 1800)	$.10(0) + .12(1800) = 216$
(800, 1600)	$.10(800) + .12(1600) = 272$ Maximum
(1400, 1000)	$.10(1400) + .12(1000) = 260$
(1600, 0)	$.1(1600) + .12(0) = 160$

Manufacture 800 Type 1 bolts and 1600 Type 2 bolts for a maximum revenue of \$272/day.

17. Let x = the number of cards from warehouse I to San Jose. Then $350 - x$ is the number of cards from warehouse II to San Jose. Let y = the number of cards from warehouse I to Memphis. Then $250 - y$ is the number of cards from warehouse II to Memphis. The constraints are represented by the inequalities

$$x \geq 0,\ y \geq 0$$
$$x \leq 350$$
$$y \leq 250$$
$$x + y \leq 500$$

$(350 - x) + (250 - y) \leq 290 \Rightarrow x + y \geq 310.$

Minimize

$.25x + .23(350 - x) + .22y + .21(250 - y)$
$= 133 + .02x + .01y.$

The graph and corner points are shown below.

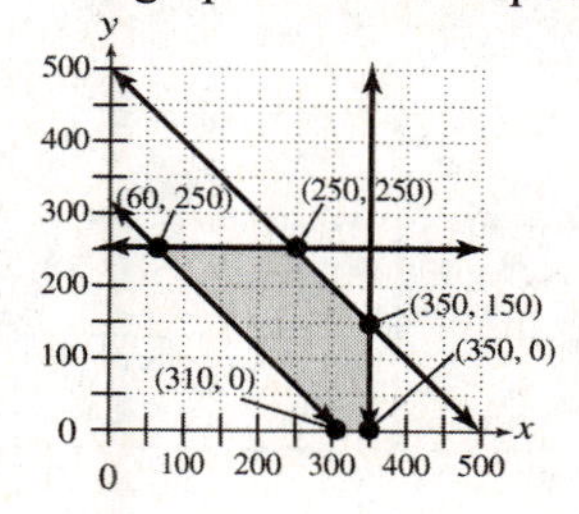

Corner Point	Value of $z = 133 + .02x + .01y$
(60, 250)	136.7 Minimum
(250, 250)	140.5
(350, 150)	141.5
(350, 0)	140
(310, 0)	139.2

The minimum cost is \$136.70. From warehouse I ship 60 boxes to San Jose and 250 boxes to Memphis. From warehouse II ship 290 boxes to San Jose and none to Memphis.

19. Let x = the amount invested in bonds and let y = the amount invested in mutual funds. (Both are in millions of dollars.)
The amount of annual interest is $.04x + .06y$.
Maximize $z = .04x + .06y$

subject to:
$$x \geq 20$$
$$y \geq 6$$
$$300x + 100y \leq 8400$$
$$x + y \leq 50$$
$$x \geq 0,\ y \geq 0$$

Graph the feasible region, and label the corner points.

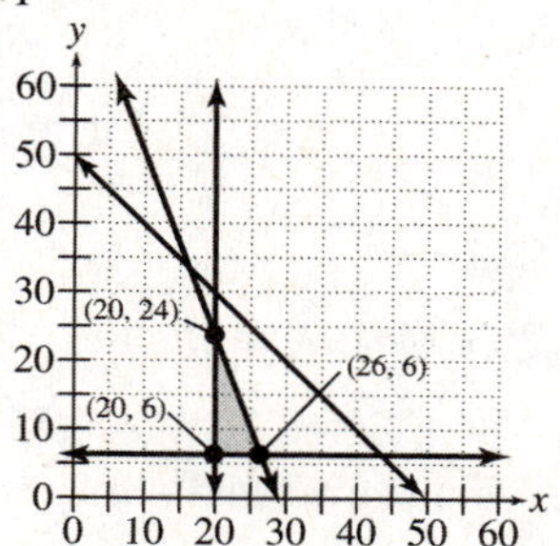

Corner Point	Value of $z = .04x + .06y$
(20, 24)	2.24 Maximum
(26, 6)	1.4
(20, 6)	1.16

He should invest \$20 million in bonds and \$24 million in mutual funds for maximum annual interest of \$2.24 million.

21. Let x = the number of humanities courses and let y = the number of science courses.

Maximize $z = 3(5y) + 2\left(\frac{1}{2}\right)(4x) + 3\left(\frac{1}{4}\right)(4x) + 4\left(\frac{1}{4}\right)(4x)$

subject to:
$$x \geq 4$$
$$4 \leq y \leq 12$$
$$4x + 5y \leq 92.$$

(*continued next page*)

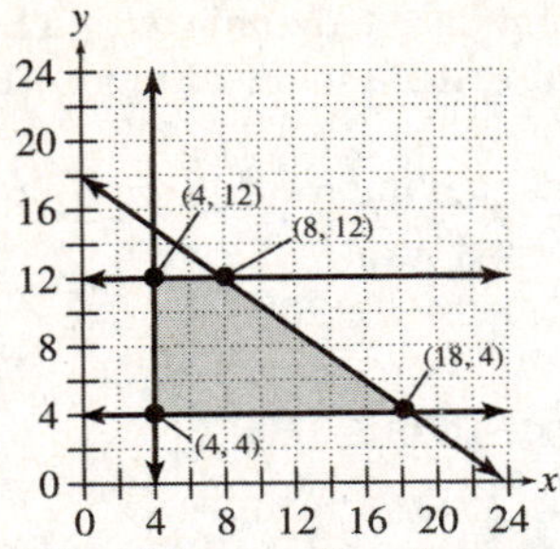

The objective function simplifies to $z = 11x + 15y$.

Corner Point	Value of $z = 11x + 15y$
(4, 4)	104
(4, 12)	224
(8, 12)	268 Maximum
(18, 4)	258

She should take 8 humanities and 12 science courses.

23.

	Number of Shares	Cost per Share	Profit	Risk
Clearbridge	x	162	30	19.4
American	y	12	2	13.7
Constraints		9000	1600	

Minimize $z = 19.4x + 13.7y$

subject to: $162x + 12y \leq 9000$

$30x + 2y \geq 1600$

Graph the feasibility region.

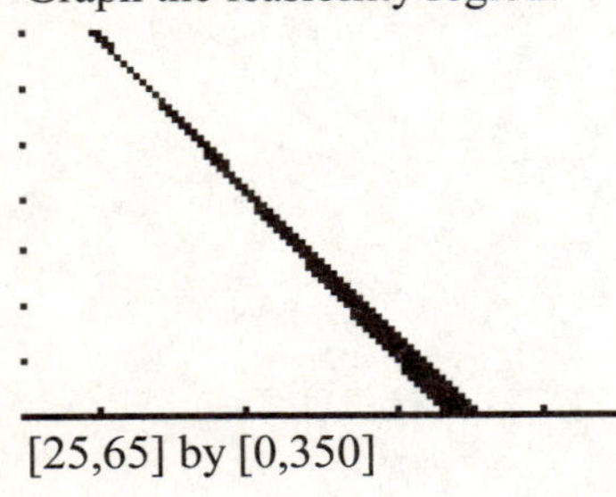

[25,65] by [0,350]

The points of intersection are summarized in the table below.

Corner Point	Value of $z = 19.4x + 13.7y$
(33.33,300)	4756.602
(55.55, 0)	1077.67
(53.33, 0)	1034.602 Minimum

Joe should buy about 53.33 shares of Clearbridge Aggressive Growth Fund and none of the American Century Mid Cap Fund.

25.

	Number of Shares	Cost per Share	Profit	Risk
Franklin	x	36	3	17.1
Delaware	y	46	8	15.5
Constraints		8000	800	

Minimize $z = 17.1x + 15.5y$

subject to: $36x + 46y \leq 8000$

$3x + 8y \geq 800$

Graph the feasibility region.

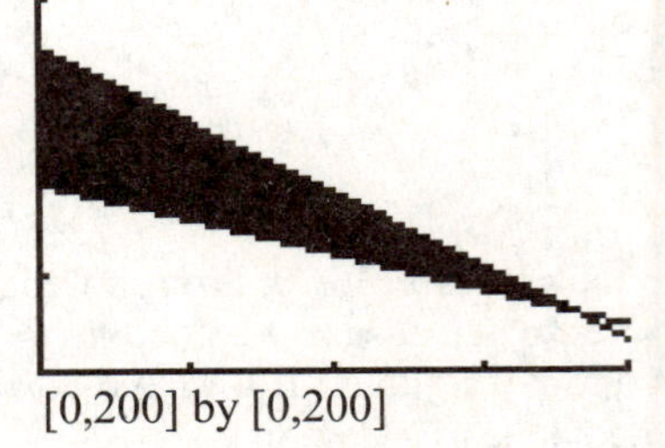

[0,200] by [0,200]

(*continued next page*)

The points of intersection are summarized in the table below.

Corner Point	Value of $z = 17.1x + 15.5y$
(0,173.91)	2695.605
(0,100)	1550 Minimum
(181.33, 32)	3596.74

Sally should buy 100 shares of Delaware Select Growth Fund and none of the Franklin MicroCap Value ADV Fund.

27. 1 Zeta + 2 Beta must not exceed 1000; thus (b) is the correct answer.

29. \$4 Zeta + \$5.25 Beta equals the total contribution margin; (c) is the correct answer.

Section 7.4 The Simplex Method: Maximization

1. Maximize $z = 32x_1 + 9x_2$

subject to:

$$\begin{aligned} 4x_1 + 2x_2 &\le 20 \\ 5x_1 + x_2 &\le 50 \\ 2x_1 + 3x_2 &\le 25 \\ x_1 \ge 0,\ x_2 &\ge 0. \end{aligned}$$

a. There are 3 constraints, so 3 slack variables are needed.

b. Use s_1, s_2, and s_3 for the slack variables.

c.

$$\begin{aligned} 4x_1 + 2x_2 + s_1 \qquad\qquad &= 20 \\ 5x_1 + x_2 \qquad + s_2 \qquad &= 50 \\ 2x_1 + 3x_2 \qquad\qquad + s_3 &= 25 \end{aligned}$$

3. Maximize $z = 8x_1 + 3x_2 + x_3$

subject to:

$$\begin{aligned} 3x_1 - x_2 + 4x_3 &\le 95 \\ 7x_1 + 6x_2 + 8x_3 &\le 118 \\ 4x_1 + 5x_2 + 10x_3 &\le 220 \\ x_1 \ge 0,\ x_2 \ge 0,\ x_3 &\ge 0. \end{aligned}$$

a. There are 3 constraints, so 3 slack variables are needed.

b. Use s_1, s_2, and s_3 for the slack variables.

c.

$$\begin{aligned} 3x_1 - x_2 + 4x_3 + s_1 \qquad\qquad &= 95 \\ 7x_1 + 6x_2 + 8x_3 \qquad + s_2 \qquad &= 118 \\ 4x_1 + 5x_2 + 10x_3 \qquad\qquad + s_3 &= 220 \end{aligned}$$

5. Maximize $z = 5x_1 + x_2$

subject to:

$$\begin{aligned} 2x_1 + 5x_2 &\le 6 \\ 4x_1 + x_2 &\le 6 \\ 5x_1 + 3x_2 &\le 15 \\ x_1 \ge 0,\ x_2 &\ge 0. \end{aligned}$$

Since there are 3 constraints, 3 slack variables are needed: s_1, s_2, and s_3.

The constraints are now

$$\begin{aligned} 2x_1 + 5x_2 + s_1 \qquad\qquad &= 6 \\ 4x_1 + x_2 \qquad + s_2 \qquad &= 6 \\ 5x_1 + 3x_2 \qquad\qquad + s_3 &= 15. \end{aligned}$$

The initial simplex tableau is

$$\begin{array}{cccccc|c} x_1 & x_2 & s_1 & s_2 & s_3 & z & \\ 2 & 5 & 1 & 0 & 0 & 0 & 6 \\ 4 & 1 & 0 & 1 & 0 & 0 & 6 \\ 5 & 3 & 0 & 0 & 1 & 0 & 15 \\ \hline -5 & -1 & 0 & 0 & 0 & 1 & 0 \end{array}.$$

7. Maximize $z = x_1 + 5x_2 + 10x_3$

subject to:

$$\begin{aligned} x_1 + 2x_2 + 3x_3 &\le 10 \\ 2x_1 + x_2 + x_3 &\le 8 \\ 3x_1 \qquad + 4x_3 &\le 6 \\ x_1 \ge 0,\ x_2 \ge 0,\ x_3 &\ge 0. \end{aligned}$$

Since there are 3 constraints, 3 slack variables are needed.: s_1, s_2, s_3.

The constraints are now

$$\begin{aligned} x_1 + 2x_2 + 3x_3 + s_1 \qquad\qquad &= 10 \\ 2x_1 + x_2 + x_3 \qquad + s_2 \qquad &= 8 \\ 3x_1 \qquad + 4x_3 \qquad\qquad + s_3 &= 6 \end{aligned}$$

The initial tableau is

$$\begin{array}{ccccccc|c} x_1 & x_2 & x_3 & s_1 & s_2 & s_3 & z & \\ 1 & 2 & 3 & 1 & 0 & 0 & 0 & 10 \\ 2 & 1 & 1 & 0 & 1 & 0 & 0 & 8 \\ 3 & 0 & 4 & 0 & 0 & 1 & 0 & 6 \\ \hline -1 & -5 & -10 & 0 & 0 & 0 & 1 & 0 \end{array}.$$

9.

$$\begin{array}{cccccc|c} x_1 & x_2 & x_3 & s_1 & s_2 & z & \\ 2 & 2 & 0 & 3 & 1 & 0 & 15 \\ 3 & 4 & 1 & 6 & 0 & 0 & 20 \\ \hline -2 & -3 & 0 & 1 & 0 & 1 & 10 \end{array}$$

The most negative indicator is –3, so the pivot column is column 2. Since $\frac{20}{4} = 5$ is the smaller quotient, the pivot row is row two. The pivot is the 4 in row two, column two.

11.

$$\begin{array}{ccccccc|c} x_1 & x_2 & x_3 & s_1 & s_2 & s_3 & z & \\ 6 & 2 & 1 & 3 & 0 & 0 & 0 & 8 \\ 0 & 2 & 0 & 1 & 0 & 1 & 0 & 7 \\ 6 & 1 & 0 & 3 & 1 & 0 & 0 & 6 \\ \hline -3 & -2 & 0 & 2 & 0 & 0 & 1 & 12 \end{array}$$

The most negative indicator is –3, so the pivot column is column one. Since $\frac{6}{6}=1$ is the smallest quotient, the pivot row is row three. Thus, the pivot is the 6 in row three, column one.

13.

$$\begin{array}{cccccc|c} x_1 & x_2 & x_3 & s_1 & s_2 & z & \\ 1 & 2 & 4 & 1 & 0 & 0 & 56 \\ 2 & \boxed{2} & 1 & 0 & 1 & 0 & 40 \\ \hline -1 & -3 & -2 & 0 & 0 & 1 & 0 \end{array}$$

Start by multiplying each entry of row 2 by $\frac{1}{2}$ in order to change the pivot to 1.

$$\begin{array}{cccccc|cl} x_1 & x_2 & x_3 & s_1 & s_2 & z & & \\ 1 & 2 & 4 & 1 & 0 & 0 & 56 & \\ 1 & 1 & \frac{1}{2} & 0 & \frac{1}{2} & 0 & 20 & \frac{1}{2}R_2 \\ \hline -1 & -3 & -2 & 0 & 0 & 1 & 0 & \end{array}$$

Now use row operations to change the entry in row one column two and the indicator –3 to 0.

$$\begin{array}{cccccc|cl} x_1 & x_2 & x_3 & s_1 & s_2 & z & & \\ -1 & 0 & 3 & 1 & -1 & 0 & 16 & -2R_2+R_1 \\ 1 & 1 & \frac{1}{2} & 0 & \frac{1}{2} & 0 & 20 & \\ \hline 2 & 0 & -\frac{1}{2} & 0 & \frac{3}{2} & 1 & 60 & 3R_2+R_3 \end{array}$$

15.

$$\begin{array}{ccccccc|c} x_1 & x_2 & x_3 & s_1 & s_2 & s_3 & z & \\ 1 & 1 & 1 & 1 & 0 & 0 & 0 & 60 \\ 3 & 1 & \boxed{2} & 0 & 1 & 0 & 0 & 100 \\ 1 & 2 & 3 & 0 & 0 & 1 & 0 & 200 \\ \hline -1 & -1 & -2 & 0 & 0 & 0 & 1 & 0 \end{array}$$

$$\begin{array}{ccccccc|cl} x_1 & x_2 & x_3 & x_4 & x_5 & x_6 & z & & \\ 1 & 1 & 1 & 1 & 0 & 0 & 0 & 60 & \\ \frac{3}{2} & \frac{1}{2} & 1 & 0 & \frac{1}{2} & 0 & 0 & 50 & \frac{1}{2}R_2 \\ 1 & 2 & 3 & 0 & 0 & 1 & 0 & 200 & \\ \hline -1 & -1 & -2 & 0 & 0 & 0 & 1 & 0 & \end{array}$$

$$\begin{array}{ccccccc|cl} x_1 & x_2 & x_3 & s_1 & s_2 & s_3 & z & & \\ -\frac{1}{2} & \frac{1}{2} & 0 & 1 & -\frac{1}{2} & 0 & 0 & 10 & -1R_2+R_1 \\ \frac{3}{2} & \frac{1}{2} & 1 & 0 & \frac{1}{2} & 0 & 0 & 50 & \\ -\frac{7}{2} & \frac{1}{2} & 0 & 0 & -\frac{3}{2} & 1 & 0 & 50 & -3R_2+R_3 \\ \hline 2 & 0 & 0 & 0 & 1 & 0 & 1 & 100 & 2R_2+R_4 \end{array}$$

17.

$$\begin{array}{cccccc|c} x_1 & x_2 & x_3 & s_1 & s_2 & z & \\ 3 & 2 & 0 & -3 & 1 & 0 & 29 \\ 4 & 0 & 1 & -2 & 0 & 0 & 16 \\ \hline -5 & 0 & 0 & -1 & 0 & 1 & 11 \end{array}$$

a. The basic variables are x_3, s_2, and z. x_1, x_2, and s_1 are nonbasic.

b. The basic feasible solution is $x_1 = 0$, $x_2 = 0$, $x_3 = 16$, $s_1 = 0$, $s_2 = 29$, $z = 11$

c. Because there are still negative indicators, this solution is not the maximum.

19.

$$\begin{array}{ccccccc|c} x_1 & x_2 & x_3 & s_1 & s_2 & s_3 & z & \\ 1 & 0 & 2 & \frac{1}{2} & 0 & \frac{1}{3} & 0 & 6 \\ 0 & 1 & -1 & 5 & 0 & -1 & 0 & 13 \\ 0 & 0 & 1 & \frac{3}{2} & 1 & -\frac{1}{3} & 0 & 21 \\ \hline 0 & 0 & 2 & \frac{1}{2} & 0 & 3 & 1 & 18 \end{array}$$

a. The basic variables are x_1, x_2, s_2, and z. x_3, s_1, and s_3 are nonbasic.

b. The basic feasible solution is $x_1 = 6$, $x_2 = 13$, $x_3 = 0$, $s_1 = 0$, $s_2 = 21$, $s_3 = 0$, $z = 18$.

c. Because there are no negative indicators, this solution is the maximum.

21. Maximize $z = x_1 + 3x_2$

subject to:

$$\begin{aligned} x_1 + x_2 &\le 10 \\ 5x_1 + 2x_2 &\le 20 \\ x_1 + 2x_2 &\le 36 \\ x_1 \ge 0,\ x_2 &\ge 0. \end{aligned}$$

Using slack variables, s_1, s_2, s_3. the constraints become:

$$\begin{aligned} x_1 + x_2 + s_1 \qquad\qquad &= 10 \\ 5x_1 + 2x_2 \qquad + s_2 \qquad &= 20 \\ x_1 + 2x_2 \qquad\qquad + s_3 &= 36. \end{aligned}$$

(continued next page)

The initial simplex tableau is

$$\begin{array}{c} \begin{matrix} x_1 & x_2 & s_1 & s_2 & s_3 & z \end{matrix} \\ \left[\begin{array}{cccccc|c} 1 & \boxed{1} & 1 & 0 & 0 & 0 & 10 \\ 5 & 2 & 0 & 1 & 0 & 0 & 20 \\ 1 & 2 & 0 & 0 & 1 & 0 & 36 \\ \hline -1 & -3 & 0 & 0 & 0 & 1 & 0 \end{array}\right]. \end{array}$$

Pivot on the 1 in row one, column two.

$$\begin{array}{c} \begin{matrix} x_1 & x_2 & s_1 & s_2 & s_3 & z \end{matrix} \\ \left[\begin{array}{cccccc|c} 1 & 1 & 1 & 0 & 0 & 0 & 10 \\ 3 & 0 & -2 & 1 & 0 & 0 & 0 \\ -1 & 0 & -2 & 0 & 1 & 0 & 16 \\ \hline 2 & 0 & 3 & 0 & 0 & 1 & 30 \end{array}\right] \begin{array}{l} \\ -2R_1 + R_2 \\ -2R_1 + R_3 \\ 3R_1 + R_4 \end{array} \end{array}$$

Since there are no negative indicators, this matrix is the final tableau. The maximum is 30 when $x_1 = 0,\ x_2 = 10,\ s_1 = 0,\ s_2 = 0,$ and $s_3 = 16.$

23. Maximize $z = 2x_1 + x_2$

subject to:

$$\begin{aligned} x_1 + 3x_2 &\le 12 \\ 2x_1 + x_2 &\le 10 \\ x_1 + x_2 &\le 4 \\ x_1 \ge 0,\ x_2 &\ge 0. \end{aligned}$$

Using slack variables $s_1,\ s_2,\ s_3$. the constraints become:

$$\begin{aligned} x_1 + 3x_2 + s_1 \qquad\qquad &= 12 \\ 2x_1 + x_2 \qquad + s_2 \qquad &= 10 \\ x_1 + x_2 \qquad\qquad + s_3 &= 4. \end{aligned}$$

The initial simplex tableau is

$$\begin{array}{c} \begin{matrix} x_1 & x_2 & s_1 & s_2 & s_3 & z \end{matrix} \\ \left[\begin{array}{cccccc|c} 1 & 3 & 1 & 0 & 0 & 0 & 12 \\ 2 & 1 & 0 & 1 & 0 & 0 & 10 \\ \boxed{1} & 1 & 0 & 0 & 1 & 0 & 4 \\ \hline -2 & -1 & 0 & 0 & 0 & 1 & 0 \end{array}\right]. \\ \uparrow \end{array}$$

Pivot on the 1 in row three, column one.

$$\begin{array}{c} \begin{matrix} x_1 & x_2 & s_1 & s_2 & s_3 & z \end{matrix} \\ \left[\begin{array}{cccccc|c} 0 & 2 & 1 & 0 & -1 & 0 & 8 \\ 0 & -1 & 0 & 1 & -2 & 0 & 2 \\ 1 & 1 & 0 & 0 & 1 & 0 & 4 \\ \hline 0 & 1 & 0 & 0 & 2 & 1 & 8 \end{array}\right] \begin{array}{l} -1R_3 + R_1 \\ -2R_3 + R_2 \\ \\ 2R_3 + R_4 \end{array} \end{array}$$

Since there are no negative indicators, this matrix is the final tableau.

The maximum is 8 when $x_1 = 4,\ x_2 = 0,\ s_1 = 8,$ $s_2 = 2,\ s_3 = 0.$

25. Maximize $z = 5x_1 + 4x_2 + x_3$

subject to:

$$\begin{aligned} -2x_1 + x_2 + 2x_3 &\le 3 \\ x_1 - x_2 + x_3 &\le 1 \\ x_1 \ge 0,\ x_2 \ge 0,\ x_3 &\ge 0. \end{aligned}$$

Using the slack variables s_1 and s_2, the constraints become:

$$\begin{aligned} -2x_1 + x_2 + 2x_3 + s_1 \qquad &= 3 \\ x_1 - x_2 + x_3 \qquad + s_2 &= 1. \end{aligned}$$

The initial simplex tableau is

$$\begin{array}{c} \begin{matrix} x_1 & x_2 & x_3 & s_1 & s_2 & z \end{matrix} \\ \left[\begin{array}{cccccc|c} -2 & 1 & 2 & 1 & 0 & 0 & 3 \\ \boxed{1} & -1 & 1 & 0 & 1 & 0 & 1 \\ \hline -5 & -4 & -1 & 0 & 0 & 1 & 0 \end{array}\right]. \\ \uparrow \end{array}$$

Pivot on the 1 in row two, column one.

$$\begin{array}{c} \begin{matrix} x_1 & x_2 & x_3 & s_1 & s_2 & z \end{matrix} \\ \left[\begin{array}{cccccc|c} 0 & -1 & 4 & 1 & 2 & 0 & 5 \\ 1 & -1 & 1 & 0 & 1 & 0 & 1 \\ \hline 0 & -9 & 4 & 0 & 5 & 1 & 5 \end{array}\right] \begin{array}{l} 2R_2 + R_1 \\ \\ 5R_2 + R_3 \end{array} \end{array}$$

There is a negative indicator in column two, but all entries in that column are negative, so there is no place to continue pivoting. Therefore, there is no maximum.

27. Maximize $z = 2x_1 + x_2 + x_3$

subject to:

$$\begin{aligned} x_1 - 3x_2 + x_3 &\le 3 \\ x_1 - 2x_2 + 2x_3 &\le 12 \\ x_1 \ge 0,\ x_2 \ge 0,\ x_3 &\ge 0. \end{aligned}$$

Using slack variables s_1 and s_2, the constraints become:

$$\begin{aligned} x_1 - 3x_2 + x_3 + s_1 \qquad &= 3 \\ x_1 - 2x_2 + 2x_3 \qquad + s_2 &= 12. \end{aligned}$$

(*continued next page*)

The initial simplex tableau is

$$\begin{array}{cccccc|c} x_1 & x_2 & x_3 & s_1 & s_2 & z & \\ \boxed{1} & -3 & 1 & 1 & 0 & 0 & 3 \\ 1 & -2 & 2 & 0 & 1 & 0 & 12 \\ \hline -2 & -1 & -1 & 0 & 0 & 1 & 0 \end{array}.$$

Pivot on the 1 in row one, column one.

$$\begin{array}{cccccc|cl} x_1 & x_2 & x_3 & s_1 & s_2 & z & & \\ 1 & -3 & 1 & 1 & 0 & 0 & 3 & \\ 0 & \boxed{1} & 1 & -1 & 1 & 0 & 9 & -R_1 + R_2 \\ \hline 0 & -7 & 1 & 2 & 0 & 1 & 6 & 2R_1 + R_3 \\ & \uparrow & & & & & & \end{array}$$

Pivot on the 1 in row two, column two.

$$\begin{array}{cccccc|cl} x_1 & x_2 & x_3 & s_1 & s_2 & z & & \\ 1 & 0 & 4 & -2 & 3 & 0 & 30 & 3R_2 + R_1 \\ 0 & 1 & 1 & -1 & 1 & 0 & 9 & \\ \hline 0 & 0 & 8 & -5 & 7 & 1 & 69 & 7R_2 + R_3 \end{array}$$

The only negative indicator is in column four, which has all negative entries, so there is no place to continue pivoting. Therefore, there is no maximum.

29. Maximize $z = 2x_1 + 2x_2 - 4x_3$

Subject to: $3x_1 + 3x_2 - 6x_3 \le 51$
$5x_1 + 5x_2 + 10x_3 \le 99$
$x_1 \ge 0,\ x_2 \ge 0,\ x_3 \ge 0.$

Using slack variables s_1 and s_2, the constraints become:

$3x_1 + 3x_2 - 6x_3 + s_1 = 51$
$5x_1 + 5x_2 + 10x_3 + s_2 = 99.$

The initial simplex tableau is

$$\begin{array}{cccccc|c} x_1 & x_2 & x_3 & s_1 & s_2 & z & \\ \boxed{3} & 3 & -6 & 1 & 0 & 0 & 51 \\ 5 & 5 & 10 & 0 & 1 & 0 & 99 \\ \hline -2 & -2 & 4 & 0 & 0 & 1 & 0 \\ \uparrow & & & & & & \end{array}.$$

Pivot on the 3 in row one, column one.

$$\begin{array}{cccccc|cl} x_1 & x_2 & x_3 & s_1 & s_2 & z & & \\ 1 & 1 & -2 & \frac{1}{3} & 0 & 0 & 17 & \frac{1}{3}R_1 \\ 5 & 5 & 10 & 0 & 1 & 0 & 99 & \\ \hline -2 & -2 & 4 & 0 & 0 & 1 & 0 & \end{array}$$

$$\begin{array}{cccccc|cl} x_1 & x_2 & x_3 & s_1 & s_2 & z & & \\ 1 & 1 & -2 & \frac{1}{3} & 0 & 0 & 17 & \\ 0 & 0 & 20 & -\frac{5}{3} & 1 & 0 & 14 & -5R_1 + R_2 \\ \hline 0 & 0 & 0 & \frac{2}{3} & 0 & 1 & 34 & 2R_1 + R_3 \end{array}$$

The maximum is 34 when
$x_1 = 17,\ x_2 = 0,\ x_3 = 0,\ s_1 = 0,\ s_2 = 14$
or when $x_1 = 0,\ x_2 = 17,\ x_3 = 0,\ s_1 = 0,$
and $s_2 = 14.$

31. Maximize $z = 300x_1 + 200x_2 + 100x_3$

subject to: $x_1 + x_2 + x_3 \le 100$
$2x_1 + 3x_2 + 4x_3 \le 320$
$2x_1 + x_2 + x_3 \le 160$
$x_1 \ge 0,\ x_2 \ge 0,\ x_3 \ge 0.$

Using slack variables $s_1,\ s_2,\ s_3$. the constraints become:

$x_1 + x_2 + x_3 + s_1 = 100$
$2x_1 + 3x_2 + 4x_3 + s_2 = 320$
$2x_1 + x_2 + x_3 + s_3 = 160.$

The initial simplex tableau is

$$\begin{array}{ccccccc|c} x_1 & x_2 & x_3 & s_1 & s_2 & s_3 & z & \\ 1 & 1 & 1 & 1 & 0 & 0 & 0 & 100 \\ 2 & 3 & 4 & 0 & 1 & 0 & 0 & 320 \\ \boxed{2} & 1 & 1 & 0 & 0 & 1 & 0 & 160 \\ \hline -300 & -200 & -100 & 0 & 0 & 0 & 1 & 0 \end{array}.$$

Pivot on the 2 in row three, column one.

$$\begin{array}{ccccccc|cl} x_1 & x_2 & x_3 & s_1 & s_2 & s_3 & z & & \\ 1 & 1 & 1 & 1 & 0 & 0 & 0 & 100 & \\ 2 & 3 & 4 & 0 & 1 & 0 & 0 & 320 & \\ \boxed{1} & \frac{1}{2} & \frac{1}{2} & 0 & 0 & \frac{1}{2} & 0 & 80 & \frac{1}{2}R_3 \\ \hline -300 & -200 & -100 & 0 & 0 & 0 & 1 & 0 & \end{array}$$

$$\begin{array}{ccccccc|cl} x_1 & x_2 & x_3 & s_1 & s_2 & s_3 & z & & \\ 0 & \frac{1}{2} & \frac{1}{2} & 1 & 0 & -\frac{1}{2} & 0 & 20 & -1R_3 + R_1 \\ 0 & 2 & 3 & 0 & 1 & -1 & 0 & 160 & -2R_3 + R_2 \\ 1 & \frac{1}{2} & \frac{1}{2} & 0 & 0 & \frac{1}{2} & 0 & 80 & \\ \hline 0 & -50 & 50 & 0 & 0 & 150 & 1 & 24{,}000 & 300R_3 + R_4 \end{array}$$

(continued on next page)

Pivot on the $\frac{1}{2}$ in row one, column two.

$$\begin{array}{ccccccc|c} x_1 & x_2 & x_3 & s_1 & s_2 & s_3 & z & \\ \hline 0 & 1 & 1 & 2 & 0 & -1 & 0 & 40 \\ 0 & 2 & 3 & 0 & 1 & -1 & 0 & 160 \\ 1 & \frac{1}{2} & \frac{1}{2} & 0 & 0 & \frac{1}{2} & 0 & 80 \\ \hline 0 & -50 & 50 & 0 & 0 & 150 & 1 & 24,000 \end{array} \begin{array}{l} 2R_1 \\ \\ \\ \\ \end{array}$$

$$\begin{array}{ccccccc|c} x_1 & x_2 & x_3 & s_1 & s_2 & s_3 & z & \\ \hline 0 & 1 & 1 & 2 & 0 & -1 & 0 & 40 \\ 0 & 0 & 1 & -4 & 1 & 1 & 0 & 80 \\ 1 & 0 & 0 & -1 & 0 & 1 & 0 & 60 \\ \hline 0 & 0 & 100 & 100 & 0 & 100 & 1 & 26,000 \end{array} \begin{array}{l} \\ -2R_1 + R_2 \\ -\frac{1}{2}R_1 + R_3 \\ 50R_1 + R_4 \end{array}$$

The maximum value is 26,000 when $x_1 = 60$, $x_2 = 40$, $x_3 = 0$, $s_1 = 0$, $s_2 = 80$, and $s_3 = 0$.

33. Maximize $z = 4x_1 - 3x_2 + 2x_3$

subject to:
$$\begin{aligned} 2x_1 - x_2 + 8x_3 &\le 40 \\ 4x_1 - 5x_2 + 6x_3 &\le 60 \\ 2x_1 - 2x_2 + 6x_3 &\le 24 \\ x_1 \ge 0,\ x_2 \ge 0,\ x_3 &\ge 0. \end{aligned}$$

Note: The third constraint simplifies to $x_1 - x_2 + 3x_3 \le 12$.

Using slack variables s_1, s_2, s_3. the constraints become:

$$\begin{aligned} 2x_1 - x_2 + 8x_3 + s_1 \qquad\qquad &= 40 \\ 4x_1 - 5x_2 + 6x_3 + \qquad s_2 \qquad &= 60 \\ x_1 - x_2 + 3x_3 + \qquad\qquad s_3 &= 12. \end{aligned}$$

The initial simplex tableau is

$$\begin{array}{ccccccc|c} x_1 & x_2 & x_3 & s_1 & s_2 & s_3 & z & \\ \hline 2 & -1 & 8 & 1 & 0 & 0 & 0 & 40 \\ 4 & -5 & 6 & 0 & 1 & 0 & 0 & 60 \\ \boxed{1} & -1 & 3 & 0 & 0 & 1 & 0 & 12 \\ \hline -4 & 3 & -2 & 0 & 0 & 0 & 1 & 0 \end{array}.$$

Pivot on the 1 in row three, column one.

$$\begin{array}{ccccccc|cl} x_1 & x_2 & x_3 & s_1 & s_2 & s_3 & z & & \\ \hline 0 & \boxed{1} & 2 & 1 & 0 & -2 & 0 & 16 & -2R_3 + R_1 \\ 0 & -1 & -6 & 0 & 1 & -4 & 0 & 12 & -4R_3 + R_2 \\ 1 & -1 & 3 & 0 & 0 & 1 & 0 & 12 & \\ \hline 0 & -1 & 10 & 0 & 0 & 4 & 1 & 48 & 4R_3 + R_4 \end{array}$$

Pivot on the 1 in row one, column two.

$$\begin{array}{ccccccc|cl} x_1 & x_2 & x_3 & s_1 & s_2 & s_3 & z & & \\ \hline 0 & 1 & 2 & 1 & 0 & -2 & 0 & 16 & \\ 0 & 0 & -4 & 1 & 1 & -6 & 0 & 28 & R_1 + R_2 \\ 1 & 0 & 5 & 1 & 0 & -1 & 0 & 28 & R_1 + R_3 \\ \hline 0 & 0 & 12 & 1 & 0 & 2 & 1 & 64 & R_1 + R_4 \end{array}$$

The maximum is 64 when $x_1 = 28,\ x_2 = 16,\ x_3 = 0,\ s_1 = 0,\ s_2 = 28$, and $s_3 = 0$.

35. Maximize $z = x_1 + 2x_2 + x_3 + 5x_4$

subject to:
$$\begin{aligned} x_1 + 2x_2 + x_3 + x_4 &\le 50 \\ 3x_1 + x_2 + 2x_3 + x_4 &\le 100 \\ x_1 \ge 0,\ x_2 \ge 0,\ x_3 \ge 0, x_4 &\ge 0. \end{aligned}$$

Using slack variables s_1 and s_2 the constraints become:

$$\begin{aligned} x_1 + 2x_2 + x_3 + x_4 + s_1 \qquad &= 50 \\ 3x_1 + x_2 + 2x_3 + x_4 \qquad + s_2 &= 100. \end{aligned}$$

The initial simplex tableau is

$$\begin{array}{ccccccc|c} x_1 & x_2 & x_3 & x_4 & s_1 & s_2 & z & \\ \hline 1 & 2 & 1 & \boxed{1} & 1 & 0 & 0 & 50 \\ 3 & 1 & 2 & 1 & 0 & 1 & 0 & 100 \\ \hline -1 & -2 & -1 & -5 & 0 & 0 & 1 & 0 \end{array}.$$

The pivot is the 1 in row one, column four.

$$\begin{array}{ccccccc|cl} x_1 & x_2 & x_3 & x_4 & s_1 & s_2 & z & & \\ \hline 1 & 2 & 1 & 1 & 1 & 0 & 0 & 50 & \\ 2 & -1 & 1 & 0 & -1 & 1 & 0 & 50 & -1R_1 + R_2 \\ \hline 4 & 8 & 4 & 0 & 5 & 0 & 1 & 250 & 5R_1 + R_3 \end{array}$$

The maximum is 250 when $x_1 = 0, x_2 = 0,\ x_3 = 0,\ x_4 = 50,\ s_1 = 0$, and $s_2 = 50$.

37.
$$\begin{array}{cccccc|c} x_1 & x_2 & x_3 & s_1 & s_2 & z & \\ \hline 1 & 1 & 1 & 1 & 0 & 0 & 12 \\ 2 & 1 & 2 & 0 & 1 & 0 & 30 \\ \hline -2 & -2 & -1 & 0 & 0 & 1 & 0 \end{array}$$

a. Pivot on the 1 in row one, column one.

$$\begin{array}{cccccc|cl} x_1 & x_2 & x_3 & s_1 & s_2 & z & & \\ \hline 1 & 1 & 1 & 1 & 0 & 0 & 12 & \\ 0 & -1 & 0 & -2 & 1 & 0 & 6 & -2R_1 + R_2 \\ \hline 0 & 0 & 1 & 2 & 0 & 1 & 24 & 2R_1 + R_3 \end{array}$$

The maximum is 24 when $x_1 = 12,\ x_2 = 0,\ x_3 = 0,\ s_1 = 0,\ s_2 = 6$.

b. Pivot on the 1 in row one, column two.

$$\begin{array}{cccccc|cl} x_1 & x_2 & x_3 & s_1 & s_2 & z & & \\ \hline 1 & 1 & 1 & 1 & 0 & 0 & 12 & \\ 1 & 0 & 1 & -1 & 1 & 0 & 18 & -1R_1 + R_2 \\ \hline 0 & 0 & 1 & 2 & 0 & 1 & 24 & 2R_1 + R_3 \end{array}$$

The maximum is 24 when $x_1 = 0,\ x_2 = 12,\ x_3 = 0,\ s_1 = 0,\ s_2 = 18$.

c. This problem has a unique maximum value of z, which is 24, but it occurs at two different basic feasible solutions.

Section 7.5 Maximization Applications

1. Let x_1 = the number of Siamese cats and let x_2 = the number of Persian cats.
The problem is to maximize $z = 12x_1 + 10x_2$ subject to:

$$\begin{aligned} 2x_1 + x_2 &\le 90 \\ x_1 + 2x_2 &\le 80 \\ x_1 + x_2 &\le 50 \end{aligned}$$

$x_1 \ge 0,\ x_2 \ge 0.$

There are three constraints to be changed into equalities, so introduce three slack variables, s_1, s_2, and s_3. The problem can now be restated as:

Find $x_1 \ge 0,\ x_2 \ge 0,\ s_1 \ge 0,\ s_2 \ge 0,\ s_3 \ge 0,$ such that

$$\begin{aligned} 2x_1 + x_2 + s_1 \qquad\qquad &= 90 \\ x_1 + 2x_2 \qquad + s_2 \qquad &= 80 \\ x_1 + x_2 \qquad\qquad + s_3 &= 50 \end{aligned}$$

and $z = 12x_1 + 10x_2$ is maximized.

The initial simplex tableau is

$$\begin{array}{c} \begin{matrix} x_1 & x_2 & s_1 & s_2 & s_3 \end{matrix} \\ \left[\begin{array}{ccccc|c} 2 & 1 & 1 & 0 & 0 & 90 \\ 1 & 2 & 0 & 1 & 0 & 80 \\ 1 & 1 & 0 & 0 & 1 & 50 \\ \hline -12 & -10 & 0 & 0 & 0 & 0 \end{array}\right]. \end{array}$$

3. Let x_1 = the number of kg of P,
x_2 = the number of kg of Q,
x_3 = the number of kg of R, and
x_4 = the number of kg of S

The constraints are

$$\begin{aligned} .375x_3 + .625x_4 &\le 500 \\ .75x_2 + .50x_3 + .375x_4 &\le 600 \\ x_1 + .25x_2 + .125x_3 \qquad &\le 300 \end{aligned}$$

$x_1 \ge 0,\ x_2 \ge 0,\ x_3 \ge 0,\ x_4 \ge 0.$

(Notice the food contents are given in *percent* of nutrient per kilogram.) The objective function to maximize is $z = 90x_1 + 70x_2 + 60x_3 + 50x_4$.

The initial tableau is

$$\begin{array}{c} \begin{matrix} x_1 & x_2 & x_3 & x_4 & s_1 & s_2 & s_3 \end{matrix} \\ \left[\begin{array}{ccccccc|c} 0 & 0 & .375 & .625 & 1 & 0 & 0 & 500 \\ 0 & .75 & .5 & .375 & 0 & 1 & 0 & 600 \\ 1 & .25 & .125 & 0 & 0 & 0 & 1 & 300 \\ \hline -90 & -70 & -60 & -50 & 0 & 0 & 0 & 0 \end{array}\right] \end{array}$$

5. a. The information is contained in the table.

	Aluminum	Steel	Profit
1 Speed	12	20	\$8
3 Speed	21	30	\$12
10 Speed	16	40	\$24
Amount Available	42,000	91,800	

Let x_1 = number of 1-speed bikes,
let x_2 = number of 3-speed bikes, and
let x_3 = number of 10-speed bikes.

The problem is to maximize
$z = 8x_1 + 12x_2 + 24x_3$
subject to:

$$\begin{aligned} 12x_1 + 21x_2 + 16x_3 &\le 42{,}000 \\ 20x_1 + 30x_2 + 40x_3 &\le 91{,}800 \end{aligned}$$

$x_1 \ge 0,\ x_2 \ge 0,\ x_3 \ge 0.$

The initial simplex tableau is

$$\begin{array}{c} \begin{matrix} x_1 & x_2 & x_3 & s_1 & s_2 \end{matrix} \\ \left[\begin{array}{ccccc|c} 12 & 21 & 16 & 1 & 0 & 42{,}000 \\ 20 & 30 & 40 & 0 & 1 & 91{,}800 \\ \hline -8 & -12 & -24 & 0 & 0 & 0 \end{array}\right]. \end{array}$$

Pivot on the 40 in row two, column three.

$$\begin{array}{c} \begin{matrix} x_1 & x_2 & x_3 & s_1 & s_2 \end{matrix} \\ \left[\begin{array}{ccccc|c} 12 & 21 & 16 & 1 & 0 & 42{,}000 \\ \frac{1}{2} & \frac{3}{4} & 1 & 0 & \frac{1}{40} & 2295 \\ \hline -8 & -12 & -24 & 0 & 0 & 0 \end{array}\right] \begin{matrix} \\ \frac{1}{40}R_2 \\ \\ \end{matrix} \end{array}$$

$$\begin{array}{c} \begin{matrix} x_1 & x_2 & x_3 & s_1 & s_2 \end{matrix} \\ \left[\begin{array}{ccccc|c} 4 & 9 & 0 & 1 & -\frac{2}{5} & 5280 \\ \frac{1}{2} & \frac{3}{4} & 1 & 0 & \frac{1}{40} & 2295 \\ \hline 4 & 6 & 0 & 0 & \frac{3}{5} & 55{,}080 \end{array}\right] \begin{matrix} -16R_2 + R_1 \\ \frac{1}{40}R_2 \\ 24R_2 + R_3 \end{matrix} \end{array}$$

From the final tableau, the maximum is 55,080 when $x_3 = 2295$, $s_1 = 5280$, and $x_1 = x_2 = s_2 = 0$. Thus, the manufacturer should make no 1-speed or 3-speed bicycles, and should make 2295 10-speed bicycles for a maximum profit of \$55,080.

b. In the optimal solution, $s_1 = 5280$ and $s_2 = 0$. $s_1 = 5280$ means 5280 units of aluminum should be left unused. $s_2 = 0$ means all of the steel is used.

7. a. Let $x_1 =$ the number of minutes allotted to the sports, let $x_2 =$ the number of minutes allotted to the news, and let $x_3 =$ the number of minutes allotted to the weather.
Maximize $z = 40x_1 + 60x_2 + 50x_3$
Subject to:

$$\begin{aligned} x_1 + x_2 + x_3 &= 27 \\ x_1 &= 2x_3 \\ x_1 + x_3 &= 2x_2 \\ x_1 \geq 0,\ x_2 \geq 0,\ x_3 &\geq 0 \end{aligned}$$

Rewrite problem in standard maximum form.
Maximize $z = 40x_1 + 60x_2 + 50x_3$

$$\begin{aligned} x_1 + x_2 + x_3 &= 27 \\ -x_1 + 2x_3 &= 0 \\ -x_1 + 2x_2 - x_3 &= 0 \\ x_1 \geq 0,\ x_2 \geq 0,\ x_3 &\geq 0 \end{aligned}$$

The initial simplex tableau is

$$\begin{array}{cccccc|c} x_1 & x_2 & x_3 & s_1 & s_2 & s_3 & \\ 1 & 1 & 1 & 1 & 0 & 0 & 27 \\ -1 & 0 & 2 & 0 & 1 & 0 & 0 \\ -1 & \boxed{2} & -1 & 0 & 0 & 1 & 0 \\ \hline -40 & -60 & -50 & 0 & 0 & 0 & 0 \end{array}$$

Pivot on the 2 in row three, column two. Multiply row three by $\frac{1}{2}$ and complete the pivot to get

$$\begin{array}{cccccc|cl} x_1 & x_2 & x_3 & s_1 & s_2 & s_3 & & \\ \frac{3}{2} & 0 & \frac{3}{2} & 1 & 0 & -\frac{1}{2} & 27 & -R_3 + R_1 \\ -1 & 0 & \boxed{2} & 0 & 1 & 0 & 0 & \\ -\frac{1}{2} & 1 & -\frac{1}{2} & 0 & 0 & \frac{1}{2} & 0 & \frac{1}{2}R_3 \\ \hline -70 & 0 & -80 & 0 & 0 & 30 & 0 & 60R_3 + R_4 \end{array}$$

Pivot on the 2 in row two, column three. Multiply row 2 by $\frac{1}{2}$ and complete the pivot to get

$$\begin{array}{cccccc|cl} x_1 & x_2 & x_3 & s_1 & s_2 & s_3 & & \\ \boxed{\frac{9}{4}} & 0 & 0 & 1 & -\frac{3}{4} & -\frac{1}{2} & 27 & -\frac{3}{2}R_2 + R_1 \\ -\frac{1}{2} & 0 & 1 & 0 & \frac{1}{2} & 0 & 0 & \frac{1}{2}R_2 \\ -\frac{3}{4} & 1 & 0 & 0 & \frac{1}{4} & \frac{1}{2} & 0 & \frac{1}{2}R_2 + R_3 \\ \hline -110 & 0 & 0 & 0 & 40 & 30 & 0 & 80R_2 + R_4 \end{array}$$

Pivot on the $\frac{9}{4}$ in row one, column one.

$$\begin{array}{cccccc|cl} x_1 & x_2 & x_3 & s_1 & s_2 & s_3 & & \\ 1 & 0 & 0 & \frac{4}{9} & -\frac{1}{3} & -\frac{2}{9} & 12 & \frac{4}{9}R_1 \\ 0 & 0 & 1 & \frac{2}{9} & \frac{1}{3} & -\frac{1}{9} & 6 & \frac{1}{2}R_1 + R_2 \\ 0 & 1 & 0 & \frac{1}{3} & 0 & \frac{1}{3} & 9 & \frac{3}{4}R_1 + R_3 \\ \hline 0 & 0 & 0 & \frac{440}{9} & \frac{10}{3} & \frac{50}{9} & 1320 & 110R_1 + R_4 \end{array}$$

The sports should get 12 minutes of airtime, the news 9 minutes and the weather 6 minutes for a maximum of 1,320,000 viewers.

b. In the optimal solution, $s_1 = 0$, $s_2 = 0$ and $s_3 = 0$. $s_1 = 0$ means that all of the 27 minutes will be used by the new show, i.e. there is no slack. $s_2 = 0$ means that the sports has exactly twice as much time as the weather. $s_3 = 0$ means that the sports and weather have exactly twice as much time as the news.

9. a. Let $x_1 =$ the number of Japanese Maples and let $x_2 =$ the number of tri-color beech trees. We want to maximize the profit function, $z = 350x_1 + 500x_2$ subject to the constraints

$$\begin{aligned} 5x_1 + 7x_2 &\leq 3600 \\ x_1 + 2x_2 &\leq 900 \\ 4x_1 + 4x_2 &\leq 2600. \end{aligned}$$

The third constraint simplifies to $x_1 + x_2 \leq 650$

The initial simplex tableau is

$$\begin{array}{ccccc|c} x_1 & x_2 & s_1 & s_2 & s_3 & \\ 5 & 7 & 1 & 0 & 0 & 3600 \\ 1 & \boxed{2} & 0 & 1 & 0 & 900 \\ 1 & 1 & 0 & 0 & 1 & 650 \\ \hline -350 & -500 & 0 & 0 & 0 & 0 \end{array}.$$

Pivot on the 2 in row two, column two. First divide each entry in row 2 by 2.

$$\begin{array}{ccccc|c} x_1 & x_2 & s_1 & s_2 & s_3 & \\ 5 & 7 & 1 & 0 & 0 & 3600 \\ \frac{1}{2} & 1 & 0 & \frac{1}{2} & 0 & 450 \\ 1 & 1 & 0 & 0 & 1 & 650 \\ \hline -350 & -500 & 0 & 0 & 0 & 0 \end{array}.$$

$$\begin{array}{ccccc|c} x_1 & x_2 & s_1 & s_2 & s_3 & \\ \boxed{\frac{3}{2}} & 0 & 1 & -\frac{7}{2} & 0 & 450 \\ \frac{1}{2} & 1 & 0 & \frac{1}{2} & 0 & 450 \\ \frac{1}{2} & 0 & 0 & -\frac{1}{2} & 1 & 200 \\ \hline -100 & 0 & 0 & 250 & 0 & 225{,}000 \end{array} \quad \begin{array}{l} -7R_2 + R_1 \\ \\ -R_2 + R_3 \\ 500R_2 + R_4 \end{array}$$

Now pivot on $\frac{3}{2}$ in row 1 column 1.First multiply each entry in row 1 by $\frac{1}{3}$.

$$\begin{array}{ccccc|c} x_1 & x_2 & s_1 & s_2 & s_3 & \\ \frac{1}{2} & 0 & \frac{1}{3} & -\frac{7}{6} & 0 & 150 \\ \frac{1}{2} & 1 & 0 & \frac{1}{2} & 0 & 450 \\ \frac{1}{2} & 0 & 0 & -\frac{1}{2} & 1 & 200 \\ \hline -100 & 0 & 0 & 250 & 0 & 225{,}000 \end{array}$$

$$\begin{array}{ccccc|c} x_1 & x_2 & s_1 & s_2 & s_3 & \\ \frac{1}{2} & 0 & \frac{1}{3} & -\frac{7}{6} & 0 & 150 \\ 0 & 1 & -\frac{1}{3} & \frac{5}{3} & 0 & 300 \\ 0 & 0 & -\frac{1}{3} & \frac{2}{3} & 1 & 50 \\ \hline 0 & 0 & \frac{200}{3} & \frac{50}{3} & 0 & 255{,}000 \end{array} \quad \begin{array}{l} \\ -R_1 + R_2 \\ -R_1 + R_3 \\ 200R_1 + R_4 \end{array}$$

$$\begin{array}{ccccc|c} x_1 & x_2 & s_1 & s_2 & s_3 & \\ 1 & 0 & \frac{2}{3} & -\frac{7}{3} & 0 & 300 \\ 0 & 1 & -\frac{1}{3} & \frac{5}{3} & 0 & 300 \\ 0 & 0 & -\frac{1}{3} & \frac{2}{3} & 1 & 50 \\ \hline 0 & 0 & \frac{200}{3} & \frac{50}{3} & 0 & 255{,}000 \end{array} \quad \begin{array}{l} 2R_1 \\ \\ \\ \\ \end{array}$$

From the final tableau, we have $x_1 = 300$ and $x_2 = 300$. Thus, the nursery should acquire 300 Japanese Maples and 300 tri-color beech trees, for a maximum profit of $255,000

b. s_1 and s_2 both are zero. $4(300) + 4(300) + s_3 = 2600 \Rightarrow s_3 = 200$
Thus, there are 200 unused hours in the delivery to the client.

11. Let x_1, x_2, and x_3 = the number of newspapers, radio, and TV ads respectively.

Maximize $z = 2000x_1 + 1200x_2 + 10{,}000x_3$

subject to:

$400x_1 + 200x_2 + 1200x_3 \le 8000$

$x_1 \le 20,\ x_2 \le 30,\ x_3 \le 6$

$x_1 \ge 0,\ x_2 \ge 0,\ x_3 \ge 0.$

The initial simplex tableau is

$$\begin{array}{c} \\ \\ \\ \\ \\ \end{array}\left[\begin{array}{ccccccc|c} x_1 & x_2 & x_3 & s_1 & s_2 & s_3 & s_4 & \\ 400 & 200 & 1200 & 1 & 0 & 0 & 0 & 8000 \\ 1 & 0 & 0 & 0 & 1 & 0 & 0 & 20 \\ 0 & 1 & 0 & 0 & 0 & 1 & 0 & 30 \\ 0 & 0 & \boxed{1} & 0 & 0 & 0 & 1 & 6 \\ \hline -2000 & -1200 & -10{,}000 & 0 & 0 & 0 & 0 & 0 \end{array}\right]$$

Pivot on the 1 in row four, column three.

$$\left[\begin{array}{ccccccc|c} x_1 & x_2 & x_3 & s_1 & s_2 & s_3 & s_4 & \\ \boxed{400} & 200 & 0 & 1 & 0 & 0 & -1200 & 800 \\ 1 & 0 & 0 & 0 & 1 & 0 & 0 & 20 \\ 0 & 1 & 0 & 0 & 0 & 1 & 0 & 30 \\ 0 & 0 & 1 & 0 & 0 & 0 & 1 & 6 \\ \hline -2000 & -1200 & 0 & 0 & 0 & 0 & 10{,}000 & 60{,}000 \end{array}\right]\begin{array}{l} \\ -1200R_4 + R_1 \\ \\ \\ \\ 10{,}000R_4 + R_5 \end{array}$$

Pivot on the 400 in row one, column one.

$$\left[\begin{array}{ccccccc|c} x_1 & x_2 & x_3 & s_1 & s_2 & s_3 & s_4 & \\ 1 & \frac{1}{2} & 0 & \frac{1}{400} & 0 & 0 & -3 & 2 \\ 1 & 0 & 0 & 0 & 1 & 0 & 0 & 20 \\ 0 & 1 & 0 & 0 & 0 & 1 & 0 & 30 \\ 0 & 0 & 1 & 0 & 0 & 0 & 1 & 6 \\ \hline -2000 & -1200 & 0 & 0 & 0 & 0 & 10{,}000 & 60{,}000 \end{array}\right]\begin{array}{l} \\ \frac{1}{400}R_1 \\ \\ \\ \\ \\ \end{array}$$

$$\left[\begin{array}{ccccccc|c} x_1 & x_2 & x_3 & s_1 & s_2 & s_3 & s_4 & \\ 1 & \boxed{\frac{1}{2}} & 0 & \frac{1}{400} & 0 & 0 & -3 & 2 \\ 0 & -\frac{1}{2} & 0 & -\frac{1}{400} & 1 & 0 & 3 & 18 \\ 0 & 1 & 0 & 0 & 0 & 1 & 0 & 30 \\ 0 & 0 & 1 & 0 & 0 & 0 & 1 & 6 \\ \hline 0 & -200 & 0 & 5 & 0 & 0 & 4000 & 64{,}000 \end{array}\right]\begin{array}{l} \\ \\ -1R_1 + R_2 \\ \\ \\ 2000R_1 + R_5 \end{array}$$

Pivot on the $\frac{1}{2}$ in row one, column two.

$$\left[\begin{array}{ccccccc|c} x_1 & x_2 & x_3 & s_1 & s_2 & s_3 & s_4 & \\ 2 & 1 & 0 & \frac{1}{200} & 0 & 0 & -6 & 4 \\ 0 & -\frac{1}{2} & 0 & -\frac{1}{400} & 1 & 0 & 3 & 18 \\ 0 & 1 & 0 & 0 & 0 & 1 & 0 & 30 \\ 0 & 0 & 1 & 0 & 0 & 0 & 1 & 6 \\ \hline 0 & -200 & 0 & 5 & 0 & 0 & 4000 & 64{,}000 \end{array}\right]\begin{array}{l} \\ 2R_1 \\ \\ \\ \\ \\ \end{array}$$

(*continued on next page*)

$$\begin{array}{c} \\ \\ \\ \\ \\ \\ \end{array}\begin{array}{ccccccc|c|l} x_1 & x_2 & x_3 & s_1 & s_2 & s_3 & s_4 & & \\ \hline 2 & 1 & 0 & \frac{1}{200} & 0 & 0 & -6 & 4 & \\ 1 & 0 & 0 & 0 & 1 & 0 & 0 & 20 & \frac{1}{2}R_1 + R_2 \\ -2 & 0 & 0 & -\frac{1}{200} & 0 & 1 & 6 & 26 & -1R_1 + R_3 \\ 0 & 0 & 1 & 0 & 0 & 0 & 1 & 6 & \\ \hline 400 & 0 & 0 & 6 & 0 & 0 & 2800 & 64,800 & 200R_1 + R_5 \end{array}$$

From the final tableau, we have $x_1 = 0$, $x_2 = 4$, and $x_3 = 6$.

No newspaper ads, 4 radio ads, and 6 TV ads will give a maximum exposure of 64,800 people.

13. a. Let $x_1 =$ the number of fund-raising parties, let $x_2 =$ the number of mailings, and let $x_3 =$ the number of dinner parties.

Maximize $z = 200,000x_1 + 100,000x_2 + 600,000x_3$

Subject to:
$$x_1 + x_2 + x_3 \le 25$$
$$3000x_1 + 1000x_2 + 12,000x_3 \le 102,000$$
$$x_1 \ge 0,\ x_2 \ge 0,\ x_3 \ge 0$$

The initial simplex tableau is

$$\begin{array}{ccccc|c} x_1 & x_2 & x_3 & s_1 & s_2 & \\ \hline 1 & 1 & 1 & 1 & 0 & 25 \\ 3000 & 1000 & \boxed{12,000} & 0 & 1 & 102,000 \\ \hline -200,000 & -100,000 & -600,000 & 0 & 0 & 0 \end{array}$$

Pivot on the 12,000, row two, column three.

$$\begin{array}{ccccc|c|l} x_1 & x_2 & x_3 & s_1 & s_2 & & \\ \hline \frac{3}{4} & \boxed{\frac{11}{12}} & 0 & 1 & -\frac{1}{12,000} & \frac{33}{2} & -R_2 + R_1 \\ \frac{1}{4} & \frac{1}{12} & 1 & 0 & \frac{1}{12,000} & \frac{17}{2} & \frac{1}{12,000}R_2 \\ \hline -50,000 & -50,000 & 0 & 0 & 50 & 5,100,000 & 600,000R_2 + R_3 \end{array}$$

Pivot on $\frac{11}{12}$ in row one, column two.

$$\begin{array}{ccccc|c|l} x_1 & x_2 & x_3 & s_1 & s_2 & & \\ \hline \boxed{\frac{9}{11}} & 1 & 0 & \frac{12}{11} & -\frac{1}{11,000} & 18 & \frac{12}{11}R_1 \\ \frac{2}{11} & 0 & 1 & -\frac{1}{11} & \frac{1}{11,000} & 7 & -\frac{1}{12}R_1 + R_2 \\ \hline -\frac{100,000}{11} & 0 & 0 & \frac{600,000}{11} & \frac{500}{11} & 6,000,000 & 50,000R_1 + R_3 \end{array}$$

Pivot on $\frac{9}{11}$ in row one, column one.

$$\begin{array}{ccccc|c|l} x_1 & x_2 & x_3 & s_1 & s_2 & & \\ \hline 1 & \frac{11}{9} & 0 & \frac{4}{3} & -\frac{1}{9000} & 22 & \frac{11}{9}R_1 \\ 0 & -\frac{2}{9} & 1 & -\frac{1}{3} & \frac{1}{9000} & 3 & -\frac{2}{11}R_1 + R_2 \\ \hline 0 & \frac{100,000}{9} & 0 & \frac{200,000}{3} & \frac{400}{9} & 6,200,000 & \frac{100,000}{11}R_1 + R_3 \end{array}$$

The party should plan 22 fund raising parties and 3 dinner parties and no mailings to raise a maximum of \$6,200,000.

b. Answers will vary.

15. Let x_1 = the number of hours running, x_2 = the number of hours biking, and x_3 = the number of hours walking. We are looking to maximize $z = 531x_1 + 472x_2 + 354x_3$ subject to the constraints

$$\begin{aligned} x_1 + x_2 + x_3 &\le 15 \\ x_1 &\le 3 \\ 2x_2 - x_3 &\le 0 \end{aligned}$$

$x_1 \ge 0, x_2 \ge 0, x_3 \ge 0.$

The initial simplex tableau is

$$\begin{array}{c} \\ \\ \\ \\ \\ \end{array}\begin{array}{cccccc|c} x_1 & x_2 & x_3 & s_1 & s_2 & s_3 & \\ \hline 1 & 1 & 1 & 1 & 0 & 0 & 15 \\ \boxed{1} & 0 & 0 & 0 & 1 & 0 & 3 \\ 0 & 2 & -1 & 0 & 0 & 1 & 0 \\ \hline -531 & -472 & -354 & 0 & 0 & 0 & 0 \end{array}$$

Pivot on the 1 in row 2, column 1.

$$\begin{array}{cccccc|c|l} x_1 & x_2 & x_3 & s_1 & s_2 & s_3 & & \\ 0 & 1 & 1 & 1 & -1 & 0 & 12 & -R_2 + R_1 \\ 1 & 0 & 0 & 0 & 1 & 0 & 3 & \\ 0 & \boxed{2} & -1 & 0 & 0 & 1 & 0 & \\ \hline 0 & -472 & -354 & 0 & 531 & 0 & 1593 & 531R_2 + R_4 \end{array}$$

Pivot on the 2 in row 3, column 2.

$$\begin{array}{cccccc|c|l} x_1 & x_2 & x_3 & s_1 & s_2 & s_3 & & \\ 0 & 0 & \boxed{3} & 2 & -2 & -1 & 24 & -R_3 + 2R_1 \\ 1 & 0 & 0 & 0 & 1 & 0 & 3 & \\ 0 & 2 & -1 & 0 & 0 & 1 & 0 & \\ \hline 0 & 0 & -590 & 0 & 531 & 236 & 1593 & 236R_2 + R_4 \end{array}$$

Finally, pivot on the 3 in row 1, column 3.

$$\begin{array}{cccccc|c|l} x_1 & x_2 & x_3 & s_1 & s_2 & s_3 & & \\ 0 & 0 & 3 & 2 & -2 & -1 & 24 & \\ 1 & 0 & 0 & 0 & 1 & 0 & 3 & \\ 0 & 6 & 0 & 2 & -2 & 2 & 24 & R_1 + 3R_3 \\ \hline 0 & 0 & 0 & 1180 & 413 & 118 & 18{,}939 & 590R_1 + 3R_4 \end{array}$$

$$\begin{array}{cccccc|c|l} x_1 & x_2 & x_3 & s_1 & s_2 & s_3 & & \\ 0 & 0 & 1 & \frac{2}{3} & -\frac{2}{3} & -\frac{1}{3} & 8 & \frac{1}{3}R_1 \\ 1 & 0 & 0 & 0 & 1 & 0 & 3 & \\ 0 & 1 & 0 & \frac{1}{3} & -\frac{1}{3} & \frac{1}{3} & 4 & \frac{1}{6}R_3 \\ \hline 0 & 0 & 0 & 1180 & 413 & 118 & 18{,}939 & \end{array}$$

Thus, $x_1 = 3$, $x_2 = 4$, and $x_3 = 8$, so

$z = 531(3) + 472(4) + 354(8) = 6313.$

Rachel should run 3 hours, bike 4 hours, and walk 8 hours for a maximum calorie expenditure of 6313 calories.

17. Using the data given, the problem should be stated as follows: Maximize $5x_1 + 4x_2 + 3x_3$ subject to:

$$\begin{aligned} 2x_1 + 3x_2 + x_3 &\le 400 \\ 4x_1 + 2x_2 + 3x_3 &\le 600 \end{aligned}$$

$x_1 \ge 0,\ x_2 \ge 0,\ x_3 \ge 0.$

when x_1 = the number of type A lamps, x_2 = the number of type B lamps, and x_3 = the number of type C lamps.

a. The coefficients of the objective function would be choice (3): 5, 4, 3

b. The constraints in the model would be choice (4): 400, 600

c. The constraint imposed by the available number of person-hours in department I could be expressed as choice (3): $2x_1 + 3x_2 + 1x_3 \le 400$

19. From Exercise 1, the initial simplex tableau is

$$\begin{array}{ccccc|c} x_1 & x_2 & s_1 & s_2 & s_3 & \\ \boxed{2} & 1 & 1 & 0 & 0 & 90 \\ 1 & 2 & 0 & 1 & 0 & 80 \\ 1 & 1 & 0 & 0 & 1 & 50 \\ \hline -12 & -10 & 0 & 0 & 0 & 0 \end{array}.$$

Pivot on the 2 in row one, column one.

$$\begin{array}{ccccc|c|l} x_1 & x_2 & s_1 & s_2 & s_3 & & \\ 1 & \frac{1}{2} & \frac{1}{2} & 0 & 0 & 45 & \frac{1}{2}R_1 \\ 1 & 2 & 0 & 1 & 0 & 80 & \\ 1 & 1 & 0 & 0 & 1 & 50 & \\ \hline -12 & -10 & 0 & 0 & 0 & 0 & \end{array}$$

(continued on next page)

$$\begin{array}{c} \begin{array}{ccccc} x_1 & x_2 & s_1 & s_2 & s_3 \end{array} \\ \left[\begin{array}{ccccc|c} 1 & \frac{1}{2} & \frac{1}{2} & 0 & 0 & 45 \\ 0 & \frac{3}{2} & -\frac{1}{2} & 1 & 0 & 35 \\ 0 & \boxed{\frac{1}{2}} & -\frac{1}{2} & 0 & 1 & 5 \\ \hline 0 & -4 & 6 & 0 & 0 & 540 \end{array}\right] \begin{array}{l} \\ -1R_1 + R_2 \\ -1R_1 + R_3 \\ 12R_1 + R_4 \end{array} \end{array}$$

Pivot on the $\frac{1}{2}$ in row three, column two.

$$\begin{array}{c} \begin{array}{ccccc} x_1 & x_2 & s_1 & s_2 & s_3 \end{array} \\ \left[\begin{array}{ccccc|c} 1 & \frac{1}{2} & \frac{1}{2} & 0 & 0 & 45 \\ 0 & \frac{3}{2} & -\frac{1}{2} & 1 & 0 & 35 \\ 0 & 1 & -1 & 0 & 2 & 10 \\ \hline 0 & -4 & 6 & 0 & 0 & 540 \end{array}\right] \begin{array}{l} \\ \\ 2R_3 \\ \\ \end{array} \end{array}$$

$$\begin{array}{c} \begin{array}{ccccc} x_1 & x_2 & s_1 & s_2 & s_3 \end{array} \\ \left[\begin{array}{ccccc|c} 1 & 0 & 1 & 0 & -1 & 40 \\ 0 & 0 & 1 & 1 & -3 & 20 \\ 0 & 1 & -1 & 0 & 2 & 10 \\ \hline 0 & 0 & 2 & 0 & 8 & 580 \end{array}\right] \begin{array}{l} -\frac{1}{2}R_3 + R_1 \\ -\frac{3}{2}R_3 + R_2 \\ \\ 4R_3 + R_4 \end{array} \end{array}$$

The breeder should raise 40 Siamese and 10 Persian cats for a maximum gross income of $580.

Exercises 21 requires the use of a computer.

21. The initial simplex tableau is

$$\begin{array}{c} \begin{array}{ccccccc} x_1 & x_2 & x_3 & x_4 & s_1 & s_2 & s_3 \end{array} \\ \left[\begin{array}{ccccccc|c} 0 & 0 & .375 & .625 & 1 & 0 & 0 & 500 \\ 0 & .75 & .5 & .375 & 0 & 1 & 0 & 600 \\ 1 & .25 & .125 & 0 & 0 & 0 & 1 & 300 \\ \hline -90 & -70 & -60 & -50 & 0 & 0 & 0 & 0 \end{array}\right] \end{array}$$

Using the SIMPLEX program on a TI-84 Plus, we get the maximum total growth value is 87,454.5 when 163.6 kg of food P, none of food Q, 1090.9 kg of food R, and 145.5 kg of food S are used.

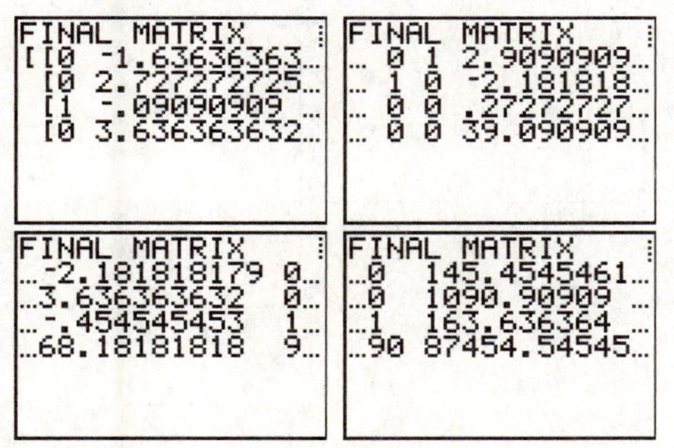

Section 7.6 The Simplex Method: Duality and Minimization

1. The transpose of
$$\begin{bmatrix} 3 & -4 & 5 \\ 1 & 10 & 7 \\ 0 & 3 & 6 \end{bmatrix} \text{ is } \begin{bmatrix} 3 & 1 & 0 \\ -4 & 10 & 3 \\ 5 & 7 & 6 \end{bmatrix}.$$

3. The transpose of

$$\begin{bmatrix} 3 & 0 & 14 & -5 & 3 \\ 4 & 17 & 8 & -6 & 1 \end{bmatrix} \text{ is } \begin{bmatrix} 3 & 4 \\ 0 & 17 \\ 14 & 8 \\ -5 & -6 \\ 3 & 1 \end{bmatrix}.$$

5. Minimize $w = 3y_1 + 5y_2$

subject to: $3y_1 + y_2 \ge 4$
$-y_1 + 2y_2 \ge 6$
$y_1 \ge 0,\ y_2 \ge 0.$

The augmented matrix is
$$\left[\begin{array}{cc|c} 3 & 1 & 4 \\ -1 & 2 & 6 \\ \hline 3 & 5 & 0 \end{array}\right].$$

The transpose of this matrix is
$$\left[\begin{array}{cc|c} 3 & -1 & 3 \\ 1 & 2 & 5 \\ \hline 4 & 6 & 0 \end{array}\right].$$

The entries in this second matrix can be used to write the following dual maximization problem:
Maximize $z = 4x_1 + 6x_2$
subject to: $3x_1 - x_2 \le 3$
$x_1 + 2x_2 \le 5$
$x_1 \ge 0,\ x_2 \ge 0.$

7. Minimize $w = 2y_1 + 8y_2$

subject to:
$$\begin{aligned} y_1 + 7y_2 &\ge 18 \\ 4y_1 + y_2 &\ge 15 \\ 5y_1 + 3y_2 &\ge 20 \\ y_1 \ge 0, y_2 &\ge 0. \end{aligned}$$

The augmented matrix is
$$\left[\begin{array}{cc|c} 1 & 7 & 18 \\ 4 & 1 & 15 \\ 5 & 3 & 20 \\ \hline 2 & 8 & 0 \end{array}\right].$$

The transpose of this matrix is
$$\left[\begin{array}{ccc|c} 1 & 4 & 5 & 2 \\ 7 & 1 & 3 & 8 \\ \hline 18 & 15 & 20 & 0 \end{array}\right].$$

The entries of this second matrix can be used to write the following dual maximization problem:

Maximize $z = 18x_1 + 15x_2 + 20x_3$

subject to:
$$\begin{aligned} x_1 + 4x_2 + 5x_3 &\le 2 \\ 7x_1 + x_2 + 3x_3 &\le 8 \\ x_1 \ge 0, x_2 \ge 0, x_3 &\ge 0. \end{aligned}$$

9. Minimize $w = 5y_1 + y_2 + 3y_3$

subject to:
$$\begin{aligned} 7y_1 + 6y_2 + 8y_3 &\ge 18 \\ 4y_1 + 5y_2 + 10y_3 &\ge 20 \\ y_1 \ge 0, y_2 \ge 0, y_3 &\ge 0. \end{aligned}$$

The augmented matrix is
$$\left[\begin{array}{ccc|c} 7 & 6 & 8 & 18 \\ 4 & 5 & 10 & 20 \\ \hline 5 & 1 & 3 & 0 \end{array}\right].$$

The transpose of this matrix is
$$\left[\begin{array}{cc|c} 7 & 4 & 5 \\ 6 & 5 & 1 \\ 8 & 10 & 3 \\ \hline 18 & 20 & 0 \end{array}\right].$$

The entries of this second matrix can be used to write the following dual maximization problem.

Maximize $z = 18x_1 + 20x_2$

subject to:
$$\begin{aligned} 7x_1 + 4x_2 &\le 5 \\ 6x_1 + 5x_2 &\le 1 \\ 8x_1 + 10x_2 &\le 3 \\ x_1 \ge 0, x_2 &\ge 0. \end{aligned}$$

11. Minimize $w = 8y_1 + 9y_2 + 3y_3$

subject to:
$$\begin{aligned} y_1 + y_2 + y_3 &\ge 5 \\ y_1 + y_2 &\ge 4 \\ 2y_1 + y_2 + 3y_3 &\ge 15 \\ y_1 \ge 0, y_2 \ge 0, y_3 &\ge 0. \end{aligned}$$

The augmented matrix is
$$\left[\begin{array}{ccc|c} 1 & 1 & 1 & 5 \\ 1 & 1 & 0 & 4 \\ 2 & 1 & 3 & 15 \\ \hline 8 & 9 & 3 & 0 \end{array}\right].$$

The transpose of this matrix is
$$\left[\begin{array}{ccc|c} 1 & 1 & 2 & 8 \\ 1 & 1 & 1 & 9 \\ 1 & 0 & 3 & 3 \\ \hline 5 & 4 & 15 & 0 \end{array}\right].$$

The entries of this second matrix can be used to write the following dual maximization problem:

Maximize $z = 5x_1 + 4x_2 + 15x_3$

subject to:
$$\begin{aligned} x_1 + x_2 + 2x_3 &\le 8 \\ x_1 + x_2 + x_3 &\le 9 \\ x_1 + 3x_3 &\le 3 \\ x_1 \ge 0, x_2 \ge 0, x_3 &\ge 0. \end{aligned}$$

13. From exercise 9, we are to minimize $w = 5y_1 + y_2 + 3y_3$ subject to the constraints
$$\begin{aligned} 7y_1 + 6y_2 + 8y_3 &\ge 18 \\ 4y_1 + 5y_2 + 10y_3 &\ge 20 \\ y_1 \ge 0, y_2 \ge 0, y_3 &\ge 0 \end{aligned}$$

We write the dual maximization problem, maximize $z = 18x_1 + 20x_2$

subject to:
$$\begin{aligned} 7x_1 + 4x_2 &\le 5 \\ 6x_1 + 5x_2 &\le 1 \\ 8x_1 + 10x_2 &\le 3 \\ x_1 \ge 0, x_2 &\ge 0. \end{aligned}$$

The initial simplex tableau is
$$\begin{array}{c} \begin{array}{ccccc} x_1 & x_2 & s_1 & s_2 & s_3 \end{array} \\ \left[\begin{array}{ccccc|c} 7 & 4 & 1 & 0 & 0 & 5 \\ 6 & \boxed{5} & 0 & 1 & 0 & 1 \\ 8 & 10 & 0 & 0 & 1 & 3 \\ \hline -18 & -20 & 0 & 0 & 0 & 0 \end{array}\right]. \end{array}$$

(*continued next page*)

Pivot on the 5 in row 2, column 2.

$$\begin{array}{c} \begin{array}{ccccc} x_1 & x_2 & s_1 & s_2 & s_3 \end{array} \\ \left[\begin{array}{rrrrr|r} 11 & 0 & 5 & -4 & 0 & 21 \\ 6 & 5 & 0 & 1 & 0 & 1 \\ -4 & 0 & 0 & -2 & 1 & 1 \\ \hline 6 & 0 & 0 & 4 & 0 & 4 \end{array}\right] \begin{array}{l} -4R_2 + 5R_1 \\ \\ -2R_2 + R_3 \\ 4R_2 + R_4 \end{array} \end{array}$$

From the final tableau, the minimum is 4 for $y_1 = 0,\ y_2 = 4,$ and $y_3 = 0.$

15. From exercise 11, we are to minimize $w = 8y_1 + 9y_2 + 3y_3$

subject to:

$$\begin{aligned} y_1 + y_2 + y_3 &\ge 5 \\ y_1 + y_2 &\ge 4 \\ 2y_1 + y_2 + 3y_3 &\ge 15 \\ y_1 \ge 0,\ y_2 \ge 0,\ y_3 &\ge 0. \end{aligned}$$

We write the dual maximization problem, maximize $z = 5x_1 + 4x_2 + 15x_3$

subject to:

$$\begin{aligned} x_1 + x_2 + 2x_3 &\le 8 \\ x_1 + x_2 + x_3 &\le 9 \\ x_1 + 3x_3 &\le 3 \\ x_1 \ge 0,\ x_2 \ge 0,\ x_3 &\ge 0. \end{aligned}$$

The initial simplex tableau is

$$\begin{array}{c} \begin{array}{cccccc} x_1 & x_2 & x_3 & s_1 & s_2 & s_3 \end{array} \\ \left[\begin{array}{rrrrrr|r} 1 & 1 & 2 & 1 & 0 & 0 & 8 \\ 1 & 1 & 1 & 0 & 1 & 0 & 9 \\ 1 & 0 & \boxed{3} & 0 & 0 & 1 & 3 \\ \hline -5 & -4 & -15 & 0 & 0 & 0 & 0 \end{array}\right]. \end{array}$$

Pivot on the 3 in row 3, column 3.

$$\begin{array}{c} \begin{array}{cccccc} x_1 & x_2 & x_3 & s_1 & s_2 & s_3 \end{array} \\ \left[\begin{array}{rrrrrr|r} 1 & 1 & 2 & 1 & 0 & 0 & 8 \\ 1 & 1 & 1 & 0 & 1 & 0 & 9 \\ \frac{1}{3} & 0 & 1 & 0 & 0 & \frac{1}{3} & 1 \\ \hline -5 & -4 & -15 & 0 & 0 & 0 & 0 \end{array}\right] \begin{array}{l} \\ \\ \frac{1}{3}R_3 \\ \\ \end{array} \end{array}$$

$$\begin{array}{c} \begin{array}{cccccc} x_1 & x_2 & x_3 & s_1 & s_2 & s_3 \end{array} \\ \left[\begin{array}{rrrrrr|r} \frac{1}{3} & \boxed{1} & 0 & 1 & 0 & -\frac{2}{3} & 6 \\ \frac{2}{3} & 1 & 0 & 0 & 1 & -\frac{1}{3} & 8 \\ \frac{1}{3} & 0 & 1 & 0 & 0 & \frac{1}{3} & 1 \\ \hline 0 & -4 & 0 & 0 & 0 & 5 & 15 \end{array}\right] \begin{array}{l} -2R_3 + R_1 \\ -R_3 + R_2 \\ \\ 15R_3 + R_4 \end{array} \end{array}$$

Now pivot on the 1 in row 1, column 2.

$$\begin{array}{c} \begin{array}{cccccc} x_1 & x_2 & x_3 & s_1 & s_2 & s_3 \end{array} \\ \left[\begin{array}{rrrrrr|r} \frac{1}{3} & \boxed{1} & 0 & 1 & 0 & -\frac{2}{3} & 6 \\ \frac{2}{3} & 1 & 0 & 0 & 1 & -\frac{1}{3} & 8 \\ \frac{1}{3} & 0 & 1 & 0 & 0 & \frac{1}{3} & 1 \\ \hline \frac{4}{3} & 0 & 0 & 4 & 0 & \frac{7}{3} & 39 \end{array}\right] \begin{array}{l} \\ \\ \\ 4R_1 + R_4 \end{array} \end{array}$$

From the final tableau, the minimum is 39 for $y_1 = 4,\ y_2 = 0,$ and $y_3 = \frac{7}{3}.$

17. Minimize $w = 2y_1 + y_2 + 3y_3$

subject to:

$$\begin{aligned} y_1 + y_2 + y_3 &\ge 100 \\ 2y_1 + y_2 &\ge 50 \\ y_1 \ge 0,\ y_2 \ge 0,\ y_3 &\ge 0. \end{aligned}$$

The augmented matrix is

$$\left[\begin{array}{rrr|r} 1 & 1 & 1 & 100 \\ 2 & 1 & 0 & 50 \\ \hline 2 & 1 & 3 & 0 \end{array}\right].$$

The transpose is

$$\left[\begin{array}{rr|r} 1 & 2 & 2 \\ 1 & 1 & 1 \\ 1 & 0 & 3 \\ \hline 100 & 50 & 0 \end{array}\right].$$

The dual maximization problem is:
Maximize $z = 100x_1 + 50x_2$

subject to:

$$\begin{aligned} x_1 + 2x_2 &\le 2 \\ x_1 + x_2 &\le 1 \\ x_1 &\le 3 \\ x_1 \ge 0,\ x_2 &\ge 0. \end{aligned}$$

The initial simplex tableau is

$$\begin{array}{c} \begin{array}{ccccc} x_1 & x_2 & s_1 & s_2 & s_3 \end{array} \\ \left[\begin{array}{rrrrr|r} 1 & 2 & 1 & 0 & 0 & 2 \\ \boxed{1} & 1 & 0 & 1 & 0 & 1 \\ 1 & 0 & 0 & 0 & 1 & 3 \\ \hline -100 & -50 & 0 & 0 & 0 & 0 \end{array}\right]. \end{array}$$

(*continued next page*)

Pivot on the 1 in row two, column one.

$$\begin{array}{ccccc|c} x_1 & x_2 & s_1 & s_2 & s_3 & \\ 0 & 1 & 1 & -1 & 0 & 1 \\ 1 & 1 & 0 & 1 & 0 & 1 \\ 0 & -1 & 0 & -1 & 1 & 2 \\ \hline 0 & 50 & 0 & 100 & 0 & 100 \end{array} \begin{array}{l} \\ -1R_2 + R_1 \\ \\ -R_2 + R_3 \\ 100R_2 + R_4 \end{array}$$

The solution to the minimization problem is found in the bottom row of the final matrix in the entries corresponding to the slack variables.
The minimum is 100 when $y_1 = 0,\ y_2 = 100,$ and $y_3 = 0.$

19. Minimize $w = 3y_1 + y_2 + 4y_3$

subject to:
$$\begin{aligned} 2y_1 + y_2 + y_3 &\ge 6 \\ y_1 + 2y_2 + y_3 &\ge 8 \\ 2y_1 + y_2 + 2y_3 &\ge 12 \\ y_1 \ge 0,\ y_2 \ge 0,\ y_3 &\ge 0 \end{aligned}$$

The augmented matrix is

$$\left[\begin{array}{ccc|c} 2 & 1 & 1 & 6 \\ 1 & 2 & 1 & 8 \\ 2 & 1 & 2 & 12 \\ \hline 3 & 1 & 4 & 0 \end{array}\right].$$

The transpose is

$$\left[\begin{array}{ccc|c} 2 & 1 & 2 & 3 \\ 1 & 2 & 1 & 1 \\ 1 & 1 & 2 & 4 \\ \hline 6 & 8 & 12 & 0 \end{array}\right].$$

The dual maximization problem is
Maximize $z = 6x_1 + 8x_2 + 12x_3$

subject to:
$$\begin{aligned} 2x_1 + x_2 + 2x_3 &\le 3 \\ x_1 + 2x_2 + x_3 &\le 1 \\ x_1 + x_2 + 2x_3 &\le 4 \\ x_1 \ge 0,\ x_2 \ge 0,\ x_3 &\ge 0. \end{aligned}$$

The initial tableau is

$$\left[\begin{array}{cccccc|c} x_1 & x_2 & x_3 & s_1 & s_2 & s_3 & \\ 2 & 1 & 2 & 1 & 0 & 0 & 3 \\ 1 & 2 & \boxed{1} & 0 & 1 & 0 & 1 \\ 1 & 1 & 2 & 0 & 0 & 1 & 4 \\ \hline -6 & -8 & -12 & 0 & 0 & 0 & 0 \end{array}\right].$$

Pivot on the 1 in row two, column three.

$$\begin{array}{cccccc|c} x_1 & x_2 & x_3 & s_1 & s_2 & s_3 & \\ 0 & -3 & 0 & 1 & -2 & 0 & 1 \\ 1 & 2 & 1 & 0 & 1 & 0 & 1 \\ -1 & -3 & 0 & 0 & -2 & 1 & 2 \\ \hline 6 & 16 & 0 & 0 & 12 & 0 & 12 \end{array} \begin{array}{l} \\ -2R_2 + R_1 \\ \\ -2R_2 + R_3 \\ 12R_2 + R_4 \end{array}$$

The minimum is 12 when $y_1 = 0,\ y_2 = 12,\ y_3 = 0.$

21. Minimize $w = 6y_1 + 4y_2 + 2y_3$

subject to:
$$\begin{aligned} 2y_1 + 2y_2 + y_3 &\ge 2 \\ y_1 + 3y_2 + 2y_3 &\ge 3 \\ y_1 + y_2 + 2y_3 &\ge 4 \\ y_1 \ge 0,\ y_2 \ge 0,\ y_3 &\ge 0. \end{aligned}$$

The augmented matrix is

$$\left[\begin{array}{ccc|c} 2 & 2 & 1 & 2 \\ 1 & 3 & 2 & 3 \\ 1 & 1 & 2 & 4 \\ \hline 6 & 4 & 2 & 0 \end{array}\right].$$

The transpose is

$$\left[\begin{array}{ccc|c} 2 & 1 & 1 & 6 \\ 2 & 3 & 1 & 4 \\ 1 & 2 & 2 & 2 \\ \hline 2 & 3 & 4 & 0 \end{array}\right].$$

The dual maximization problem is
Maximize $z = 2x_1 + 3x_2 + 4x_3$

subject to:
$$\begin{aligned} 2x_1 + x_2 + x_3 &\le 6 \\ 2x_1 + 3x_2 + x_3 &\le 4 \\ x_1 + 2x_2 + 2x_3 &\le 2 \\ x_1 \ge 0,\ x_2 \ge 0,\ x_3 &\ge 0. \end{aligned}$$

(*continued next page*)

The initial simplex tableau is

$$\begin{array}{c} \begin{array}{cccccc} x_1 & x_2 & x_3 & s_1 & s_2 & s_3 \end{array} \\ \left[\begin{array}{cccccc|c} 2 & 1 & 1 & 1 & 0 & 0 & 6 \\ 2 & 3 & 1 & 0 & 1 & 0 & 4 \\ 1 & 2 & \boxed{2} & 0 & 0 & 1 & 2 \\ \hline -2 & -3 & -4 & 0 & 0 & 0 & 0 \end{array}\right] \end{array}.$$

Pivot on the 2 in row three, column three.

$$\begin{array}{c} \begin{array}{cccccc} x_1 & x_2 & x_3 & s_1 & s_2 & s_3 \end{array} \\ \left[\begin{array}{cccccc|c} 2 & 1 & 1 & 1 & 0 & 0 & 6 \\ 2 & 3 & 1 & 0 & 1 & 0 & 4 \\ \frac{1}{2} & 1 & 1 & 0 & 0 & \frac{1}{2} & 1 \\ \hline -2 & -3 & -4 & 0 & 0 & 0 & 0 \end{array}\right] \end{array} \begin{array}{l} \\ \\ \frac{1}{2}R_3 \\ \\ \end{array}$$

$$\begin{array}{c} \begin{array}{cccccc} x_1 & x_2 & x_3 & s_1 & s_2 & s_3 \end{array} \\ \left[\begin{array}{cccccc|c} \frac{3}{2} & 0 & 0 & 1 & 0 & -\frac{1}{2} & 5 \\ \frac{3}{2} & 2 & 0 & 0 & 1 & -\frac{1}{2} & 3 \\ \frac{1}{2} & 1 & 1 & 0 & 0 & \frac{1}{2} & 1 \\ \hline 0 & 1 & 0 & 0 & 0 & 2 & 4 \end{array}\right] \end{array} \begin{array}{l} -R_3 + R_1 \\ -R_3 + R_2 \\ \\ 4R_3 + R_4 \end{array}$$

The minimum is 4 when $y_1 = 0$, $y_2 = 0$, and $y_3 = 2$.

23. Minimize $w = 20y_1 + 12y_2 + 40y_3$

subject to:

$$\begin{aligned} y_1 + y_2 + 5y_3 &\geq 20 \\ 2y_1 + y_2 + y_3 &\geq 30 \\ y_1 \geq 0,\ y_2 \geq 0,\ y_3 &\geq 0. \end{aligned}$$

The augmented matrix is

$$\left[\begin{array}{ccc|c} 1 & 1 & 5 & 20 \\ 2 & 1 & 1 & 30 \\ \hline 20 & 12 & 40 & 0 \end{array}\right].$$

The transpose is

$$\left[\begin{array}{cc|c} 1 & 2 & 20 \\ 1 & 1 & 12 \\ 5 & 1 & 40 \\ \hline 20 & 30 & 0 \end{array}\right].$$

The dual maximization problem is

Maximize $z = 20x_1 + 30x_2$

subject to:

$$\begin{aligned} x_1 + 2x_2 &\leq 20 \\ x_1 + x_2 &\leq 12 \\ 5x_1 + x_2 &\leq 40 \\ x_1 \geq 0,\ x_2 &\geq 0. \end{aligned}$$

The initial simplex tableau is

$$\begin{array}{c} \begin{array}{ccccc} x_1 & x_2 & s_1 & s_2 & s_3 \end{array} \\ \left[\begin{array}{ccccc|c} 1 & \boxed{2} & 1 & 0 & 0 & 20 \\ 1 & 1 & 0 & 1 & 0 & 12 \\ 5 & 1 & 0 & 0 & 1 & 40 \\ \hline -20 & -30 & 0 & 0 & 0 & 0 \end{array}\right] \end{array}.$$

Pivot on the 2 in row one, column two.

$$\begin{array}{c} \begin{array}{ccccc} x_1 & x_2 & s_1 & s_2 & s_3 \end{array} \\ \left[\begin{array}{ccccc|c} \frac{1}{2} & 1 & \frac{1}{2} & 0 & 0 & 10 \\ 1 & 1 & 0 & 1 & 0 & 12 \\ 5 & 1 & 0 & 0 & 1 & 40 \\ \hline -20 & -30 & 0 & 0 & 0 & 0 \end{array}\right] \end{array} \begin{array}{l} \frac{1}{2}R_1 \\ \\ \\ \\ \end{array}$$

$$\begin{array}{c} \begin{array}{ccccc} x_1 & x_2 & s_1 & s_2 & s_3 \end{array} \\ \left[\begin{array}{ccccc|c} \frac{1}{2} & 1 & \frac{1}{2} & 0 & 0 & 10 \\ \boxed{\frac{1}{2}} & 0 & -\frac{1}{2} & 1 & 0 & 2 \\ \frac{9}{2} & 0 & -\frac{1}{2} & 0 & 1 & 30 \\ \hline -5 & 0 & 15 & 0 & 0 & 300 \end{array}\right] \end{array} \begin{array}{l} \\ -R_1 + R_2 \\ -R_1 + R_3 \\ 30R_1 + R_4 \end{array}$$

Pivot on the $\frac{1}{2}$ in row two, column one.

$$\begin{array}{c} \begin{array}{ccccc} x_1 & x_2 & s_1 & s_2 & s_3 \end{array} \\ \left[\begin{array}{ccccc|c} \frac{1}{2} & 1 & \frac{1}{2} & 0 & 0 & 10 \\ 1 & 0 & -1 & 2 & 0 & 4 \\ \frac{9}{2} & 0 & -\frac{1}{2} & 0 & 1 & 30 \\ \hline -5 & 0 & 15 & 0 & 0 & 300 \end{array}\right] \end{array} \begin{array}{l} \\ 2R_2 \\ \\ \\ \end{array}$$

$$\begin{array}{c} \begin{array}{ccccc} x_1 & x_2 & s_1 & s_2 & s_3 \end{array} \\ \left[\begin{array}{ccccc|c} 0 & 1 & 1 & -1 & 0 & 8 \\ 1 & 0 & -1 & 2 & 0 & 4 \\ 0 & 0 & 4 & -9 & 1 & 12 \\ \hline 0 & 0 & 10 & 10 & 0 & 320 \end{array}\right] \end{array} \begin{array}{l} -\frac{1}{2}R_2 + R_1 \\ \\ -\frac{9}{2}R_2 + R_3 \\ 5R_2 + R_4 \end{array}$$

The minimum is 320 when $y_1 = 10$, $y_2 = 10$, and $y_3 = 0$.

25. Minimize $w = 4y_1 + 2y_2 + y_3$

subject to:
$$\begin{aligned} y_1 + y_2 + y_3 &\ge 4 \\ 3y_1 + y_2 + 3y_3 &\ge 6 \\ y_1 + y_2 + 3y_3 &\ge 5 \\ y_1 \ge 0,\ y_2 \ge 0,\ y_3 &\ge 0. \end{aligned}$$

The augmented matrix is

$$\left[\begin{array}{ccc|c} 1 & 1 & 1 & 4 \\ 3 & 1 & 3 & 6 \\ 1 & 1 & 3 & 5 \\ \hline 4 & 2 & 1 & 0 \end{array}\right].$$

The transpose is

$$\left[\begin{array}{ccc|c} 1 & 3 & 1 & 4 \\ 1 & 1 & 1 & 2 \\ 1 & 3 & 3 & 1 \\ \hline 4 & 6 & 5 & 0 \end{array}\right].$$

The dual maximization problem is

Maximize $z = 4x_1 + 6x_2 + 5x_3$

subject to:
$$\begin{aligned} x_1 + 3x_2 + x_3 &\le 4 \\ x_1 + x_2 + x_3 &\le 2 \\ x_1 + 3x_2 + 3x_3 &\le 1 \\ x_1 \ge 0,\ x_2 > 0,\ x_3 &\ge 0. \end{aligned}$$

The initial simplex tableau is

$$\begin{array}{c} \begin{array}{cccccc} x_1 & x_2 & x_3 & s_1 & s_2 & s_3 \end{array} \\ \left[\begin{array}{cccccc|c} 1 & 3 & 1 & 1 & 0 & 0 & 4 \\ 1 & 1 & 1 & 0 & 1 & 0 & 2 \\ 1 & \boxed{3} & 3 & 0 & 0 & 1 & 1 \\ \hline -4 & -6 & -5 & 0 & 0 & 0 & 0 \end{array}\right]. \end{array}$$

Pivot on the 3 in row three, column two.

$$\begin{array}{c} \begin{array}{cccccc} x_1 & x_2 & x_3 & s_1 & s_2 & s_3 \end{array} \\ \left[\begin{array}{cccccc|c} 1 & 3 & 1 & 1 & 0 & 0 & 4 \\ 1 & 1 & 1 & 0 & 1 & 0 & 2 \\ \frac{1}{3} & 1 & 1 & 0 & 0 & \frac{1}{3} & \frac{1}{3} \\ \hline -4 & -6 & -5 & 0 & 0 & 0 & 0 \end{array}\right] \begin{array}{l} \\ \\ \frac{1}{3}R_3 \\ \\ \end{array} \end{array}$$

$$\begin{array}{c} \begin{array}{cccccc} x_1 & x_2 & x_3 & s_1 & s_2 & s_3 \end{array} \\ \left[\begin{array}{cccccc|c} 0 & 0 & -2 & 1 & 0 & -1 & 3 \\ \frac{2}{3} & 0 & 0 & 0 & 1 & -\frac{1}{3} & \frac{5}{3} \\ \boxed{\frac{1}{3}} & 1 & 1 & 0 & 0 & \frac{1}{3} & \frac{1}{3} \\ \hline -2 & 0 & 1 & 0 & 0 & 2 & 2 \end{array}\right] \begin{array}{l} -3R_3 + R_1 \\ -1R_3 + R_2 \\ \\ 6R_3 + R_4 \end{array} \end{array}$$

Pivot on the $\frac{1}{3}$ in row three, column one.

$$\begin{array}{c} \begin{array}{cccccc} x_1 & x_2 & x_3 & s_1 & s_2 & s_3 \end{array} \\ \left[\begin{array}{cccccc|c} 0 & 0 & -2 & 1 & 0 & -1 & 3 \\ \frac{2}{3} & 0 & 0 & 0 & 1 & -\frac{1}{3} & \frac{5}{3} \\ 1 & 3 & 3 & 0 & 0 & 1 & 1 \\ \hline -2 & 0 & 1 & 0 & 0 & 2 & 2 \end{array}\right] \begin{array}{l} \\ \\ 3R_3 \\ \\ \end{array} \end{array}$$

$$\begin{array}{c} \begin{array}{cccccc} x_1 & x_2 & x_3 & s_1 & s_2 & s_3 \end{array} \\ \left[\begin{array}{cccccc|c} 0 & 0 & -2 & 1 & 0 & -1 & 3 \\ 0 & -2 & -2 & 0 & 1 & -1 & 1 \\ 1 & 3 & 3 & 0 & 0 & 1 & 1 \\ \hline 0 & 6 & 7 & 0 & 0 & 4 & 4 \end{array}\right] \begin{array}{l} \\ -\frac{2}{3}R_3 + R_2 \\ \\ 2R_3 + R_4 \end{array} \end{array}$$

The minimum is 4 when $y_1 = 0$, $y_2 = 0$, and $y_3 = 4$.

27. Let y_1 = the amount of product A and let y_2 = the amount of product B.

The given information can be expressed as the following minimization problem:

Minimize $w = 24y_1 + 40y_2$

subject to:
$$\begin{aligned} 4y_1 + 2y_2 &\ge 20 \\ 2y_1 + 5y_2 &\ge 18 \\ y_1 \ge 0,\ y_2 &\ge 0. \end{aligned}$$

The augmented matrix is

$$\left[\begin{array}{cc|c} 4 & 2 & 20 \\ 2 & 5 & 18 \\ \hline 24 & 40 & 0 \end{array}\right].$$

The transpose is

$$\left[\begin{array}{cc|c} 4 & 2 & 24 \\ 2 & 5 & 40 \\ \hline 20 & 18 & 0 \end{array}\right].$$

The dual maximization problem is

Maximize $z = 20x_1 + 18x_2$

subject to:
$$\begin{aligned} 4x_1 + 2x_2 &\le 24 \\ 2x_1 + 5x_2 &\le 40 \\ x_1 \ge 0,\ x_2 &\ge 0. \end{aligned}$$

The initial simplex tableau is

$$\begin{array}{c} \begin{array}{cccc} x_1 & x_2 & s_1 & s_2 \end{array} \\ \left[\begin{array}{cccc|c} \boxed{4} & 2 & 1 & 0 & 24 \\ 2 & 5 & 0 & 1 & 40 \\ \hline -20 & -18 & 0 & 0 & 0 \end{array}\right]. \end{array}$$

(*continued next page*)

Pivot on the 4 in row one, column one.

$$\begin{array}{cccc|c} x_1 & x_2 & x_3 & x_4 & x_5 \\ \end{array}$$
$$\left[\begin{array}{cccc|c} 1 & \frac{1}{2} & \frac{1}{4} & 0 & 6 \\ 2 & 5 & 0 & 1 & 40 \\ \hline -20 & -18 & 0 & 0 & 0 \end{array}\right] \begin{array}{l} \frac{1}{4}R_1 \\ \\ \\ \end{array}$$

$$\begin{array}{cccc} x_1 & x_2 & s_1 & s_2 \end{array}$$
$$\left[\begin{array}{cccc|c} 1 & \frac{1}{2} & \frac{1}{4} & 0 & 6 \\ 0 & \boxed{4} & -\frac{1}{2} & 1 & 28 \\ \hline 0 & -8 & 5 & 0 & 120 \end{array}\right] \begin{array}{l} \\ -2R_1 + R_2 \\ 20R_1 + R_3 \end{array}$$

Pivot on the 4 in row two, column two.

$$\begin{array}{ccccc} x_1 & x_2 & x_3 & x_4 & x_5 \end{array}$$
$$\left[\begin{array}{cccc|c} 1 & \frac{1}{2} & \frac{1}{4} & 0 & 6 \\ 0 & 1 & -\frac{1}{8} & \frac{1}{4} & 7 \\ \hline 0 & -8 & 5 & 0 & 120 \end{array}\right] \begin{array}{l} \\ \frac{1}{4}R_2 \\ \\ \end{array}$$

$$\begin{array}{cccc} x_1 & x_2 & x_3 & x_4 \end{array}$$
$$\left[\begin{array}{cccc|c} 1 & 0 & \frac{5}{16} & -\frac{1}{8} & \frac{5}{2} \\ 0 & 1 & -\frac{1}{8} & \frac{1}{4} & 7 \\ \hline 0 & 0 & 4 & 2 & 176 \end{array}\right] \begin{array}{l} -\frac{1}{2}R_2 + R_1 \\ \\ 8R_2 + R_3 \end{array}$$

Glenn should use 4 servings of product *A* and 2 servings of product *B* for a minimum cost of \$1.76.

29. Let $y_1 =$ the number of additional units of regular beer to produce, and let $y_2 =$ the number of additional units of light beer to produce

The sale of 12 units of regular beer and 10 units of light beer already generates $12 \cdot 100{,}000 + 10 \cdot 300{,}000 = \$4{,}200{,}000$ in revenue. Additional units need only to generate $7{,}000{,}000 - 4{,}200{,}000 = \$2{,}800{,}000$ in revenue.

Minimize $w = 36{,}000y_1 + 48{,}000y_2$

subject to: $100{,}000y_1 + 300{,}000y_2 \geq 2{,}800{,}000$

$y_1 + y_2 \geq 20$

$y_1 \geq 0,\ y_2 \geq 0$

The augmented matrix is

$$\left[\begin{array}{cc|c} 100{,}000 & 300{,}000 & 2{,}800{,}000 \\ 1 & 1 & 20 \\ \hline 36{,}000 & 48{,}000 & 0 \end{array}\right]$$

The transpose is

$$\left[\begin{array}{cc|c} 100{,}000 & 1 & 36{,}000 \\ 300{,}000 & 1 & 48{,}000 \\ \hline 2{,}800{,}000 & 20 & 0 \end{array}\right]$$

The dual maximization problem is

Maximize $z = 2{,}800{,}000x_1 + 20x_2$

subject to: $100{,}000x_1 + x_2 \leq 36{,}000$

$300{,}000x_1 + x_2 \leq 48{,}000$

$x_1 \geq 0,\ x_2 \geq 0$

The initial simplex tableau is

$$\begin{array}{cccc} x_1 & x_2 & s_1 & s_2 \end{array}$$
$$\left[\begin{array}{cccc|c} 100{,}000 & 1 & 1 & 0 & 36{,}000 \\ \boxed{300{,}000} & 1 & 0 & 1 & 48{,}000 \\ \hline -2{,}800{,}000 & -20 & 0 & 0 & 0 \end{array}\right]$$

Pivot on the 300,000 in row two, column one.

$$\begin{array}{cccc} x_1 & x_2 & s_1 & s_2 \end{array}$$
$$\left[\begin{array}{cccc|c} 0 & \boxed{\frac{2}{3}} & 1 & -\frac{1}{3} & 20{,}000 \\ 1 & \frac{1}{300{,}000} & 0 & \frac{1}{300{,}000} & \frac{16}{100} \\ \hline 0 & -\frac{32}{3} & 0 & \frac{28}{3} & 448{,}000 \end{array}\right] \begin{array}{l} -100{,}000R_2 + R_1 \\ \frac{1}{300{,}000}R_2 \\ 2{,}800{,}000R_2 + R_3 \end{array}$$

Pivot on the $\frac{2}{3}$ in row one, column two.

$$\left[\begin{array}{cccc|c} 0 & 1 & \frac{3}{2} & -\frac{1}{2} & 30{,}000 \\ 1 & 0 & -\frac{1}{200{,}000} & \frac{1}{200{,}000} & \frac{3}{50} \\ \hline 0 & 0 & 16 & 4 & 768{,}000 \end{array}\right] \begin{array}{l} \frac{3}{2}R_1 \\ -\frac{1}{300{,}000}R_1 + R_2 \\ \frac{32}{3}R_1 + R_3 \end{array}$$

The total cost is the cost of the original 12 and 10 units of beer plus the cost of the additional units of beer.

$768{,}000 + 12 \cdot 36{,}000 + 10 \cdot 48{,}000 = \$1{,}680{,}000$

The brewery should make 16 additional units of regular beer and 4 additional units of light beer for a total of 28 units of regular beer and 14 units of light beer at a minimum cost of \$1,680,000.

31. The linear program (P)
Minimize $z = x_1 + 2x_2$

$$-2x_1 + x_2 \geq 1$$
$$x_1 - 2x_2 \geq 1$$
$$x_1 \geq 0,\ x_2 \geq 0$$

has no feasible solution. Graphing the constraints of (P) shows that the constraints do not form a feasible space. To determine the dual of (P), (D), we have the augmented matrix

$$\left[\begin{array}{cc|c} -2 & 1 & 1 \\ 1 & -2 & 1 \\ \hline 1 & 2 & 0 \end{array}\right]$$

the transpose

$$\left[\begin{array}{cc|c} -2 & 1 & 1 \\ 1 & -2 & 2 \\ \hline 1 & 1 & 0 \end{array}\right]$$

and the dual maximization problem
Maximize $w = y_1 + y_2$
subject to: $-2y_1 + y_2 \leq 1$

$$y_1 - 2y_2 \leq 2$$
$$y_1 \geq 0,\ y_2 \geq 0$$

Graphing the constraints of (D) does show a feasible space but the objective function is unbounded.

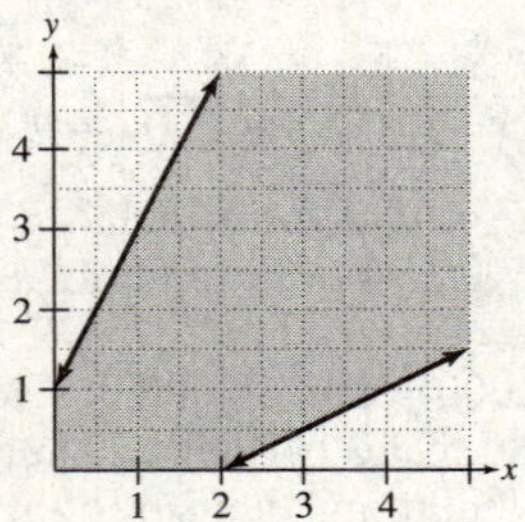

The answer is (a).

33. Let x_1 = the number of toy bears and let x_2 = the number of monkeys.
Maximize $z = x_1 + 1.5x_2$
subject to: $x_1 + 2x_2 \leq 200$

$$4x_1 + 3x_2 \leq 600$$
$$x_2 \leq 90$$
$$x_1 \geq 0,\ x_2 \geq 0.$$

a. The augmented matrix is

$$\left[\begin{array}{cc|c} 1 & 2 & 200 \\ 4 & 3 & 600 \\ 0 & 1 & 90 \\ \hline 1 & 1.5 & 0. \end{array}\right].$$

The transpose is

$$\left[\begin{array}{ccc|c} 1 & 4 & 0 & 1 \\ 2 & 3 & 1 & 1.5 \\ \hline 200 & 600 & 90 & 0 \end{array}\right].$$

The dual minimization problem is
Minimize $w = 200y_1 + 600y_2 + 90y_3$
subject to: $y_1 + 4y_2 \geq 1$

$$2y_1 + 3y_2 + y_3 \geq 1.5$$
$$y_1 \geq 0,\ y_2 \geq 0,\ y_3 \geq 0.$$

b. The solution to the dual minimization problem is found in the last row of the final simplex. tableau of the original problem. The minimum is 180 when $y_1 = .6$, $y_2 = .1$, and $y_3 = 0$.

c. The shadow cost for felt is in row four, column three of the final matrix, \$.60/unit. So, an increase of 10 squares of felt would increase the profit by \$6. The profit would be \$186.

d. The shadow cost for stuffing is \$.10/unit; the shadow cost for trim is \$0/unit. So, a decrease of 10 oz of stuffing and 10 ft of trim would decrease the profit by \$1. The profit would be \$179.

Section 7.7 The Simplex Method: Nonstandard Problems

1. Maximize $z = -5x_1 + 4x_2 - 2x_3$
subject to
$$-2x_2 + 5x_3 \geq 8$$
$$4x_1 - x_2 + 3x_3 \leq 12$$
$$x_1 \geq 0,\ x_2 \geq 0,\ x_3 \geq 0.$$

a. Insert a surplus variable in the first constraint and slack variables in the last constraint. The problem then becomes:
Maximize $z = -5x_1 + 4x_2 - 2x_3$
subject to:
$$-2x_2 + 5x_3 - s_1 = 8$$
$$4x_1 - x_2 + 3x_3 + s_2 = 12$$
$$x_1 \geq 0,\ x_2 \geq 0,\ x_3 \geq 0,\ s_1 \geq 0,\ s_2 \geq 0.$$

b. The initial simplex tableau is

x_1	x_2	x_3	s_1	s_2	
0	−2	5	−1	0	8
4	−1	3	0	1	12
5	−4	2	0	0	0

3. Maximize $z = 2x_1 - 3x_2 + 4x_3$
subject to:
$$x_1 + x_2 + x_3 \leq 100$$
$$x_1 + x_2 + x_3 \geq 75$$
$$x_1 + x_2 \geq 27$$
$$x_1 \geq 0,\ x_2 \geq 0,\ x_3 \geq 0.$$

a. Insert a slack variable in the first constraint and surplus variables in the last two constraints. The problem then becomes:
Maximize $z = 2x_1 - 3x_2 + 4x_3$
subject to:
$$x_1 + x_2 + x_3 + s_1 = 100$$
$$x_1 + x_2 + x_3 - s_2 = 75$$
$$x_1 + x_2 - s_3 = 27$$
$$x_1 \geq 0,\ x_2 \geq 0,\ x_3 \geq 0,\ s_1 \geq 0,\ s_2 \geq 0,\ s_3 \geq 0.$$

b. The initial simplex tableau is

x_1	x_2	x_3	s_1	s_2	s_3	
1	1	1	1	0	0	100
1	1	1	0	−1	0	75
1	1	0	0	0	−1	27
−2	3	−4	0	0	0	0

5. Minimize $w = 2y_1 + 5y_2 - 3y_3$
subject to:
$$y_1 + 2y_2 + 3y_3 \geq 115$$
$$2y_1 + y_2 + y_3 \leq 200$$
$$y_1 + y_3 \geq 50$$
$$y_1 \geq 0,\ y_2 \geq 0,\ y_3 \geq 0.$$
Rewrite the objective function to get a maximization problem.
Maximize $z = -2y_1 - 5y_2 + 3y_3$
subject to:
$$y_1 + 2y_2 + 3y_3 \geq 115$$
$$2y_1 + y_2 + y_3 \leq 200$$
$$y_1 + y_3 \geq 50$$
$$y_1 \geq 0,\ y_2 \geq 0,\ y_3 \geq 0.$$
Insert surplus variables in the first and third constraints and a slack variable in the second to get the initial simplex tableau.

y_1	y_2	y_3	s_1	s_2	s_3	
1	2	3	−1	0	0	115
2	1	1	0	1	0	200
1	0	1	0	0	−1	50
2	5	−3	0	0	0	0

7. Minimize $w = 10y_1 + 8y_2 + 15y_3$
subject to:
$$y_1 + y_2 + y_3 \geq 12$$
$$5y_1 + 4y_2 + 9y_3 \geq 48$$
$$y_1 \geq 0,\ y_2 \geq 0,\ y_3 \geq 0.$$
Rewrite the objective function to get a maximization problem.
Maximize $z = -10y_1 - 8y_2 - 15y_3$
subject to:
$$y_1 + y_2 + y_3 \geq 12$$
$$5y_1 + 4y_2 + 9y_3 \geq 48$$
$$y_1 \geq 0,\ y_2 \geq 0,\ y_3 \geq 0.$$
Insert surplus variables in both constraints to get the initial simplex tableau.

y_1	y_2	y_3	s_1	s_2	
1	1	1	−1	0	12
5	4	9	0	−1	48
10	8	15	0	0	0

9. Maximize $z = 12x_1 + 10x_2$

subject to: $x_1 + 2x_2 \ge 24$

$x_1 + x_2 \le 40$

$x_1 \ge 0,\ x_2 \ge 0.$

The initial simplex tableau is

$$\begin{array}{cccc} x_1 & x_2 & s_1 & s_2 \end{array}$$
$$\left[\begin{array}{cccc|c} \boxed{1} & 2 & -1 & 0 & 24 \\ 1 & 1 & 0 & 1 & 40 \\ \hline -12 & -10 & 0 & 0 & 0 \end{array}\right]$$

For Stage I pivoting, pivot on the 1 in row one, column one.

$$\begin{array}{cccc} x_1 & x_2 & s_1 & s_2 \end{array}$$
$$\left[\begin{array}{cccc|c} 1 & 2 & -1 & 0 & 24 \\ 0 & -1 & \boxed{1} & 1 & 16 \\ \hline 0 & 14 & -12 & 0 & 288 \end{array}\right] \begin{array}{l} \\ -R_1 + R_2 \\ 12R_2 + R_3 \end{array}$$

For Stage II pivoting, pivot on the 1 in row two, column three.

$$\begin{array}{cccc} x_1 & x_2 & s_1 & s_2 \end{array}$$
$$\left[\begin{array}{cccc|c} 1 & 1 & 0 & 1 & 40 \\ 0 & -1 & 1 & 1 & 16 \\ \hline 0 & 2 & 0 & 12 & 480 \end{array}\right] \begin{array}{l} R_2 + R_1 \\ \\ 12R_1 + R_3 \end{array}$$

The maximum is 480 when $x_1 = 40$ and $x_2 = 0$.

11. Find $x_1 \ge 0$, $x_2 \ge 0$, and $x_3 \ge 0$ such that

$x_1 + x_2 + 2x_3 \le 38$

$2x_1 + x_2 + x_3 \ge 24$

and $z = 3x_1 + 2x_2 + 2x_3$ is maximized.

The initial simplex tableau is

$$\begin{array}{ccccc} x_1 & x_2 & x_3 & s_1 & s_2 \end{array}$$
$$\left[\begin{array}{ccccc|c} 1 & 1 & 2 & 1 & 0 & 38 \\ \boxed{2} & 1 & 1 & 0 & -1 & 24 \\ \hline -3 & -2 & -2 & 0 & 0 & 0 \end{array}\right]$$

For Stage I pivoting, pivot on the 2 in row two, column one.

$$\begin{array}{ccccc} x_1 & x_2 & x_3 & s_1 & s_2 \end{array}$$
$$\left[\begin{array}{ccccc|c} 1 & 1 & 2 & 1 & 0 & 38 \\ 1 & \frac{1}{2} & \frac{1}{2} & 0 & -\frac{1}{2} & 12 \\ \hline -3 & -2 & -2 & 0 & 0 & 0 \end{array}\right] \begin{array}{l} \\ \frac{1}{2}R_2 \\ \\ \end{array}$$

$$\begin{array}{ccccc} x_1 & x_2 & x_3 & s_1 & s_2 \end{array}$$
$$\left[\begin{array}{ccccc|c} 0 & \frac{1}{2} & \frac{3}{2} & 1 & \boxed{\frac{1}{2}} & 26 \\ 1 & \frac{1}{2} & \frac{1}{2} & 0 & -\frac{1}{2} & 12 \\ \hline 0 & -\frac{1}{2} & -\frac{1}{2} & 0 & -\frac{3}{2} & 36 \end{array}\right] \begin{array}{l} -1R_2 + R_1 \\ \\ 3R_2 + R_3 \end{array}$$

For Stage II pivoting, pivot on the $\frac{1}{2}$ in row one, column five.

$$\begin{array}{ccccc} x_1 & x_2 & x_3 & s_1 & s_2 \end{array}$$
$$\left[\begin{array}{ccccc|c} 0 & 1 & 3 & 2 & 1 & 52 \\ 1 & \frac{1}{2} & \frac{1}{2} & 0 & -\frac{1}{2} & 12 \\ \hline 0 & -\frac{1}{2} & -\frac{1}{2} & 0 & -\frac{3}{2} & 36 \end{array}\right] \begin{array}{l} 2R_1 \\ \\ \\ \end{array}$$

$$\begin{array}{ccccc} x_1 & x_2 & x_3 & s_1 & s_2 \end{array}$$
$$\left[\begin{array}{ccccc|c} 0 & 1 & 3 & 2 & 1 & 52 \\ 1 & 1 & 2 & 1 & 0 & 38 \\ \hline 0 & 1 & 4 & 3 & 0 & 114 \end{array}\right] \begin{array}{l} \\ \frac{1}{2}R_1 + R_2 \\ \frac{3}{2}R_1 + R_3 \end{array}$$

The maximum is 114 when $x_1 = 38$, $x_2 = 0$, and $x_3 = 0$.

13. Find $x_1 \geq 0$ and $x_2 \geq 0$ such that

$$\begin{aligned} x_1 + 2x_2 &\leq 18 \\ x_1 + 3x_2 &\geq 12 \\ 2x_1 + 2x_2 &\leq 30 \end{aligned}$$

and $z = 5x_1 + 10x_2$ is maximized.

The initial simplex tableau is

$$\begin{array}{ccccc|c} x_1 & x_2 & s_1 & s_2 & s_3 & \\ \hline 1 & 2 & 1 & 0 & 0 & 18 \\ \boxed{1} & 3 & 0 & -1 & 0 & 12 \\ 2 & 2 & 0 & 0 & 1 & 30 \\ \hline -5 & -10 & 0 & 0 & 0 & 0 \end{array}.$$

For Stage I pivoting, we can pivot on the 1 in row two, column one or the 3 in row two, column two. Choosing the 1 in row two, column one as the pivot, we proceed as follows.

$$\begin{array}{ccccc|c|l} x_1 & x_2 & s_1 & s_2 & s_3 & & \\ \hline 0 & -1 & 1 & 1 & 0 & 6 & -R_2 + R_1 \\ 1 & 3 & 0 & -1 & 0 & 12 & \\ 0 & -4 & 0 & \boxed{2} & 1 & 6 & -2R_2 + R_3 \\ \hline 0 & 5 & 0 & -5 & 0 & 60 & 5R_2 + R_4 \end{array}$$

For Stage II, pivot on the 2 in row three, column four.

$$\begin{array}{ccccc|c|l} x_1 & x_2 & s_1 & s_2 & s_3 & & \\ \hline 0 & -1 & 1 & 1 & 0 & 6 & \\ 1 & 3 & 0 & -1 & 0 & 12 & \\ 0 & -2 & 0 & 1 & \frac{1}{2} & 3 & \frac{1}{2}R_3 \\ \hline 0 & 5 & 0 & -5 & 0 & 60 & \end{array}$$

$$\begin{array}{ccccc|c|l} x_1 & x_2 & s_1 & s_2 & s_3 & & \\ \hline 0 & \boxed{1} & 1 & 0 & -\frac{1}{2} & 3 & -R_3 + R_1 \\ 1 & 1 & 0 & 0 & \frac{1}{2} & 15 & R_3 + R_2 \\ 0 & -2 & 0 & 1 & \frac{1}{2} & 3 & \\ \hline 0 & -5 & 0 & 0 & \frac{5}{2} & 75 & 5R_3 + R_4 \end{array}$$

Now pivot on the 1 in row one, column two.

$$\begin{array}{ccccc|c|l} x_1 & x_2 & s_1 & s_2 & s_3 & & \\ \hline 0 & 1 & 1 & 0 & -\frac{1}{2} & 3 & \\ 1 & 0 & -1 & 0 & 1 & 12 & -R_1 + R_2 \\ 0 & 0 & 2 & 1 & -\frac{1}{2} & 9 & 2R_1 + R_3 \\ \hline 0 & 0 & 5 & 0 & 0 & 90 & 5R_1 + R_4 \end{array}$$

The above tableau gives the solution:
The maximum is 90 when $x_1 = 12$ and $x_2 = 3$.

However, in Stage I, we could also choose the 3 in row two, column two as the pivot and proceed as follows.

$$\begin{array}{ccccc|c|l} x_1 & x_2 & s_1 & s_2 & s_3 & & \\ \hline 1 & 2 & 1 & 0 & 0 & 18 & \\ \frac{1}{3} & \boxed{1} & 0 & -\frac{1}{3} & 0 & 4 & \frac{1}{3}R_2 \\ 2 & 2 & 0 & 0 & 1 & 30 & \\ \hline -5 & -10 & 0 & 0 & 0 & 0 & \end{array}$$

$$\begin{array}{ccccc|c|l} x_1 & x_2 & s_1 & s_2 & s_3 & & \\ \hline \frac{1}{3} & 0 & 1 & \boxed{\frac{2}{3}} & 0 & 10 & -2R_2 + R_1 \\ \frac{1}{3} & 1 & 0 & -\frac{1}{3} & 0 & 4 & \\ \frac{4}{3} & 0 & 0 & \frac{2}{3} & 1 & 22 & -2R_2 + R_3 \\ \hline -\frac{5}{3} & 0 & 0 & -\frac{10}{3} & 0 & 40 & 10R_2 + R_4 \end{array}$$

For Stage II, pivot on the $\frac{2}{3}$ in row one, column four.

$$\begin{array}{ccccc|c|l} x_1 & x_2 & s_1 & s_2 & s_3 & & \\ \hline \frac{1}{2} & 0 & \frac{3}{2} & 1 & 0 & 15 & \frac{3}{2}R_1 \\ \frac{1}{3} & 1 & 0 & -\frac{1}{3} & 0 & 4 & \\ \frac{4}{3} & 0 & 0 & \frac{2}{3} & 1 & 22 & \\ \hline -\frac{5}{3} & 0 & 0 & -\frac{10}{3} & 0 & 40 & \end{array}$$

$$\begin{array}{ccccc|c|l} x_1 & x_2 & s_1 & s_2 & s_3 & & \\ \hline \frac{1}{2} & 0 & \frac{3}{2} & 1 & 0 & 15 & \\ \frac{1}{2} & 1 & \frac{1}{2} & 0 & 0 & 9 & \frac{1}{3}R_1 + R_2 \\ 1 & 0 & -1 & 0 & 1 & 12 & -\frac{2}{3}R_1 + R_3 \\ \hline 0 & 0 & 5 & 0 & 0 & 90 & \frac{10}{3}R_1 + R_4 \end{array}$$

This tableau gives the solution:
The maximum is 90 when $x_1 = 0$ and $x_2 = 9$.
Thus, the maximum is 90 when $x_1 = 12$ and $x_2 = 3$ or when $x_1 = 0$ and $x_2 = 9$.

15. Minimize $w = 3y_1 + 2y_2$

subject to: $2y_1 + 3y_2 \ge 60$
$y_1 + 4y_2 \ge 40$
$y_1 \ge 0,\ y_2 \ge 0.$

The initial simplex tableau is

$$\begin{array}{cccc|c} y_1 & y_2 & s_1 & s_2 & \\ \boxed{2} & 3 & -1 & 0 & 60 \\ 1 & 4 & 0 & -1 & 40 \\ \hline 3 & 2 & 0 & 0 & 0 \end{array}.$$

For Stage I, pivot on the 2 in row one, column one.

$$\begin{array}{cccc|cl} y_1 & y_2 & s_1 & s_2 & & \\ 1 & \frac{3}{2} & -\frac{1}{2} & 0 & 30 & \frac{1}{2}R_1 \\ 0 & \boxed{\frac{5}{2}} & \frac{1}{2} & -1 & 10 & -R_1 + R_2 \\ \hline 0 & -\frac{5}{2} & \frac{3}{2} & 0 & -90 & -3R_1 + R_3 \end{array}$$

To continue Stage I, pivot on the $\frac{5}{2}$ in row two, column two.

$$\begin{array}{cccc|cl} y_1 & y_2 & s_1 & s_2 & & \\ 1 & 0 & -\frac{4}{5} & \boxed{\frac{3}{5}} & 24 & -\frac{3}{5}R_2 + R_1 \\ 0 & 1 & \frac{1}{5} & -\frac{2}{5} & 4 & \frac{2}{5}R_2 \\ \hline 0 & 0 & 2 & -1 & -80 & \frac{5}{2}R_2 + R_3 \end{array}$$

For Stage II, pivot on the $\frac{3}{5}$ in row one, column four.

$$\begin{array}{cccc|cl} y_1 & y_2 & s_1 & s_2 & & \\ \frac{5}{3} & 0 & -\frac{4}{3} & 1 & 40 & \frac{5}{3}R_1 \\ \frac{2}{3} & 1 & -\frac{1}{3} & 0 & 20 & \frac{2}{5}R_1 + R_2 \\ \hline \frac{5}{3} & 0 & \frac{2}{3} & 0 & -40 & R_1 + R_3 \end{array}$$

The minimum is 40 when $y_1 = 0$ and $y_2 = 20$.

17. Maximize $z = 3x_1 + 2x_2$

subject to: $x_1 + x_2 = 50$
$4x_1 + 2x_2 \ge 120$
$5x_1 + 2x_2 \le 200$
$x_1 \ge 0,\ x_2 \ge 0$

The initial simplex tableau is

$$\begin{array}{cccccc|c} x_1 & x_2 & s_1 & s_2 & s_3 & s_4 & \\ 1 & 1 & -1 & 0 & 0 & 0 & 50 \\ 1 & 1 & 0 & 1 & 0 & 0 & 50 \\ \boxed{4} & 2 & 0 & 0 & -1 & 0 & 120 \\ 5 & 2 & 0 & 0 & 0 & 1 & 200 \\ \hline -3 & -2 & 0 & 0 & 0 & 0 & 0 \end{array}$$

For Stage I, pivot on the 4 in row three, column one.

$$\begin{array}{cccccc|cl} x_1 & x_2 & s_1 & s_2 & s_3 & s_4 & & \\ 0 & \boxed{\frac{1}{2}} & -1 & 0 & \frac{1}{4} & 0 & 20 & -R_3 + R_1 \\ 0 & \frac{1}{2} & 0 & 1 & \frac{1}{4} & 0 & 20 & -R_3 + R_2 \\ 1 & \frac{1}{2} & 0 & 0 & -\frac{1}{4} & 0 & 30 & \frac{1}{4}R_3 \\ 0 & -\frac{1}{2} & 0 & 0 & \frac{5}{4} & 1 & 50 & -5R_3 + R_4 \\ \hline 0 & -\frac{1}{2} & 0 & 0 & -\frac{3}{4} & 0 & 90 & 3R_3 + R_5 \end{array}$$

To continue Stage I, pivot on the $\frac{1}{2}$ in row one column two.

$$\begin{array}{cccccc|cl} x_1 & x_2 & s_1 & s_2 & s_3 & s_4 & & \\ 0 & 1 & -2 & 0 & \frac{1}{2} & 0 & 40 & 2R_1 \\ 0 & 0 & \boxed{1} & 1 & 0 & 0 & 0 & -\frac{1}{2}R_1 + R_2 \\ 1 & 0 & 1 & 0 & -\frac{1}{2} & 0 & 10 & -\frac{1}{2}R_1 + R_3 \\ 0 & 0 & -1 & 0 & \frac{3}{2} & 1 & 70 & \frac{1}{2}R_1 + R_4 \\ \hline 0 & 0 & -1 & 0 & -\frac{1}{2} & 0 & 110 & \frac{1}{2}R_1 + R_5 \end{array}$$

For Stage II, pivot on the 1 in row two, column three.

$$\begin{array}{cccccc|cl} x_1 & x_2 & s_1 & s_2 & s_3 & s_4 & & \\ 0 & 1 & 0 & 2 & \frac{1}{2} & 0 & 40 & 2R_2 + R_1 \\ 0 & 0 & 1 & 1 & 0 & 0 & 0 & \\ 1 & 0 & 0 & -1 & -\frac{1}{2} & 0 & 10 & -R_2 + R_3 \\ 0 & 0 & 0 & 1 & \boxed{\frac{3}{2}} & 1 & 70 & R_2 + R_4 \\ \hline 0 & 0 & 0 & 1 & -\frac{1}{2} & 0 & 110 & R_2 + R_5 \end{array}$$

(*continued next page*)

Pivot on the $\frac{3}{2}$ in row four, column five.

$$\begin{array}{c} \\ \\ \\ \\ \\ \end{array}\begin{matrix} x_1 & x_2 & s_1 & s_2 & s_3 & s_4 & \end{matrix}$$

$$\left[\begin{array}{cccccc|c} 0 & 1 & 0 & \frac{5}{3} & 0 & -\frac{1}{3} & \frac{50}{3} \\ 0 & 0 & 1 & 1 & 0 & 0 & 0 \\ 1 & 0 & 0 & -\frac{2}{3} & 0 & \frac{1}{3} & \frac{100}{3} \\ 0 & 0 & 0 & \frac{2}{3} & 1 & \frac{2}{3} & \frac{140}{3} \\ \hline 0 & 0 & 0 & \frac{4}{3} & 0 & \frac{1}{3} & \frac{400}{3} \end{array}\right]\begin{array}{l} -\frac{1}{2}R_4 + R_1 \\ \\ \frac{1}{2}R_4 + R_3 \\ \frac{2}{3}R_4 \\ \frac{1}{2}R_4 + R_5 \end{array}$$

The maximum is $133\frac{1}{3}$ when $x_1 = 33\frac{1}{3}$ and $x_2 = 16\frac{2}{3}$.

19. Minimize $w = 32y_1 + 40y_2$

Maximize $z = -w = -32y_1 - 40y_2$.

subject to: $20y_1 + 10y_2 = 200$

$25y_1 + 40y_2 \le 500$

$18y_1 + 24y_2 \ge 300$

$y_1 \ge 0,\ y_2 \ge 0$

The initial simplex tableau is

$$\begin{matrix} y_1 & y_2 & s_1 & s_2 & s_3 & s_4 & \end{matrix}$$

$$\left[\begin{array}{cccccc|c} \boxed{20} & 10 & -1 & 0 & 0 & 0 & 200 \\ 20 & 10 & 0 & 1 & 0 & 0 & 200 \\ 25 & 40 & 0 & 0 & 1 & 0 & 500 \\ 18 & 24 & 0 & 0 & 0 & -1 & 300 \\ \hline 32 & 40 & 0 & 0 & 0 & 0 & 0 \end{array}\right]$$

For Stage I, pivot on the 20 in row one, column one.

$$\begin{matrix} y_1 & y_2 & s_1 & s_2 & s_3 & s_4 & \end{matrix}$$

$$\left[\begin{array}{cccccc|c} 1 & \frac{1}{2} & -\frac{1}{20} & 0 & 0 & 0 & 10 \\ 0 & 0 & 1 & 1 & 0 & 0 & 0 \\ 0 & \frac{55}{2} & \frac{5}{4} & 0 & 1 & 0 & 250 \\ 0 & \boxed{15} & \frac{9}{10} & 0 & 0 & -1 & 120 \\ \hline 0 & 24 & \frac{8}{5} & 0 & 0 & 0 & -320 \end{array}\right]\begin{array}{l} \frac{1}{20}R_1 \\ -20R_1 + R_2 \\ -25R_1 + R_3 \\ -18R_1 + R_4 \\ -32R_1 + R_5 \end{array}$$

Pivot on the 15 in row four, column two.

$$\begin{matrix} y_1 & y_2 & s_1 & s_2 & s_3 & s_4 & \end{matrix}$$

$$\left[\begin{array}{cccccc|c} 1 & 0 & -\frac{4}{50} & 0 & 0 & \frac{1}{30} & 6 \\ 0 & 0 & 1 & 1 & 0 & 0 & 0 \\ 0 & 0 & -\frac{2}{5} & 0 & 1 & \frac{11}{6} & 30 \\ 0 & 1 & \frac{3}{50} & 0 & 0 & -\frac{1}{15} & 8 \\ \hline 0 & 0 & \frac{4}{25} & 0 & 0 & \frac{8}{5} & -512 \end{array}\right]\begin{array}{l} -\frac{1}{2}R_4 + R_1 \\ \\ -\frac{55}{2}R_4 + R_3 \\ \frac{1}{15}R_4 \\ -24R_4 + R_5 \end{array}$$

The program is optimal after Stage I.
The minimum is 512 when $y_1 = 6$ and $y_2 = 8$.

21. Maximize $z = -5x_1 + 4x_2 - 2x_3$

subject to
$$\begin{aligned} -2x_2 + 5x_3 &\ge 8 \\ 4x_1 - x_2 + 3x_3 &\le 12 \\ x_1 \ge 0,\ x_2 \ge 0,\ x_3 &\ge 0. \end{aligned}$$

Insert a surplus variable in the first constraint and slack variables in the last two constraints. The problem then becomes:

Maximize $z = -5x_1 + 4x_2 - 2x_3$

subject to:
$$\begin{aligned} -2x_2 + 5x_3 - s_1 \quad &= 8 \\ 4x_1 - x_2 + 3x_3 \quad + s_2 &= 12 \\ x_1 \ge 0,\ x_2 \ge 0,\ x_3 \ge 0,\ s_1 \ge 0,\ s_2 &\ge 0. \end{aligned}$$

The initial simplex tableau is

$$\begin{matrix} x_1 & x_2 & x_3 & s_1 & s_2 & \end{matrix}$$

$$\left[\begin{array}{ccccc|c} 0 & -2 & \boxed{5} & -1 & 0 & 8 \\ 4 & -1 & 3 & 0 & 1 & 12 \\ \hline 5 & -4 & 2 & 0 & 0 & 0 \end{array}\right].$$

For Stage I, pivot on the 5 in row 1, column 3.

$$\begin{matrix} x_1 & x_2 & x_3 & s_1 & s_2 & \end{matrix}$$

$$\left[\begin{array}{ccccc|c} 0 & -\frac{2}{5} & 1 & -\frac{1}{5} & 0 & \frac{8}{5} \\ 4 & -1 & 3 & 0 & 1 & 12 \\ \hline 5 & -4 & 2 & 0 & 0 & 0 \end{array}\right]\begin{array}{l} \frac{1}{5}R_1 \\ \\ \\ \end{array}$$

$$\begin{matrix} x_1 & x_2 & x_3 & s_1 & s_2 & \end{matrix}$$

$$\left[\begin{array}{ccccc|c} 0 & -\frac{2}{5} & 1 & -\frac{1}{5} & 0 & \frac{8}{5} \\ 4 & \boxed{\frac{1}{5}} & 0 & \frac{3}{5} & 1 & \frac{36}{5} \\ \hline 5 & -\frac{16}{5} & 0 & \frac{2}{5} & 0 & -\frac{16}{5} \end{array}\right]\begin{array}{l} \\ -3R_1 + R_2 \\ -2R_1 + R_3 \end{array}$$

This completes Stage I because. the solution given in the usual way from the matrix has nonnegative values for all variables. Since there are negative indicators in the objective row, we continue with Stage II. Now pivot on the $\frac{1}{5}$ in row 1, column 2.

$$\begin{matrix} x_1 & x_2 & x_3 & s_1 & s_2 & \end{matrix}$$

$$\left[\begin{array}{ccccc|c} 8 & 0 & 1 & 1 & 2 & 16 \\ 4 & \frac{1}{5} & 0 & \frac{3}{5} & 1 & \frac{36}{5} \\ \hline 69 & 0 & 0 & 10 & 16 & 112 \end{array}\right]\begin{array}{l} R_1 + 2R_2 \\ \\ R_3 + 16R_2 \end{array}$$

$$\begin{matrix} x_1 & x_2 & x_3 & s_1 & s_2 & \end{matrix}$$

$$\left[\begin{array}{ccccc|c} 8 & 0 & 1 & 1 & 2 & 16 \\ 20 & 1 & 0 & 3 & 5 & 36 \\ \hline 69 & 0 & 0 & 10 & 16 & 112 \end{array}\right]\begin{array}{l} \\ 5R_2 \\ \\ \end{array}$$

The maximum is 112 when $x_1 = 0,\ x_2 = 36,\ x_3 = 16.$

23. Maximize $z = 2x_1 - 3x_2 + 4x_3$

subject to:
$$x_1 + x_2 + x_3 \le 100$$
$$x_1 + x_2 + x_3 \ge 75$$
$$x_1 + x_2 \ge 27$$
$$x_1 \ge 0,\ x_2 \ge 0,\ x_3 \ge 0.$$

The initial simplex tableau is

$$\begin{array}{cccccc|c} x_1 & x_2 & x_3 & s_1 & s_2 & s_3 & \\ 1 & 1 & 1 & 1 & 0 & 0 & 100 \\ 1 & 1 & \boxed{1} & 0 & -1 & 0 & 75 \\ 1 & 1 & 0 & 0 & 0 & -1 & 27 \\ \hline -2 & 3 & -4 & 0 & 0 & 0 & 0 \end{array}.$$

For Stage I, pivot on the 1 in row 2, column 3.

$$\begin{array}{cccccc|cl} x_1 & x_2 & x_3 & s_1 & s_2 & s_3 & & \\ 0 & 0 & 0 & 1 & 1 & 0 & 25 & -R_2 + R_1 \\ 1 & 1 & 1 & 0 & -1 & 0 & 75 & \\ \boxed{1} & 1 & 0 & 0 & 0 & -1 & 27 & \\ \hline 2 & 7 & 0 & 0 & -4 & 0 & 300 & 4R_2 + R_4 \end{array}$$

Now pivot on the 1 in row 3 column 1.

$$\begin{array}{cccccc|cl} x_1 & x_2 & x_3 & s_1 & s_2 & s_3 & & \\ 0 & 0 & 0 & 1 & 1 & 0 & 25 & \\ 0 & 0 & 1 & 0 & -1 & 1 & 48 & -R_3 + R_2 \\ 1 & 1 & 0 & 0 & 0 & -1 & 27 & \\ \hline 0 & 5 & 0 & 0 & -4 & 2 & 246 & -2R_3 + R_4 \end{array}$$

This completes Stage I because the solution given in the usual way from the matrix has nonnegative values for all variables. Since there is a negative indicator in the objective row, we continue with Stage II.

Pivot on the 1 in row 1 column 5.

$$\begin{array}{cccccc|cl} x_1 & x_2 & x_3 & s_1 & s_2 & s_3 & & \\ 0 & 0 & 0 & 1 & \boxed{1} & 0 & 25 & \\ 0 & 0 & 1 & 1 & 0 & 1 & 73 & R_1 + R_2 \\ 1 & 1 & 0 & 0 & 0 & -1 & 27 & \\ \hline 0 & 5 & 0 & 4 & 0 & 2 & 346 & 4R_1 + R_4 \end{array}$$

The maximum is 346 when $x_1 = 27,\ x_2 = 0,\ x_3 = 73.$

In exercises 25–27, the two-stage program in Appendix A was used to produce the matrix giving a feasible solution. (We used the LINPROG program on a TI84 Plus.) A different program might produce a different feasible solution, but will produce the same final solution.

25. Minimize $w = 2y_1 + 5y_2 - 3y_3$

subject to:
$$y_1 + 2y_2 + 3y_3 \ge 115$$
$$2y_1 + y_2 + y_3 \le 200$$
$$y_1 + y_3 \ge 50$$
$$y_1 \ge 0,\ y_2 \ge 0,\ y_3 \ge 0.$$

The initial simplex tableau is

$$\begin{array}{cccccc|c} y_1 & y_2 & y_3 & s_1 & s_2 & s_3 & \\ 1 & 2 & 3 & -1 & 0 & 0 & 115 \\ 2 & 1 & 1 & 0 & 1 & 0 & 200 \\ \boxed{1} & 0 & 1 & 0 & 0 & -1 & 50 \\ \hline 2 & 5 & -3 & 0 & 0 & 0 & 0 \end{array}.$$

Stage I of the two-stage program produces the matrix

$$\begin{array}{cccccc|c} y_1 & y_2 & y_3 & s_1 & s_2 & s_3 & \\ 0 & 1 & 1 & -\frac{1}{2} & 0 & \frac{1}{2} & \frac{65}{2} \\ 0 & 0 & -2 & \frac{1}{2} & 1 & \frac{3}{2} & \frac{135}{2} \\ 1 & 0 & 1 & 0 & 0 & -1 & 50 \\ \hline 0 & 0 & -10 & \frac{5}{2} & 0 & -\frac{1}{2} & -\frac{525}{2} \end{array}.$$

This gives the feasible solution $y_1 = 50$, $y_2 = \frac{65}{2}$, $y_3 = 0$, $s_1 = 0$, $s_2 = \frac{135}{2}$, and $s_3 = 0$.

Stage II produces the final matrix

$$\begin{array}{cccccc|c} y_1 & y_2 & y_3 & s_1 & s_2 & s_3 & \\ 2 & 1 & 1 & 0 & 1 & 0 & 200 \\ 1 & 1 & 0 & 0 & 1 & 1 & 150 \\ 5 & 1 & 0 & 1 & 3 & 0 & 485 \\ \hline 8 & 8 & 0 & 0 & 3 & 0 & 600 \end{array}.$$

The minimum is –600 when $y_1 = 0,\ y_2 = 0,$ and $y_3 = 200.$

27. Minimize $w = 10y_1 + 8y_2 + 15y_3$

subject to: $\begin{aligned} y_1 + y_2 + y_3 &\geq 12 \\ 5y_1 + 4y_2 + 9y_3 &\geq 48 \\ y_1 \geq 0,\ y_2 \geq 0,\ y_3 &\geq 0. \end{aligned}$

Insert surplus variables in both constraints to get the initial simplex tableau.

$$\begin{array}{ccccc|c} y_1 & y_2 & y_3 & s_1 & s_2 & \\ 1 & 1 & 1 & -1 & 0 & 12 \\ 5 & 4 & 9 & 0 & -1 & 48 \\ \hline 10 & 8 & 15 & 0 & 0 & 0 \end{array}$$

Stage I of the two-stage program produces the matrix

$$\begin{array}{ccccc|c} y_1 & y_2 & y_3 & s_1 & s_2 & \\ 0 & 1 & -4 & -5 & 1 & 12 \\ 1 & 0 & 5 & 4 & -1 & 0 \\ \hline 0 & 0 & -3 & 0 & 2 & -96 \end{array}.$$

This gives the feasible solution $y_1 = 0,\ y_2 = 12,$ $y_3 = 0,\ s_1 = 0,$ and $s_2 = 0.$

Stage II produces the final matrix

$$\begin{array}{ccccc|c} y_1 & y_2 & y_3 & s_1 & s_2 & \\ \frac{4}{5} & 1 & 0 & -\frac{9}{5} & \frac{1}{5} & 12 \\ \frac{1}{5} & 0 & 1 & \frac{4}{5} & -\frac{1}{5} & 0 \\ \hline \frac{3}{5} & 0 & 0 & \frac{12}{5} & \frac{7}{5} & -96 \end{array}.$$

The minimum is 96 when $y_1 = 0,\ y_2 = 12,$ and $y_3 = 0.$

29. Let y_1 = the amount of ingredient I per barrel of gasoline, let y_2 = the amount of ingredient II per barrel of gasoline, let y_3 = the amount of ingredient III per barrel of gasoline.

Minimize $w = .30y_1 + .09y_2 + .27y_3$

Maximize $z = -w = -.30y_1 - .09y_2 - .27y_3$

subject to: $\begin{aligned} y_1 + y_2 + y_3 &\geq 10 \\ y_1 + y_2 + y_3 &\leq 15 \\ y_1 - \tfrac{1}{4}y_2 \quad &\geq 0 \quad \left(\text{Since } y_1 \geq \tfrac{1}{4}y_2\right) \\ -y_1 \quad + y_3 &\geq 0 \quad (\text{Since } y_3 \geq y_1) \\ y_1 \geq 0,\ y_2 \geq 0,\ y_3 &\geq 0 \end{aligned}$

The initial simplex tableau is

$$\begin{array}{ccccccc|c} y_1 & y_2 & y_3 & s_1 & s_2 & s_3 & s_4 & \\ 1 & 1 & 1 & -1 & 0 & 0 & 0 & 10 \\ 1 & 1 & 1 & 0 & 1 & 0 & 0 & 15 \\ 1 & -\frac{1}{4} & 0 & 0 & 0 & -1 & 0 & 0 \\ -1 & 0 & 1 & 0 & 0 & 0 & -1 & 0 \\ \hline .30 & .09 & .27 & 0 & 0 & 0 & 0 & 0 \end{array}.$$

31. Let y_1 = the number of computers from W_1 to D_1; let y_2 = the number of computers from W_2 to D_1; let y_3 = the number of computers from W_1 to D_2; let y_4 = the number of computers from W_2 to D_2.

Minimize $w = 14y_1 + 12y_2 + 22y_3 + 10y_4$

Maximize $z = -w = -14y_1 - 12y_2 - 22y_3 - 10y_4$

subject to:

$$\begin{aligned} y_1 + y_2 \qquad\qquad &= 32 \\ y_3 + y_4 &= 20 \\ y_1 + \qquad y_3 \qquad &\le 25 \\ y_2 + \qquad y_4 &\le 30 \end{aligned}$$

$y_1 \ge 0,\ y_2 \ge 0, y_3 \ge 0,\ y_4 \ge 0.$

The initial simplex tableau is

$$\begin{array}{cccccccccc|c} y_1 & y_2 & y_3 & y_4 & s_1 & s_2 & s_3 & s_4 & s_5 & s_6 & \\ \hline 1 & 1 & 0 & 0 & -1 & 0 & 0 & 0 & 0 & 0 & 32 \\ 1 & 1 & 0 & 0 & 0 & 1 & 0 & 0 & 0 & 0 & 32 \\ 0 & 0 & 1 & 1 & 0 & 0 & -1 & 0 & 0 & 0 & 20 \\ 0 & 0 & 1 & 1 & 0 & 0 & 0 & 1 & 0 & 0 & 20 \\ 1 & 0 & 1 & 0 & 0 & 0 & 0 & 0 & 1 & 0 & 25 \\ 0 & 1 & 0 & 1 & 0 & 0 & 0 & 0 & 0 & 1 & 30 \\ \hline 14 & 12 & 22 & 10 & 0 & 0 & 0 & 0 & 0 & 0 & 0 \end{array}.$$

33. Let x_1 = the number of barrels of oil supplied by S_1 to D_1.
Let x_2 = the number of barrels of oil supplied by S_2 to D_1.
Let x_3 = the number of barrels of oil supplied by S_1 to D_2.
Let x_4 = the number of barrels of oil supplied by S_2 to D_2.

Minimize $w = 30x_1 + 25x_2 + 20x_3 + 22x_4$

subject to:

$$\begin{aligned} x_1 + x_2 \qquad\qquad &\ge 3000 \\ x_3 + x_4 &\ge 5000 \\ x_1 \qquad + x_3 \qquad &\le 5000 \\ x_2 \qquad + x_4 &\le 5000 \\ 2x_1 + 5x_2 + 6x_3 + 4x_4 &\le 40{,}000 \end{aligned}$$

The initial simplex tableau is

$$\begin{array}{ccccccccc|c} x_1 & x_2 & x_3 & x_4 & s_1 & s_2 & s_3 & s_4 & s_5 & \\ \hline \boxed{1} & 1 & 0 & 0 & -1 & 0 & 0 & 0 & 0 & 3000 \\ 0 & 0 & 1 & 1 & 0 & -1 & 0 & 0 & 0 & 5000 \\ 1 & 0 & 1 & 0 & 0 & 0 & 1 & 0 & 0 & 5000 \\ 0 & 1 & 0 & 1 & 0 & 0 & 0 & 1 & 0 & 5000 \\ 2 & 5 & 6 & 4 & 0 & 0 & 0 & 0 & 1 & 40{,}000 \\ \hline 30 & 25 & 20 & 22 & 0 & 0 & 0 & 0 & 0 & 0 \end{array}$$

(continued next page)

For Stage I, pivot on the 1 in row one, column one.

$$\begin{array}{ccccccccc|c} x_1 & x_2 & x_3 & x_4 & s_1 & s_2 & s_3 & s_4 & s_5 & \\ 1 & 1 & 0 & 0 & -1 & 0 & 0 & 0 & 0 & 3000 \\ 0 & 0 & 1 & 1 & 0 & -1 & 0 & 0 & 0 & 5000 \\ 0 & -1 & \boxed{1} & 0 & 1 & 0 & 1 & 0 & 0 & 2000 \\ 0 & 1 & 0 & 1 & 0 & 0 & 0 & 1 & 0 & 5000 \\ 0 & 3 & 6 & 4 & 2 & 0 & 0 & 0 & 1 & 34,000 \\ \hline 0 & -5 & 20 & 22 & 30 & 0 & 0 & 0 & 0 & -90,000 \end{array} \begin{array}{l} \\ \\ \\ -R_1 + R_3 \\ \\ -2R_1 + R_5 \\ -30R_1 + R_6 \end{array}$$

To continue Stage I, pivot on the 1 in row three, column three.

$$\begin{array}{ccccccccc|c} x_1 & x_2 & x_3 & x_4 & s_1 & s_2 & s_3 & s_4 & s_5 & \\ 1 & 1 & 0 & 0 & -1 & 0 & 0 & 0 & 0 & 3000 \\ 0 & 1 & 0 & \boxed{1} & -1 & -1 & -1 & 0 & 0 & 3000 \\ 0 & -1 & 1 & 0 & 1 & 0 & 1 & 0 & 0 & 2000 \\ 0 & 1 & 0 & 1 & 0 & 0 & 0 & 1 & 0 & 5000 \\ 0 & 9 & 0 & 4 & -4 & 0 & -6 & 0 & 1 & 22,000 \\ \hline 0 & 15 & 0 & 22 & 10 & 0 & -20 & 0 & 0 & -130,000 \end{array} \begin{array}{l} \\ \\ -R_3 + R_2 \\ \\ \\ -6R_3 + R_5 \\ -20R_3 + R_6 \end{array}$$

To continue Stage I, pivot on the 1 in row two, column four.

$$\begin{array}{ccccccccc|c} x_1 & x_2 & x_3 & x_4 & s_1 & s_2 & s_3 & s_4 & s_5 & \\ 1 & 1 & 0 & 0 & -1 & 0 & 0 & 0 & 0 & 3000 \\ 0 & 1 & 0 & 1 & -1 & -1 & -1 & 0 & 0 & 3000 \\ 0 & -1 & 1 & 0 & 1 & 0 & 1 & 0 & 0 & 2000 \\ 0 & 0 & 0 & 0 & 1 & 1 & 1 & 1 & 0 & 2000 \\ 0 & \boxed{5} & 0 & 0 & 0 & 4 & -2 & 0 & 1 & 10,000 \\ \hline 0 & -7 & 0 & 0 & 32 & 22 & 2 & 0 & 0 & -196,000 \end{array} \begin{array}{l} \\ \\ \\ \\ -R_2 + R_4 \\ -4R_2 + R_5 \\ -22R_2 + R_6 \end{array}$$

We have completed Stage I. For Stage II, pivot on the 5 in row five, column two.

$$\begin{array}{ccccccccc|c} x_1 & x_2 & x_3 & x_4 & s_1 & s_2 & s_3 & s_4 & s_5 & \\ 1 & 1 & 0 & 0 & -1 & 0 & 0 & 0 & 0 & 3000 \\ 0 & 1 & 0 & 1 & -1 & -1 & -1 & 0 & 0 & 3000 \\ 0 & -1 & 1 & 0 & 1 & 0 & 1 & 0 & 0 & 2000 \\ 0 & 0 & 0 & 0 & 1 & 1 & 1 & 1 & 0 & 2000 \\ 0 & 1 & 0 & 0 & 0 & \frac{4}{5} & -\frac{2}{5} & 0 & \frac{1}{5} & 2000 \\ \hline 0 & -7 & 0 & 0 & 32 & 22 & 2 & 0 & 0 & -196,000 \end{array} \begin{array}{l} \\ \\ \\ \\ \\ \frac{1}{5}R_5 \\ \\ \end{array}$$

$$\begin{array}{ccccccccc|c} x_1 & x_2 & x_3 & x_4 & s_1 & s_2 & s_3 & s_4 & s_5 & \\ 1 & 0 & 0 & 0 & -1 & -\frac{4}{5} & \frac{2}{5} & 0 & -\frac{1}{5} & 1000 \\ 0 & 0 & 0 & 1 & -1 & -\frac{9}{5} & -\frac{3}{5} & 0 & -\frac{1}{5} & 1000 \\ 0 & 0 & 1 & 0 & 1 & \frac{4}{5} & \frac{3}{5} & 0 & \frac{1}{5} & 4000 \\ 0 & 0 & 0 & 0 & 1 & 1 & \boxed{1} & 1 & 0 & 2000 \\ 0 & 1 & 0 & 0 & 0 & \frac{4}{5} & -\frac{2}{5} & 0 & \frac{1}{5} & 2000 \\ \hline 0 & 0 & 0 & 0 & 32 & \frac{138}{5} & -\frac{4}{5} & 0 & \frac{7}{5} & -182,000 \end{array} \begin{array}{l} \\ -R_5 + R_1 \\ -R_5 + R_2 \\ R_5 + R_3 \\ \\ \\ 7R_5 + R_6 \end{array}$$

(*continued next page*)

Finally, pivot on the 1 in row four, column seven.

$$\begin{array}{c} \\ \\ \\ \\ \\ \\ \end{array}\begin{array}{cccccccccc} x_1 & x_2 & x_3 & x_4 & s_1 & s_2 & s_3 & s_4 & s_5 & \end{array}$$

$$\left[\begin{array}{ccccccccc|c} 1 & 0 & 0 & 0 & -\frac{7}{5} & -\frac{6}{5} & 0 & -\frac{2}{5} & -\frac{1}{5} & 200 \\ 0 & 0 & 0 & 1 & -\frac{2}{5} & -\frac{6}{5} & 0 & \frac{3}{5} & -\frac{1}{5} & 2200 \\ 0 & 0 & 1 & 0 & \frac{2}{5} & \frac{1}{5} & 0 & -\frac{3}{5} & \frac{1}{5} & 2800 \\ 0 & 0 & 0 & 0 & 1 & 1 & 1 & 1 & 0 & 2000 \\ 0 & 1 & 0 & 0 & \frac{2}{5} & \frac{6}{5} & 0 & \frac{2}{5} & \frac{1}{5} & 2800 \\ \hline 0 & 0 & 0 & 0 & \frac{164}{5} & \frac{142}{5} & 0 & \frac{4}{5} & \frac{7}{5} & -180{,}400 \end{array}\right]\begin{array}{l} -\frac{2}{5}R_4 + R_1 \\ \frac{3}{5}R_4 + R_2 \\ -\frac{3}{5}R_4 + R_3 \\ \\ \frac{2}{5}R_4 + R_5 \\ \frac{4}{5}R_4 + R_6 \end{array}$$

This gives $x_1 = 200$, $x_2 = 2800$, $x_3 = 2800$, and $x_4 = 2200$. This means the company should ship 200 barrels from supplier S_1 to distributor D_1, 2800 barrels from supplier S_2 to distributor D_1, 2800 barrels from supplier S_1 to distributor D_2 and 2200 barrels from supplier S_2 to distributor D_2 at a minimum cost of \$180,400.

35. Let y_1 = the amount of bluegrass seed, y_2 = the amount of rye seed, and y_3 = the amount of Bermuda seed. The problem is to minimize $w = .12y_1 + .15y_2 + .05y_3$ subject to

$$y_1 \ge .2(y_1 + y_2 + y_3)$$
$$y_3 \le \frac{2}{3}y_2$$
$$y_1 + y_2 + y_3 \ge 5000$$

Or, maximize: $z = -w = -.12y_1 - .15y_2 - .05y_3$ subject to

$$.8y_1 - .2y_2 - .2y_3 \ge 0$$
$$2y_2 - 3y_3 \ge 0$$
$$y_1 + y_2 + y_3 \ge 5000$$

Adding surplus variables s_1, s_2, and s_3, the initial tableau is

$$\begin{array}{cccccc} y_1 & y_2 & y_3 & s_1 & s_2 & s_3 \end{array}$$

$$\left[\begin{array}{cccccc|c} \boxed{.8} & -.2 & -.2 & -1 & 0 & 0 & 0 \\ 0 & 2 & -3 & 0 & -1 & 0 & 0 \\ 1 & 1 & 1 & 0 & 0 & -1 & 5000 \\ \hline .12 & .15 & .05 & 0 & 0 & 0 & 0 \end{array}\right]$$

Since y_4 through y_6 are negative, this does not have a feasible solution. For Stage I, use the .8 in row one, column one.

$$\begin{array}{cccccc} y_1 & y_2 & y_3 & s_1 & s_2 & s_3 \end{array}$$

$$\left[\begin{array}{cccccc|c} 1 & -.25 & -.25 & -1.25 & 0 & 0 & 0 \\ 0 & \boxed{2} & -3 & 0 & -1 & 0 & 0 \\ 0 & 1.25 & 1.25 & 1.25 & 0 & -1 & 5000 \\ \hline 0 & .18 & .08 & .15 & 0 & 0 & 0 \end{array}\right]\begin{array}{l} 1.25R_1 \\ \\ -R_1 + R_3 \\ -.12R_1 + R_4 \end{array}$$

(*continued next page*)

Continue Stage 1 by pivoting on the 2 in row two, column two.

$$\begin{array}{c} \\ \\ \\ \\ \\ \end{array}\begin{array}{cccccc|c} y_1 & y_2 & y_3 & s_1 & s_2 & s_3 & \\ \hline 1 & 0 & -.625 & -1.25 & -.125 & 0 & 0 \\ 0 & 1 & -1.5 & 0 & -.5 & 0 & 0 \\ 0 & 0 & \boxed{3.125} & 1.25 & .625 & -1 & 5000 \\ \hline 0 & 0 & .35 & .15 & .09 & 0 & 0 \end{array} \begin{array}{l} \\ .25R_2 + R_1 \\ .5R_2 \\ -1.25R_1 + R_3 \\ -.18R_1 + R_4 \end{array}$$

To finish Stage 1, pivot on the 3.125 in row three, column three.

$$\begin{array}{cccccc|c} y_1 & y_2 & y_3 & s_1 & s_2 & s_3 & \\ \hline 1 & 0 & 0 & -1 & 0 & -.2 & 1000 \\ 0 & 1 & 0 & .6 & -.2 & -.48 & 2400 \\ 0 & 0 & 1 & .4 & .2 & -.32 & 1600 \\ \hline 0 & 0 & 0 & .01 & .02 & .112 & -560 \end{array} \begin{array}{l} \\ .625R_3 + R_1 \\ 1.5R_3 + R_2 \\ .32R_3 \\ -.35R_3 + R_4 \end{array}$$

Stage II pivoting is not necessary. The solution is $y_1 = 1000$, $y_2 = 2400$, $y_3 = 1600$ and $w = 560$. In other words, Topgrade Turf should use 1000 lbs.of bluegrass seed, 2400 lbs. of rye seed, and 1600 lbs. of Bermuda seed for a minimum cost of $560.

37. Let x_1 = amount allotted to commercial loans (in millions) and x_2 = amount allotted to home loans (in millions)

Maximize $z = .10x_1 + .12x_2$

subject to:

$$\begin{aligned} x_1 + x_2 &\le 25 \\ 2x_1 + 3x_2 &\le 72 \\ -4x_1 + x_2 &\ge 0 \\ x_1 + x_2 &\ge 10 \\ x_1 \ge 0,\ x_2 &\ge 0 \end{aligned}$$

The initial simplex tableau is

$$\begin{array}{cccccc|c} x_1 & x_2 & s_1 & s_2 & s_3 & s_4 & \\ \hline 1 & 1 & 1 & 0 & 0 & 0 & 25 \\ 2 & 3 & 0 & 1 & 0 & 0 & 72 \\ -4 & \boxed{1} & 0 & 0 & -1 & 0 & 0 \\ 1 & 1 & 0 & 0 & 0 & -1 & 10 \\ \hline -.10 & -.12 & 0 & 0 & 0 & 0 & 0 \end{array}$$

For Stage I, pivot on the 1 in row three, column two.

$$\begin{array}{cccccc|c} x_1 & x_2 & s_1 & s_2 & s_3 & s_4 & \\ \hline 5 & 0 & 1 & 0 & 1 & 0 & 25 \\ 14 & 0 & 0 & 1 & 3 & 0 & 72 \\ -4 & 1 & 0 & 0 & -1 & 0 & 0 \\ \boxed{5} & 0 & 0 & 0 & 1 & -1 & 10 \\ \hline -\frac{29}{50} & 0 & 0 & 0 & -\frac{3}{25} & 0 & 0 \end{array} \begin{array}{l} \\ -R_3 + R_1 \\ -3R_3 + R_2 \\ \\ -R_3 + R_4 \\ \frac{3}{25}R_3 + R_5 \end{array}$$

To continue Stage I, pivot on the 5 in row four, column one.

$$\begin{array}{cccccc|c} x_1 & x_2 & s_1 & s_2 & s_3 & s_4 & \\ \hline 0 & 0 & 1 & 0 & 0 & \boxed{1} & 15 \\ 0 & 0 & 0 & 1 & \frac{1}{5} & \frac{14}{5} & 44 \\ 0 & 1 & 0 & 0 & -\frac{1}{5} & -\frac{4}{5} & 8 \\ 1 & 0 & 0 & 0 & \frac{1}{5} & -\frac{1}{5} & 2 \\ \hline 0 & 0 & 0 & 0 & -\frac{1}{250} & -\frac{29}{250} & \frac{29}{25} \end{array} \begin{array}{l} \\ -5R_4 + R_1 \\ -14R_4 + R_2 \\ 4R_4 + R_3 \\ \frac{1}{5}R_4 \\ \frac{29}{50}R_4 + R_5 \end{array}$$

For Stage II, pivot on the 1 in row one, column six.

$$\begin{array}{cccccc|c} x_1 & x_2 & s_1 & s_2 & s_3 & s_4 & \\ \hline 0 & 0 & 1 & 0 & 0 & 1 & 15 \\ 0 & 0 & -\frac{14}{5} & 1 & \boxed{\frac{1}{5}} & 0 & 2 \\ 0 & 1 & \frac{4}{5} & 0 & -\frac{1}{5} & 0 & 20 \\ 1 & 0 & \frac{1}{5} & 0 & \frac{1}{5} & 0 & 5 \\ \hline 0 & 0 & \frac{29}{250} & 0 & -\frac{1}{250} & 0 & \frac{29}{10} \end{array} \begin{array}{l} \\ \\ -\frac{14}{5}R_1 + R_2 \\ \frac{4}{5}R_1 + +R_3 \\ \frac{1}{5}R_1 + R_4 \\ \frac{29}{250}R_1 + R_5 \end{array}$$

(continued next page)

To continue stage II, pivot on the $\frac{1}{5}$ in row two, column five.

$$\begin{array}{cccccc|c|l}
x_1 & x_2 & s_1 & s_2 & s_3 & s_4 & & \\
0 & 0 & 1 & 0 & 0 & 1 & 15 & \\
0 & 0 & -14 & 5 & 1 & 0 & 10 & 5R_2 \\
0 & 1 & -2 & 1 & 0 & 0 & 22 & \frac{1}{5}R_2 + R_3 \\
1 & 0 & 3 & -1 & 0 & 0 & 3 & -\frac{1}{5}R_2 + R_4 \\
\hline
0 & 0 & \frac{3}{50} & \frac{1}{50} & 0 & 0 & \frac{147}{50} & \frac{1}{250}R_2 + R_5
\end{array}$$

This gives $x_1 = 3$, $x_2 = 22$, and the maximum is $\frac{147}{50} = 2.94$. Allot \$3,000,000 for commercial loans and \$22,000,000 for home loans for a maximum interest income of \$2,940,000.

39. Let y_1 = the amount of regular beer and y_2 = the amount of light beer.

Minimize $36{,}000y_1 + 48{,}000y_2$

subject to:

$$\begin{aligned}
y_1 &\ge 12 \\
y_2 &\ge 10 \\
-2y_1 + y_2 &\le 0 \quad (\text{Since } y_2 \le 2y_1) \\
y_1 + y_2 &\ge 42.
\end{aligned}$$

(Since the company already produces at least $12 + 10 = 22$ units and can produce at least 20 additional units: $y_1 + y_2 \ge 22 + 20 = 42$.)

The initial simplex tableau is

$$\begin{array}{cccccc|c}
y_1 & y_2 & s_1 & s_2 & s_3 & s_4 & \\
\boxed{1} & 0 & -1 & 0 & 0 & 0 & 12 \\
0 & 1 & 0 & -1 & 0 & 0 & 10 \\
-2 & 1 & 0 & 0 & 1 & 0 & 0 \\
1 & 1 & 0 & 0 & 0 & -1 & 42 \\
\hline
36 & 48 & 0 & 0 & 0 & 0 & 0
\end{array}.$$

For Stage I, pivot on the 1 in row one, column one.

$$\begin{array}{cccccc|c|l}
y_1 & y_2 & s_1 & s_2 & s_3 & s_4 & & \\
1 & 0 & -1 & 0 & 0 & 0 & 12 & \\
0 & \boxed{1} & 0 & -1 & 0 & 0 & 10 & \\
0 & 1 & -2 & 0 & 1 & 0 & 24 & 2R_1 + R_3 \\
0 & 1 & 1 & 0 & 0 & -1 & 30 & -R_1 + R_4 \\
\hline
0 & 48 & 36 & 0 & 0 & 0 & -432 & -36R_1 + R_5
\end{array}$$

To continue Stage I, pivot on the 1 in row two, column two.

$$\begin{array}{cccccc|c|l}
y_1 & y_2 & s_1 & s_2 & s_3 & s_4 & & \\
1 & 0 & -1 & 0 & 0 & 0 & 12 & \\
0 & 1 & 0 & -1 & 0 & 0 & 10 & \\
0 & 0 & -2 & 1 & 1 & 0 & 14 & -R_2 + R_3 \\
0 & 0 & 1 & \boxed{1} & 0 & -1 & 20 & -R_2 + R_4 \\
\hline
0 & 0 & 36 & 48 & 0 & 0 & -912 & -48R_2 + R_5
\end{array}$$

Now pivot on the 1 in row four, column four.

$$\begin{array}{cccccc|c|l}
y_1 & y_2 & s_1 & s_2 & s_3 & s_4 & & \\
1 & 0 & -1 & 0 & 0 & 0 & 12 & \\
0 & 1 & 1 & 0 & 0 & -1 & 30 & R_4 + R_2 \\
0 & 0 & \boxed{-3} & 0 & 1 & 1 & -6 & -R_4 + R_3 \\
0 & 0 & 1 & 1 & 0 & -1 & 20 & \\
\hline
0 & 0 & -12 & 0 & 0 & 48 & -1872 & -48R_4 + R_5
\end{array}$$

For Stage II, pivot on the –3 in row three, column three.

$$\begin{array}{cccccc|c|l}
y_1 & y_2 & s_1 & s_2 & s_3 & s_4 & & \\
1 & 0 & 0 & 0 & -\frac{1}{3} & -\frac{1}{3} & 14 & R_3 + R_1 \\
0 & 1 & 0 & 0 & \frac{1}{3} & -\frac{2}{3} & 28 & -R_3 + R_2 \\
0 & 0 & 1 & 0 & -\frac{1}{3} & -\frac{1}{3} & 2 & -\frac{1}{3}R_3 \\
0 & 0 & 0 & 1 & \boxed{\frac{1}{3}} & -\frac{2}{3} & 18 & -R_3 + R_4 \\
\hline
0 & 0 & 0 & 0 & -4 & 44 & -1848 & 12R_3 + R_5
\end{array}$$

Finally, pivot on the $\frac{1}{3}$ in row four, column five.

$$\begin{array}{cccccc|c|l}
y_1 & y_2 & s_1 & s_2 & s_3 & s_4 & & \\
1 & 0 & 0 & 1 & 0 & -1 & 32 & \frac{1}{3}R_4 + R_1 \\
0 & 1 & 0 & -1 & 0 & 0 & 10 & -\frac{1}{3}R_4 + R_2 \\
0 & 0 & 1 & 1 & 0 & -1 & 20 & \frac{1}{3}R_4 + R_3 \\
0 & 0 & 0 & 3 & 1 & -2 & 54 & 3R_4 \\
\hline
0 & 0 & 0 & 12 & 0 & 36 & -1632 & 4R_4 + R_5
\end{array}$$

Make 32 units of regular beer and 10 units of light beer for a minimum cost of \$1,632,000.

41. From Exercise 29, we know the initial simplex tableau is

$$\begin{array}{c} \\ \\ \\ \\ \\ \\ \end{array}\begin{array}{ccccccc|c} y_1 & y_2 & y_3 & s_1 & s_2 & s_3 & s_4 & \\ 1 & 1 & 1 & -1 & 0 & 0 & 0 & 10 \\ 1 & 1 & 1 & 0 & 1 & 0 & 0 & 15 \\ 1 & -\frac{1}{4} & 0 & 0 & 0 & -1 & 0 & 0 \\ -1 & 0 & 1 & 0 & 0 & 0 & -1 & 0 \\ \hline .30 & .09 & .27 & 0 & 0 & 0 & 0 & 0 \end{array}.$$

Using the LINPROG program on a TI-84 Plus, we find the optimal solution is

$y_1 = 1\frac{2}{3},\ y_2 = 6\frac{2}{3},$ and $y_3 = 1\frac{2}{3}.$

For each barrel of gasoline, the mixture should contain $1\frac{2}{3}$ ounces of ingredient I, $6\frac{2}{3}$ ounces of ingredient II, and $1\frac{2}{3}$ ounces of ingredient III at a minimum cost of $1.55 per barrel.

```
FINAL MATRIX
[[0 1 0 -.66666...
 [0 0 0 1      ...
 [1 0 0 -.16666...
 [0 0 1 -.16666...
 [0 0 0 31/200 ...
```

```
FINAL MATRIX
...0 1.333333333 ...
...1 0           ...
...0 -.666666667 ...
...0 -.666666667 ...
...0 13/50       ...
```

```
FINAL MATRIX
....666666667  6....
...0           5 ...
....166666667  1....
...-.833333333 1....
...23/200      -3...
```

```
FINAL MATRIX
...   6.666666667]
...   5          ]
...   1.666666667]
...3  1.666666667]
...   -31/20     ]]
```

43. From Exercise 31, we know the initial simplex tableau is

$$\begin{array}{cccccccccc|c} y_1 & y_2 & y_3 & y_4 & s_1 & s_2 & s_3 & s_4 & s_5 & s_6 & \\ 1 & 1 & 0 & 0 & -1 & 0 & 0 & 0 & 0 & 0 & 32 \\ 1 & 1 & 0 & 0 & 0 & 1 & 0 & 0 & 0 & 0 & 32 \\ 0 & 0 & 1 & 1 & 0 & 0 & -1 & 0 & 0 & 0 & 20 \\ 0 & 0 & 1 & 1 & 0 & 0 & 0 & 1 & 0 & 0 & 20 \\ 1 & 0 & 1 & 0 & 0 & 0 & 0 & 0 & 1 & 0 & 25 \\ 0 & 1 & 0 & 1 & 0 & 0 & 0 & 0 & 0 & 1 & 30 \\ \hline 14 & 12 & 22 & 10 & 0 & 0 & 0 & 0 & 0 & 0 & 0 \end{array}.$$

Using the LINPROG program on a TI-84 Plus, we find the optimal solution is $y_1 = 22,\ y_2 = 0,\ y_3 = 10,$ and $y_4 = 20.$ The manufacturer should send 22 computers from W_1 to D_1, 0 from W_1 to D_2, 10 from W_2 to D_1, and 20 computers from W_2 to D_2 for a minimum cost of $628.

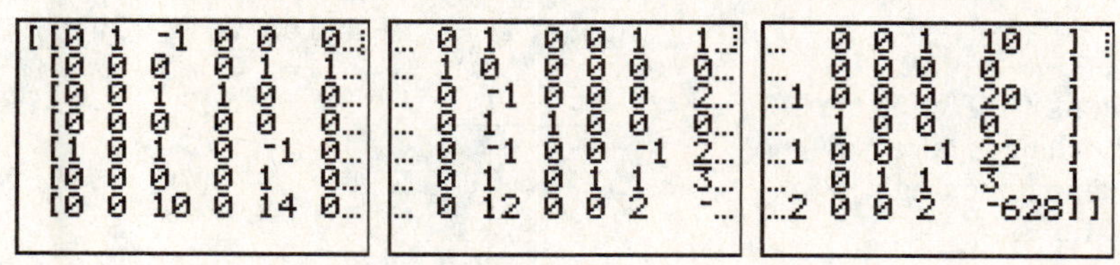

Chapter 7 Review Exercises

1. $y \le 3x + 2$
   Graph the solid boundary line $y = 3x + 2$ which goes through the points (0, 2) and (2, 8). The inequality is $\le$, so shade the half-plane below the line.

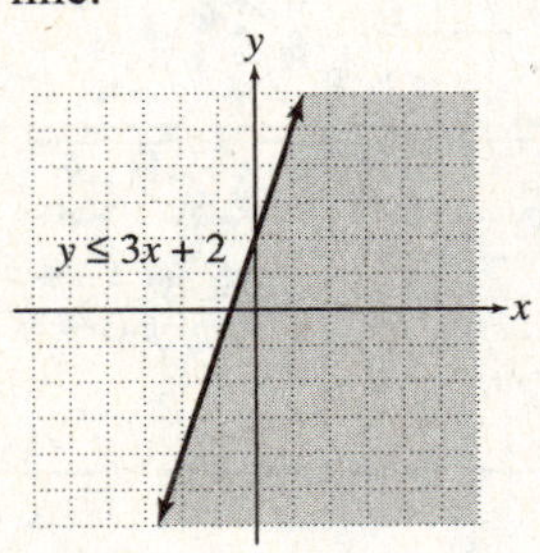

2. $2x - y \ge 6$
   Graph the solid boundary line $2x - y = 6$, which goes through the points (0, –6) and (3, 0). Testing the origin gives the statement $2(0) - 0 \ge 6$, which is false, so shade the half-plane not containing the origin, which is the region below the line.

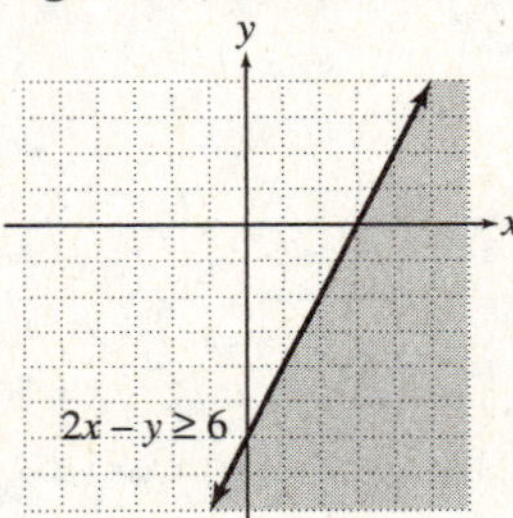

3. $3x + 4y \ge 12$
   Graph the solid boundary line $3x + 4y = 12$, which goes through (0, 3) and (4, 0). Testing the origin gives $3(0) + 4(0) \ge 12$, which is false, so shade the half-plane not containing the origin, which is the region above the line.

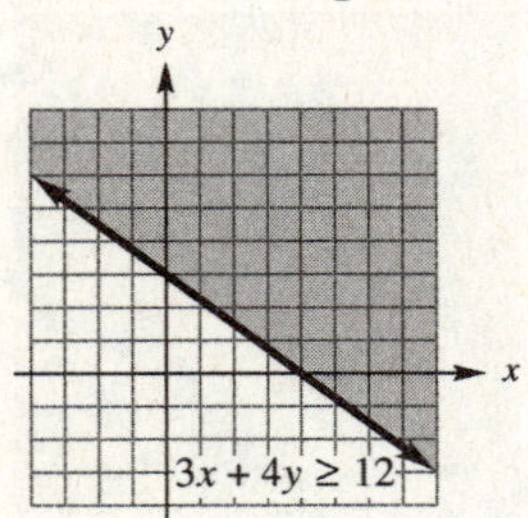

4. $y \le 4$
   Graph the solid boundary line $y = 4$, which is the horizontal line crossing the y-axis at (0, 4). Shade the half-plane below the line.

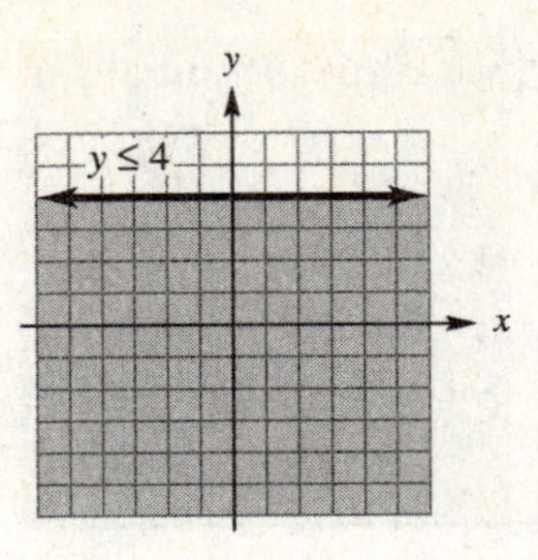

5. $x + y \le 6$
   $2x - y \ge 3$
   $x + y \le 6$ is the region on or below the line $x + y = 6$; $2x - y \ge 3$ is the region on or below the line $2x - y = 3$. The system of inequalities must meet both conditions so we shade the overlap of the two half-planes.

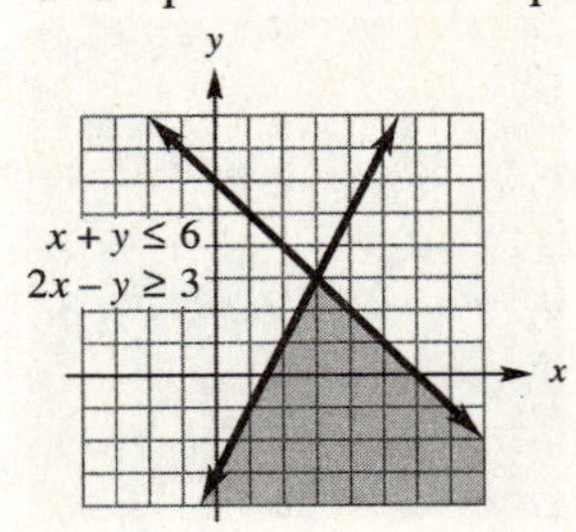

6. $4x + y \ge 8$
   $2x - 3y \le 6$
   Graph $4x + y = 8$ as a solid line using (2, 0) and (0, 8), and $2x - 3y = 6$ as a solid line using (3, 0) and (0, –2). The test point (0, 0) gives $0 + 0 \ge 8$, which is false and $0 - 0 \le 6$, which is true. Shade all points above $4x + y = 8$ and above $2x - 3y = 6$. The solution region is to the right of $4x + y = 8$ and above $2x - 3y = 6$.

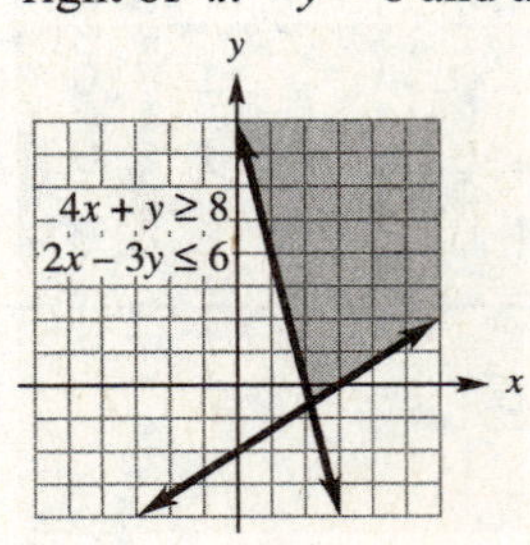

7. $2 \le x \le 5$
   $1 \le y \le 6$
   $x - y \le 3$
   The graph of $2 \le x \le 5$ is the region lying on or between the two vertical lines $x = 2$ and $x = 5$. The graph of $1 \le y \le 6$ is the region lying on or between the two horizontal lines $y = 1$ and $y = 6$. The graph of $x - y \le 3$ is the region lying on or below the line $x - y = 3$. Shade the region that is common to all three graphs.

(continued on next page)

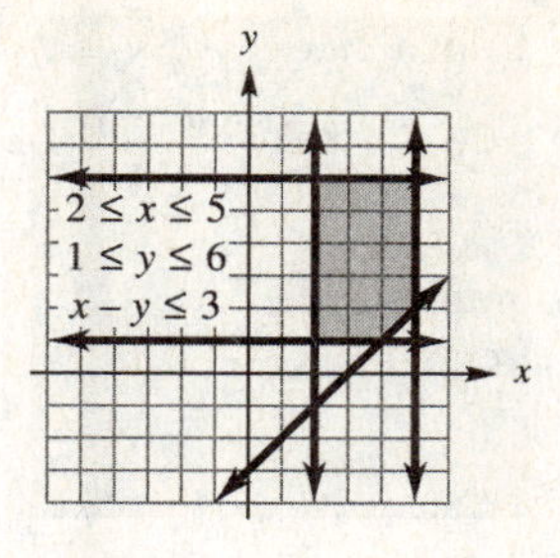

8. $x + 2y \le 4$
$2x - 3y \le 6$
$x \ge 0$
$y \ge 0$

Graph $x + 2y = 4$ and $2x - 3y = 6$ as solid lines. $x \ge 0$ and $y \ge 0$ restrict the region to quadrant I. Use (0, 0) as a test point to get $0 + 0 \le 4$ and $0 + 0 \le 6$, which are true. The region is all points on or below $x + 2y = 4$, and on or above $2x - 3y = 6$ in quadrant I.

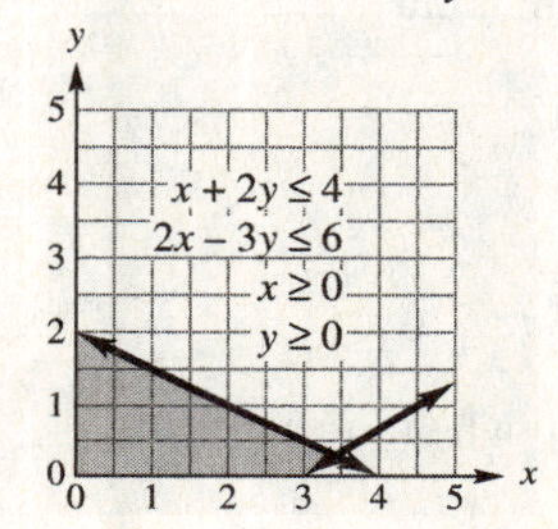

9. Let x = the time spent on a summary report and let y = the time spent on an inference report. Then we have the following inequalities:

$2x + 4y \le 15$ Analysis time
$1.5x + 2y \le 9$ Writing time
$x \ge 0$
$y \ge 0.$

The solution of this system of inequalities is the graph of the feasible region.

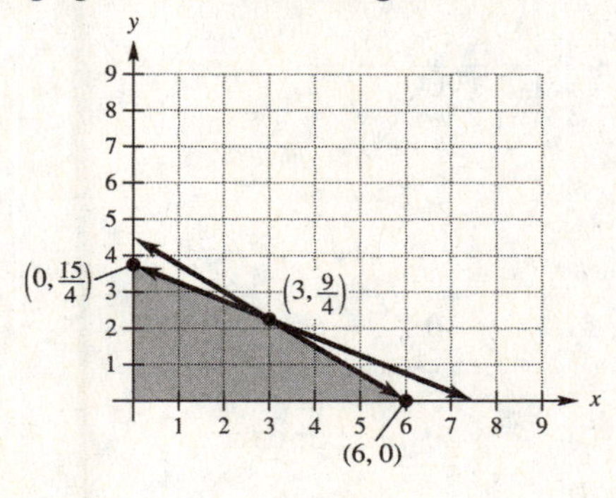

10. Let x = the number of margherita pizzas sold and let y = the number of basic pizzas sold

Then we have the following inequalities:

$5x + 4y \le 1000$
$2x + y \le 320$
$x \ge 60$
$y \ge 40$

The solution of this system is the graph of the feasible region.

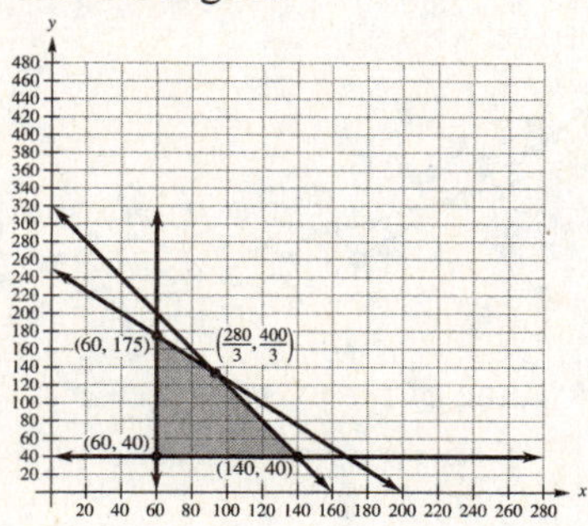

11.

Corner Point	Value of $z = 3x + 4y$
(1, 6)	27
(6 ,7)	46 Maximum
(7, 3)	33
$(1, 2\frac{1}{2})$	13
(2, 1)	10 Minimum

The maximum value of 46 occurs at (6, 7) the minimum value of 10 occurs at (2, 1).

12.

Corner Point	Value of $z = 3x + 4y$
(0, 8)	32
(8, 8)	56 Maximum
(5, 2)	23
(2, 0)	6 Minimum

The maximum value is 56 at (8, 8); the minimum value is 6 at (2, 0).

13. Maximize $z = 6x + 2y$
subject to $2x + 7y \leq 14$
$2x + 3y \leq 10$
$x \geq 0,\ y \geq 0.$

Graph the feasible region.

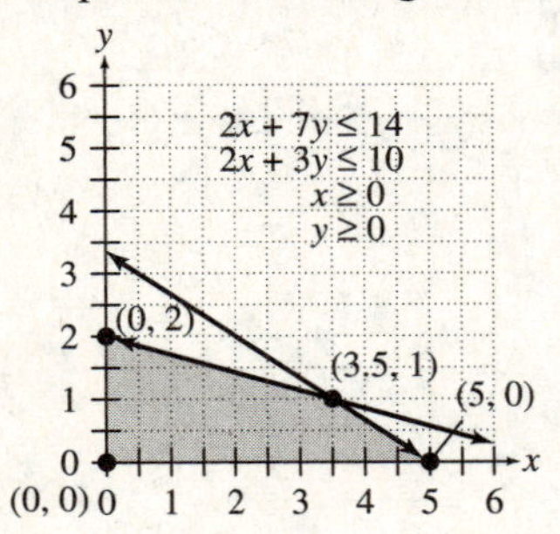

Corner Point	Value of $z = 6x + 2y$
(0, 0)	0
(0, 2)	4
(3.5, 1)	23
(5, 0)	30 Maximum

The maximum is 30 when $x = 5, y = 0$.

14. Find $x \geq 0$ and $y \geq 0$ such that
$8x + 9y \geq 72$
$6x + 8y \geq 72$
and $w = 2x + 10y$ is minimized.
Graph the feasible region.

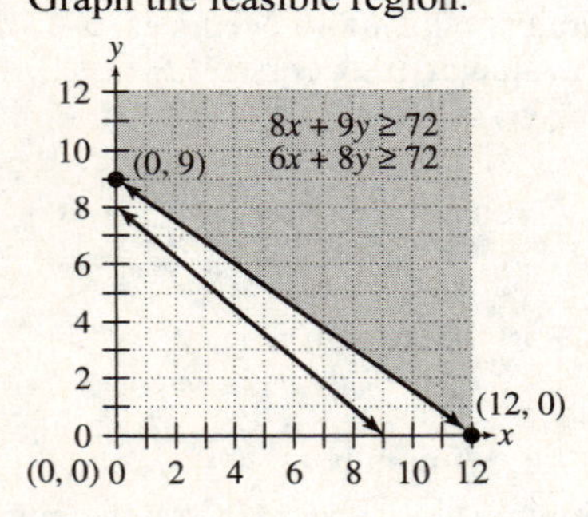

Corner Point	Value of $z = 2x + 10y$
(0, 9)	90
(12, 0)	24 Minimum

The minimum is 24 when $x = 12, y = 0$.

15. Find $x \geq 0$ and $y \geq 0$ such that
$x + y \leq 50$
$2x + y \geq 20$
$x + 2y \geq 30$
and $w = 5x + 2y$ is minimized.
Graph the feasible region.

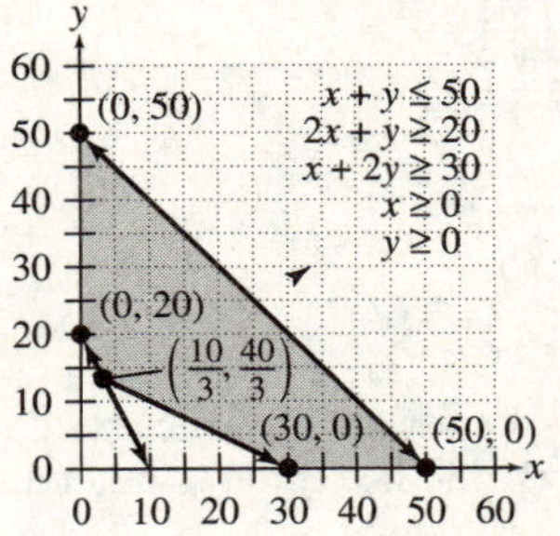

Corner Point	Value of $w = 5x + 2y$
(0, 20)	40 Minimum
(0, 50)	100
(50, 0)	250
(30, 0)	150
$\left(\frac{10}{3}, \frac{40}{3}\right)$	$\frac{130}{3} = 43\frac{1}{3}$

The minimum is 40 when $x = 0, y = 20$.

16. Maximize $z = 5x - 2y$
subject to: $3x + 2y \leq 12$
$5x + y \geq 5$
$x \geq 0,\ y \geq 0.$

Graph the feasible region.

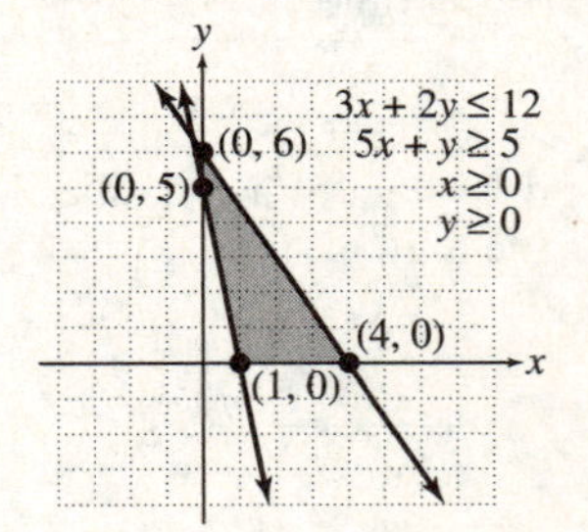

Corner Point	Value of $z = 5x - 2y$
(0, 5)	–10
(0, 6)	–12
(4, 0)	20 Maximum
(1, 0)	5

The maximum is 20 when $x = 4, y = 0$.

17. From the graph for Exercise 9, the corner points are (0, 3.75), (3, 2.25), (6, 0), and (0, 0). Since x was the number of summary reports and y the number of inference reports, the revenue function is $z = 500x + 750y$. Evaluate this objective function at each corner point.

Corner Point	Value of $z = 500x + 750y$
(0, 3.75)	2812.5
(3, 2.25)	3187.5 Maximum
(6,0)	3000
(0, 0)	0

Linda should complete 3 summary reports and 2.25 inference reports to produce a maximum profit of $3187.50.

18. The profit function is $z = 20x + 15y$. Refer to the graph in Exercise 10, and evaluate at the corner points.

Corner Point	Value of $z = 20x + 15y$
(60, 40)	1800
(60, 175)	3825
(93, 133)	3855
(140, 40)	3400

A maximum profit of $3855 is obtained by making 133 basic pizzas and 93 margherita pizzas.

19.

	Number of Shares	Cost per Share	Profit	Risk
Blackrock	x	22	3	13.2
Columbia	y	17	2	12.8
Constraints		10,000	1200	

Minimize $z = 13.2x + 12.8y$
subject to: $22x + 17y \le 10{,}000$
$3x + 2y \ge 1200$

Graph the feasibility region

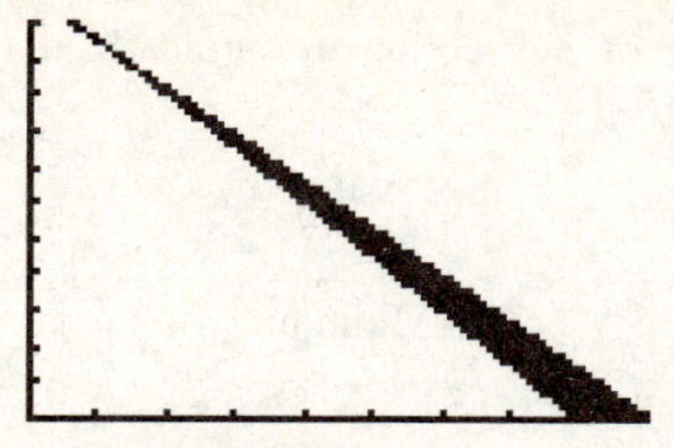

[0,500] by [0,550]

The points of intersection are summarized in the table below.

Corner Point	Value of $z = 13.2x + 12.8y$
(57, 514)	7331.6
(400,0)	5280 Minimum
(454, 0)	5992.8

Donald should buy 400 shares of Blackrock Equity Dividend Fund and none of the Columbia Dividend Income Fund. The risk has a value of 5280 for Donald.

20.

	Number of Shares	Cost per Share	Profit	Risk
Alger	x	15	2	17
Dynamic	y	22	3.50	20
Constraints		5016	700	

Minimize $z = 17x + 20y$
subject to: $15x + 22y \le 5016$
$2x + 3.5y \ge 700$

Graph the feasibility region

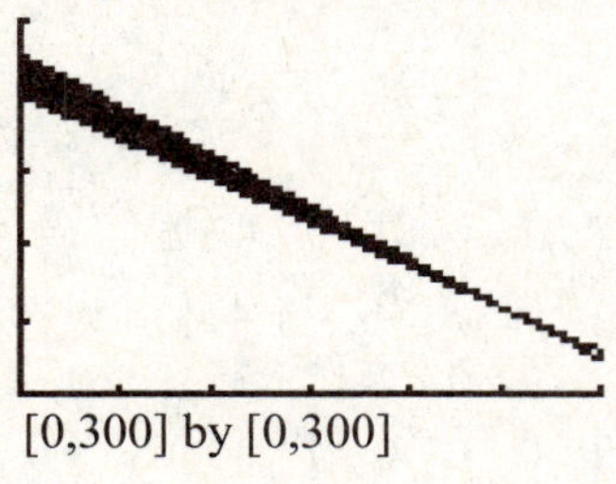

[0,300] by [0,300]

(continued next page)

The points of intersection are summarized in the table below.

Corner Point	Value of $z = 17x + 20y$
(0, 200)	4000 Minimum
(254, 55)	5418
(0, 228)	4560

Kimberly should buy 200 shares of Dynamic US Growth Fund and none of the Alger Spectra Z Fund. The risk has a value of 4000 for Kimberly.

21. Maximize $z = 5x_1 + 6x_2 + 3x_3$

subject to:
$$\begin{aligned} x_1 + x_2 + x_3 &\le 100 \\ 2x_1 + 3x_2 &\le 500 \\ x_1 + 2x_3 &\le 350 \\ x_1 \ge 0,\ x_2 \ge 0,\ x_3 &\ge 0. \end{aligned}$$

a.
$$\begin{aligned} x_1 + x_2 + x_3 + s_1 &= 100 \\ 2x_1 + 3x_2 + s_2 &= 500 \\ x_1 + 2x_3 + s_3 &= 350 \end{aligned}$$

b.
$$\begin{array}{c} \begin{matrix} x_1 & x_2 & x_3 & s_1 & s_2 & s_3 \end{matrix} \\ \left[\begin{array}{rrrrrr|r} 1 & 1 & 1 & 1 & 0 & 0 & 100 \\ 2 & 3 & 0 & 0 & 1 & 0 & 500 \\ 1 & 0 & 2 & 0 & 0 & 1 & 350 \\ \hline -5 & -6 & -3 & 0 & 0 & 0 & 0 \end{array}\right] \end{array}$$

22. Maximize $z = 2x_1 + 9x_2$

subject to:
$$\begin{aligned} 3x_1 + 5x_2 &\le 47 \\ x_1 + x_2 &\le 25 \\ 5x_1 + 2x_2 &\le 35 \\ 2x_1 + x_2 &\le 30 \\ x_1 \ge 0,\ x_2 &\ge 0. \end{aligned}$$

a.
$$\begin{aligned} 3x_1 + 5x_2 + s_1 &= 47 \\ x_1 + x_2 + s_2 &= 25 \\ 5x_1 + 2x_2 + s_3 &= 35 \\ 2x_1 + x_2 + s_4 &= 30 \end{aligned}$$

b.
$$\begin{array}{c} \begin{matrix} x_1 & x_2 & s_1 & s_2 & s_3 & s_4 \end{matrix} \\ \left[\begin{array}{rrrrrr|r} 3 & 5 & 1 & 0 & 0 & 0 & 47 \\ 1 & 1 & 0 & 1 & 0 & 0 & 25 \\ 5 & 2 & 0 & 0 & 1 & 0 & 35 \\ 2 & 1 & 0 & 0 & 0 & 1 & 30 \\ \hline -2 & -9 & 0 & 0 & 0 & 0 & 0 \end{array}\right] \end{array}$$

23. Maximize $z = x_1 + 8x_2 + 2x_3$

subject to:
$$\begin{aligned} x_1 + x_2 + x_3 &\le 90 \\ 2x_1 + 5x_2 + x_3 &\le 120 \\ x_1 + 3x_2 &\le 80 \\ x_1 \ge 0,\ x_2 \ge 0,\ x_3 &\ge 0. \end{aligned}$$

a.
$$\begin{aligned} x_1 + x_2 + x_3 + s_1 &= 90 \\ 2x_1 + 5x_2 + x_3 + s_2 &= 120 \\ x_1 + 3x_2 + s_3 &= 80 \end{aligned}$$

b.
$$\begin{array}{c} \begin{matrix} x_1 & x_2 & x_3 & s_1 & s_2 & s_3 \end{matrix} \\ \left[\begin{array}{rrrrrr|r} 1 & 1 & 1 & 1 & 0 & 0 & 90 \\ 2 & 5 & 1 & 0 & 1 & 0 & 120 \\ 1 & 3 & 0 & 0 & 0 & 1 & 80 \\ \hline -1 & -8 & -2 & 0 & 0 & 0 & 0 \end{array}\right] \end{array}$$

24. Maximize $z = 15x_1 + 12x_2$

subject to:
$$\begin{aligned} 2x_1 + 5x_2 &\le 50 \\ x_1 + 3x_2 &\le 25 \\ 4x_1 + x_2 &\le 18 \\ x_1 + x_2 &\le 12 \\ x_1 \ge 0,\ x_2 &\ge 0. \end{aligned}$$

a.
$$\begin{aligned} 2x_1 + 5x_2 + s_1 &= 50 \\ x_1 + 3x_2 + s_2 &= 25 \\ 4x_1 + x_2 + s_3 &= 18 \\ x_1 + x_2 + s_4 &= 12 \end{aligned}$$

b.
$$\begin{array}{c} \begin{matrix} x_1 & x_2 & s_1 & s_2 & s_3 & s_4 \end{matrix} \\ \left[\begin{array}{rrrrrr|r} 2 & 5 & 1 & 0 & 0 & 0 & 50 \\ 1 & 3 & 0 & 1 & 0 & 0 & 25 \\ 4 & 1 & 0 & 0 & 1 & 0 & 18 \\ 1 & 1 & 0 & 0 & 0 & 1 & 12 \\ \hline -15 & -12 & 0 & 0 & 0 & 0 & 0 \end{array}\right] \end{array}$$

25.

$$\begin{array}{ccccc} x_1 & x_2 & x_3 & s_1 & s_2 \end{array}$$
$$\left[\begin{array}{ccccc|c} 1 & 2 & 3 & 1 & 0 & 28 \\ \boxed{2} & 4 & 8 & 0 & 1 & 32 \\ \hline -5 & -2 & -3 & 0 & 0 & 0 \end{array}\right]$$

Pivot on the 2 in row two, column one.

$$\begin{array}{ccccc} x_1 & x_2 & x_3 & s_1 & s_2 \end{array}$$
$$\left[\begin{array}{ccccc|c} 1 & 2 & 3 & 1 & 0 & 28 \\ 1 & 2 & 4 & 0 & \frac{1}{2} & 16 \\ \hline -5 & -2 & -3 & 0 & 0 & 0 \end{array}\right]\begin{array}{l} \\ \frac{1}{2}R_2 \\ \\ \end{array}$$

$$\begin{array}{ccccc} x_1 & x_2 & x_3 & s_1 & s_2 \end{array}$$
$$\left[\begin{array}{ccccc|c} 0 & 0 & -1 & 1 & -\frac{1}{2} & 12 \\ 1 & 2 & 4 & 0 & \frac{1}{2} & 16 \\ \hline 0 & 8 & 17 & 0 & \frac{5}{2} & 80 \end{array}\right]\begin{array}{l} -R_2 + R_1 \\ \\ 5R_2 + R_3 \end{array}$$

The maximum is 80 when
$x_1 = 16,\ x_2 = 0,\ x_3 = 0,\ s_1 = 12,\ s_2 = 0.$

26.

$$\begin{array}{cccc} x_1 & x_2 & s_1 & s_2 \end{array}$$
$$\left[\begin{array}{cccc|c} 2 & 1 & 1 & 0 & 10 \\ 9 & \boxed{3} & 0 & 1 & 15 \\ \hline -2 & -3 & 0 & 0 & 0 \end{array}\right]$$

Pivot on the 3 in row two, column two.

$$\begin{array}{cccc} x_1 & x_2 & s_1 & s_2 \end{array}$$
$$\left[\begin{array}{cccc|c} 2 & 1 & 1 & 0 & 10 \\ 3 & 1 & 0 & \frac{1}{3} & 5 \\ \hline -2 & -3 & 0 & 0 & 0 \end{array}\right]\begin{array}{l} \\ \frac{1}{3}R_2 \\ \\ \end{array}$$

$$\begin{array}{cccc} x_1 & x_2 & s_1 & s_2 \end{array}$$
$$\left[\begin{array}{cccc|c} -1 & 0 & 1 & -\frac{1}{3} & 5 \\ 3 & 1 & 0 & \frac{1}{3} & 5 \\ \hline 7 & 0 & 0 & 1 & 15 \end{array}\right]\begin{array}{l} -R_2 + R_1 \\ \\ 3R_2 + R_3 \end{array}$$

The maximum is 15 when $x_1 = 0,\ x_2 = 5,\ s_1 = 5,$ and $s_2 = 0.$

27.

$$\begin{array}{cccccc} x_1 & x_2 & x_3 & s_1 & s_2 & s_3 \end{array}$$
$$\left[\begin{array}{cccccc|c} 1 & 2 & 2 & 1 & 0 & 0 & 50 \\ \boxed{4} & 24 & 0 & 0 & 1 & 0 & 20 \\ 1 & 0 & 2 & 0 & 0 & 1 & 15 \\ \hline -5 & -3 & -2 & 0 & 0 & 0 & 0 \end{array}\right]$$

Pivot on the 4 in row two, column one.

$$\begin{array}{cccccc} x_1 & x_2 & x_3 & s_1 & s_2 & s_3 \end{array}$$
$$\left[\begin{array}{cccccc|c} 1 & 2 & 2 & 1 & 0 & 0 & 50 \\ 1 & 6 & 0 & 0 & \frac{1}{4} & 0 & 5 \\ 1 & 0 & 2 & 0 & 0 & 1 & 15 \\ \hline -5 & -3 & -2 & 0 & 0 & 0 & 0 \end{array}\right]\begin{array}{l} \\ \frac{1}{4}R_2 \\ \\ \\ \end{array}$$

$$\begin{array}{cccccc} x_1 & x_2 & x_3 & s_1 & s_2 & s_3 \end{array}$$
$$\left[\begin{array}{cccccc|c} 0 & -4 & 2 & 1 & -\frac{1}{4} & 0 & 45 \\ 1 & 6 & 0 & 0 & \frac{1}{4} & 0 & 5 \\ 0 & -6 & \boxed{2} & 0 & -\frac{1}{4} & 1 & 10 \\ \hline 0 & 27 & -2 & 0 & \frac{5}{4} & 0 & 25 \end{array}\right]\begin{array}{l} -R_2 + R_1 \\ \\ -R_2 + R_3 \\ 5R_2 + R_4 \end{array}$$

Pivot on the 2 in row three, column three.

$$\begin{array}{cccccc} x_1 & x_2 & x_3 & s_1 & s_2 & s_3 \end{array}$$
$$\left[\begin{array}{cccccc|c} 0 & -4 & 2 & 1 & -\frac{1}{4} & 0 & 45 \\ 1 & 6 & 0 & 0 & \frac{1}{4} & 0 & 5 \\ 0 & -3 & 1 & 0 & -\frac{1}{8} & \frac{1}{2} & 5 \\ \hline 0 & 27 & -2 & 0 & \frac{5}{4} & 0 & 25 \end{array}\right]\begin{array}{l} \\ \\ \frac{1}{2}R_3 \\ \\ \end{array}$$

$$\begin{array}{cccccc} x_1 & x_2 & x_3 & s_1 & s_2 & s_3 \end{array}$$
$$\left[\begin{array}{cccccc|c} 0 & 2 & 0 & 1 & 0 & -1 & 35 \\ 1 & 6 & 0 & 0 & \frac{1}{4} & 0 & 5 \\ 0 & -3 & 1 & 0 & -\frac{1}{8} & \frac{1}{2} & 5 \\ \hline 0 & 21 & 0 & 0 & 1 & 1 & 35 \end{array}\right]\begin{array}{l} -2R_3 + R_1 \\ \\ \\ 2R_3 + R_4 \end{array}$$

The maximum is 35 when
$x_1 = 5,\ x_2 = 0,\ x_3 = 5,\ s_1 = 35,\ s_2 = 0,$ and $s_3 = 0.$

28.

$$\begin{array}{c} \\ \\ \\ \\ \\ \end{array}\begin{array}{cccccc} x_1 & x_2 & s_1 & s_2 & s_3 & \\ \end{array}$$

$$\left[\begin{array}{rrrrr|r} 1 & -2 & 1 & 0 & 0 & 38 \\ 1 & -1 & 0 & 1 & 0 & 12 \\ 2 & \underline{1} & 0 & 0 & 1 & 30 \\ \hline -1 & -2 & 0 & 0 & 0 & 0 \end{array}\right]$$

Pivot on the 1 in row three, column two.

$$\begin{array}{c} x_1\ x_2\ s_1\ s_2\ s_3 \end{array}$$

$$\left[\begin{array}{rrrrr|r} 5 & 0 & 1 & 0 & 2 & 98 \\ 3 & 0 & 0 & 1 & 1 & 42 \\ 2 & 1 & 0 & 0 & 1 & 30 \\ \hline 3 & 0 & 0 & 0 & 2 & 60 \end{array}\right]\begin{array}{l} 2R_3 + R_1 \\ R_3 + R_2 \\ \\ 2R_3 + R_4 \end{array}$$

The maximum is 60 when $x_1 = 0, x_2 = 30,\ s_1 = 98,\ s_2 = 42,$ and $s_3 = 0.$

29. From exercise 21, we have

$$\begin{array}{c} x_1\ x_2\ x_3\ s_1\ s_2\ s_3 \end{array}$$

$$\left[\begin{array}{rrrrrr|r} 1 & 1 & 1 & 1 & 0 & 0 & 100 \\ 2 & 3 & 0 & 0 & 1 & 0 & 500 \\ 1 & 0 & 2 & 0 & 0 & 1 & 350 \\ \hline -5 & -6 & -3 & 0 & 0 & 0 & 0 \end{array}\right].$$

Pivot on the 1 in row 1 column 2.

$$\begin{array}{c} x_1\ x_2\ x_3\ s_1\ s_2\ s_3 \end{array}$$

$$\left[\begin{array}{rrrrrr|r} 1 & \boxed{1} & 1 & 1 & 0 & 0 & 100 \\ -1 & 0 & -3 & -3 & 1 & 0 & 200 \\ 1 & 0 & 2 & 0 & 0 & 1 & 350 \\ \hline 1 & 0 & 3 & 6 & 0 & 0 & 600 \end{array}\right]\begin{array}{l} \\ -3R_1 + R_2 \\ \\ 6R_1 + R_4 \end{array}$$

The maximum is 600 when $x_1 = 0,\ x_2 = 100,$ and $x_3 = 0.$

30. From exercise 20, we have

$$\begin{array}{c} x_1\ x_2\ s_1\ s_2\ s_3\ s_4 \end{array}$$

$$\left[\begin{array}{rrrrrr|r} 3 & 5 & 1 & 0 & 0 & 0 & 47 \\ 1 & 1 & 0 & 1 & 0 & 0 & 25 \\ 5 & 2 & 0 & 0 & 1 & 0 & 35 \\ 2 & 1 & 0 & 0 & 0 & 1 & 30 \\ \hline -2 & -9 & 0 & 0 & 0 & 0 & 0 \end{array}\right].$$

Pivot on the 5 in row 1 column 2. First divide row 1 by 5, then continue the process.

$$\begin{array}{c} x_1\ x_2\ s_1\ s_2\ s_3\ s_4 \end{array}$$

$$\left[\begin{array}{rrrrrr|r} \frac{3}{5} & 1 & \frac{1}{5} & 0 & 0 & 0 & \frac{47}{5} \\ 1 & 1 & 0 & 1 & 0 & 0 & 25 \\ 5 & 2 & 0 & 0 & 1 & 0 & 35 \\ 2 & 1 & 0 & 0 & 0 & 1 & 30 \\ \hline -2 & -9 & 0 & 0 & 0 & 0 & 0 \end{array}\right]\begin{array}{l} \frac{1}{5}R_1 \\ \\ \\ \\ \\ \end{array}$$

$$\begin{array}{c} x_1\ x_2\ s_1\ s_2\ s_3\ s_4 \end{array}$$

$$\left[\begin{array}{rrrrrr|r} \frac{3}{5} & 1 & \frac{1}{5} & 0 & 0 & 0 & \frac{47}{5} \\ \frac{2}{5} & 0 & -\frac{1}{5} & 1 & 0 & 0 & \frac{78}{5} \\ \frac{19}{5} & 0 & -\frac{2}{5} & 0 & 1 & 0 & \frac{81}{5} \\ \frac{7}{5} & 0 & -\frac{1}{5} & 0 & 0 & 1 & \frac{103}{5} \\ \hline \frac{17}{5} & 0 & \frac{9}{5} & 0 & 0 & 0 & \frac{423}{5} \end{array}\right]\begin{array}{l} \\ -R_1 + R_2 \\ -2R_1 + R_3 \\ -R_1 + R_4 \\ 9R_1 + R_5 \end{array}$$

The maximum is $\frac{423}{5}$ when $x_1 = 0$ and $x_2 = \frac{47}{5}.$

31. From exercise 21, we have

$$\begin{array}{c} x_1\ x_2\ x_3\ s_1\ s_2\ s_3 \end{array}$$

$$\left[\begin{array}{rrrrrr|r} 1 & 1 & 1 & 1 & 0 & 0 & 90 \\ 2 & \boxed{5} & 1 & 0 & 1 & 0 & 120 \\ 1 & 3 & 0 & 0 & 0 & 1 & 80 \\ \hline -1 & -8 & -2 & 0 & 0 & 0 & 0 \end{array}\right].$$

Pivot on the 5 in row 2 column 2. First divide row 2 by 5, then continue the process.

$$\begin{array}{c} x_1\ x_2\ x_3\ s_1\ s_2\ s_3 \end{array}$$

$$\left[\begin{array}{rrrrrr|r} 1 & 1 & 1 & 1 & 0 & 0 & 90 \\ \frac{2}{5} & 1 & \frac{1}{5} & 0 & \frac{1}{5} & 0 & 24 \\ 1 & 3 & 0 & 0 & 0 & 1 & 80 \\ \hline -1 & -8 & -2 & 0 & 0 & 0 & 0 \end{array}\right]\begin{array}{l} \\ \frac{1}{5}R_1 \\ \\ \\ \end{array}$$

$$\begin{array}{c} x_1\ x_2\ x_3\ s_1\ s_2\ s_3 \end{array}$$

$$\left[\begin{array}{rrrrrr|r} \frac{3}{5} & 0 & \boxed{\frac{4}{5}} & 1 & -\frac{1}{5} & 0 & 66 \\ \frac{2}{5} & 1 & \frac{1}{5} & 0 & \frac{1}{5} & 0 & 24 \\ -\frac{1}{5} & 0 & -\frac{3}{5} & 0 & -\frac{3}{5} & 1 & 8 \\ \hline \frac{11}{5} & 0 & -\frac{2}{5} & 0 & \frac{8}{5} & 0 & 192 \end{array}\right]\begin{array}{l} -R_2 + R_1 \\ \\ -3R_2 + R_3 \\ \\ \end{array}$$

Now pivot on the $\frac{4}{5}$ in row 1 column 3. First multiply row 1 by $\frac{5}{4}$, then continue the process.

$$\begin{array}{c} x_1\ x_2\ x_3\ s_1\ s_2\ s_3 \end{array}$$

$$\left[\begin{array}{rrrrrr|r} \frac{3}{4} & 0 & 1 & \frac{5}{4} & -\frac{1}{4} & 0 & \frac{165}{2} \\ \frac{2}{5} & 1 & \frac{1}{5} & 0 & \frac{1}{5} & 0 & 24 \\ -\frac{1}{5} & 0 & -\frac{3}{5} & 0 & -\frac{3}{5} & 1 & 8 \\ \hline \frac{11}{5} & 0 & -\frac{2}{5} & 0 & \frac{8}{5} & 0 & 192 \end{array}\right]\begin{array}{l} \frac{5}{4}R_1 \\ \\ \\ \\ \end{array}$$

(*continued on next pgae*)

$$\begin{array}{c} \\ \\ \\ \\ \end{array}\begin{array}{cccccc} x_1 & x_2 & x_3 & s_1 & s_2 & s_3 \end{array}$$

$$\left[\begin{array}{cccccc|c} \frac{3}{4} & 0 & 1 & \frac{5}{4} & -\frac{1}{4} & 0 & \frac{165}{2} \\ \frac{1}{4} & 1 & 0 & -\frac{1}{4} & \frac{1}{4} & 0 & \frac{15}{2} \\ \frac{1}{4} & 0 & 0 & \frac{3}{4} & -\frac{3}{4} & 1 & \frac{115}{2} \\ \hline \frac{5}{2} & 0 & 0 & \frac{1}{2} & \frac{3}{2} & 0 & 225 \end{array}\right]\begin{array}{l} \\ -\frac{1}{5}R_1 + R_2 \\ \frac{3}{5}R_1 + R_3 \\ \frac{2}{5}R_1 + R_4 \end{array}$$

The maximum is 225 when $x_1 = 0$, $x_2 = \frac{15}{2}$, and $x_3 = \frac{165}{2}$.

32. From exercise 22, we have

$$\begin{array}{cccccc} x_1 & x_2 & s_1 & s_2 & s_3 & s_4 \end{array}$$

$$\left[\begin{array}{cccccc|c} 2 & 5 & 1 & 0 & 0 & 0 & 50 \\ 1 & 3 & 0 & 1 & 0 & 0 & 25 \\ \boxed{4} & 1 & 0 & 0 & 1 & 0 & 18 \\ 1 & 1 & 0 & 0 & 0 & 1 & 12 \\ \hline -15 & -12 & 0 & 0 & 0 & 0 & 0 \end{array}\right].$$

Pivot on the 4 in row 3 column 1. First divide row 3 by 4, then continue the process.

$$\begin{array}{cccccc} x_1 & x_2 & s_1 & s_2 & s_3 & s_4 \end{array}$$

$$\left[\begin{array}{cccccc|c} 2 & 5 & 1 & 0 & 0 & 0 & 50 \\ 1 & 3 & 0 & 1 & 0 & 0 & 25 \\ 1 & \frac{1}{4} & 0 & 0 & \frac{1}{4} & 0 & \frac{9}{2} \\ 1 & 1 & 0 & 0 & 0 & 1 & 12 \\ \hline -15 & -12 & 0 & 0 & 0 & 0 & 0 \end{array}\right]\begin{array}{l} \\ \\ \frac{1}{4}R_3 \\ \\ \\ \end{array}$$

$$\begin{array}{cccccc} x_1 & x_2 & s_1 & s_2 & s_3 & s_4 \end{array}$$

$$\left[\begin{array}{cccccc|c} 0 & \frac{9}{2} & 1 & 0 & -\frac{1}{2} & 0 & 41 \\ 0 & \boxed{\frac{11}{4}} & 0 & 1 & -\frac{1}{4} & 0 & \frac{41}{2} \\ 1 & \frac{1}{4} & 0 & 0 & \frac{1}{4} & 0 & \frac{9}{2} \\ 0 & \frac{3}{4} & 0 & 0 & -\frac{1}{4} & 1 & \frac{15}{2} \\ \hline 0 & -\frac{33}{4} & 0 & 0 & \frac{15}{4} & 0 & \frac{135}{2} \end{array}\right]\begin{array}{l} -2R_3 + R_1 \\ -R_3 + R_2 \\ \\ -R_3 + R_4 \\ 15R_3 + R_5 \end{array}$$

Now pivot on the $\frac{11}{4}$ in row 2 column 2. First multiply row 2 by $\frac{4}{11}$, then continue the process.

$$\begin{array}{cccccc} x_1 & x_2 & s_1 & s_2 & s_3 & s_4 \end{array}$$

$$\left[\begin{array}{cccccc|c} 0 & \frac{9}{2} & 1 & 0 & -\frac{1}{2} & 0 & 41 \\ 0 & 1 & 0 & \frac{4}{11} & -\frac{1}{11} & 0 & \frac{82}{11} \\ 1 & \frac{1}{4} & 0 & 0 & \frac{1}{4} & 0 & \frac{9}{2} \\ 0 & \frac{3}{4} & 0 & 0 & -\frac{1}{4} & 1 & \frac{15}{2} \\ \hline 0 & -\frac{33}{4} & 0 & 0 & \frac{15}{4} & 0 & \frac{135}{2} \end{array}\right]\begin{array}{l} \\ \frac{11}{4}R_2 \\ \\ \\ \\ \end{array}$$

$$\begin{array}{cccccc} x_1 & x_2 & s_1 & s_2 & s_3 & s_4 \end{array}$$

$$\left[\begin{array}{cccccc|c} 0 & 0 & 1 & -\frac{18}{11} & -\frac{1}{11} & 0 & \frac{82}{11} \\ 0 & 1 & 0 & \frac{4}{11} & -\frac{1}{11} & 0 & \frac{82}{11} \\ 1 & 0 & 0 & -\frac{1}{11} & \frac{3}{11} & 0 & \frac{29}{11} \\ 0 & 0 & 0 & -\frac{3}{11} & -\frac{2}{11} & 1 & \frac{21}{11} \\ \hline 0 & 0 & 0 & 3 & 3 & 0 & 129 \end{array}\right]\begin{array}{l} -\frac{9}{2}R_2 + R_1 \\ \\ -\frac{1}{4}R_2 + R_3 \\ -\frac{3}{4}R_2 + R_4 \\ \frac{33}{4}R_2 + R_5 \end{array}$$

The maximum is 129 when $x_1 = \frac{29}{11}$ and $x_2 = \frac{82}{11}$.

33. a. Let x_1 = the number of item *A*, let x_2 = the number of item *B*, and x_3 = the number of item *C*.

b. The objective function is $z = 4x_1 + 3x_2 + 3x_3$.

c. The constraints are

$$\begin{aligned} 2x_1 + 3x_2 + 6x_3 &\le 1200 \\ x_1 + 2x_2 + 2x_3 &\le 800 \\ 2x_1 + 2x_2 + 4x_3 &\le 500 \\ x_1 \ge 0,\ x_2 \ge 0,\ x_3 &\ge 0. \end{aligned}$$

34. a. Let x_1 = the amount invested in oil leases, x_2 = the amount in bonds, and x_3 = the amount invested in stock.

b. We want to maximize the objective function $z = .15x_1 + .09x_2 + .05x_3$.

c.

$$\begin{aligned} x_1 + x_2 + x_3 &\le 50{,}000 \\ x_1 + x_2 \qquad &\le 15{,}000 \\ x_1 \qquad + x_3 &\le 25{,}000 \\ x_1 \ge 0,\ x_2 \ge 0,\ x_3 &\ge 0 \end{aligned}$$

35. a. Let x_1 = the number of gallons of Fruity wine, and x_2 = the number of gallons of Crystal wine to be made.

b. The objective function is $z = 12x_1 + 15x_2$.

c. The ingredients available are the limitations. The constraints are

$$\begin{aligned} 2x_1 + x_2 &\le 110 \\ 2x_1 + 3x_2 &\le 125 \\ 2x_1 + x_2 &\le 90 \\ x_1 \ge 0,\ x_2 &\ge 0. \end{aligned}$$

36. a. Let x_1 = the number of 5-gallon bags, x_2 = the number of 10-gallon bags, and x_3 = the number of 20-gallon bags.

b. The objective function is $z = x_1 + .9x_2 + .95x_3$.

c.
$$\begin{aligned} x_1 + 1.1x_2 + 1.5x_3 &\le 8 \\ x_1 + 1.2x_2 + 1.3x_3 &\le 8 \\ 2x_1 + 3x_2 + 4x_3 &\le 8 \\ x_1 \ge 0,\ x_2 \ge 0,\ x_3 &\ge 0 \end{aligned}$$

37. It is necessary to use the simplex method when there are more than two variables.

38. Problem with constraints involving "≤" can be solved with the use of slack variables, constraints involving "≥" can be solved with the use of surplus variables, while constraints involving "=" require both slack and surplus variables.

39. Any standard minimization problem can be solved using the method of duals.

40. a. Maximize $z = 6x_1 + 7x_2 + 5x_3$
subject to:
$$\begin{aligned} 4x_1 + 2x_2 + 3x_3 &\le 9 \\ 5x_1 + 4x_2 + x_3 &\le 10 \\ x_1 \ge 0,\ x_2 \ge 0,\ x_3 &\ge 0 \end{aligned}$$

b. The first constraint would become $4x_1 + 2x_2 + 3x_3 \ge 9$

c. $z = 22.7$ when $x_1 = 0,\ x_2 = 2.1,\ x_3 = 1.6$

d. Minimize $w = 9y_1 + 10y_2$
subject to:
$$\begin{aligned} 4y_1 + 5y_2 &\ge 6 \\ 2y_1 + 4y_2 &\ge 7 \\ 3y_1 + y_2 &\ge 5 \\ y_1 \ge 0,\ y_2 &\ge 0 \end{aligned}$$

e. The minimum value is $w = 22.7$, when $y_1 = 1.3$, and $y_2 = 1.1$.

41. Using the method of duals,
$$\left[\begin{array}{cccccc|c} 1 & 0 & 0 & 3 & 1 & 2 & 12 \\ 0 & 0 & 1 & 4 & 5 & 3 & 5 \\ 0 & 1 & 0 & -2 & 7 & -6 & 8 \\ \hline 0 & 0 & 0 & 5 & 7 & 3 & 172 \end{array}\right]$$
indicates a minimum value of 172 at (5, 7, 3, 0, 0, 0).

42. Using the method of duals,
$$\left[\begin{array}{cccccc|c} 0 & 0 & 1 & 6 & 3 & 1 & 2 \\ 1 & 0 & 0 & 4 & -2 & 2 & 8 \\ 0 & 1 & 0 & 10 & 7 & 0 & 12 \\ \hline 0 & 0 & 0 & 9 & 5 & 8 & 62 \end{array}\right]$$
indicates a minimum of 62 at (9, 5, 8, 0, 0, 0).

43. Using the method of duals,
$$\left[\begin{array}{cccc|c} 1 & 0 & 7 & -1 & 100 \\ 0 & 1 & 1 & 3 & 27 \\ \hline 0 & 0 & 7 & 2 & 640 \end{array}\right]$$
indicates a minimum of 640 at (7, 2, 0, 0).

44. Minimize $w = 5y_1 + 2y_2$
subject to
$$\begin{aligned} 2y_1 + 3y_2 &\ge 6 \\ 2y_1 + y_2 &\ge 7 \\ y_1 \ge 0,\ y_2 &\ge 0 \end{aligned}$$
Rewrite the objective function by forming the augmented matrix of the system and then form its transpose.
$$\left[\begin{array}{cc|c} 2 & 3 & 6 \\ 2 & 1 & 7 \\ \hline 5 & 2 & 0 \end{array}\right] \rightarrow \left[\begin{array}{cc|c} 2 & 2 & 5 \\ 3 & 1 & 2 \\ \hline 6 & 7 & 0 \end{array}\right].$$
This gives the problem
Maximize $z = 6x_1 + 7x_2$
subject to
$$\begin{aligned} 2x_1 + 2x_2 &\le 5 \\ 3x_1 + x_2 &\le 2 \\ x_1 \ge 0,\ x_2 &\ge 0 \end{aligned}$$
The initial simplex tableau is
$$\begin{array}{c} \begin{array}{cccc} x_1 & x_2 & s_1 & s_2 \end{array} \\ \left[\begin{array}{cccc|c} 2 & 2 & 1 & 0 & 5 \\ 3 & \boxed{1} & 0 & 1 & 2 \\ \hline -6 & -7 & 0 & 0 & 0 \end{array}\right] \end{array}.$$
Pivot on the 1 in row 2 column 2.
$$\begin{array}{c} \begin{array}{cccc} x_1 & x_2 & s_1 & s_2 \end{array} \\ \left[\begin{array}{cccc|c} -4 & 0 & 1 & -2 & 1 \\ 3 & 1 & 0 & 1 & 2 \\ \hline 15 & 0 & 0 & 7 & 14 \end{array}\right] \begin{array}{l} -2R_2 + R_1 \\ \\ 7R_2 + R_3 \end{array} \\ \begin{array}{ccc} \uparrow & \uparrow & \uparrow \\ y_1 & y_2 & w \end{array} \end{array}$$
The minimum is 14 when $y_1 = 0$ and $y_2 = 7$.

45. Minimize $w = 18y_1 + 10y_2$
subject to: $y_1 + y_2 \ge 17$
$5y_1 + 8y_2 \ge 42$
$y_1 \ge 0,\ y_2 \ge 0.$

Rewrite the objective function by forming the augmented matrix of the system and then form its transpose.

$$\left[\begin{array}{cc|c} 1 & 1 & 17 \\ 5 & 8 & 42 \\ \hline 18 & 10 & 0 \end{array}\right] \to \left[\begin{array}{cc|c} 1 & 5 & 18 \\ 1 & 8 & 10 \\ \hline 17 & 42 & 0 \end{array}\right].$$

This gives the problem
Maximize $z = 17x_1 + 42x_2$
subject to $x_1 + 5x_2 \le 18$.
$x_1 + 8x_2 \le 10$
$x_1 \ge 0,\ x_2 \ge 0$

The initial simplex tableau is

$$\begin{array}{c} \begin{array}{cccc} x_1 & x_2 & s_1 & s_2 \end{array} \\ \left[\begin{array}{cccc|c} 1 & 5 & 1 & 0 & 18 \\ 1 & 8 & 0 & 1 & 10 \\ \hline -17 & -42 & 0 & 0 & 0 \end{array}\right] \end{array}.$$

Pivot on the 8 in row 2 column 2.

$$\begin{array}{c} \begin{array}{cccc} x_1 & x_2 & s_1 & s_2 \end{array} \\ \left[\begin{array}{cccc|c} 1 & 5 & 1 & 0 & 18 \\ \frac{1}{8} & 1 & 0 & \frac{1}{8} & \frac{5}{4} \\ \hline -17 & -42 & 0 & 0 & 0 \end{array}\right] \begin{array}{l} \\ \frac{1}{8}R_2 \\ \\ \end{array} \end{array}$$

$$\begin{array}{c} \begin{array}{cccc} x_1 & x_2 & s_1 & s_2 \end{array} \\ \left[\begin{array}{cccc|c} \frac{3}{8} & 0 & 1 & -\frac{5}{8} & \frac{47}{4} \\ \frac{1}{8} & 1 & 0 & \frac{1}{8} & \frac{5}{4} \\ \hline -\frac{47}{4} & 0 & 0 & \frac{21}{4} & \frac{105}{2} \end{array}\right] \begin{array}{l} -5R_2 + R_1 \\ \\ 42R_2 + R_3 \end{array} \end{array}$$

Now pivot on the $\frac{1}{8}$ in row 2 column 1.

$$\begin{array}{c} \begin{array}{cccc} x_1 & x_2 & s_1 & s_2 \end{array} \\ \left[\begin{array}{cccc|c} 0 & -3 & 1 & -1 & 8 \\ \frac{1}{8} & 1 & 0 & \frac{1}{8} & \frac{5}{4} \\ \hline 0 & 94 & 0 & 17 & 170 \end{array}\right] \begin{array}{l} -3R_2 + R_1 \\ \\ 94R_2 + R_3 \end{array} \\ \begin{array}{ccc} \uparrow & \uparrow & \uparrow \\ y_1 & y_2 & w \end{array} \end{array}$$

The minimum is 170 when $y_1 = 0$ and $y_2 = 17$.

46. Minimize $w = 4y_1 + 5y_2$
subject to $10y_1 + 5y_2 \ge 100$
$20y_1 + 10y_2 \ge 150$
$y_1 \ge 0,\ y_2 \ge 0$

Rewrite the objective function by forming the augmented matrix of the system and then form its transpose.

$$\left[\begin{array}{cc|c} 10 & 5 & 100 \\ 20 & 10 & 150 \\ \hline 4 & 5 & 0 \end{array}\right] \to \left[\begin{array}{cc|c} 10 & 20 & 4 \\ 5 & 10 & 5 \\ \hline 100 & 150 & 0 \end{array}\right].$$

This gives the problem
Maximize $z = 100x_1 + 150x_2$
subject to $10x_1 + 20x_2 \le 4$.
$5x_1 + 10x_2 \le 5$
$x_1 \ge 0,\ x_2 \ge 0$

The initial simplex tableau is

$$\begin{array}{c} \begin{array}{cccc} x_1 & x_2 & s_1 & s_2 \end{array} \\ \left[\begin{array}{cccc|c} 10 & 20 & 1 & 0 & 4 \\ 5 & 10 & 0 & 1 & 5 \\ \hline -100 & -150 & 0 & 0 & 0 \end{array}\right] \end{array}.$$

Pivot on the 20 in row 1 column 2.

$$\begin{array}{c} \begin{array}{cccc} x_1 & x_2 & s_1 & s_2 \end{array} \\ \left[\begin{array}{cccc|c} \frac{1}{2} & 1 & \frac{1}{20} & 0 & \frac{1}{5} \\ 5 & 10 & 0 & 1 & 5 \\ \hline -100 & -150 & 0 & 0 & 0 \end{array}\right] \begin{array}{l} \frac{1}{20}R_1 \\ \\ \\ \end{array} \end{array}$$

$$\begin{array}{c} \begin{array}{cccc} x_1 & x_2 & s_1 & s_2 \end{array} \\ \left[\begin{array}{cccc|c} \frac{1}{2} & 1 & \frac{1}{20} & 0 & \frac{1}{5} \\ 0 & 0 & -\frac{1}{2} & 1 & 3 \\ \hline -25 & 0 & \frac{15}{2} & 0 & 30 \end{array}\right] \begin{array}{l} \\ -10R_1 + R_2 \\ 150R_1 + R_3 \end{array} \end{array}$$

Now pivot on the $\frac{1}{2}$ in row 1 column 1.

$$\begin{array}{c} \begin{array}{cccc} x_1 & x_2 & s_1 & s_2 \end{array} \\ \left[\begin{array}{cccc|c} 1 & 2 & \frac{1}{10} & 0 & \frac{2}{5} \\ 0 & 0 & -\frac{1}{2} & 1 & 3 \\ \hline -25 & 0 & \frac{15}{2} & 0 & 30 \end{array}\right] \begin{array}{l} 2R_1 \\ \\ \\ \end{array} \end{array}$$

$$\begin{array}{c} \begin{array}{cccc} x_1 & x_2 & s_1 & s_2 \end{array} \\ \left[\begin{array}{cccc|c} 1 & 2 & \frac{1}{10} & 0 & \frac{2}{5} \\ 0 & 0 & -\frac{1}{2} & 1 & 3 \\ \hline 0 & 50 & 10 & 0 & 40 \end{array}\right] \begin{array}{l} 2R_1 \\ \\ 25R_1 + R_3 \end{array} \\ \begin{array}{ccc} \uparrow & \uparrow & \uparrow \\ y_1 & y_2 & w \end{array} \end{array}$$

The minimum is 40 when $y_1 = 10$ and $y_2 = 0$.

47. Maximize $z = 20x_1 + 30x_2$
subject to $5x_1 + 10x_2 \le 120$
$10x_1 + 15x_2 \ge 200$
$x_1 \ge 0,\ x_2 \ge 0$

The initial simplex tableau is

$$\begin{array}{c} \begin{array}{cccc} x_1 & x_2 & s_1 & s_2 \end{array} \\ \left[\begin{array}{cccc|c} 5 & 10 & 1 & 0 & 120 \\ 10 & 15 & 0 & -1 & 200 \\ \hline -20 & -30 & 0 & 0 & 0 \end{array}\right] \end{array}.$$

48. Minimize $w = 4y_1 + 2y_2$
subject to $y_1 + 3y_2 \ge 6$
$2y_1 + 8y_2 \le 21$
$y_1 \ge 0,\ y_2 \ge 0$

The initial simplex tableau is

$$\begin{array}{c} \begin{array}{cccc} y_1 & y_2 & s_1 & s_2 \end{array} \\ \left[\begin{array}{cccc|c} 1 & 3 & -1 & 0 & 6 \\ 2 & 8 & 0 & 1 & 21 \\ \hline 4 & 2 & 0 & 0 & 0 \end{array}\right] \end{array}.$$

49. Minimize $w = 12y_1 + 20y_2 - 8y_3$
subject to: $y_1 + y_2 + 2y_3 \ge 48$
$y_1 + y_2 \le 12$
$y_3 \ge 10$
$3y_1 + y_3 \ge 30$
$y_1 \ge 0,\ y_2 \ge 0, y_3 \ge 0.$

The initial simplex tableau is

$$\begin{array}{c} \begin{array}{ccccccc} y_1 & y_2 & y_3 & s_1 & s_2 & s_3 & s_4 \end{array} \\ \left[\begin{array}{ccccccc|c} 1 & 1 & 2 & -1 & 0 & 0 & 0 & 48 \\ 1 & 1 & 0 & 0 & 1 & 0 & 0 & 12 \\ 0 & 0 & 1 & 0 & 0 & -1 & 0 & 10 \\ 3 & 0 & 1 & 0 & 0 & 0 & -1 & 30 \\ \hline 12 & 20 & -8 & 0 & 0 & 0 & 0 & 0 \end{array}\right] \end{array}.$$

50. Maximize $w = 6x_1 - 3x_2 + 4x_3$
subject to: $2x_1 + x_2 + x_3 \le 112$
$x_1 + x_2 + x_3 \ge 80$
$x_1 + x_2 \le 45$
$x_1 \ge 0,\ x_2 \ge 0,\ x_3 \ge 0.$

The initial simplex tableau is

$$\begin{array}{c} \begin{array}{cccccc} x_1 & x_2 & x_3 & s_1 & s_2 & s_3 \end{array} \\ \left[\begin{array}{cccccc|c} 2 & 1 & 1 & 1 & 0 & 0 & 112 \\ 1 & 1 & 1 & 0 & -1 & 0 & 80 \\ 1 & 1 & 0 & 0 & 0 & 1 & 45 \\ \hline -6 & 3 & -4 & 0 & 0 & 0 & 0 \end{array}\right] \end{array}.$$

51. If $w = -z$

$$\left[\begin{array}{cccccc|c} 0 & 1 & 0 & 2 & 5 & 0 & 17 \\ 0 & 0 & 1 & 3 & 1 & 1 & 25 \\ 1 & 0 & 0 & 4 & 2 & \frac{1}{2} & 8 \\ \hline 0 & 0 & 0 & 2 & 5 & 0 & -427 \end{array}\right]$$

indicates a minimum value of 427 at (8, 17, 25, 0 ,0, 0).

52.
$$\left[\begin{array}{ccccccc|c} 0 & 0 & 2 & 1 & 0 & 6 & 6 & 92 \\ 1 & 0 & 3 & 0 & 0 & 0 & 2 & 47 \\ 0 & 1 & 0 & 0 & 0 & 1 & 0 & 68 \\ 0 & 0 & 4 & 0 & 1 & 0 & 3 & 35 \\ \hline 0 & 0 & 5 & 0 & 0 & 2 & 9 & -1957 \end{array}\right]$$

The minimum value is 1957 at (47, 68, 0, 92, 35, 0, 0).

53. Maximize $z = 20x_1 + 30x_2$
subject to $5x_1 + 10x_2 \le 120$
$10x_1 + 15x_2 \ge 200$
$x_1 \ge 0,\ x_2 \ge 0$

The initial simplex tableau is

$$\begin{array}{c} \begin{array}{cccc} x_1 & x_2 & s_1 & s_2 \end{array} \\ \left[\begin{array}{cccc|c} 5 & 10 & 1 & 0 & 120 \\ \boxed{10} & 15 & 0 & -1 & 200 \\ \hline -20 & -30 & 0 & 0 & 0 \end{array}\right] \end{array}.$$

Pivot on the 10 in row 2 column 1.

$$\begin{array}{c} \begin{array}{cccc} x_1 & x_2 & s_1 & s_2 \end{array} \\ \left[\begin{array}{cccc|c} 0 & 5 & 2 & 1 & 40 \\ 10 & 15 & 0 & -1 & 200 \\ \hline 0 & 0 & 0 & -2 & 400 \end{array}\right] \end{array} \begin{array}{l} -R_2 + 2R_1 \\ \\ 2R_2 + R_3 \end{array}$$

This completes Stage I because the solution given in the usual way from the matrix has nonnegative values for all variables. Since there is a negative indicator in the objective row, we continue with Stage II. Pivot on the 1 in row 1 column 4.

$$\begin{array}{c} \begin{array}{cccc} x_1 & x_2 & s_1 & s_2 \end{array} \\ \left[\begin{array}{cccc|c} 0 & 5 & 2 & \boxed{1} & 40 \\ 10 & 20 & 2 & 0 & 240 \\ \hline 0 & 10 & 4 & 0 & 480 \end{array}\right] \end{array} \begin{array}{l} \\ \\ 2R_1 + R_3 \end{array}$$

$$\begin{array}{c} \begin{array}{cccc} x_1 & x_2 & s_1 & s_2 \end{array} \\ \left[\begin{array}{cccc|c} 0 & 5 & 2 & 1 & 40 \\ 1 & 2 & \frac{1}{5} & 0 & 24 \\ \hline 0 & 10 & 4 & 0 & 480 \end{array}\right] \end{array} \begin{array}{l} \\ \frac{1}{10}R_2 \\ \\ \end{array}$$

The maximum is 480 when $x_1 = 24$ and $x_2 = 0$.

54. Minimize $w = 4y_1 + 2y_2$
subject to $y_1 + 3y_2 \ge 6$
$2y_1 + 8y_2 \le 21$
$y_1 \ge 0,\ y_2 \ge 0$

The initial simplex tableau is

$$\begin{array}{c} \begin{array}{cccc} y_1 & y_2 & s_1 & s_2 \end{array} \\ \left[\begin{array}{cccc|c} \boxed{1} & 3 & -1 & 0 & 6 \\ 2 & 8 & 0 & 1 & 21 \\ \hline 4 & 2 & 0 & 0 & 0 \end{array}\right] \end{array}.$$

Pivot on the 1 in row 1 column 1.

$$\begin{array}{c} \begin{array}{cccc} y_1 & y_2 & s_1 & s_2 \end{array} \\ \left[\begin{array}{cccc|c} 1 & 3 & -1 & 0 & 6 \\ 0 & 2 & 2 & 1 & 9 \\ \hline 0 & -10 & 4 & 0 & -24 \end{array}\right] \end{array} \begin{array}{l} \\ -2R_1 + R_2 \\ -4R_1 + R_3 \end{array}$$

This completes Stage I because the solution given in the usual way from the matrix has nonnegative values for all variables. Since there is a negative indicator in the objective row, we continue with Stage II. Pivot on the 3 in row 1 column 2.

$$\begin{array}{c} \begin{array}{cccc} y_1 & y_2 & s_1 & s_2 \end{array} \\ \left[\begin{array}{cccc|c} \frac{1}{3} & 1 & -\frac{1}{3} & 0 & 2 \\ 0 & 2 & 2 & 1 & 9 \\ \hline 0 & -10 & 4 & 0 & -24 \end{array}\right] \end{array} \begin{array}{l} \frac{1}{3}R_1 \\ \\ \\ \end{array}$$

$$\begin{array}{c} \begin{array}{cccc} y_1 & y_2 & s_1 & s_2 \end{array} \\ \left[\begin{array}{cccc|c} \frac{1}{3} & 1 & -\frac{1}{3} & 0 & 2 \\ -\frac{2}{3} & 0 & \frac{8}{3} & 1 & 5 \\ \hline \frac{10}{3} & 0 & \frac{2}{3} & 0 & -4 \end{array}\right] \end{array} \begin{array}{l} \\ -2R_1 + R_2 \\ 10R_1 + R_3 \end{array}$$

The minimum is 4 when $y_1 = 0$ and $y_2 = 2$.

55. Minimize $w = 4y_1 - 8y_2$
subject to: $y_1 + y_2 \le 50$
$2y_1 - 4y_2 \ge 20$
$y_1 - y_2 \le 22$
$y_1 \ge 0,\ y_2 \ge 0.$

The initial simplex tableau is

$$\begin{array}{c} \begin{array}{ccccc} y_1 & y_2 & s_1 & s_2 & s_3 \end{array} \\ \left[\begin{array}{ccccc|c} 1 & 1 & 1 & 0 & 0 & 50 \\ \boxed{2} & -4 & 0 & -1 & 0 & 20 \\ 1 & -1 & 0 & 0 & 1 & 22 \\ \hline 4 & -8 & 0 & 0 & 0 & 0 \end{array}\right] \end{array}.$$

For Stage I, pivot on the 2 in row two, column one.

$$\begin{array}{c} \begin{array}{ccccc} y_1 & y_2 & s_1 & s_2 & s_3 \end{array} \\ \left[\begin{array}{ccccc|c} 1 & 1 & 1 & 0 & 0 & 50 \\ 1 & -2 & 0 & -\frac{1}{2} & 0 & 10 \\ 1 & -1 & 0 & 0 & 1 & 22 \\ \hline 4 & -8 & 0 & 0 & 0 & 0 \end{array}\right] \end{array} \begin{array}{l} \\ \frac{1}{2}R_2 \\ \\ \\ \end{array}$$

$$\begin{array}{c} \begin{array}{ccccc} y_1 & y_2 & s_1 & s_2 & s_3 \end{array} \\ \left[\begin{array}{ccccc|c} 0 & 3 & 1 & \frac{1}{2} & 0 & 40 \\ 1 & -2 & 0 & -\frac{1}{2} & 0 & 10 \\ 0 & 1 & 0 & \frac{1}{2} & 1 & 12 \\ \hline 0 & 0 & 0 & 2 & 0 & -40 \end{array}\right] \end{array} \begin{array}{l} -R_2 + R_1 \\ \\ -R_2 + R_3 \\ -4R_2 + R_4 \end{array}$$

The minimum is 40 when $y_1 = 10$ and $y_2 = 0$.

56. Maximize $z = 2x_1 + 4x_2$
subject to: $3x_1 + 2x_2 \le 12$
$5x_1 + x_2 \ge 5$
$x_1 \ge 0,\ x_2 \ge 0.$

The initial simplex tableau is

$$\begin{array}{c} \begin{array}{cccc} x_1 & x_2 & s_1 & s_2 \end{array} \\ \left[\begin{array}{cccc|c} 3 & 2 & 1 & 0 & 12 \\ 5 & \boxed{1} & 0 & -1 & 5 \\ \hline -2 & -4 & 0 & 0 & 0 \end{array}\right] \end{array}.$$

For Stage I, pivot on the 1 in row 2 column 1.

$$\begin{array}{c} \begin{array}{cccc} x_1 & x_2 & s_1 & s_2 \end{array} \\ \left[\begin{array}{cccc|c} -7 & 0 & 1 & \boxed{2} & 2 \\ 5 & 1 & 0 & -1 & 5 \\ \hline 18 & 0 & 0 & -4 & 20 \end{array}\right] \end{array} \begin{array}{l} -2R_2 + R_1 \\ \\ 4R_2 + R_3 \end{array}$$

For Stage II, pivot on the 2 in row 1 column 4.

$$\begin{array}{c} \begin{array}{cccc} x_1 & x_2 & s_1 & s_2 \end{array} \\ \left[\begin{array}{cccc|c} -\frac{7}{2} & 0 & \frac{1}{2} & 1 & 1 \\ \frac{3}{2} & 1 & \frac{1}{2} & 0 & 6 \\ \hline 4 & 0 & 2 & 0 & 24 \end{array}\right] \end{array} \begin{array}{l} \frac{1}{2}R_1 \\ R_1 + R_2 \\ 4R_1 + R_3 \end{array}$$

The maximum is 24 when $x_1 = 0$ and $x_2 = 6$.

57. From Exercise 33, the initial simplex tableau is

$$\begin{array}{c} \begin{array}{cccccc} x_1 & x_2 & x_3 & s_1 & s_2 & s_3 \end{array} \\ \left[\begin{array}{cccccc|c} 2 & 3 & 6 & 1 & 0 & 0 & 1200 \\ 1 & 2 & 2 & 0 & 1 & 0 & 800 \\ \boxed{2} & 2 & 4 & 0 & 0 & 1 & 500 \\ \hline -4 & -3 & -3 & 0 & 0 & 0 & 0 \end{array}\right] \end{array}.$$

Pivot on the 2 in row three, column one.

$$\begin{array}{c} \begin{array}{cccccc} x_1 & x_2 & x_3 & s_1 & s_2 & s_3 \end{array} \\ \left[\begin{array}{cccccc|c} 2 & 3 & 6 & 1 & 0 & 0 & 1200 \\ 1 & 2 & 2 & 0 & 1 & 0 & 800 \\ 1 & 1 & 2 & 0 & 0 & \frac{1}{2} & 250 \\ \hline -4 & -3 & -3 & 0 & 0 & 0 & 0 \end{array}\right] \end{array} \begin{array}{l} \\ \\ \frac{1}{2}R_3 \\ \\ \end{array}$$

$$\begin{array}{c} \begin{array}{cccccc} x_1 & x_2 & x_3 & s_1 & s_2 & s_3 \end{array} \\ \left[\begin{array}{cccccc|c} 0 & 1 & 2 & 1 & 0 & -1 & 700 \\ 0 & 1 & 0 & 0 & 1 & -\frac{1}{2} & 550 \\ 1 & 1 & 2 & 0 & 0 & \frac{1}{2} & 250 \\ \hline 0 & 1 & 5 & 0 & 0 & 2 & 1000 \end{array}\right] \end{array} \begin{array}{l} -2R_3 + R_1 \\ -1R_3 + R_2 \\ \\ 4R_3 + R_4 \end{array}$$

Roberta should get 250 units of item A, none of item B, and none of item C for a maximum profit of $1000.

58. From Exercise 34, we have the initial simplex tableau (after multiplying the fourth row by 100 to clear the decimals):

$$\begin{array}{c} \begin{array}{cccccc} x_1 & x_2 & x_3 & s_1 & s_2 & s_3 \end{array} \\ \left[\begin{array}{cccccc|c} 1 & 1 & 1 & 1 & 0 & 0 & 50,000 \\ \boxed{1} & 1 & 0 & 0 & 1 & 0 & 15,000 \\ 1 & 0 & 1 & 0 & 0 & 1 & 25,000 \\ \hline -15 & -9 & -5 & 0 & 0 & 0 & 0 \end{array}\right] \end{array}.$$

$$\begin{array}{c} \begin{array}{cccccc} x_1 & x_2 & x_3 & s_1 & s_2 & s_3 \end{array} \\ \left[\begin{array}{cccccc|c} 0 & 0 & 1 & 1 & -1 & 0 & 35,000 \\ 1 & 1 & 0 & 0 & 1 & 0 & 15,000 \\ 0 & -1 & \boxed{1} & 0 & -1 & 1 & 10,000 \\ \hline 0 & 6 & -5 & 0 & 15 & 0 & 225,000 \end{array}\right] \end{array} \begin{array}{l} -1R_2 + R_1 \\ \\ -1R_2 + R_3 \\ 15R_2 + R_4 \end{array}$$

$$\begin{array}{c} \begin{array}{cccccc} x_1 & x_2 & x_3 & s_1 & s_2 & s_3 \end{array} \\ \left[\begin{array}{cccccc|c} 0 & 1 & 0 & 1 & 0 & -1 & 25,000 \\ 1 & 1 & 0 & 0 & 1 & 0 & 15,000 \\ 0 & -1 & 1 & 0 & -1 & 1 & 10,000 \\ \hline 0 & 1 & 0 & 0 & 10 & 5 & 275,000 \end{array}\right] \end{array} \begin{array}{l} -1R_3 + R_1 \\ \\ \\ 5R_3 + R_4 \end{array}$$

Since $x_1 = 15,000$ and $x_3 = 10,000$, he should invest $15,000 in oil leases and $10,000 in stock for a maximum return of .01(275,000) = $2750.

59. From Exercise 35, we have the initial simplex tableau:

$$\begin{array}{c} \begin{array}{ccccc} x_1 & x_2 & s_1 & s_2 & s_3 \end{array} \\ \left[\begin{array}{ccccc|c} 2 & 1 & 1 & 0 & 0 & 110 \\ 2 & \boxed{3} & 0 & 1 & 0 & 125 \\ 2 & 1 & 0 & 0 & 1 & 90 \\ \hline -12 & -15 & 0 & 0 & 0 & 0 \end{array}\right] \end{array}$$

$$\begin{array}{c} \begin{array}{ccccc} x_1 & x_2 & s_1 & s_2 & s_3 \end{array} \\ \left[\begin{array}{ccccc|c} 2 & 1 & 1 & 0 & 0 & 110 \\ \frac{2}{3} & \boxed{1} & 0 & \frac{1}{3} & 0 & \frac{125}{3} \\ 2 & 1 & 0 & 0 & 1 & 90 \\ \hline -12 & -15 & 0 & 0 & 0 & 0 \end{array}\right] \end{array} \begin{array}{l} \\ \frac{1}{3}R_2 \\ \\ \\ \end{array}$$

$$\begin{array}{c} \begin{array}{ccccc} x_1 & x_2 & s_1 & s_2 & s_3 \end{array} \\ \left[\begin{array}{ccccc|c} \frac{4}{3} & 0 & 1 & -\frac{1}{3} & 0 & \frac{205}{3} \\ \frac{2}{3} & 1 & 0 & \frac{1}{3} & 0 & \frac{125}{3} \\ \boxed{\frac{4}{3}} & 0 & 0 & -\frac{1}{3} & 1 & \frac{145}{3} \\ \hline -2 & 0 & 0 & 5 & 0 & 625 \end{array}\right] \end{array} \begin{array}{l} -R_2 + R_1 \\ \\ -R_2 + R_3 \\ 15R_2 + R_4 \end{array}$$

$$\begin{array}{c} \begin{array}{ccccc} x_1 & x_2 & s_1 & s_2 & s_3 \end{array} \\ \left[\begin{array}{ccccc|c} \frac{4}{3} & 0 & 1 & -\frac{1}{3} & 0 & \frac{205}{3} \\ \frac{2}{3} & 1 & 0 & \frac{1}{3} & 0 & \frac{125}{3} \\ \boxed{1} & 0 & 0 & -\frac{1}{4} & \frac{3}{4} & \frac{145}{4} \\ \hline -2 & 0 & 0 & 5 & 0 & 625 \end{array}\right] \end{array} \begin{array}{l} \\ \\ \frac{4}{3}R_3 \\ \\ \end{array}$$

$$\begin{array}{c} \begin{array}{ccccc} x_1 & x_2 & s_1 & s_2 & s_3 \end{array} \\ \left[\begin{array}{ccccc|c} 0 & 0 & 1 & 0 & -1 & 20 \\ 0 & 1 & 0 & \frac{1}{2} & -\frac{1}{2} & \frac{35}{2} \\ 1 & 0 & 0 & -\frac{1}{4} & \frac{3}{4} & \frac{145}{4} \\ \hline 0 & 0 & 0 & \frac{9}{2} & \frac{3}{2} & \frac{1395}{2} \end{array}\right] \end{array} \begin{array}{l} -\frac{4}{3}R_3 + R_1 \\ -\frac{2}{3}R_3 + R_2 \\ \\ 2R_3 + R_4 \end{array}$$

The winery should make 17.5 gallons of Crystal wine and 36.25 gallons of Fruity wine for a maximum profit of $697.50.

60. From Exercise 36, we have the initial tableau:

$$\begin{array}{cccccc|c} x_1 & x_2 & x_3 & s_1 & s_2 & s_3 & \\ \hline 1 & 1.1 & 1.5 & 1 & 0 & 0 & 8 \\ 1 & 1.2 & 1.3 & 0 & 1 & 0 & 8 \\ \boxed{2} & 3 & 4 & 0 & 0 & 1 & 8 \\ \hline -1 & -.9 & -.95 & 0 & 0 & 0 & 0 \end{array}.$$

$$\begin{array}{cccccc|c|l} x_1 & x_2 & x_3 & s_1 & s_2 & s_3 & & \\ \hline 1 & 1.1 & 1.5 & 1 & 0 & 0 & 8 & \\ 1 & 1.2 & 1.3 & 0 & 1 & 0 & 8 & \\ \boxed{1} & 1.5 & 2 & 0 & 0 & .5 & 4 & \frac{1}{2}R_3 \\ \hline -1 & -.9 & -.95 & 0 & 0 & 0 & 0 & \end{array}$$

$$\begin{array}{cccccc|c|l} x_1 & x_2 & x_3 & s_1 & s_2 & s_3 & & \\ \hline 0 & -.4 & -.5 & 1 & 0 & -.5 & 4 & -1R_3 + R_1 \\ 0 & -.3 & -.7 & 0 & 1 & -.5 & 4 & -1R_3 + R_2 \\ 1 & 1.5 & 2 & 0 & 0 & .5 & 4 & \\ \hline 0 & .6 & 1.05 & 0 & 0 & .5 & 4 & R_3 + R_4 \end{array}$$

The final tableau gives $x_1 = 4$, $x_2 = 0$, $x_3 = 0$, and $z = 4$. Therefore, 4 units of 5-gallon bags (and none of the others) should be made for a maximum profit of $4 per unit.

61. Let y_1 = the number of cases of corn
y_2 = the number of cases of beans
y_3 = the number of cases of carrots

Minimize $w = 10y_1 + 15y_2 + 25y_3$

Maximize $z = -w = -10y_1 - 15y_2 - 25y_3$

subject to:
$$\begin{aligned} y_1 + y_2 + y_3 &\geq 1000 \\ y_1 - 2y_2 &\geq 0 \\ y_3 &\geq 340 \\ y_1 \geq 0,\ y_2 \geq 0,\ y_3 &\geq 0 \end{aligned}$$

The initial simplex tableau is

$$\begin{array}{cccccc|c} y_1 & y_2 & y_3 & s_1 & s_2 & s_3 & \\ \hline 1 & 1 & 1 & -1 & 0 & 0 & 1000 \\ \boxed{1} & -2 & 0 & 0 & -1 & 0 & 0 \\ 0 & 0 & 1 & 0 & 0 & -1 & 340 \\ \hline 10 & 15 & 25 & 0 & 0 & 0 & 0 \end{array}$$

For Stage I, pivot on the 1 in row two, column one.

$$\begin{array}{cccccc|c|l} y_1 & y_2 & y_3 & s_1 & s_2 & s_3 & & \\ \hline 0 & \boxed{3} & 1 & -1 & 1 & 0 & 1000 & -R_2 + R_1 \\ 1 & -2 & 0 & 0 & -1 & 0 & 0 & \\ 0 & 0 & 1 & 0 & 0 & -1 & 340 & \\ \hline 0 & 35 & 25 & 0 & 10 & 0 & 0 & -10R_2 + R_4 \end{array}$$

Continuing Stage I, pivot on the 3 in row one, column two.

$$\begin{array}{cccccc|c|l} y_1 & y_2 & y_3 & s_1 & s_2 & s_3 & & \\ \hline 0 & 1 & \frac{1}{3} & -\frac{1}{3} & \frac{1}{3} & 0 & \frac{1000}{3} & \frac{1}{3}R_1 \\ 1 & 0 & \frac{2}{3} & -\frac{2}{3} & -\frac{1}{3} & 0 & \frac{2000}{3} & 2R_1 + R_2 \\ 0 & 0 & \boxed{1} & 0 & 0 & -1 & 340 & \\ \hline 0 & 0 & \frac{40}{3} & \frac{35}{3} & -\frac{5}{3} & 0 & -\frac{35,000}{3} & -35R_1 + R_4 \end{array}$$

Pivot on the 1 in row three, column three.

$$\begin{array}{cccccc|c|l} y_1 & y_2 & y_3 & s_1 & s_2 & s_3 & & \\ \hline 0 & 1 & 0 & -\frac{1}{3} & \boxed{\frac{1}{3}} & \frac{1}{3} & 220 & -\frac{1}{3}R_3 + R_1 \\ 1 & 0 & 0 & -\frac{2}{3} & -\frac{1}{3} & \frac{2}{3} & 440 & -\frac{2}{3}R_3 + R_2 \\ 0 & 0 & 1 & 0 & 0 & -1 & 340 & \\ \hline 0 & 0 & 0 & \frac{35}{3} & -\frac{5}{3} & \frac{40}{3} & -16,200 & -\frac{40}{3}R_3 + R_4 \end{array}$$

For Stage II, pivot on the $\frac{1}{3}$ in row one, column five.

$$\begin{array}{cccccc|c|l} y_1 & y_2 & y_3 & s_1 & s_2 & s_3 & & \\ \hline 0 & 3 & 0 & -1 & 1 & 1 & 660 & 3R_1 \\ 1 & 1 & 0 & -1 & 0 & 1 & 660 & \frac{1}{3}R_1 + R_2 \\ 0 & 0 & 1 & 0 & 0 & -1 & 340 & \\ \hline 0 & 5 & 0 & 10 & 0 & 15 & -15,100 & \frac{5}{3}R_1 + R_4 \end{array}$$

The Cauchy Canners should produce 660 cases of corn, no cases of beans, and 340 cases of carrots for a minimum cost of $15,100.

62. First put the data into a table.

	Lumber	Concrete	Advertising	Total Spent
Atlantic	1000	3000	2000	$3000
Pacific	2000	3000	3000	$4000
Minimum Use	8000	18000	15000	

Let y_1 = the number of Atlantic boats and let y_2 = the number of Pacific boats.

The problem is to minimize $w = 3000y_1 + 4000y_2$

subject to:
$$\begin{aligned} 1000y_1 + 2000y_2 &\ge 8000 \\ 3000y_1 + 3000y_2 &\ge 18,000 \\ 2000y_1 + 3000y_2 &\ge 15,000 \\ y_1 \ge 0,\ y_2 &\ge 0. \end{aligned}$$

The matrix for this problem is

$$\left[\begin{array}{cc|c} 1000 & 2000 & 8000 \\ 3000 & 3000 & 18,000 \\ 2000 & 3000 & 15,000 \\ \hline 3000 & 4000 & 0 \end{array}\right].$$

The transpose of this matrix is

$$\left[\begin{array}{ccc|c} 1000 & 3000 & 2000 & 3000 \\ 2000 & 3000 & 3000 & 4000 \\ \hline 8000 & 18,000 & 15,000 & 0 \end{array}\right].$$

The dual problem is as follows.

Maximize $z = 8000x_1 + 18,000x_2 + 15,000x_3$

subject to:
$$\begin{aligned} 1000x_1 + 3000x_2 + 2000x_3 &\le 3000 \\ 2000x_1 + 3000x_2 + 3000x_3 &\le 4000 \\ x_1 \ge 0,\ x_2 \ge 0,\ x_3 &\ge 0. \end{aligned}$$

The initial simplex tableau is

$$\begin{array}{c} \begin{array}{ccccc} x_1 & x_2 & x_3 & s_1 & s_2 \end{array} \\ \left[\begin{array}{ccccc|c} 1000 & \boxed{3000} & 2000 & 1 & 0 & 3000 \\ 2000 & 3000 & 3000 & 0 & 1 & 4000 \\ \hline -8000 & -18,000 & -15,000 & 0 & 0 & 0 \end{array}\right]. \end{array}$$

Pivot on the 3000 in row one, column two.

$$\begin{array}{c} \begin{array}{ccccc} x_1 & x_2 & x_3 & s_1 & s_2 \end{array} \\ \left[\begin{array}{ccccc|c} \frac{1}{3} & \boxed{1} & \frac{2}{3} & \frac{1}{3000} & 0 & 1 \\ 2000 & 3000 & 3000 & 0 & 1 & 4000 \\ \hline -8000 & -18,000 & -15,000 & 0 & 0 & 0 \end{array}\right] \frac{1}{3000}R_1 \end{array}$$

$$\begin{array}{c} \begin{array}{ccccc} x_1 & x_2 & x_3 & s_1 & s_2 \end{array} \\ \left[\begin{array}{ccccc|c} \frac{1}{3} & 1 & \frac{2}{3} & \frac{1}{3000} & 0 & 1 \\ 1000 & 0 & \boxed{1000} & -1 & 1 & 1000 \\ \hline -2000 & 0 & -3000 & 6 & 0 & 18,000 \end{array}\right] \begin{array}{l} \\ -3000R_1 + R_2 \\ 18,000R_1 + R_3 \end{array} \end{array}$$

(*continued on next page*)

Pivot on the 1000 in row two, column three.

$$\begin{array}{ccccc|c} x_1 & x_2 & x_3 & x_4 & x_5 & \\ \frac{1}{3} & 1 & \frac{2}{3} & \frac{1}{3000} & 0 & 1 \\ 1 & 0 & 1 & -\frac{1}{1000} & \frac{1}{1000} & 1 \\ \hline -2000 & 0 & -3000 & 6 & 0 & 18,000 \end{array} \quad \frac{1}{1000}R_2$$

$$\begin{array}{ccccc|c} x_1 & x_2 & x_3 & x_4 & x_5 & \\ -\frac{1}{3} & 1 & 0 & \frac{1}{1000} & -\frac{1}{1500} & \frac{1}{3} \\ 1 & 0 & 1 & -\frac{1}{1000} & \frac{1}{1000} & 1 \\ \hline 1000 & 0 & 0 & 3 & 3 & 21,000. \end{array} \quad \begin{array}{l} -\frac{2}{3}R_2 + R_1 \\ \\ 3000R_2 + R_3 \end{array}$$

The contractor should build 3 Atlantic and 3 Pacific models for a minimum cost of \$21,000.

63. Let y_1 = kg of whole tomatoes and let y_2 = kg of tomato sauce.

Minimize $w = 4y_1 + 3.25y_2$

subject to:
$$\begin{aligned} y_1 + y_2 &\le 3,000,000 \\ y_1 &\ge 800,000 \\ y_2 &\ge 80,000 \\ \frac{6}{60}y_1 + \frac{3}{60}y_2 &\ge 110,000 \\ y_1 \ge 0,\ y_2 &\ge 0 \end{aligned}$$

The initial simplex tableau is

$$\begin{array}{cccccc|c} y_1 & y_2 & s_1 & s_2 & s_3 & s_4 & \\ 1 & 1 & 1 & 0 & 0 & 0 & 3,000,000 \\ \boxed{1} & 0 & 0 & -1 & 0 & 0 & 800,000 \\ 0 & 1 & 0 & 0 & -1 & 0 & 80,000 \\ .1 & .05 & 0 & 0 & 0 & -1 & 110,000 \\ \hline 4 & 3.25 & 0 & 0 & 0 & 0 & 0 \end{array}$$

For Stage I, pivot on the 1 in row two, column one.

$$\begin{array}{cccccc|c} y_1 & y_2 & s_1 & s_2 & s_3 & s_4 & \\ 0 & 1 & 1 & 1 & 0 & 0 & 2,200,000 \\ 1 & 0 & 0 & -1 & 0 & 0 & 800,000 \\ 0 & \boxed{1} & 0 & 0 & -1 & 0 & 80,000 \\ 0 & .05 & 0 & .1 & 0 & -1 & 30,000 \\ \hline 0 & 3.25 & 0 & 4 & 0 & 0 & -3,200,000 \end{array} \quad \begin{array}{l} -R_2 + R_1 \\ \\ \\ -.1R_2 + R_4 \\ -4R_2 + R_5 \end{array}$$

Pivot on the 1 in row three, column two.

$$\begin{array}{cccccc|c} y_1 & y_2 & s_1 & s_2 & s_3 & s_4 & \\ 0 & 0 & 1 & 1 & 1 & 0 & 2,120,000 \\ 1 & 0 & 0 & -1 & 0 & 0 & 800,000 \\ 0 & 1 & 0 & 0 & -1 & 0 & 80,000 \\ 0 & 0 & 0 & \boxed{.1} & .05 & -1 & 26,000 \\ \hline 0 & 0 & 0 & 4 & 3.25 & 0 & -3,460,000 \end{array} \quad \begin{array}{l} -R_3 + R_1 \\ \\ \\ -.05R_3 + R_4 \\ -3.25R_3 + R_5 \end{array}$$

(*continued on next page*)

Pivot on the .1 in row four, column four.

$$\begin{array}{c} \\ \\ \\ \\ \\ \\ \end{array}\begin{array}{cccccc|c} y_1 & y_2 & s_1 & s_2 & s_3 & s_4 & \\ \hline 0 & 0 & 1 & 0 & .5 & 10 & 1,860,000 \\ 1 & 0 & 0 & 0 & .5 & -10 & 1,060,000 \\ 0 & 1 & 0 & 0 & -1 & 0 & 80,000 \\ 0 & 0 & 0 & 1 & .5 & -10 & 260,000 \\ \hline 0 & 0 & 0 & 0 & 1.25 & 40 & -4,500,000 \end{array}\begin{array}{l} \\ -R_4 + R_1 \\ R_4 + R_2 \\ \\ 10R_4 \\ -4R_4 + R_5 \end{array}$$

This program is optimal after Stage I.
1,060,000 kg of tomatoes should be used for canned whole tomatoes and 80,000 kg of tomatoes should be used for sauce at a minimum cost of \$4,500,000.

64. Let y_1 = the number of runs of type I and let y_2 = the number of runs of type II.

Minimize $w = 15,000y_1 + 6000y_2$

Maximize $z = -w = -15,000y_1 - 6000y_2$

subject to:
$$\begin{aligned} 3000y_1 + 3000y_2 &= 18,000 \\ 2000y_1 + 1000y_2 &= 7000 \\ 2000y_1 + 3000y_2 &\geq 14,000 \\ y_1 \geq 0,\ y_2 &\geq 0 \end{aligned}$$

The initial simplex tableau is

$$\begin{array}{ccccccc|c} y_1 & y_2 & s_1 & s_2 & s_3 & s_4 & s_5 & \\ \hline 3000 & 3000 & -1 & 0 & 0 & 0 & 0 & 18,000 \\ 3000 & 3000 & 0 & 1 & 0 & 0 & 0 & 18,000 \\ \boxed{2000} & 1000 & 0 & 0 & -1 & 0 & 0 & 7000 \\ 2000 & 1000 & 0 & 0 & 0 & 1 & 0 & 7000 \\ 2000 & 3000 & 0 & 0 & 0 & 0 & -1 & 14,000 \\ \hline 15,000 & 6000 & 0 & 0 & 0 & 0 & 0 & 0 \end{array}$$

For Stage I, pivot on the 2000 in row three, column one.

$$\begin{array}{ccccccc|c|l} y_1 & y_2 & s_1 & s_2 & s_3 & s_4 & s_5 & & \\ \hline 0 & 1500 & -1 & 0 & 1.5 & 0 & 0 & 7500 & -3000R_3 + R_1 \\ 0 & 1500 & 0 & 1 & 1.5 & 0 & 0 & 7500 & -3000R_3 + R_2 \\ 1 & .5 & 0 & 0 & -.0005 & 0 & 0 & 3.5 & .0005R_3 \\ 0 & 0 & 0 & 0 & 1 & 1 & 0 & 0 & -2000R_3 + R_4 \\ 0 & \boxed{2000} & 0 & 0 & 1 & 0 & -1 & 7000 & -2000R_3 + R_5 \\ \hline 0 & -1500 & 0 & 0 & 7.5 & 0 & 0 & -52,500 & -15,000R_3 + R_6 \end{array}$$

Pivot on the 2000 in row five, column two.

$$\begin{array}{ccccccc|c|l} y_1 & y_2 & s_1 & s_2 & s_3 & s_4 & s_5 & & \\ \hline 0 & 0 & -1 & 0 & .75 & 0 & \boxed{.75} & 2250 & -1500R_5 + R_1 \\ 0 & 0 & 0 & 1 & .75 & 0 & .75 & 2250 & -1500R_5 + R_2 \\ 1 & 0 & 0 & 0 & -.00075 & 0 & .00025 & 1.75 & -.5R_5 + R_3 \\ 0 & 0 & 0 & 0 & 1 & 1 & 0 & 0 & \\ 0 & 1 & 0 & 0 & .0005 & 0 & -.0005 & 3.5 & .0005R_5 \\ \hline 0 & 0 & 0 & 0 & 8.25 & 0 & -.75 & -47,250 & 1500R_5 + R_6 \end{array}$$

(*continued next page*)

Pivot on the .75 in row one column seven.

$$\begin{array}{ccccccc|c|l} y_1 & y_2 & s_1 & s_2 & s_3 & s_4 & s_5 & & \\ 0 & 0 & -\frac{4}{3} & 0 & 1 & 0 & 1 & 3000 & \frac{4}{3}R_1 \\ 0 & 0 & \boxed{1} & 1 & 0 & 0 & 0 & 0 & -.75R_1+R_2 \\ 1 & 0 & \frac{1}{3000} & 0 & -.001 & 0 & 0 & 1 & -.00025R_1+R_3 \\ 0 & 0 & 0 & 0 & 1 & 1 & 0 & 0 & \\ 0 & 1 & -\frac{1}{1500} & 0 & .001 & 0 & 0 & 5 & .0005R_1+R_5 \\ \hline 0 & 0 & -1 & 0 & 9 & 0 & 0 & -45,000 & .75R_1+R_6 \end{array}$$

For Stage II, pivot on the 1 in row two, column three.

$$\begin{array}{ccccccc|c|l} y_1 & y_2 & y_3 & y_4 & y_5 & y_6 & y_7 & & \\ 0 & 0 & 0 & \frac{4}{3} & 1 & 0 & 1 & 3000 & \frac{4}{3}R_2+R_1 \\ 0 & 0 & 1 & 1 & 0 & 0 & 0 & 0 & \\ 1 & 0 & 0 & -\frac{1}{3000} & -.001 & 0 & 0 & 1 & -\frac{1}{3000}R_2+R_3 \\ 0 & 0 & 0 & 0 & 1 & 1 & 0 & 0 & \\ 0 & 1 & 0 & \frac{1}{1500} & .001 & 0 & 0 & 5 & \frac{1}{1500}R_2+R_5 \\ \hline 0 & 0 & 0 & 1 & 9 & 0 & 0 & -45,000 & R_2+R_6 \end{array}$$

The company should produce 1 run of type I and 5 runs of type II for a minimum cost of $45,000.

Case 7 Cooking with Linear Programming

1. Let x_1 = the number of 100 gram units of feta cheese, x_2 = the number of 100 gram units of lettuce, x_3 = the number of 100 gram units of salad dressing, and x_4 = the number of 100 gram units of tomato.

Maximize $z = 4.09x_1 + 2.37x_2 + 2.5x_3 + 4.64x_4$

subject to:

$$\begin{aligned} 263x_1 + 14x_2 + 448.8x_3 + 21x_4 &< 260 \\ 492.5x_1 + 36x_2 + 5x_4 &> 210 \\ 10.33x_1 + 1.62x_2 + .85x_4 &> 6 \\ x_1 + x_2 + x_3 + x_4 &< 4 \\ x_3 &\geq .3125 \end{aligned}$$

$x_1 \geq 0,\ x_2 \geq 0,\ x_3 \geq 0,\ x_4 \geq 0$

Using a computer with linear programming software, the optimal solution is:

$x_1 = .243037,\ x_2 = 2.35749,\ x_3 = .3125,$

$x_4 = 1.08698$

Converting to kitchen units gives approximately $\frac{1}{6}$ cup feta cheese $4\frac{1}{4}$ cups of lettuce, $\frac{1}{8}$ cup of salad dressing and $\frac{7}{8}$ of a tomato for a salad with a maximum of about 12.41 g of carbohydrates.

2. Let y_1 = the number of 100 gram units of beef, y_2 = the number of 100 gram units of oil, y_3 = the number of 100 gram units of onion, and y_4 = the number of 100 gram units of soy sauce.

Minimize $215y_1 + 884y_2 + 38y_3 + 60y_4$

subject to:

$$\begin{aligned} y_2 + 8.63y_3 + 5.57y_4 &< 10 \\ 26y_1 + 1.16y_3 + 10.51y_4 &> 50 \\ 6.4y_3 &> 3.5 \\ y_2 &\geq .045 \end{aligned}$$

$y_1 \geq 0,\ y_2 \geq 0,\ y_3 \geq 0,\ y_4 \geq 0$

Using a computer with linear programming software, the optimal solution is

$y_1 = 1.51873,\ y_2 = .045,\ y_3 = .546875,$

$y_4 = .939941$

Converting to kitchen units gives approximately $5\frac{1}{3}$ ounces of beef, $\frac{1}{3}$ tablespoon of oil, $\frac{1}{2}$ an onion, and $5\frac{1}{4}$ tablespoons of soy sauce for a stir-fry with a minimum 443.48 calories.

Chapter 8 Sets and Probability

Section 8.1 Sets

1. $3 \in \{2, 5, 7, 9, 10\}$
This statement is false. The number 3 is not an element of the set.

3. $9 \notin \{2, 1, 5, 8\}$
The statement is true. 9 is not an element of the set.

5. $\{2, 5, 8, 9\} = \{2, 5, 9, 8\}$
The statement is true. The sets contain exactly the same elements, so they are equal. The ordering of the elements in a set is unimportant.

7. {all whole numbers greater than 7 and less than 10} = {8, 9}
The statement is true. 8 and 9 are the only such numbers.

9. $\{x \mid x \text{ is an odd integer}, 6 \le x \le 18\}$
$= \{7, 9, 11, 15, 17\}$
The statement is false. The number 13 should be included.

11. Answers vary.
Possible answer:
$\{0\}$ is a set containing 1 element, namely zero.
$\{\varnothing\}$ is a set containing 1 element, namely, a mathematical symbol representing the empty set.

13. Since every element of A is also an element of U, $A \subseteq U$.

15. $A \not\subseteq E$ since A contains elements that do not belong to E, namely –3, 0, 3.

17. $\varnothing \subseteq A$ since the empty set is a subset of every set.

19. $D \subseteq B$, since every element of D is also an element of B.

21. $\{A, B, C\}$ contains 3 elements. Therefore, it has $2^3 = 8$ subsets.

23. $\{x \mid x \text{ is an integer strictly between 0 and 8}\}$
$= \{1, 2, 3, 4, 5, 6, 7\}$. This set contains 7 elements. Therefore it has $2^7 = 128$ subsets.

25. $\left\{ x \middle| x \text{ is an integer less than or equal to 0 or greater than or equal to 8} \right\}$

27. Answers vary.

29. $\{8, 11, 15\} \cap \{8, 11, 19, 20\} = \{8, 11\}$

31. $\{6, 12, 14, 16\} \cap \{6, 14, 19\} = \{6, 14\}$
$\{6, 14\}$ is the set of all elements belonging to both of the sets, so it is the intersection of those sets.

33. $\{3, 5, 9, 10\} \cup \varnothing = \{3, 5, 9, 10\}$.

35. $\{1, 2, 4\} \cup \{1, 2\} = \{1, 2, 4\}$
The answer set $\{1, 2, 4\}$ consists of all elements belonging to the first set, to the second set, or to both sets, and therefore it is the union of the first two sets.

37. $X \cap Y = \{b, 1, 3\}$ since only these elements are contained in both.

39. $X' = \{d, e, f, 4, 5, 6\}$ since these are the elements that are in U but not in X.

41. $X' \cap Y'$
$= \{d, e, f, 4, 5, 6\} \cap \{a, c, e, 2, 4, 6\}$
$= \{e, 4, 6\}$

43. $X \cup (Y \cap Z)$
$Y \cap Z = \{b, d, f, 1, 3, 5\} \cap \{b, d, 2, 3, 5\}$
$= \{b, d, 3, 5\}$
$X \cup (Y \cap Z) = \{a, b, c, 1, 2, 3\} \cup \{b, d, 3, 5\}$
$= \{a, b, c, d, 1, 2, 3, 5\}$

45. M' is the set of all students in this school not taking this course.

47. $N \cap P$ is the set of all students in this school taking both accounting and philosophy.

49. A pair of sets is disjoint if the two sets have no elements in common. The pairs of these sets that are disjoint are C and D, A and E, C and E, and D and E.

51. A' is the set of all stocks with a high price less than or equal to \$50. A' = {First Solar, Ford}

53. $A' \cap B'$ is the set of all stocks with a high price less than or equal to \$50 and a last price either less than or equal to \$25 or greater than or equal to \$80. $A' \cap B' =$ {Ford}.

55. $M \cap E$ is the set of all male employed clients.

57. $M' \cup S'$ is the set of all female or married clients.

59. {Television, Magazines}

61. $\varnothing$

63. $F =$ {Comcast, Direct TV, Dish Network, Time Warner}

65. $H =$ {Dish Network, Time Warner, Verizon, Cox}

67. $F \cup G =$ {Comcast, Direct TV, Dish Network, Time Warner} $\cup$ {Dish Network, Time Warner} ={Comcast, Direct TV, Dish Network, Time Warner}

69. $I' =$ {Time Warner, Verizon, Cox}

71. {Soybeans, Rice, Cotton}

73. {Wheat, Corn, Soybeans, Rice, Cotton}

Section 8.2 Applications of Venn Diagrams

1. $A \cap B'$
Shade the region inside B that is outside A.

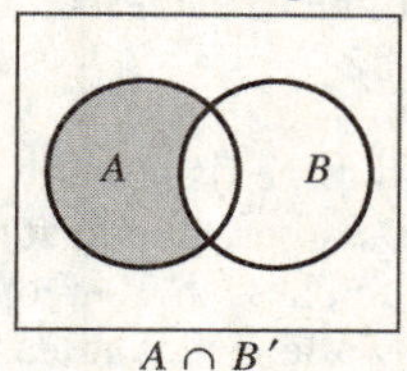

$A \cap B'$

3. $B' \cup A'$
Shade the region outside B and the region outside of region A.

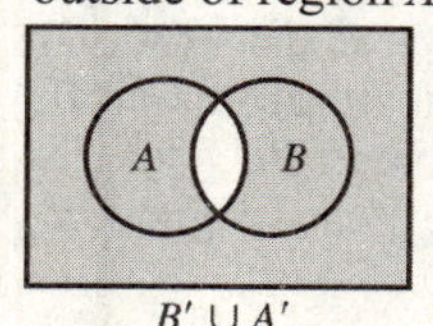

$B' \cup A'$

5. $B' \cup (A \cap B')$
First shade the common region that is in A and outside B. Then shade the rest of the region outside B.

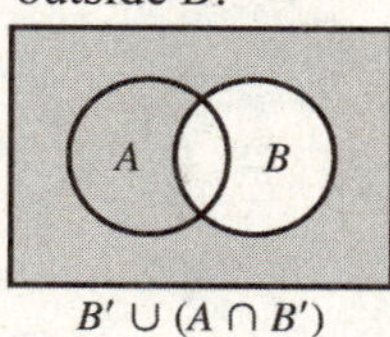

$B' \cup (A \cap B')$

7. $U' = \varnothing$
There are no elements outside the universal set. Therefore, there is no region to be shaded.

9. Three sets divide the universal set into at most <u>8</u> regions.

11. $(A \cap C') \cup B$
First find $A \cap C'$, the region in A *and* not in C.

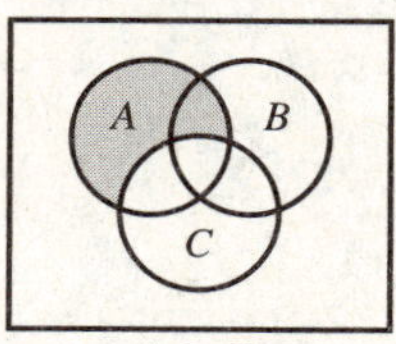

$A \cap C'$

For the union, we want the region in $(A \cap C')$ *or* in B, or both.

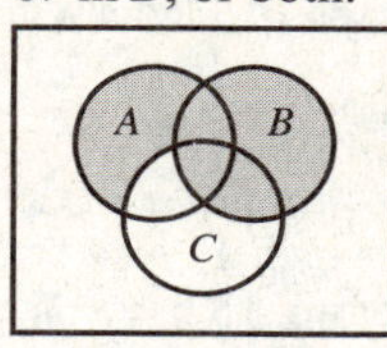

$(A \cap C') \cup B$

13. $A' \cap (B \cap C)$
First find $B \cap C$, the region in B *and* in C.

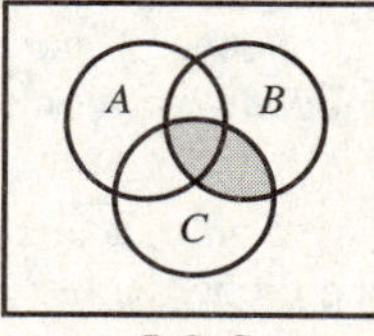

$B \cap C$

Now find A', the region not in A.

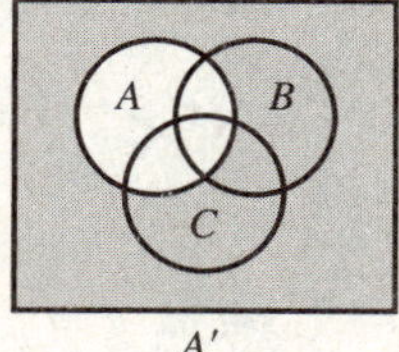

A'

(*continued next page*)

For the intersection, we want the region in A' *and* in $(B \cap C)$.

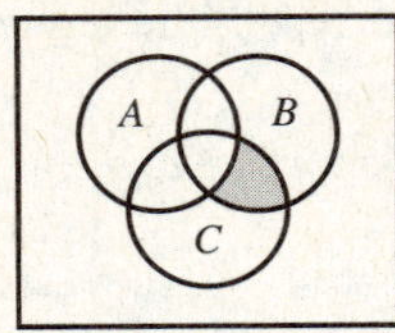

$A' \cap (B \cap C)$

15. $(A \cap B') \cup C$

First find $A \cap B'$, the region in *A and* not in *B*.

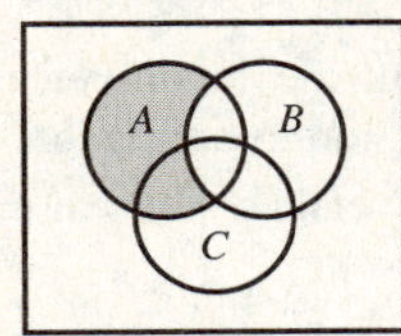

$A \cap B'$

For the union, we want the region in $(A \cap B')$ *or* in *C*, or both.

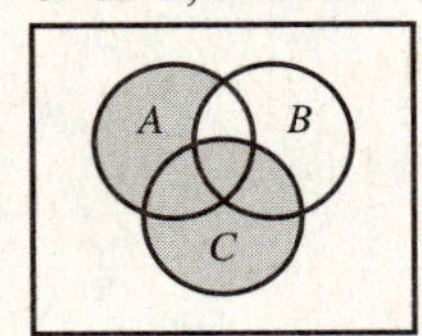

$(A \cap B') \cup C$

17. Percentage of assisted living residents with only dementia:
100% – 24% – 33% – 25% = 18%

19. There are 26 applicants with sales experience and a college degree, of which 11 also have a real estate license. Therefore, 15 have only sales experience and a college degree. There are 16 applicants with sales experience and a real estate license, of which 11 also have a college degree. Therefore, there are 5 applicants with only sales experience and a real estate license. There are 66 – 15 – 11 – 5 = 35 applicants with only sales experience.

Using similar reasoning, we find that there are 10 applicants with only a college degree, 4 applicants with a only a college degree and a real estate license, and 3 applicants with only a real estate license. This is illustrated in the Venn diagram below.

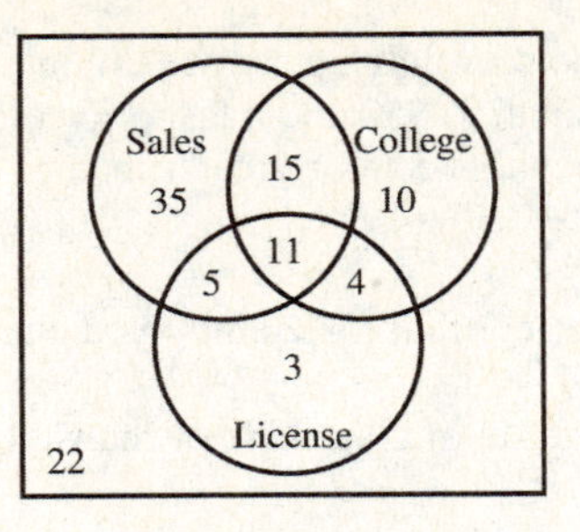

a. There were 35 + 15 + 11 + 5 + 10 + 4 + 3 + 22 = 105 applicants.

b. 22 + 3 + 4 + 10 =39 applicants did not have sales experience.

c. 15 applicants had sales experience and a college degree, but not a real estate license.

d. Three applicants had only a real estate license.

21. a. The total number of people surveyed is the sum of all those who caught at least one fish minus the sum of those who caught at least two fish minus the number of those who caught all three fish.
(124 + 133 + 146) – (75 + 67 + 79 – 45)
= 227

b. The total number of people who caught at least one walleye or at least one smallmouth bass equals the sum of the number of those who caught at least one walleye and the number of those who caught at least one smallmouth bass minus the number of those who caught at least one of each fish.
124 + 133 – 75 = 182

c. 45 people caught all three fish, while 75 people caught at least one walleye and at least one smallmouth bass, so 30 people caught at least one walleye and at least one smallmouth bass only. 67 people caught at least one walleye and at least one yellow perch, so 67 – 45 = 22 people caught at least one walleye and at least one yellow perch only. Therefore, 124 – (30 + 45 + 22) = 27 people caught at least one walleye only. The information in the exercise is summarized in the Venn diagram on the next page.

(*continued next page*)

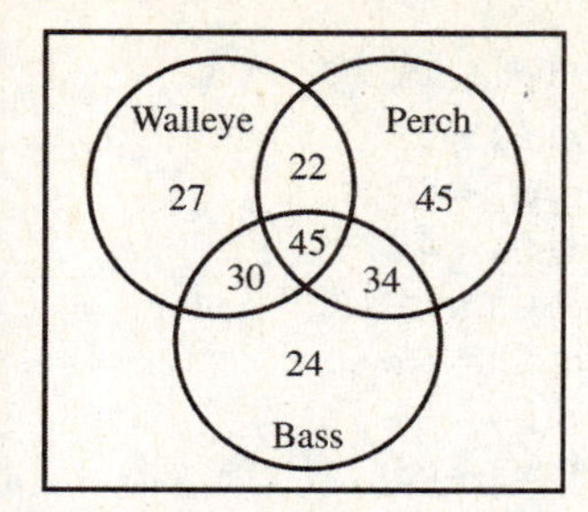

23. In the diagram, the regions are labeled (a) to (g). First, the number for each region is found by starting with innermost region (d). 15 had all three; then proceed to work outward as follows:
Region (a):
Since 25 had A, 25 – (2 + 15 + 1) = 7.
Region (b):
Since 17 had A and B, 17 – 15 = 2.
Region (c):
Since 27 had B, 27 – (2 + 15 + 7) = 3.
Region (d):
15 had all three.
Region (e):
Since 22 had B and Rh, 22 – 15 = 7.
Region (f):
Since 30 had Rh, 30 – (1 + 15 + 7) = 7.
Region (g):
12 had none.
Region with no label:
Since 16 had A and Rh, 16 – 15 = 1.

a. 7 + 2 + 3 + 15 + 7 + 7 + 12 + 1 = 54 patients were represented.

b. 7 + 3 + 7 = 17 patients had exactly one antigen.

c. 2 + 1 + 7 = 10 patients had exactly two antigens.

d. Since a person having only the Rh antigen has type O-positive blood, 7 had type O-positive blood.

e. Since a person having A, B, and Rh antigens is AB-positive, 15 had AB-positive blood.

f. Since a person having only the B antigen is B-negative, 3 had B-negative blood.

g. Since a person having neither A, B, nor Rh antigens is O-negative, 12 had O-negative blood.

h. Since a person having A and Rh antigens is A-positive, 1 had A-positive blood.

25. a.
$$n(B \cup C) = n(B) + n(C) - n(B \cap C)$$
$$= 218,000 + 182,000 - 92,000$$
$$= 308,000$$

b.
$$n[F'] = 90,000 + 23,000 + 92,000 + 22,000$$
$$= 227,000$$

c. $n(C \cap B) = 92,000$

d.
$$n((C \cup W) \cap A)$$
$$= 90,000 + 23,000$$
$$= 113,000$$

27. a. $n(A \cap E) = 23,682$ thousand

b.
$$n(A \cup H)$$
$$= 45,873 + 37,089 - 7801$$
$$= 75,161 \text{ thousand}$$

c.
$$n[B \cap H'] = 20,223 - 5824$$
$$= 14,399 \text{ thousand}$$

d.
$$n[(A \cup B) \cap G] = 7350 + 4054$$
$$= 11,404 \text{ thousand}$$

29. Answers will vary.

31.
$$n(A \cup B) = n(A) + n(B) - n(A \cap B)$$
$$30 = 12 + 27 - n(A \cap B)$$
$$30 = 39 - n(A \cap B)$$
$$n(A \cap B) = 9$$

33.
$$n(A \cup B) = n(A) + n(B) - n(A \cap B)$$
$$35 = 13 + n(B) - 5$$
$$n(B) = 27$$

35. $n(A) = 28$, $n(B) = 12$, $n(A \cup B) = 30$, $n(A') = 19$

This gives

$n(A \cup B) = n(A) + n(B) - n(A \cap B) \Rightarrow$
$30 = 28 + 12 - n(A \cap B)$
$n(A \cap B) = 28 + 12 - 30 = 10$

$n(A) = 28$ and $n(A \cap B) = 10$, so $n(A \cap B') = 18$

$n(B) = 12$ and $n(A \cap B) = 10$, so $n(B \cap A') = 2$

Since $n(A') = 19$, 2 of which are accounted for in $B \cap A'$, 17 remain in $A' \cap B'$.

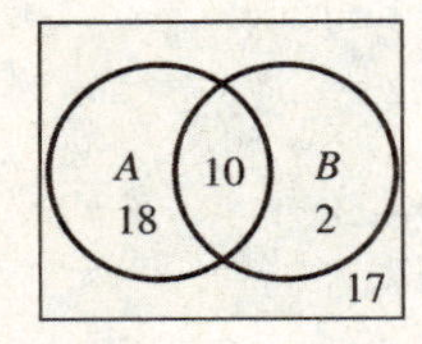

37. $n(A') = 28$, $n(B) = 25$, $n(A' \cup B') = 45$, $n(A \cap B) = 12$

$n(B) = 25$ and $n(A \cap B) = 12$, so $n(B \cap A') = 13$

Since $n(A') = 28$, of which 13 are accounted for, 15 are in $A' \cap B'$.

$$n(A' \cup B') = n(A') + n(B') - n(A' \cap B')$$
$$45 = 28 + n(B') - 15$$
$$45 = 13 + n(B')$$
$$32 = n(B')$$

15 are in $A' \cap B'$, so the rest are in $A \cap B'$, and $n(A \cap B') = 17$.

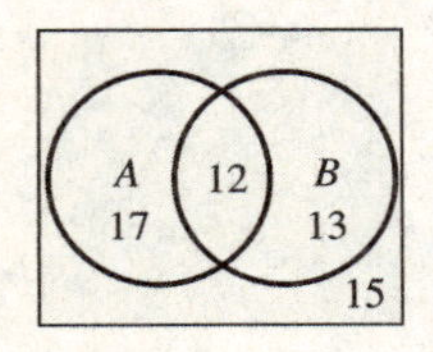

39.

$$n(A) = 54$$
$$n(A \cap B) = 22$$
$$n(A \cup B) = 85$$
$$n(A \cap B \cap C) = 4$$
$$n(A \cap C) = 15$$
$$n(B \cap C) = 16$$
$$n(C) = 44$$
$$n(B') = 63$$

Start with $A \cap B \cap C$. Now $n(A \cap C) = 15$, of which 4 are in $A \cap B \cap C$, so $n(A \cap B' \cap C) = 11$. $n(B \cap C) = 16$, of which 4 are in $A \cap B \cap C$, so $n(B \cap C \cap A') = 12$.

$n(C) = 44$, so 17 are in $C \cap A' \cap B'$.

$n(A \cap B) = 22$, so 18 are in $A \cap B \cap C'$.

$n(A) = 54$, so $54 - 11 - 18 - 4 = 21$ are in $A \cap B' \cap C'$.

$$n(A \cup B) = n(A) + n(B) - n(A \cap B)$$
$$85 = 54 + n(B) - 22$$
$$53 = n(B)$$

This leaves 19 in $B \cap A' \cap C'$. $n(B') = 63$, of which $21 + 11 + 17 = 49$ are accounted for, leaving 14 in $A' \cap B' \cap C'$.

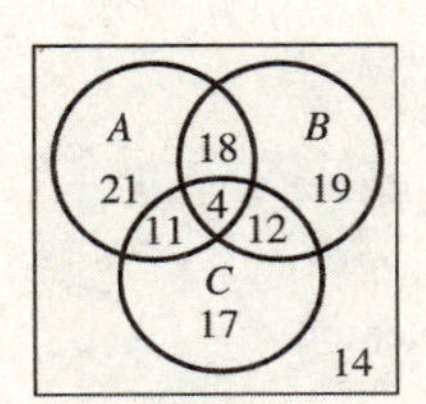

41. $(A \cap B)'$ is the complement of the intersection of A and B; hence it contains all elements not in $A \cap B$.

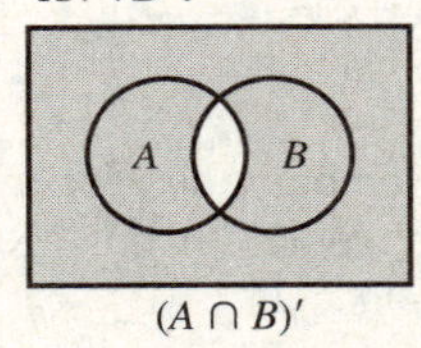

$(A \cap B)'$

$A' \cup B'$ is the union of the complements of A and B; hence it contains any element that is either not in A or not in B.

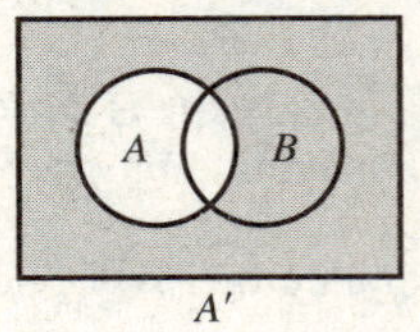

A'

(*continued on next page*)

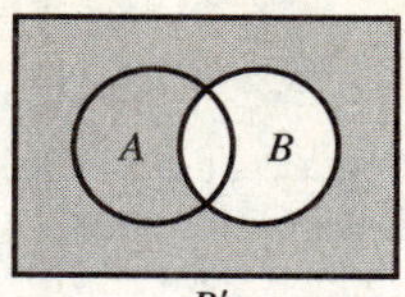

B'

Note that $(A \cap B)' = A' \cup B'$, as claimed.

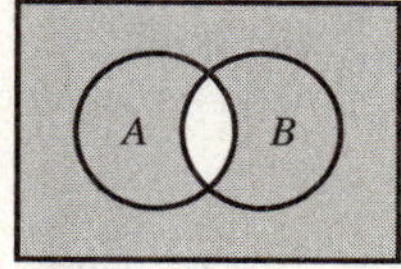

$A' \cup B'$

43. $A \cup (B \cap C)$ contains all points in A and the points where B and C overlap.

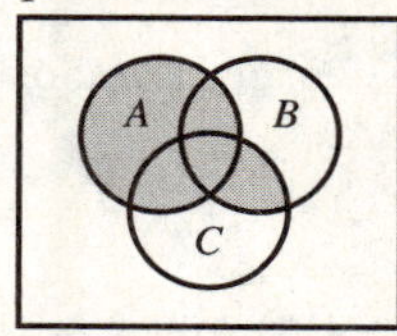

$A \cup (B \cap C)$

$(A \cup B) \cap (A \cup C)$ contains the points where $A \cup B$ and $A \cup C$ overlap.

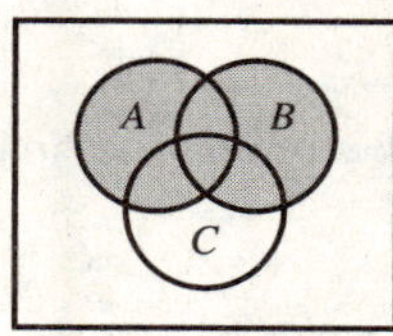

$A \cup B$

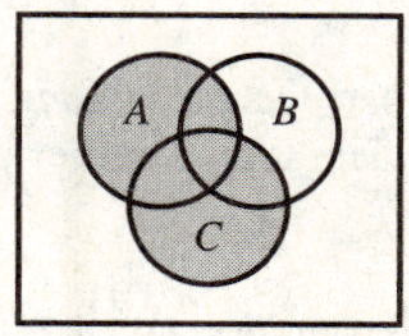

$A \cup C$

Note that $A \cup (B \cap C) = (A \cup B) \cap (A \cup C)$, as claimed.

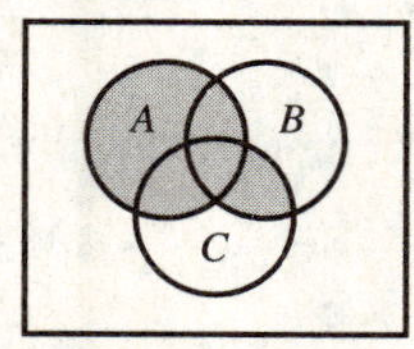

$(A \cup B) \cap (A \cup C)$

45. The complement of *A* and (intersection) *B* equals the union of the complement of *A* and the complement of *B*.

47. A union (*B* intersect *C*) equals (*A* union *B*) intersect (*A* union *C*).

Section 8.3 Introduction to Probability

1. Answers vary. Possible answer:
A coin or die is fair if the probability of any result is the same, that is, all results are equally likely.

3. There are 12 months in a year. The sample space is the set
{January, February, March, …, December}.

5. There are 80 points on the test. The sample space is the set {0, 1, 2, …, 80}.

7. There are only 2 choices the management can make. The sample space is the set
{go ahead, cancel}.

9. There are four possibilities, so the sample space is the set {Q1, Q2, Q3, Q4}.

11. Answers vary. Possible answer: Disjoint events are events that cannot occur at the same time.

13. Wearing a hat and wearing glasses are not disjoint, because it is possible to wear both at the same time.

15. Being a doctor and being under 5 years old are disjoint, since it is impossible to be under 5 years old and be a doctor.

17. Being a female and being a pilot are not disjoint, since there are many female pilots.

19. *S* = {(Connie, Kate), (Connie, Lindsey), (Connie, Jackie), (Connie, Taisa), (Connie, Nicole), (Kate, Lindsey), (Kate, Jackie), (Kate, Taisa), (Kate, Nicole), (Lindsey, Jackie), (Lindsey, Taisa), (Lindsey, Nicole), (Jackie, Taisa), (Jackie, Nicole), (Taisa, Nicole)}

a. Taisa is selected.
{(Connie, Taisa),(Kate, Taisa), (Lindsey, Taisa), (Jackie, Taisa), (Taisa, Nicole)}

b. The two names selced have the same number of letters.
{(Connie, Jackie), (Connie, Nicole), (Jackie, Nicole)}

21. The sample space is {(forest sage, rag painting), (forest sage, colorwash), (evergreen whisper, rag painting), (evergreen whisper, colorwash), (opaque emerald, rag painting), (opaque emerald, colorwash)}

a. There are 3 combinations with colorwash: {(colorwash, forest sage), (colorwash, evergreen whisper), (colorwash, opaque emerald)}.

$$P_C = \frac{n(C)}{n(S)} = \frac{3}{6} = \frac{1}{2}.$$

b. There are 4 combinations with opaque emerald or rag painting: {(opaque emerald, rag painting), (opaque emerald, colorwash), (forest sage, rag painting), (evergreen whisper, rag painting)}.

$$P_{O \cup R} = \frac{n(O \cup R)}{n(S)} = \frac{4}{6} = \frac{2}{3}$$

23. There are 2 possibilities for size and 3 possibilities for color. The sample space is {(10-foot, beige), (10-foot, forest green) (10-foot, rust), (12-foot, beige), (12-foot, forest green), (12-foot, rust)}

a. Doug buys a 12-foot forest green umbrella. There is 1 possibility: {(12-foot, forest green)}.

$$P_{12' \cap G} = \frac{n(12' \cap G)}{n(S)} = \frac{1}{6}$$

b. Doug buys a 10-foot umbrella. There are 3 possibilities: {(10-foot, beige), (10-foot, forest green), (10-foot, rust)}

$$P_{10'} = \frac{n(10')}{n(S)} = \frac{3}{6} = \frac{1}{2}$$

c. Doug buys a rust-colored umbrella. There are 2 possibilities: {(10-foot, rust), (12-foot, rust)}

$$P_R = \frac{n(R)}{n(S)} = \frac{2}{6} = \frac{1}{3}$$

25. Use the sample space $S = \{1, 2, 3, 4, 5, 6\}$. Let E be the event "getting a number less than 4." So, $E = \{1, 2, 3\}$. Since S contains six elements,

$$P(E) = \frac{3}{6} = \frac{1}{2}.$$

27. Use the sample space $S = \{1, 2, 3, 4, 5, 6\}$. Let E be the event "getting a 2 or a 5." So, $E = \{2, 5\}$. Since S contains 6 elements,

$$P(E) = \frac{2}{6} = \frac{1}{3}.$$

29. {male beagle, male boxer, male collie, male Labrador, female beagle, female boxer, female collie, female Labrador}

31. There are four possibilities, {male beagle, male boxer, male collie, male Labrador}.

$$P(E) = \frac{4}{8} = \frac{1}{2}.$$

33. There is one possibility, {female Labrador}.

$$P(E) = \frac{1}{8}.$$

35. There are six possibilities, {male beagle, male boxer, male collie, female beagle, female boxer, female collie}.

$$P(E) = \frac{6}{8} = \frac{3}{4}.$$

For exercises 37, the total number of fatalities in 2010 is 4690.

37. $P(E) = \frac{646}{4690} \approx .1377$

39. The table lists religious service attendance as number of respondents. The total number of respondents is 1974.

a. Several times a year: $P(E) = \frac{191}{1974} \approx .0968$

b. 2–3 times a month: $P(E) = \frac{168}{1974} \approx .0851$

c. nearly every week or more frequently:

$$P(E) = \frac{83 + 380 + 140}{1974} = \frac{603}{1974} \approx .3055$$

41. a. G' = The person is not overweight.

b. $F \cap G$ = The person has a family history of heart disease and is overweight.

c. $E \cup G'$ = The person smokes or is not overweight.

43. The total number of females who were surveyed is 2826.

a. six feet tall or taller: $\frac{3+1}{2826}=\frac{4}{2826}\approx .0014.$

b. less than 63 inches tall:
$\frac{354+924}{2826}=\frac{1278}{2826}\approx .4522.$

c. between 63 and 68.99 inches tall:
$\frac{1058+421}{2826}=\frac{1479}{2826}\approx .5234.$

45. This experiment is possible, since all probabilities are non-negative and .09 + .32 + .21 + .25 + .13 = 1

47. This experiment is not possible, since the sum of the probabilities is $\frac{39}{40}$, which is less than 1.

49. This experiment is not possible, since a probability cannot be negative.

Section 8.4 Basic Concepts of Probability

1. Let G be the event of the marble landing in a green slot, and let B be the event of the marble landing in a black slot. There are 2 slots that are green and 18 slots that are black, so

$$P(G\cup B)=P(G)+P(B)-P(G\cap B)$$
$$=\frac{2}{38}+\frac{18}{38}-\frac{0}{38}=\frac{20}{38}=\frac{10}{19}.$$

3. Let O be the event of the marble landing in an odd slot, and let B be the event of the marble landing in a black slot. There are 18 slots that are odd, 18 slots that are black, and 8 slots that are odd and black, so

$$P(O\cup B)=P(O)+P(B)-P(O\cap B)$$
$$=\frac{18}{38}+\frac{18}{38}-\frac{8}{38}=\frac{28}{38}=\frac{14}{19}.$$

5. Let E be the event of the marble landing in a slot numbered 0, 00, 1, 2, or 3. There are 5 such slots, so $P(E)=\frac{5}{38}.$

7. Let E be the event of the marble landing in a slot numbered 25–36. There are 12 such slots, so $P(E)=\frac{12}{38}=\frac{6}{19}.$

For Exercises 9–13, count outcomes by referring to Figure 8.22 in the text.

9. a. $P(\text{sum is }8)=\frac{5}{36}$

b. $P(\text{sum is }9)=\frac{4}{36}=\frac{1}{9}$

c. $P(\text{sum is }10)=\frac{3}{36}=\frac{1}{12}$

d. $P(\text{sum is }13)=\frac{0}{36}=0$

11. a. $P(\text{not more than }5)=\frac{10}{36}=\frac{5}{18}$

b. $P(\text{not less than }8)=\frac{15}{36}=\frac{5}{12}$

c. $P(\text{between 3 and 7 (exclusive)})=\frac{12}{36}=\frac{1}{3}$

13. The shoes come in two shades of beige (light and dark) and black, so $P(\text{shoes are black})=\frac{1}{3}.$

15. a. $P(\text{a sister or an aunt})=\frac{2}{10}+\frac{3}{10}$
$=\frac{5}{10}=\frac{1}{2}$

b. $P(\text{a sister or a cousin})=\frac{2}{10}+\frac{2}{10}$
$=\frac{4}{10}=\frac{2}{5}$

c. $P(\text{a sister or her mother})=\frac{2}{10}+\frac{1}{10}$
$=\frac{3}{10}$

17.

E F
.11 .19 .32
.38

a. $P(E\cup F)=P(E)+P(F)-P(E\cap F)$
$=.30+.51-.19=.62$

b. $P(E'\cap F)=P(F)-P(E\cap F)$
$=.51-.19=.32$

c. $P(E \cap F') = P(E) - P(E \cap F)$
$= .30 - .19 = .11$

d. $P(E' \cup F') = P(E') + P(F') - P(E' \cap F')$
$= .70 + .49 - .38 = .81$

19. 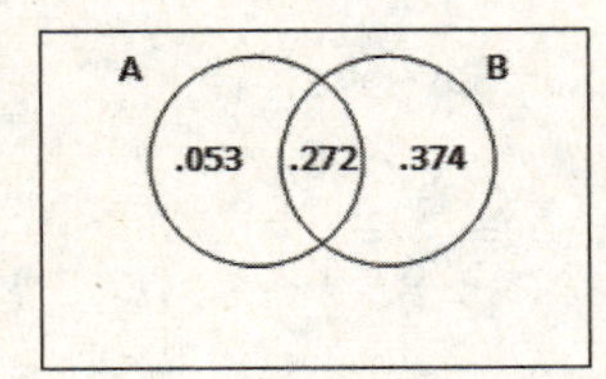

a. $P(A \cup B) = P(A) + P(B) - P(A \cap B)$
$= .325 + .646 - .272 = .699$

b. $P(A' \cap B) = P(B) - P(A \cap B)$
$= .646 - .272 = .374$

c. $P(A' \cap B') = P(A \cup B)' = 1 - P(A \cup B)$
$= 1 - .699 = .301$

d. $P(A' \cup B') = P(A') + P(B') - P(A' \cap B')$
$= (1 - .325) + (1 - .646) - .301$
$= .728$

21. 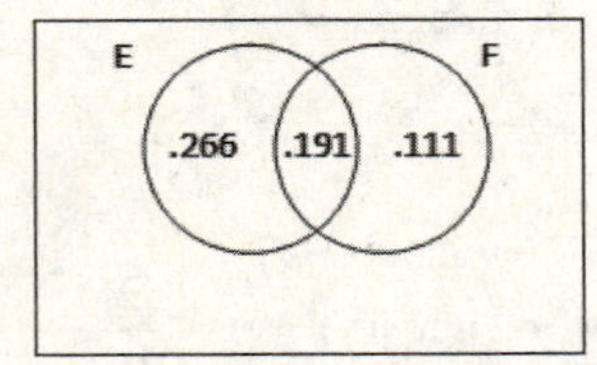

a. $P(E \cap F') = P(E) - P(E \cap F)$
$= .457 - .191 = .266$

b. $P(E' \cap F') = 1 - [P(E) + P(F) - P(E \cap F)]$
$= 1 - [.457 + .302 - .191]$
$= .432$

c. $P(E' \cup F') = 1 - P(E \cap F) = 1 - .191$
$= .809$

23. When rolling a die, there are 6 equally likely outcomes. The sample space is $S = \{1, 2, 3, 4, 5, 6\}$. Let E be the event "2 is rolled." Then $P(E) = \frac{1}{6}$ and $P(E') = \frac{5}{6}$. The odds in favor of rolling a 5 are $\frac{P(E)}{P(E')} = \frac{\frac{1}{6}}{\frac{5}{6}} = \frac{1}{5}$, written 1 to 5.

25. Let E be the event " 2, 3, 5, or 6 is rolled." Then $P(E) = \frac{4}{6} = \frac{2}{3}$ and $P(E') = \frac{1}{3}$. The odds in favor of rolling 2, 3, 5, or 6 are $\frac{P(E)}{P(E')} = \frac{\frac{2}{3}}{\frac{1}{3}} = 2$, written 2 to 1.

27. There are 3 yellow, 4 white, and 8 blue marbles.

a. Yellow: There are 3 ways to win and 12 ways to lose. The odds are 3 to 12 or 1 to 4.

b. Blue: There are 8 ways to win and 7 ways to lose; the odds are 8 to 7.

c. White: There are 4 ways to win and 11 ways to lose; the odds are 4 to 11.

29. Let E be the event "rolling a 7 or 11." Then $P(E) = \frac{8}{36}$ and $P(E') = \frac{28}{36}$. The odds in favor of rolling a 7 or 11 are $\frac{\frac{8}{36}}{\frac{28}{36}} = \frac{8}{28} = \frac{2}{7}$, written 2 to 7.

31. 21:79

33. 6:94

35. $\frac{43}{50}$

37. $\frac{6}{25}$

39. Using a graphing calculator,

a. P(the sum is 9 or more) $\approx .2778$

b. P(the sum is less than 7) $\approx .4167$

The probabilities compare very well to the results in Exercise 10.

41. Using a graphing calculator,

a. $P(\text{the sum is 5 or less}) \approx .0463$

b. $P(\text{neither a 1 nor a 6 is rolled}) \approx .2963$

43. $P(H) = \frac{192+459+498+544}{2023} = \frac{1693}{2023} \approx .8369$

45. $P(G \cap H) = \frac{459}{2023} \approx .2269$

47. $P(B' \cup H') = \frac{2023-544}{2023} = \frac{1479}{2023} \approx .7311$

49. $P(U \cap M) = \frac{12}{100} = .12$

51. $P(U' \cap B) = \frac{7+1}{100} = .08$

53. $P(O \cup M) = \frac{45+3-12}{100} = .60$

In exercises 55–57, let A = the event "working full time," let B = the event "working part time," let C indicate "working 0–19 hours," let D indicate "working 40–49 hours," and let E indicate "working less than 30 hours".

55. $P(A) = \frac{1021}{1352} \approx .7552$

57. $P(A \cap D) = \frac{468}{1352} \approx .3462$

59.

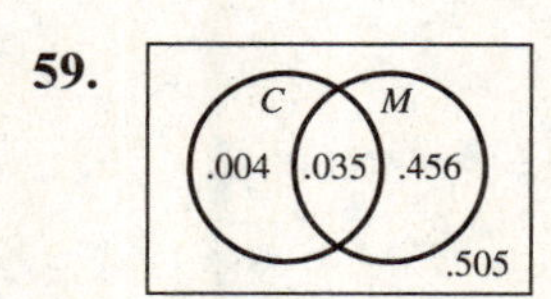

a. $P(C') = .456 + .505 = .961$

b. $P(M) = .035 + .456 = .491$

c. $P(M') = .004 + .505 = .509$

d. $P(M' \cap C') = 1 - (.004 + .035 + .456) = .505$

e. $P(C \cap M') = .004$ (inside C and outside M)

f. $P(C \cup M') = .004 + .035 + .505 = .544$

Section 8.5 Conditional Probability and Independent Events

1. Roll a fair die.

$$P(3 \mid \text{odd}) = \frac{P(3 \cap \text{odd})}{P(\text{odd})} = \frac{\frac{1}{6}}{\frac{1}{2}} = \frac{1}{3}$$

3. $P(\text{odd} \mid 3) = \frac{P(\text{odd} \cap 3)}{P(3)} = \frac{\frac{1}{6}}{\frac{1}{6}} = 1$

5. There are 6 doubles, 1 of which has a sum of 6.

$P(\text{sum of 6} \mid \text{double}) = \frac{1}{6}$

7. Since the first card is a heart, there are 51 cards remaining, 12 of them hearts.

$P(\text{second is heart} \mid \text{first is heart}) = \frac{12}{51} = \frac{4}{17}$

9. $P(\text{jack and 10}) = \frac{8 \cdot 4}{52 \cdot 51} \approx .012$

There are 8 possibilities for the first card (4 jacks and 4 tens), but for the second card there are only 4 (the 4 tens if a jack was picked or the 4 jacks if a 10 was picked).

11. Answers vary.

13. Answers vary.

15. No. Information about the event that it snows tomorrow does affect the probability that an instructor will be late for class. So, the events are dependent.

17. Yes. Knowledge of the event that it rains in the Amazon jungle gives no information about the occurrence or nonoccurrence of the event that an instructor in New York will write a difficult exam. So, the events are independent.

19. No, for a two-child family, the knowledge that each child is the same sex influences the probability of the event that there is at most one male. Eliminating the "one of each" possibility lowers the probability of "at most one male" from .75 to .5. However, the events are independent for a three-child family because the first event does not influence the probability of the second event. The probability of at most one male is .5 whether male-female mixes are allowed or not.

21. P(P | L)

$$= \frac{P(P \cap L)}{P(L)} = \frac{(.551)(.352)}{(.551)(.352) + (.258)(.648)}$$

$$= \frac{.1940}{.3612} \approx .537$$

23. Let M represent "male" and L represent "not in labor force." We are seeking

$$P(L \mid M) = \frac{P(L \cap M)}{P(M)}.$$

$$P(L \cap M) = P(L) + P(M) - P(L \cup M)$$
$$= .34 + .485 - .701 = .124$$

$$P(L \mid M) = \frac{P(L \cap M)}{P(M)} = \frac{.124}{.485} \approx .256.$$

The probability of not being in the labor force, given that the person is male is .256.

25. $$P(\text{ins} \mid \text{not grad}) = \frac{P(\text{ins} \cap \text{not grad})}{P(\text{not grad})} = \frac{192}{279} \approx .6882$$

27. $$P(\text{hs grad} \mid \text{no ins}) = \frac{P(\text{hs grad} \cap \text{no ins})}{P(\text{no ins})} = \frac{113}{330} \approx .3424$$

29. $P(\text{gd con in banks} \mid \text{ha con in corp})$

$$= \frac{P(\text{gd con in banks} \cap \text{ha con in corp})}{P(\text{ha con in corp})}$$

$$= \frac{14}{271} \approx .0517$$

31. $P(\text{gd con in corp} \mid \text{some con in banks})$

$$= \frac{P(\text{gd con in corp} \cap \text{some con in banks})}{P(\text{some con in banks})}$$

$$= \frac{123}{662} \approx .1858$$

33. $P(M) = .527$ (directly from the chart)

35. $P(M \cap C) = P(M \text{ and } C) = .042$

37. $$P(M \mid C) = \frac{P(M \cap C)}{P(C)} = \frac{.042}{.049} = \frac{6}{7} \approx .857$$

39. $$P(C \mid M) = \frac{P(M \cap C)}{P(M)} = \frac{.042}{.527} \approx .0797$$

$P(C) = .049$

Since $P(C \mid M) \neq P(C)$, they are dependent.

41. $P(C) = .0800$; $P(D) = .0050$; $P(C \cap D) = .0004$;

$$P(C \mid D) = \frac{P(C \cap D)}{P(D)} = \frac{.0004}{.0050} = .0800; \text{ Yes,}$$

$P(C) = P(C \mid D)$, therefore the two events are independent.

43. Complete the probability tree.

1st try — 2nd try — 3rd try

- .75 pass
- .25 fail
  - .80 pass
  - .20 fail
    - .70 pass
    - .30 fail

P(fails 1st and 2nd test)
$= P$(fails 1st) $\cdot$ P(fails 2nd | fails 1st)
$= (.25)(.20) = .05$

45. $P(\text{American flight is domestic})$

$$= \frac{517{,}971}{638{,}947} \approx .8107$$

47. $P(\text{Domestic flight is Delta})$

$$= \frac{729{,}997}{1{,}775{,}737} \approx .4111$$

49. $P(\text{never has stress} \mid \text{earns} \leq \$50{,}000)$

$$= \frac{35}{564} \approx .0621$$

51. $P(\text{never has stress} \mid \text{earns} > \$100{,}000)$

$$= \frac{5}{142} \approx .0352$$

53. $P(\text{earns} < \$50{,}000 \mid \text{sometimes has stress})$

$$= \frac{238}{440} \approx .5409$$

55. $P(\text{at least 1 hit}) = 1 - P(\text{no hit})$
$P(\text{no hit}) = P(\text{no hits in the first four})$
$= (1 - .32)(1 - .16)(1 - .08)(1 - .04)$
$= .5045$
$P(\text{at least 1 hit}) = 1 - .5045 = .4955$

57. $P(\text{have computer service})$
$= 1 - P(\text{no computer service})$
$= 1 - (.003)(.005)$
$= 1 - .000015$
$= .999985$
Answers vary. It is fairly realistic to assume independence because the chance of a failure of one computer does not usually depend on the failure of another, so long as the cause of failure does not lie with something the two systems have in common, like a power source.

59. a. The probability of success with one component is
$1 - .03 = .97;$
with 2, $1 - (.03)^2 = .9991;$
with 3, $1 - (.03)^3 = .999973;$
with 4, $1 - (.03)^4 = .99999919.$
Therefore, 4 (the original and 3 backups) will do the job.

b. Answers vary. It is probably reasonable to assume independence here so long as the cause of failure is some internal defect and the components are from different manufacture lots.

61. Let A be the event "student studies" and B be the event "student gets a good grade." We are told that
$P(A) = .6, P(B) = .7$, and $P(A \cap B) = .52.$
$P(A) \cdot P(B) = (.6)(.7) = .42$
Since $P(A) \cdot P(B)$ is not equal to $P(A \cap B)$, A and B are not independent. Rather, they are dependent events.

Section 8.6 Bayes' Formula

For Exercises 1:

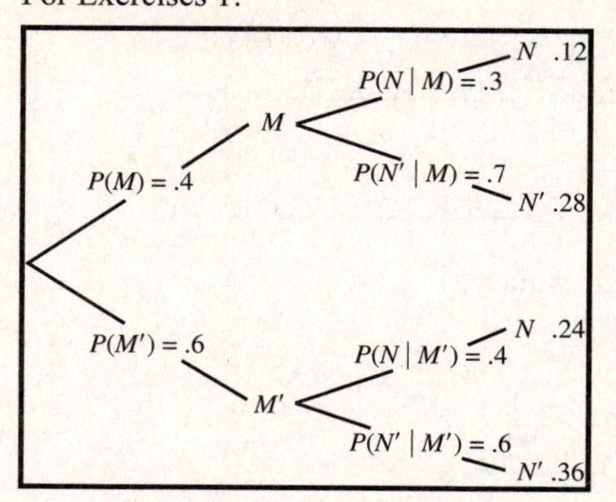

1. $P(M \mid N) = \dfrac{.12}{.12 + .24} = \dfrac{.12}{.36} = \dfrac{1}{3}$

For Exercises 3–5:

$P(R_1) = .05$
$P(Q \mid R_1) = .4$ — Q .02
$P(Q' \mid R_1) = .6$ — Q' .03
$P(R_2) = .6$
$P(Q \mid R_2) = .3$ — Q .18
$P(Q' \mid R_2) = .7$ — Q' .42
$P(R_3) = .35$
$P(Q \mid R_3) = .6$ — Q .21
$P(Q' \mid R_3) = .4$ — Q' .14

3. $P(R_1 \mid Q) = \dfrac{.02}{.02 + .18 + .21} = \dfrac{.02}{.41} = \dfrac{2}{41} \approx .0488$

5. $P(R_3 \mid Q) = \dfrac{.21}{.02 + .18 + .21} = \dfrac{.21}{.41} \approx .5122$

For Exercises 7:

$P(\text{jar 1}) = 1/2$ — jar 1 — $P(\text{white} \mid \text{jar 1}) = 1/3$ — white 1/6

$P(\text{jar 2}) = 1/3$ — jar 2 — $P(\text{white} \mid \text{jar 2}) = 2/3$ — white 2/9

$P(\text{jar 3}) = 1/6$ — jar 3 — $P(\text{white} \mid \text{jar 3}) = 1/2$ — white 1/12

7. $P(\text{jar 2} \mid \text{white})$

$$= \frac{\frac{2}{9}}{\frac{1}{6}+\frac{2}{9}+\frac{1}{12}} = \frac{8}{17} \approx .4706$$

9. $P(\geq \$35 \text{ million} \mid \text{under } 60)$

$$= \frac{P(\text{under } 60)P(\geq \$35 \text{ million}|<60)}{P(<60)P(\geq \$35 \text{ m.}|<60) + P(\geq 60)P(\geq \$35 \text{ m}|\geq 60)}$$

$$= \frac{(.52)(.2308)}{(.52)(.2308)+(.48)(.1667)}$$

$$= \frac{.120016}{.200032} \approx .6$$

11. $P(\text{some college or assos} \mid \geq \$75\text{K})$

$$= \frac{P(\text{some col.})P(\geq \$75\text{K}|\text{some col})}{[P(\text{HS})P(\geq \$75\text{K}|\text{HS}) + P(\text{some col.})P(\geq \$75\text{K}|\text{some col}) + P(\text{Bach.})P(\geq \$75\text{K}|\text{Bach})]}$$

$$= \frac{(.277)(.314)}{(.408)(.170)+(.277)(.314)+(.315)(.550)}$$

$$= \frac{.086978}{.329588} \approx .264$$

13. $P(\text{HS} \mid <\$75\text{K})$

$$= \frac{P(\text{HS})P(<\$75\text{K}|\text{HS})}{[P(\text{HS})P(<\$75\text{K}|\text{HS}) + P(\text{some col.})P(<\$75\text{K}|\text{some col}) + P(\text{Bach.})P(<\$75\text{K}|\text{Bach})]}$$

$$= \frac{(.408)(.830)}{(.408)(.830)+(.277)(.686)+(.315)(.450)}$$

$$= \frac{.33864}{.670412} \approx .505$$

15. $P(\text{Public} \mid \geq \$400\text{K})$

$$= \frac{P(\text{Public})P(\geq \$400\text{K}|\text{Public})}{P(\text{Public})P(\geq \$400\text{K}|\text{Public}) + P(\text{Priv.})P(\geq \$400\text{K}|\text{Priv.})}$$

$$= \frac{(.5111)(.2643)}{(.5111)(.2643)+(.4889)(.2906)}$$

$$= \frac{.13508373}{.27715807} \approx .4874$$

17. $P(\text{male} \mid \text{Not too Happy})$

$$= \frac{P(\text{male})P(\text{Not too Happy}|\text{male})}{P(\text{m})P(\text{not too happy}|\text{male}) + P(\text{f})P(\text{not too happy}|\text{female})}$$

$$= \frac{(.493)(.139)}{(.493)(.139)+(.507)(.142)}$$

$$= \frac{.068527}{.140521} \approx .488$$

19. Let L be the event "the object was shipped by land," A be the event "the object was shipped by air," S be the event "the object was shipped by sea," and E be the event "an error occurred."

$P(L \mid E)$

$$= \frac{P(L)\cdot P(E \mid L)}{P(L)\cdot P(E \mid L) + P(A)\cdot P(E \mid A) + P(S)\cdot P(E \mid S)}$$

$$= \frac{(.50)(.02)}{(.50)(.02)+(.40)(.04)+(.10)(.14)}$$

$$= \frac{.0100}{.0400} = .25$$

The correct response is (c).

For Exercises 21:

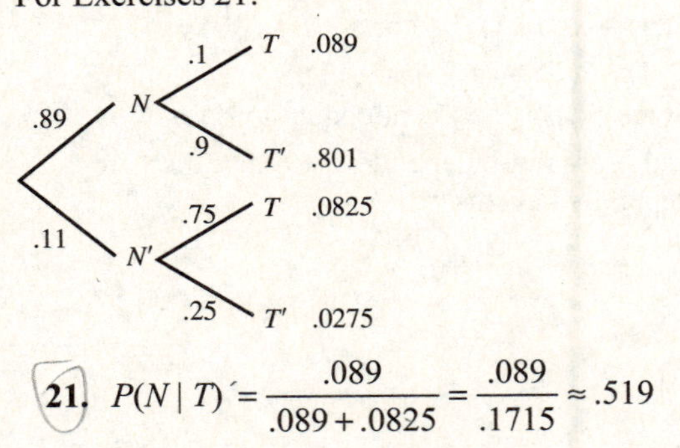

21. $P(N \mid T)' = \frac{.089}{.089+.0825} = \frac{.089}{.1715} \approx .519$

23. $P(\text{GM} \mid \text{car})$

$$= \frac{P(\text{GM})P(\text{car}|\text{GM})}{[P(\text{GM})P(\text{car}|\text{GM}) + P(\text{Ford})P(\text{car}|\text{Ford}) + P(\text{Chrys.})P(\text{car}|\text{Chrys})]}$$

$$= \frac{(.392)(.395)}{(.392)(.395)+(.350)(.370)+(.258)(.332)}$$

$$= \frac{.15484}{.369996} \approx .418$$

25. $P(\text{Ford} \mid \text{truck})$

$$= \frac{P(\text{Ford})P(\text{truck}|\text{Ford})}{[P(\text{GM})P(\text{truck}|\text{GM}) + P(\text{Ford})P(\text{truck}|\text{Ford}) + P(\text{Chrys.})P(\text{truck}|\text{Chrys})]}$$

$$= \frac{(.350)(.630)}{(.392)(.605)+(.350)(.630)+(.258)(.668)}$$

$$= \frac{.2205}{.630004} \approx .350$$

27. $P(\text{adv.} \mid \text{NYSE})$

$$= \frac{P(\text{NYSE})P(\text{adv.}|\text{NYSE})}{P(\text{NYSE})P(\text{adv}|\text{NYSE}) + P(\text{NASDAQ})P(\text{adv}|\text{NASDAQ})}$$

$$= \frac{(.547)(.692)}{(.547)(.692)+(.453)(.657)}$$

$$= \frac{.378524}{.676145} \approx .560$$

29. $P(\text{unch.} \mid \text{NYSE})$

$$= \frac{P(\text{NYSE})P(\text{unch}|\text{NYSE})}{P(\text{NYSE})P(\text{unch}|\text{NYSE}) + P(\text{NASDAQ})P(\text{unch}|\text{NASDAQ})}$$

$$= \frac{(.547)(.041)}{(.547)(.041)+(.453)(.046)}$$

$$= \frac{.022427}{.043265} \approx .518$$

31. $P(45\text{–}64 \mid \text{lives alone})$

$$= \frac{(.332)(.152)}{(.173)(.034)+(.170)(.100)+(.163)(.089)+(.332)(.152)+(.161)(.289)} = \frac{.050464}{.134382} \approx .376$$

33. $P(\text{not living alone})$

$$= 1 - [(.173)(.034)+(.170)(.100)+(.163)(.089)+(.332)(.152)+(.161)(.289)] = 1 - .134382 \approx .865$$

Chapter 8 Review Exercises

1. $9 \in \{8, 4, -3, -9, 6\}$
Because 9 is not an element of the given set, the statement is false.

2. $4 \in \{3, 9, 7\}$
Because 4 is not an element of the given set, the statement is false.

3. $2 \notin \{0, 1, 2, 3, 4\}$
Because 2 is an element of the given set, the statement is false.

4. $0 \notin \{0, 1, 2, 3, 4\}$
Because 0 is an element of the given set, the statement is false.

5. $\{3, 4, 5\} \subseteq \{2, 3, 4, 5, 6\}$
The statement is true because every member of the first set is in the second set.

6. $(1, 2, 5, 8) \subseteq \{1, 2, 5, 10, 11\}$
This statement is false because 8 is an element of the first set but not of the second.

7. $\{1,5,9\} \subset \{1,5,6,9,10\}$
This statement is true because every member of the first set is a member of the second set and the second set contains at least one element not in the first set.

8. $0 \subseteq \varnothing$
This statement is false because the empty set has no subsets except itself. Also, 0 is an element, not a set, so it cannot be a subset.

9. $\{x \mid x$ is a national holiday$\}$ = {New Year's Day, Martin Luther King's Birthday, Presidents' Day, Memorial Day, Independence Day, Labor Day, Columbus Day, Veterans' Day, Thanksgiving, Christmas}

10. $\{x \mid x \text{ is an integer}, -3 \le x < 1\} = \{-3, -2, -1, 0\}$

11. {all counting numbers less than 5} = {1, 2, 3, 4}

12. $\{x \mid x \text{ is a leap year between 1989 and 2006}\}$
$= \{1992, 1996, 2000, 2004\}$

13. M' contains all the elements of U not in M.
$M' = \{B_1, B_2, B_3, B_6, B_{12}\}$

14. N' contains all the elements of U not in N.
$N' = \{B_3, B_6, B_{12}, D\}$

15. $M \cap N$ contains all the elements that are common to M and N.
$M \cap N = \{A, C, E\}$

16. $M \cup N$ contains all the elements in either M or N or both.
$M \cup N = \{A, B_1, B_2, C, D, E\}$

17. $M \cup N'$ contains all the elements either in M or not in N, or both.
$M \cup N' = \{A, B_3, B_6, B_{12}, C, D, E\}$

18. $M' \cap N$ contains all the elements in N and not in M.
$M' \cap N = \{B_1, B_2\}$

19. $A \cap C$ is the set of all female students older than 22.

20. $B \cap D$ is the set of all finance majors with a GPA > 3.5.

21. $A \cup D$ is the set of females or students with a GPA > 3.5.

22. $A' \cap D$ is the set of male students with a GPA > 3.5.

23. $B' \cap C'$ is the set of non-finance majors who are 22 or younger

24. $B \cup A'$
Shade the region inside B as well as all of the region outside of A.

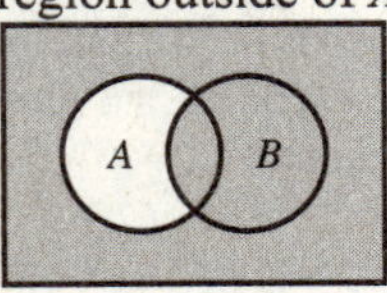

$B \cup A'$

25. $A' \cap B$
Shade all the region inside B that is also outside of A.

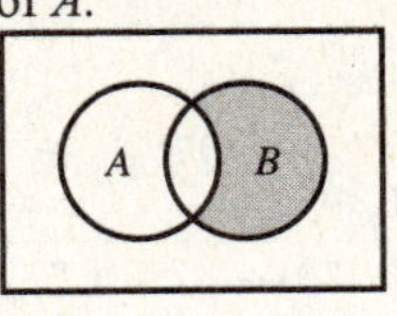

26. $A' \cap (B' \cap C)$
First choose the regions that are inside C and outside of B. From those regions, then shade the region outside of A.

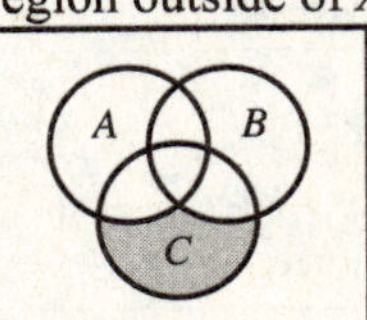

For Exercises 27–30, use the following Venn diagram. Let A be action movies, P be movies rated PG-13, and M be movies made after 2000. Since 21 movies were action, rated PG-13, and made after 2000, and 28 were rated PG-13 and made after 2000, 7 movies were rated PG-13 and made after 2000 only. Since 23 movies were action and made after 2000, then 2 were action and made after 2000 only. 22 movies were action and PG-13, so 22 – 21 = 1 was action and PG-13 only. 27 movies were action, so 27 – (21 + 1 + 2) = 3 were action only. Similar reasoning shows that there were 5 movies that were PG-13 only and 8 movies that were made after 2000 only.

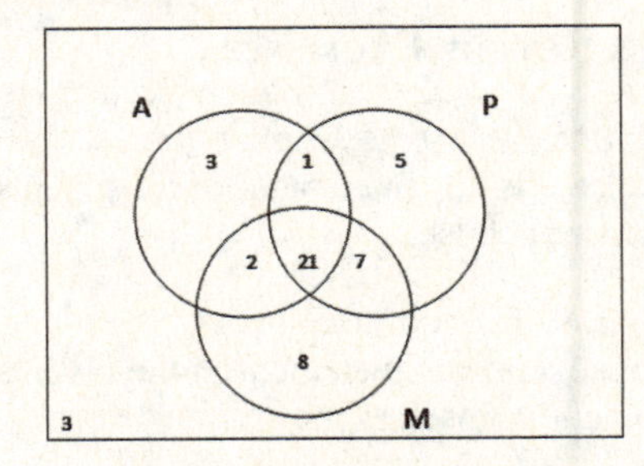

27. $3 + 2 = 5$
5 movies were action movies and were not rated PG–13.

28. $50 - 8 - 3 = 39$
39 movies were action or rated PG-13.

29. $50 - 3 - 3 = 44$
44 movies were made after 2000 or rated PG-13.

30. $50 - 3 - 1 - 5 - 2 - 21 - 7 - 8 = 3$
3 movies were neither action or rated PG-13, and made before 2000.

31. The sample space for rolling a die is $\{1, 2, 3, 4, 5, 6\}$.

32. The sample space for choosing a color and then a number is {(red, 10), (red, 20), (red, 30), (blue, 10), (blue, 20), (blue, 30), (green, 10), (green, 20), (green, 30)}.

33. {(rock, rock), (rock, pop), (rock, alternative), (pop, rock), (pop, pop), (pop, alternative), (alternative, rock), (alternative, pop), (alternative, alternative)}

34. No, the outcomes in the sample space are not equally likely because the probabilities for each type of music are not equal.

35. {(Dell, Epson), (Dell, HP), (Gateway, Epson), (Gateway, HP), (HP, Epson), (HP, HP)}

36. No, the outcomes in the sample space are not equally likely because the probabilities for each type of computer and printer are not equal.

37. "A customer buys neither" is written $E' \cap F'$. (This event can also be written as $(E \cup F)'$.)

38. "A customer buys at least one" is written $E \cup F$.

39. Answers vary. Possible answer: This answer must be incorrect because any probability must be between 0 and 1 inclusive.

40. Answers vary. Possible answer: Disjoint sets are sets which have no elements in common, for example, the set of all females and the set of all people who have been President of the United States.

41. Answers vary. Possible answer: Two events are mutually exclusive if their intersection is the empty set. An example is the events "rolling a die and getting a 3" and "rolling a die and getting an even number."

42. Answers vary. Possible answer: Disjoint sets and mutually exclusive events both have no elements in common, that is, their intersection is the empty set. Mutually exclusive events are in fact disjoint subsets of the sample space.

43. P(consumer discretionary or consumer staples) $= .1187 + .1084 = .2271$

44. P(information tech or telecommservices) $= .1793 + .0293 = .2086$

45. P(Materials or Utilities) $= .0357 + .0035 = .0392$

46. P(Not in Health Care) $= 1 - .1350 = .8650$

47.
$$P(\text{Not Associate's}) = \frac{3586-895}{3586} = \frac{2691}{3586} \approx .7504$$

48.
$$P(\text{Bachelor's or Master's}) = \frac{1781+730}{3586} = \frac{2511}{3586} \approx .7002$$

49.
$$P(\$1 \text{ billion or greater in assets}) = \frac{531}{7245} \approx .0733$$

50.
$$P(\$25\text{-}49.99 \text{ mil in assets and savings}) = \frac{99}{7245} \approx .0137$$

51.
$$P(\$300\text{-}499.99 \text{ mil in assets and comm.}) = \frac{613}{7245} \approx .0846$$

52. P(Less than \$25 million or savings)
$$= \frac{367}{7245} + \frac{1023}{7245} - \frac{48}{7245} = \frac{1342}{7245} \approx .1852$$

53. P(\$1 billion or greater or commercial)
$$= \frac{531}{7245} + \frac{6222}{7245} - \frac{429}{7245} = \frac{6324}{7245} \approx .8729$$

54.
$$P(\$500-999.99 \text{ million} \mid \text{savings}) = \frac{P(\$500-999.99 \text{ million} \cap \text{savings})}{P(\text{savings})} = \frac{122}{1023} \approx .1193$$

55.
$$P(\$500-999.99 \text{ million} \mid \text{comm}) = \frac{P(\$500-999.99 \text{ million} \cap \text{comm})}{P(\text{comm})} = \frac{434}{6222} \approx .0698$$

56. $P(\text{savings} \mid \$1 \text{ billion or greater})$
$= \dfrac{P(\text{savings} \cap \$1 \text{ billion or greater})}{P(\$1 \text{ billion or greater})}$
$= \dfrac{102}{531} \approx .1921$

57. $P(\text{commercial} \mid \text{less than } \$25 \text{ million})$
$= \dfrac{P(\text{commercial} \cap \text{less than } \$25 \text{ million})}{P(\text{less than } \$25 \text{ million})}$
$= \dfrac{319}{367} \approx .8692$

58. $P(\text{less than } \$25 \text{ million} \mid \text{commercial})$
$= \dfrac{P(\text{less than } \$25 \text{ million} \cap \text{commercial})}{P(\text{commercial})}$
$= \dfrac{319}{6222} \approx .0513$

59.

		2nd Parent	
		N_2	T_2
1st	N_1	N_1N_2	N_1T_2
Parent	T_1	T_1N_2	T_1T_2

60. P(child has disease) $= P(T_1T_2) = \frac{1}{4}$

61. There are 4 possible combinations, but only 2 have a normal cell combined with a trait cell (N_1T_2, T_1N_2).

P(child is carrier) $= \frac{2}{4} = \frac{1}{2}$

62. P(child is neither carrier nor has disease)
$= P(N_1N_2) = \frac{1}{4}$

63. There are 36 possibilities, with 5 having a sum of 8:
(4, 4), (3, 5), (5, 3), (2, 6) and (6, 2).
$P(8) = \frac{5}{36} \approx .139$

64. P(no more than 4)
$= P(4) + P(3) + P(2)$
$= \frac{3}{36} + \frac{2}{36} + \frac{1}{36} = \frac{6}{36} \approx .167$

65. P(at least 9)
$= P(9) + P(10) + P(11) + P(12)$
$= \frac{4}{36} + \frac{3}{36} + \frac{2}{36} + \frac{1}{36} = \frac{10}{36} \approx .278$

66. P(odd and greater than 8)
$= P(9) + P(11)$
$= \frac{4}{36} + \frac{2}{36} = \frac{6}{36} \approx .167$

67. A roll less than 4 means 3 or 2. There are 2 ways to get 3 and 1 way to get 2. Hence,
$P(2 \mid \text{less than } 4) = \frac{1}{3}$.

68. $P(7 \mid \text{at least one is a } 4) = \frac{2}{11} \approx .182$, since there are 11 possibilities with at least one of the dice being a 4 {(4, 1), (1, 4), (4, 2), (2, 4), (4, 3), (3, 4), (4, 4), (5, 4), (4, 5), (6, 4), (4, 6)} with only (4, 3) and (3, 4) having a sum of 7.

For Exercises 69–72, draw a Venn diagram and use the given information to fill in the probabilities for each of the regions.

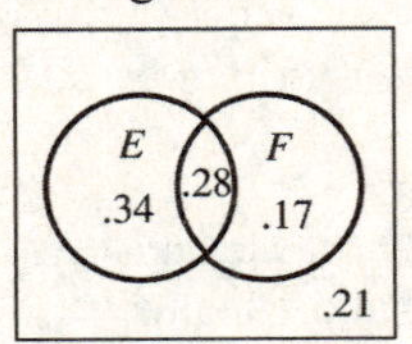

69. $P(E \cup F) = .34 + .28 + .17 = .79$

70. $P(E \cap F') = .34$

71. $P(E' \cup F) = .28 + .17 + .21 = .66$

72. $P(E' \cap F') = 1 - (.34 + .28 + .17) = .21$

73. $P(E' \mid F) = 1 - P(E \mid F) = 1 - .3 = .7$

74. $P(E \mid F') = \frac{.08}{.6} = \frac{2}{15} \approx .1333$

75. Answers vary.

76. Answers vary.

77. a. $\frac{1}{40}$

b. 942:58 = 471:29

78. a. 732:268 = 183:67

b. $\frac{373}{500} = .746$

Use the following probability tree for Exercises 79–82.

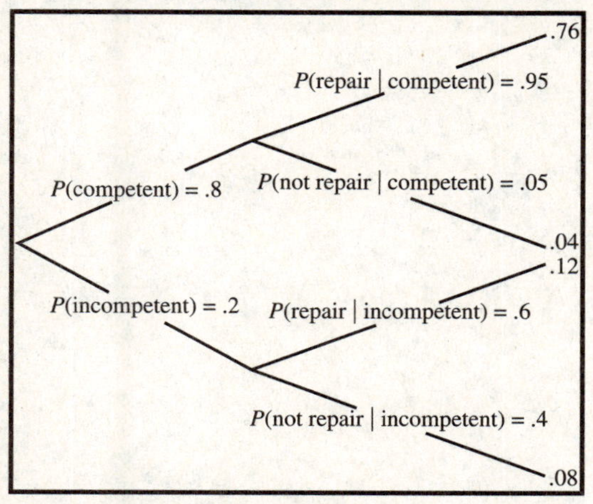

79. $P(\text{competent} \mid \text{repaired})$
$= \frac{.76}{.76+.12} = \frac{.76}{.88} = \frac{19}{22}$

80. $P(\text{incompetent} \mid \text{repaired})$
$= \frac{.12}{.76+.12} = \frac{12}{88} = \frac{3}{22}$

81. $P(\text{competent} \mid \text{not repaired})$
$= \frac{.04}{.04+.08} = \frac{.04}{.12} = \frac{1}{3}$

82. $P(\text{incompetent} \mid \text{not repaired})$
$= \frac{.08}{.04+.08} = \frac{8}{12} = \frac{2}{3}$

83. Let D mean defective.
$P(D) = .17(.04) + .39(.02) + .35(.07) + .09(.03)$
$= .0418$

$P(4 \mid D) = \frac{P(4 \cap D)}{P(D)} = \frac{.09(.03)}{.0418}$
$= \frac{.0027}{.0418} \approx .0646$

84. $P(2 \mid D) = \frac{P(2 \cap D)}{P(D)} = \frac{.39(.02)}{.0418}$
$= \frac{.0078}{.0418} \approx .1866$

85. $P(\text{second class}) = \frac{357}{1316} \approx .271$

86. $P(\text{surviving}) = \frac{499}{1316} \approx .379$

87. $P(\text{surviving}|\text{first class}) = \frac{203}{325} \approx .625$

88. $P(\text{surviving}|\text{third class child}) = \frac{27}{79} \approx .342$

89. $P(\text{female}|\text{first class survivor}) = \frac{140}{203} \approx .690$

90. $P(\text{third class}|\text{male survivor}) = \frac{75}{146} \approx .514$

91. No. Answers vary.

92. No. Answers vary.

93. $P(\text{commute} < 15 \text{ min})$
$= P(\text{male})P(< 15 \text{ min}|\text{male})$
$+ P(\text{female})P(< 15 \text{ min}|\text{female})$
$= (.528)(.259) + (.472)(.299) \approx .2779$

94. $P(\text{commute} \geq 60 \text{ min})$
$= P(\text{male})P(\geq 60 \text{ min}|\text{male})$
$+ P(\text{female})P(\geq 60 \text{ min}|\text{female})$
$= (.528)(.096) + (.472)(.066) \approx .0818$

95. $P(\text{female and commute 30-44 min})$
$= P(\text{female})P(\text{30-44 min}|\text{female})$
$= (.472)(.192) \approx .0906$

96. $P(\text{male and commute 45-59 min})$
$= P(\text{male})P(\text{45-59 min}|\text{male})$
$= (.528)(.082) \approx .0433$

97. $P(\text{m} \mid \text{commute} \geq 60 \text{ min})$
$= \frac{P(\text{male})P(\geq 60 \text{ min}|\text{male})}{P(\text{m})P(\geq 60 \text{ min}|\text{male}) + P(\text{f})P(\geq 60 \text{ min}|\text{female})}$
$= \frac{(.528)(.096)}{(.528)(.096) + (.472)(.066)}$
$= \frac{.050688}{.08184} \approx .6194$

98. $P(\text{f} \mid \text{commute 30-44 min})$

$$= \frac{P(\text{female})P(\text{30-44 min}|\text{female})}{P(\text{m})P(\text{30-44 min}|\text{male}) + P(\text{f})P(\text{30-44 min}|\text{female})}$$

$$= \frac{(.472)(.192)}{(.528)(.210)+(.472)(.192)}$$

$$= \frac{.090624}{.201504} \approx .4497$$

99. $P(\text{f} \mid \text{commute } \leq 29 \text{ min})$

$$= \frac{P(\text{female})P(\leq 29 \text{ min}|\text{female})}{P(\text{m})P(\leq 29 \text{ min}|\text{male}) + P(\text{f})P(\leq 29 \text{ min}|\text{female})}$$

$$= \frac{(.472)(.674)}{(.528)(.613)+(.472)(.674)}$$

$$= \frac{.318128}{.641792} \approx .4957$$

100. $P(\text{m} \mid \text{commute } \geq 45 \text{ min})$

$$= \frac{P(\text{male})P(\geq 45 \text{ min}|\text{male})}{P(\text{m})P(\geq 45 \text{ min}|\text{male}) + P(\text{f})P(\geq 45 \text{ min}|\text{female})}$$

$$= \frac{(.528)(.178)}{(.528)(.178)+(.472)(.134)}$$

$$= \frac{.093984}{.157232} \approx .5977$$

Case 8 Medical Diagnosis

1. $P(T' \mid D') = .999 \Rightarrow P(T \mid D') = 1 - .999 = .001$

2. $P(D) = .005 \Rightarrow P(D') = 1 - .005 = .995$

$$P(D \mid T) = \frac{P(D)P(T \mid D)}{P(D)P(T \mid D) + P(D')P(T \mid D')}$$

$$= \frac{.005(.99)}{.005(.99) + .995(.001)}$$

$$= \frac{.00495}{.005945} \approx .833$$

3. $$P(D \mid T) = \frac{P(D)P(T \mid D)}{P(D)P(T \mid D) + P(D')P(T \mid D')}$$

$$= \frac{.0005(.99)}{.0005(.99) + .9995(.001)}$$

$$= \frac{.000495}{.0014945} \approx .331$$

Chapter 9 Counting, Probability Distributions, and Further Topics in Probability

Section 9.1 Probability Distributions and Expected Value

1. The number of possible samples is 16.

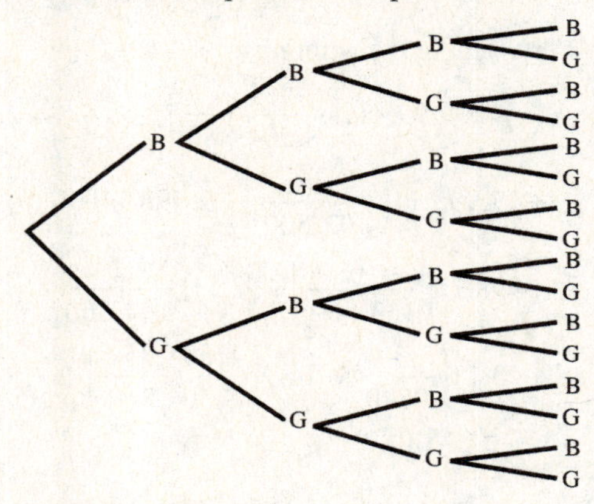

Number of Boys	$P(x)$
0	$\frac{1}{16} \approx .063$
1	$\frac{4}{16} = .25$
2	$\frac{6}{16} = .375$
3	$\frac{4}{16} = .25$
4	$\frac{1}{16} \approx .063$

3. Let x be the number of queens drawn. Then x can take on values 0, 1, 2, or 3.

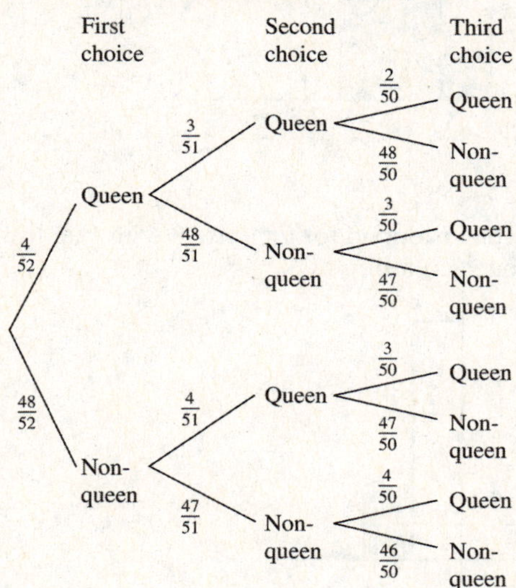

Number of queens	$P(x)$
0	$\frac{48}{52} \cdot \frac{47}{51} \cdot \frac{46}{50} \approx .7826$
1	$\frac{4}{52} \cdot \frac{48}{51} \cdot \frac{47}{50} + \frac{48}{52} \cdot \frac{4}{51} \cdot \frac{47}{50} + \frac{48}{52} \cdot \frac{47}{51} \cdot \frac{4}{50} \approx .2042$
2	$\frac{4}{52} \cdot \frac{3}{51} \cdot \frac{48}{50} + \frac{4}{52} \cdot \frac{48}{51} \cdot \frac{3}{50} + \frac{48}{52} \cdot \frac{4}{51} \cdot \frac{3}{50} \approx .0130$
3	$\frac{4}{52} \cdot \frac{3}{51} \cdot \frac{2}{50} \approx .0002$

5.

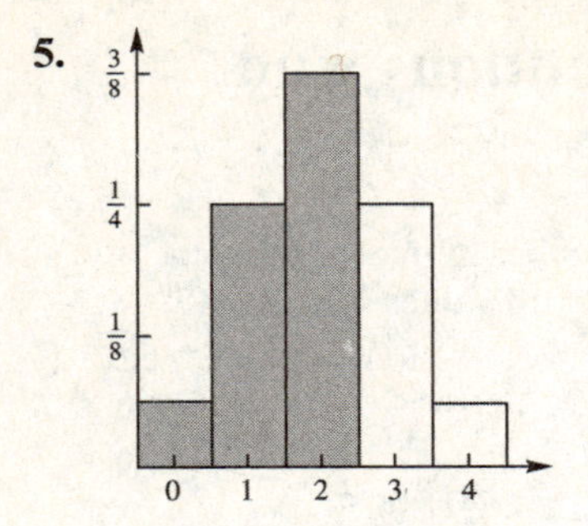

7. The histogram for Exercise 3 with P(at least one queens) follows.

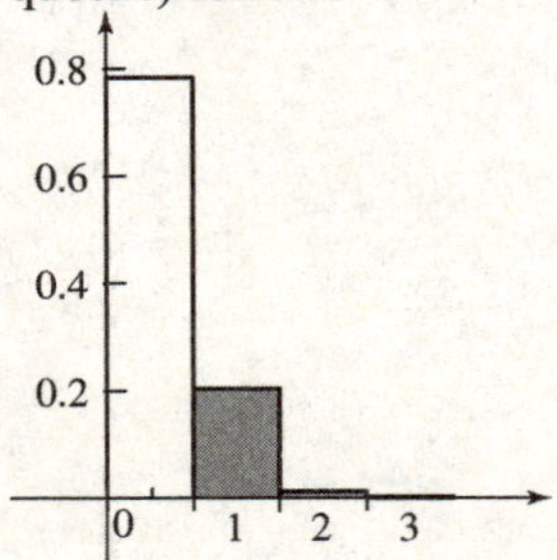

9. expected value $= 1(.1) + 3(.5) + 5(.2) + 7(.2)$
$= 4$

11. expected value
$= 0(.21) + 2(.24) + 4(.21) + 8(.17) + 16(.17)$
$= 5.4$

13. $E(x) = 1(.2) + 2(.3) + 3(.1) + 4(.4) = 2.7$

15. $E(x) = 1(.3) + 2(.25) + 3(.2) + 4(.15) + 5(.1)$
$= 2.5$

17. $E(x) = 1\left(\frac{18}{38}\right) - 1\left(\frac{20}{38}\right) = -\frac{2}{38} = -\frac{1}{19} \approx -0.05$

19. You have one chance in a thousand of winning \$500 on a \$1 bet for a net return of \$499. In the 999 other outcomes you lose your dollar

$$E(x) = 499\left(\frac{1}{1000}\right) + (-1)\left(\frac{999}{1000}\right)$$
$$= -\frac{500}{1000} = -\$.50 \text{ or } -50¢$$

21. Let x denote the winnings.

$$E(x) = 49{,}999\left(\frac{1}{2{,}000{,}000}\right) + 9{,}999\left(\frac{2}{2{,}000{,}000}\right) - 1\left(\frac{1{,}999{,}997}{2{,}000{,}000}\right)$$
$$= -\$.965 = -96.5¢ \approx -97¢$$

23. Let x denote the winnings.

$$E(x) = 99{,}998.11\left(\frac{1}{8{,}504{,}860}\right) + 49{,}998.11\left(\frac{1}{302{,}500}\right) + 9{,}998.11\left(\frac{1}{282{,}735}\right) + 998.11\left(\frac{1}{153{,}560}\right) + 98.11\left(\frac{1}{104{,}560}\right) + 23.11\left(\frac{1}{9540}\right) - 1.89\left(\frac{8{,}504{,}854}{8{,}504{,}860}\right)$$
$$\approx -\$1.67$$

25. $E(x) = 0(.2643) + 1(.4173) + 2(.2470) + 3(.0650) + 4(.0064) = 1.1319$

27. The distribution is not valid since no probability can be less than 0.

29. The sum of the probabilities total 1.0, so this is a valid distribution.

31. Let x denote the missing probability. Then,
$.01 + .09 + .25 + .45 + .05 + x = 1.0 \Rightarrow$
$.85 + x = 1.0 \Rightarrow x = 1.0 - .85 = .15$

33. Let x denote the missing probability. Then
$.20 + x + .25 + .30 = 1.0 \Rightarrow x + .75 = 1.0 \Rightarrow$
$x = 1.0 - .75 = .25$

35. Let x and y denote the missing probabilities. Then
$.10 + .10 + .20 + .25 + .05 + x + y = 1.0 \Rightarrow$
$.70 + x + y = 1.0 \Rightarrow x + y = 1.0 - .7 = .3$
Answers may vary. One possible answer would be .15 and .15.

37. $E(x) = .0007[100(15{,}000) + 250(10{,}000) + 500(5{,}000)]$
$= .0007(6{,}500{,}000) = \$4550$

39. $E(x) = 0(.6427) + 1(.3061) + 2(.0486) + 3(.0026)$
$= .4111$

41.

Account Number	Expected value	Exist. vol. + exp. value	Class
3	2000	22,000	C
4	1000	51,000	B
5	25,000	30,000	C
6	60,000	60,000	A
7	16,000	46,000	B

43. **a.** Let x denote the cost of using each antibiotic. For Drug A,
$E(x) = .7(160) + .3(350) = \$217$.
For Drug B,
$E(x) = .90(210) + .10(400) = \$229$.

b. Drug A, since the total expected cost is less.

45. **a.** $E(x) = 750,000(.5) + 350,000(.5)$
$= 550,000$ pounds

b. $E(x) = 750,000(.67) + 350,000(.33)$
$= 618,000$ pounds

Section 9.2 The Multiplication Principle, Permutations and Combinations

1. ${}_4P_2 = \dfrac{4!}{(4-2)!} = \dfrac{4!}{2!} = 4 \cdot 3 = 12$

3. ${}_8C_5 = \dfrac{8!}{5!(8-5)!} = \dfrac{8!}{3!5!} = \dfrac{8 \cdot 7 \cdot 6}{3 \cdot 2 \cdot 1} = 56$

5. ${}_8P_1 = \dfrac{8!}{(8-1)!} = \dfrac{8!}{7!} = 8$

7. $4! = 4 \cdot 3 \cdot 2 \cdot 1 = 24$

9. ${}_9C_6 = \dfrac{9!}{6!(9-6)!}$
$= \dfrac{9!}{6!3!} = \dfrac{9 \cdot 8 \cdot 7}{3 \cdot 2 \cdot 1} = 84$

11. ${}_{13}P_3 = \dfrac{13!}{(13-3)!} = \dfrac{13!}{10!} = 13 \cdot 12 \cdot 11 = 1716$

13. ${}_{25}P_5 = 6,375,600$

15. ${}_{14}P_5 = 240,240$

17. ${}_{18}C_5 = \dfrac{18!}{5!13!} = \dfrac{18 \cdot 17 \cdot 16 \cdot 15 \cdot 14}{5 \cdot 4 \cdot 3 \cdot 2 \cdot 1} = 8568$

19. ${}_{28}C_{14} = 40,116,600$

21. If $0! = 0$, ${}_4P_4 = \dfrac{4!}{(4-4)!} = \dfrac{24}{0} =$ undefined

23. **a.** There are two possibilities for each line, and, by the multiplication principle, $2 \cdot 2 \cdot 2 = 8$ trigrams.

b. Since each hexagram is made up of two trigrams, and there are 8 possible trigrams, it follows that there are $8 \cdot 8 = 64$ possible hexagrams.

25. $6 \cdot 8 \cdot 4 \cdot 3 = 576$
There are 576 varieties of autos available.

27. Yes; Since a social security number has 9 digits with no restrictions, there are $10^9 = 1,000,000,000$ (1 billion) different social security numbers. This is enough for every one of the people in the United States to have a social security number.

29. Since a zip code has nine digits with no restrictions, there are 10^9 or 1,000,000,000 different 9-digit zip codes.

31. There are $12 \cdot 10 = 120$ different ways Sherri can buy 1 Pantene shampoo and 1 Pantene conditioner.

33. **a.** There are 8 possibilities for the first digit, 2 possibilities for the second digit, and 10 possibilities for the last digit. The total number of possible area codes is
$8 \cdot 2 \cdot 10 = 160$
There are 8 possibilities for the first digit and 10 possibilities for each of the next six digits. The total number of phone numbers is
$8 \cdot 10 \cdot 10 \cdot 10 \cdot 10 \cdot 10 \cdot 10 = 8 \cdot 10^6$
$= 8,000,000$

b. Some numbers, like 911, 800, and 900, are reserved for special purposes.

35. In this new plan, there would be 8 possibilities for the first digit, 2 possibilities for the second digit, and 10 possibilities for each of the last 2 digits. The total number of area codes is $8 \cdot 2 \cdot 10 \cdot 10 = 1600$.

37. Answers vary. Possible answer:
A permutation of a elements ($a \geq 1$) from a set of b elements is any arrangement, without repetition, of the a elements.

39. Since order makes a difference, the number of arrangements is

$${}_{12}P_5 = \frac{12!}{(10-5)!} = \frac{12!}{5!} = 95{,}040.$$

41. Order is important, and there are no repeats, so there are ${}_{24}P_3 = \frac{24!}{(24-3)!} = 24 \cdot 23 \cdot 22 = 12{,}144$ ways to name a fraternity using 3 Greek letters.

43. Four people are being selected. Since each receives a different job, order is important. The total number of different officer selections is

$${}_{32}P_4 = \frac{32!}{(32-4)!} = \frac{32!}{28!} = 32 \cdot 31 \cdot 30 \cdot 29 = 863{,}040$$

45. $${}_{17}P_2 = \frac{17!}{(17-2)!} = \frac{17!}{15!} = 17 \cdot 16 = 272$$

47. This is a combinations problem.

a. $${}_{10}C_4 = \frac{10!}{4!6!} = 210$$

b. $${}_{10}C_6 = \frac{10!}{6!4!} = 210$$

49. Order is not important, so use a combination.

$${}_{11}C_4 = \frac{11!}{(11-4)!4!} = 330$$

There are 330 ways to select a group of 4 co-captains from a group of 11.

51. Answers vary. Possible answer: Combinations are not ordered, while permutations are ordered.

53. a. With repetition permitted, the tree diagram shows 9 different pairs.

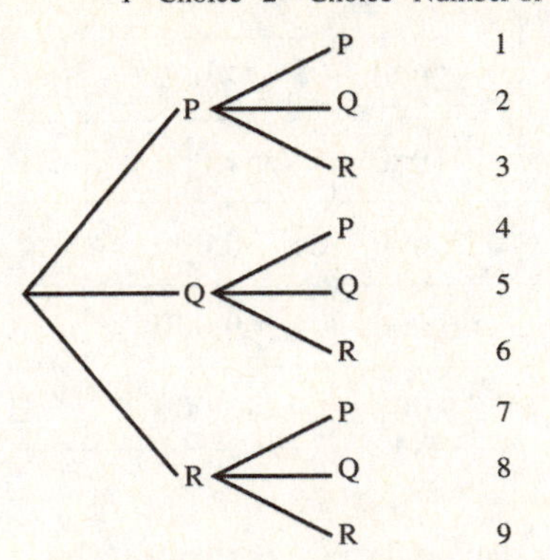

b. If repetition is not permitted, one branch is missing from each of the clusters of second branches, for a total of 6 different pairs.

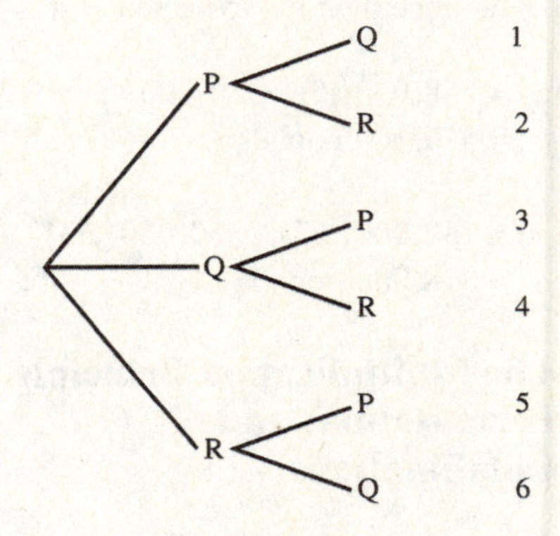

c. Find the number of combinations of 3 elements taken 2 at a time.

$${}_3C_2 = \frac{3!}{2!1!} = 3$$

No repetitions are allowed, so the answer cannot equal that for part a. However, since order does not matter, our answer is only half of the answer for part b. For example PQ and QP are distinct in b. but not in c. Thus, the answer differs from both a. and b.

55. These are combination problems since order is not important.

a. ${}_{18}C_5 = 8568$ delegations

b. ${}_{8}C_5 = 56$ delegations

c. ${}_{10}C_3 \cdot {}_{8}C_2 = 120 \cdot 28 = 3360$ delegations

d. At least one means one or two or three or four or five.

$$\begin{aligned} &{}_{10}C_1 \cdot {}_{8}C_4 + {}_{10}C_2 \cdot {}_{8}C_3 + {}_{10}C_3 \cdot {}_{8}C_2 \\ &\quad + {}_{10}C_4 \cdot {}_{8}C_1 + {}_{10}C_5 \\ &= 700 + 2520 + 3360 + 1680 + 252 \\ &= 8512 \end{aligned}$$

8512 delegations would have at least one Democrat.

57. a. The number of different double-scoops will be ${}_{21}C_2 = 210$.

b. The number of different triple-scoops will be ${}_{21}P_3 = 7980$ since order matters.

59. a. Since order is not important, this is a combination problem.

${}_{99}C_6 = 1{,}120{,}529{,}256$

b. Since order is important, this is a permutation problem.

${}_{99}P_6 = 806{,}781{,}064{,}320$

61. Using the multiplication principle, we have $3 \cdot 3 \cdot 3 \cdot 3 = 81$

63. It is not possible since $26^3 = 17{,}576$ (the number of different 3-initial names) and the biologist needs 52,000 names. Actually, 4-initial names would do the biologist's job since $26^4 = 456{,}976$.

65. Since order is not important, these are combination problems.

a. ${}_{5}C_3 = 10$

b. Since you are taking 3 Diet Coke, and only 1 Diet Coke exists, this situation is impossible. The answer is 0.

c. ${}_{3}C_3 = 1$

d. ${}_{5}C_2 \cdot {}_{1}C_1 = 10 \cdot 1 = 10$

e. ${}_{5}C_2 \cdot {}_{3}C_1 = 10 \cdot 3 = 30$

f. ${}_{3}C_2 \cdot {}_{5}C_1 = 3 \cdot 5 = 15$

g. Again, since you are picking 2 Diet Coke and only 1 Diet Coke exists, this situation is impossible. The answer is 0.

67. Since order is not important, we use combinations. ${}_{35}C_{20} = 3{,}247{,}943{,}160$

There are 3,247,943,160 ways to select 20 volunteers from the group of 35.

69. Since the order in each column is important, we use permutations.

$\left({}_{15}P_5\right)^4 \cdot \left({}_{15}P_4\right) = 5.524 \times 10^{26}$

71. a. martini

There are 7 total letters, $1m$, $1a$, $1r$, $1t$, $2i$'s, and $1n$.

$$\frac{7!}{1!1!1!1!2!1!} = 2520$$

b. nunnery

There are 7 letters, $3n$'s, $1u$, $1e$, $1r$, and $1y$.

$$\frac{7!}{3!1!1!1!1!} = 840$$

c. grinding

There are 8 letters, $2g$'s, $1r$, $2i$'s, $2n$'s, and $1d$.

$$\frac{8!}{2!1!2!2!1!} = 5040$$

73. a. The key word is "distinguishable", so use permutations. Total of 12 dinners:

${}_{12}P_{12} = 479{,}001{,}600$ ways

b. Since dinners of the same company are considered identical, the problem is to find the number of different arrangements of the three colors. ${}_{3}P_3 = 6$

c. $\dfrac{12!}{4!3!5!} = 27{,}720$

Section 9.3 Applications of Counting

1. There are ${}_{24}C_3 = 2024$ ways to select 3 tiles from the batch of 24. There are ${}_{5}C_1 = 5$ ways to select one defective tile from the batch, so there are ${}_{19}C_2 = 171$ ways to choose 2 good tiles from the 19 good tiles in the batch. Thus,
$P(1 \text{ defective}) = \frac{{}_{5}C_1 \cdot {}_{19}C_2}{{}_{24}C_3} = \frac{5 \cdot 171}{2024} \approx .422.$

3. There are 8 computers, 5 non-defective, and a sample of size 1 is chosen.
$P(\text{no defectives}) = \frac{{}_{5}C_1}{{}_{8}C_1} = \frac{5}{8}$

5. There are 8 computers, 5 non-defective, and a sample of size 3 is chosen.
$P(\text{no defectives}) = \frac{{}_{5}C_3}{{}_{8}C_3} = \frac{10}{56} = \frac{5}{28}$

7. There are 42 prizes, 10 of which are \$100 prizes; 3 are drawn.
$P(\text{all are \$100 prizes}) = \frac{{}_{10}C_3}{{}_{242}C_3} = \frac{120}{2{,}332{,}880} \approx .00005$

9. There are 42 prizes, 20 of which are \$25 prizes and 200 "dummy" tickets; 3 are drawn.
$P(\text{two \$25 prizes}) = \frac{{}_{20}C_2 \cdot {}_{200}C_1}{{}_{242}C_3} = \frac{190 \cdot 1 \cdot 200}{2{,}332{,}880} \approx .0163$

11. There are 200 dummy tickets, and all 3 must come from that group.
$P(\text{no winning ticket})$
$= \frac{{}_{200}C_3}{{}_{242}C_3} = \frac{1{,}313{,}400}{2{,}332{,}880} \approx .5630$

13. Since order does not make a difference, the number of 2-card hands is ${}_{52}C_2 = 1326.$

15. There are 48 non-deuces (2's) in a deck of cards.
$P(\text{no deuces}) = \frac{{}_{48}C_2}{{}_{52}C_2} = \frac{1128}{1326} \approx .851$

17. From Exercise 13, we know there are 1326 different 2-card hands. To see how many hands have the same suit, there are ${}_{13}C_2 = 78$ ways to get a 2-card hand of one particular suit. Because there are 4 different suits, there are $4 \cdot 78 = 312$ hands of the same suit. Therefore, there are $1326 - 312 = 1014$ hands of different suits.
$P(\text{different suits}) = \frac{1014}{1326} \approx .765$

19. $P(\text{no more than 1 diamond})$
$= P(\text{no diamonds}) + P(1 \text{ diamond})$
Because there are 39 non-diamonds in a deck,
$P(\text{no more than 1 diamond}) = \frac{741}{1326} + \frac{39 \cdot 13}{1326} = \frac{1248}{1326} \approx .941.$

Alternate solution:
$P(\text{at least 1 black card}) = 1 - P(2 \text{ diamonds}) = 1 - \frac{{}_{13}C_2}{{}_{52}C_2} = \frac{78}{1326} \approx .941$

21. Answer varies. Possible answer:
The advantage of using this rule is that it is many times easier to calculate the probability of the complement of an event than of the given event. For example in dealing a hand of 5 cards, if E is the event to get at least one heart, it is much easier to calculate $P(E)'$, the probability of getting no hearts and subtracting it from 1.

23. There are 22 rocks in all, so there are ${}_{22}C_5 = 26{,}334$ ways to select 5 rocks. There are ${}_{7}C_3 = 35$ ways to select 3 sedimentary rocks and ${}_{15}C_2 = 105$ ways to select the remaining 2 rocks. Therefore, the probability of selecting 5 rocks of which 3 are sedimentary is
$\frac{{}_{7}C_3 \cdot {}_{15}C_2}{{}_{22}C_5} = \frac{35 \cdot 105}{26{,}334} \approx .1396.$

25. a. We must pick all 6 of our numbers from the 99 total numbers.
$P(\text{all } 6) = \frac{{}_{6}C_6}{{}_{99}C_6} \approx 8.9 \times 10^{-10}$

b. $P(\text{all } 6) = \frac{1}{{}_{99}P_6} \approx 1.2 \times 10^{-12}$

27. The probability of two individuals independently selecting the winning numbers is

$$\left(\frac{1}{195{,}249{,}054}\right)^2 \approx 2.62\times10^{-17}.$$

29. a. $P(\text{3 women and 1 man}) = \frac{{}_4C_3 \cdot {}_{11}C_1}{{}_{15}C_4} = \frac{4\cdot 11}{1365} \approx .0322$

b. $P(\text{all men}) = \frac{{}_{11}C_4}{{}_{15}C_4} = \frac{330}{1365} \approx .2418$

c. $P(\text{at least 1 woman})$
$= 1 - P(\text{no women})$
$= 1 - P(\text{all men})$
$= 1 - .2418 = .7582$

31. There are ${}_{20}C_7 = 77{,}520$ ways to select 7 volunteers from the pool of 20.

a. $P(\text{all 20-39}) = \frac{{}_{10}C_7}{{}_{20}C_7} = \frac{120}{77{,}520} \approx .0015$

b. $P(\text{5 20-39, 2 60 or older}) = \frac{{}_{10}C_5 \cdot {}_2C_2}{{}_{20}C_7} = \frac{252\cdot 1}{77{,}520} \approx .0033$

c. $P(\text{3 40-59, 4 other}) = \frac{{}_8C_3 \cdot {}_{12}C_4}{{}_{20}C_7} = \frac{56\cdot 495}{77{,}520} \approx .3576$

33. The probability that at least 2 of the 100 U.S. Senators have the same birthday is

$$1 - \frac{{}_{365}P_{100}}{(365)^{100}} \approx 1.$$

35. There are ${}_{20}C_4 = 4845$ ways to pick the correct 4 numbers out of the 20 the state picks. There are ${}_{80}C_4 = 1{,}581{,}580$ ways to pick 4 numbers out of 80. The probability of winning $55 is

$$\frac{{}_{20}C_4}{{}_{80}C_4} = \frac{4845}{1581580} \approx .0031.$$

37. There are 60 losing numbers from which all 4 must be picked. There are 80 numbers from which 4 are picked. The probability of losing is

$$\frac{{}_{20}C_0 \cdot {}_{60}C_4}{{}_{80}C_4} = \frac{1\cdot 487{,}635}{1{,}581{,}580} \approx .3083$$

39. Answers vary. Theoretical answers: This exercise should be solved by computer methods. The solution will vary according to the computer program that is used. The answers are (a) .0399, (b) .5191, (c) .0226.

Section 9.4 Binomial Probability

1. $n = 10, p = .7, x = 6, 1 - p = .3$
$P(\text{exactly 6}) = {}_{10}C_6(.7)^6(.3)^4 \approx .2001$

3. $n = 10, p = .7, x = 0, 1 - p = 1 - .7 = .3$
$P(\text{none}) = {}_{10}C_0(.7)^0(.3)^{10} \approx .000006$

5. $P(\text{at least 1}) = 1 - P(\text{none})$
From Exercise 3, we have $P(\text{none}) \approx .000006$, so
$P(\text{at least 1}) = 1 - .000006 \approx .999994$.

For exercises 7–11, we have $n = 8,\ p = .8$, and $1 - p = .2$.

7. $x = 2$; $P(\text{exactly 2}) = {}_8C_2(.8)^2(.2)^8 \approx .0011$.

9. $x = 0$; $P(\text{none}) = {}_8C_0(.8)^0(.2)^{10} \approx .000003$

11. $P(\text{at least 1}) = 1 - P(\text{none})$. From Exercise 9, we have $P(\text{none}) \approx .000003$, so
$P(\text{at least 1}) \approx 1 - .000003 \approx .999997$.

For exercises 13–15, we have $n = 5,\ p = \frac{1}{2}$, and $1 - p = \frac{1}{2}$.

13. $x = 5$; $P(\text{all heads}) = {}_5C_5\left(\frac{1}{2}\right)^5\left(\frac{1}{2}\right)^0 = \frac{1}{32}$

15. "No more than 3 heads" means 0 heads, 1 head, 2 heads, or 3 heads.

$$P(\text{0 heads}) = {}_5C_0\left(\frac{1}{2}\right)^0\left(\frac{1}{2}\right)^5 = \frac{1}{32}$$

$$P(\text{1 head}) = {}_5C_1\left(\frac{1}{2}\right)^1\left(\frac{1}{2}\right)^4 = \frac{5}{32}$$

$$P(\text{2 heads}) = {}_5C_2\left(\frac{1}{2}\right)^2\left(\frac{1}{2}\right)^3 = \frac{10}{32}$$

$$P(\text{3 heads}) = {}_5C_3\left(\frac{1}{2}\right)^3\left(\frac{1}{2}\right)^2 = \frac{10}{32}$$

$P(\text{no more than 3 heads})$

$$= \frac{1}{32}+\frac{5}{32}+\frac{10}{32}+\frac{10}{32}=\frac{26}{32}=\frac{13}{16}$$

17. Answer varies. Possible answer:
A problem involves a binomial experiment if the experiment is repeated several times, there are only two possible outcomes, and the repeated trials are independent.

For exercises 19–21, we have $n = 15, p = .48$, and $1 - p = .52$.

19. $P(3) = {}_{15}C_3(.48)^3(.52)^{12} \approx .0197$

21. $P(\text{at most }2) = P(0)+P(1)+P(2)$

$$= {}_{15}C_0(.48)^0(.52)^{15} + {}_{15}C_1(.48)^1(.52)^{14} + {}_{15}C_2(.48)^2(.52)^{13}$$

$$\approx .00005+.00076+.00491 \approx .0057$$

For exercises 23–27, we have $n = 9, p = .54$, and $1 - p = .46$.

23. $P(9) = {}_9C_9(.54)^9(.46)^0 \approx .0039$

25. $P(\text{at most }3)$

$$= P(0)+P(1)+P(2)+P(3)$$

$$= {}_9C_0(.54)^0(.46)^9 + {}_9C_1(.54)^1(.46)^8 + {}_9C_2(.54)^2(.46)^7 + {}_9C_3(.54)^3(.46)^6$$

$$\approx .0009+.0097+.0458+.1253 \approx .1817$$

27. Since 54% of small businesses chose the cost of health care as hurting the operating environment of their business, we would expect that $.54 \cdot 500 = 270$ businesses chose the cost of health care as hurting their business "a lot".

29. $n = 100,\ p = .027$, and $1 - p = .973$

$P(\text{exactly 2 sets of twins})$

$$= {}_{100}C_2(.027)^2(.973)^{98} \approx .247$$

31. $n = 9,\ p = .11,\ 1 - p = .89$

$$P(2) = {}_9C_2(.11)^2(.89)^7 \approx .193$$

33. $P(\text{none}) = {}_9C_0(.11)^0(.89)^9 \approx .350$

35. Since 11% of Americans are left-handed, $.11 \times 35 = 3.85$ would be expected to be left-handed from the sample of 35.

For exercises 37–39, we have $n = 16, p = .11$, and $1 - p = .89$.

37. $P(\text{at most }3) = P(0)+P(1)+P(2)+P(3)$

$$= {}_{16}C_0(.11)^0(.89)^{16} + {}_{16}C_1(.11)^1(.89)^{15} + {}_{16}C_2(.11)^2(.89)^{14} + {}_{16}C_3(.11)^3(.89)^{13}$$

$$\approx .15497+.30645+.28407+.16385 \approx .9093$$

39. We would expect that $.11 \cdot 300 = 33$ US residents would have a great deal of confidence in banks and financial institutions.

41. In order to find the probability, we will use the TI – 83 distribution mode. Enter the distribution mode by hitting the 2nd key followed by the VARS key (distribution mode on the secondary level). Since we are looking for the probability of 20 or less, use the binomcdf function. Enter the number of trial, probability of success and stopping number, all separated by commas to get the answer. Therefore the P(at most 20) = binomcdf (100, .23, 20) $\approx .2811$.

43. We would expect that $.23 \cdot 100 = 23$ registered vehicles would be pickup trucks.

Section 9.5 Markov Chains

1. $\begin{bmatrix} \frac{1}{4} & \frac{3}{4} \end{bmatrix}$ could be a probability vector because it is a matrix with only one row containing only nonnegative entries whose sum is $\frac{1}{4}+\frac{3}{4}=1$.

3. $\begin{bmatrix} 0 & 1 \end{bmatrix}$ could be a probability vector because it has only one row, the entries are nonnegative, and $0 + 1 = 1$.

5. $\begin{bmatrix} .3 & -.1 & .6 \end{bmatrix}$ cannot be a probability vector because it has a negative entry, $-.1$.

7. $\begin{bmatrix} .7 & .2 \\ .5 & .5 \end{bmatrix}$ cannot be a transition matrix because the sum of the entries in the first row is $.7 + .2 = .9 \neq 1$.

9. $\begin{bmatrix} \frac{4}{9} & \frac{1}{3} \\ \frac{1}{5} & \frac{7}{10} \end{bmatrix}$

This cannot be a transition matrix because the sum of the entries in row 1 is $\frac{4}{9}+\frac{1}{3}=\frac{7}{9}\neq 1$ and the sum of the entries in row 2 is $\frac{1}{5}+\frac{7}{10}=\frac{9}{10}\neq 1$.

11. $\begin{bmatrix} \frac{1}{2} & \frac{1}{4} & 1 \\ \frac{2}{3} & 0 & \frac{1}{3} \\ \frac{1}{3} & 1 & 0 \end{bmatrix}$

This could not be a transition matrix because the sum of the entries in row 1 is $\frac{1}{2}+\frac{1}{4}+1=\frac{7}{4}\neq 1$, and the sum of the entries in row 3 is $\frac{1}{3}+1=\frac{4}{3}\neq 1$.

13. This is not a transition diagram because the sum of the probabilities for changing from state A to states A, B, and C is $\frac{1}{3}+\frac{1}{2}+1\neq 1$.

15. This is a transition diagram. The information given in this diagram can also be given by the following matrix.

$$\begin{array}{c} \\ A \\ B \\ C \end{array} \begin{array}{c} \begin{array}{ccc} A & B & C \end{array} \\ \begin{bmatrix} .6 & .20 & .20 \\ .9 & .02 & .08 \\ .4 & .0 & .6 \end{bmatrix} \end{array}$$

17. Let $A = \begin{bmatrix} .2 & .8 \\ .9 & .1 \end{bmatrix}$

A is a regular transition matrix since A^2 contains all positive entries.

19. Let $P = \begin{bmatrix} 0 & 1 & 0 \\ .3 & .3 & .4 \\ 1 & 0 & 0 \end{bmatrix}$.

$$P^2 = \begin{bmatrix} 0 & 1 & 0 \\ .3 & .3 & .4 \\ 1 & 0 & 0 \end{bmatrix}\begin{bmatrix} 0 & 1 & 0 \\ .3 & .3 & .4 \\ 1 & 0 & 0 \end{bmatrix} = \begin{bmatrix} .3 & .3 & .4 \\ .49 & .39 & .12 \\ 0 & 1 & 0 \end{bmatrix}$$

$$P^3 = \begin{bmatrix} 0 & 1 & 0 \\ .3 & .3 & .4 \\ 1 & 0 & 0 \end{bmatrix}\begin{bmatrix} .3 & .3 & .4 \\ .49 & .39 & .12 \\ 0 & 1 & 0 \end{bmatrix} = \begin{bmatrix} .49 & .39 & .12 \\ .237 & .607 & .156 \\ .3 & .3 & .4 \end{bmatrix}$$

P is a regular transition matrix since P^3 contains all positive entries.

21. Let $A = \begin{bmatrix} .23 & .41 & 0 & .36 \\ 0 & .27 & .21 & .52 \\ 0 & 0 & 1 & 0 \\ .48 & 0 & .39 & .13 \end{bmatrix}$

$$A^2 = \begin{bmatrix} .2257 & .205 & .2265 & .3428 \\ .2496 & .0729 & .4695 & .208 \\ 0 & 0 & 1 & 0 \\ .1728 & .1968 & .4407 & .1897 \end{bmatrix}$$

A is not regular. Any power of A will have zero entries in the 1st, 2nd, and 4th column of row 3; thus, it cannot have all positive entries.

23. $[v_1 \quad v_2]\begin{bmatrix} .55 & .45 \\ .19 & .81 \end{bmatrix} = [v_1 \quad v_2].$

We obtain the two equations
$.55v_1 + .19v_2 = v_1,\ .45v_1 + .81v_2 = v_2.$
Simplify these equations to get the system
$-.45v_1 + .19v_2 = 0,$
$.45v_1 - .19v_2 = 0.$
These equations are dependent. Since **V** is a probability vector, $v_1 + v_2 = 1.$
Solve the system
$-.45v_1 + .19v_2 = 0.$
$v_1 + \quad v_2 = 1$
by the substitution method.
$-.45(1 - v_2) + .19v_2 = 0 \Rightarrow$
$-.45 + .45v_2 + .19v_2 = 0 \Rightarrow .64v_2 = .45 \Rightarrow$
$v_2 = \frac{.45}{.64} = \frac{45}{64}$
$v_1 = 1 - \frac{45}{64} = \frac{19}{64}$
Thus, the equilibrium vector is
$\left[\frac{19}{64}, \frac{45}{64}\right].$

25. $[v_1 \quad v_2]\begin{bmatrix} \frac{2}{3} & \frac{1}{3} \\ \frac{1}{8} & \frac{7}{8} \end{bmatrix} = [v_1 \quad v_2].$

We obtain the two equations
$\frac{2}{3}v_1 + \frac{1}{8}v_2 = v_1$
$\frac{1}{3}v_1 + \frac{7}{8}v_2 = v_2.$
Multiply both equations by 24 to eliminate fractions.
$16v_1 + 3v_2 = 24v_1$
$8v_1 + 21v_2 = 24v_2$
Simplify both equations.
$-8v_1 + 3v_2 = 0$
$8v_1 - 3v_2 = 0$
These equations are dependent.
$v_1 + v_2 = 1 \Rightarrow v_1 = 1 - v_2.$

Substituting $1 - v_2$ for v_1 in the first equation gives
$-8(1 - v_2) + 3v_2 = 0 \Rightarrow -8 + 8v_2 + 3v_2 = 0 \Rightarrow$
$11v_2 = 8 \Rightarrow v_2 = \frac{8}{11}$ and $v_1 = 1 - v_2 = \frac{3}{11}.$
The equilibrium vector is
$\left[\frac{3}{11} \quad \frac{8}{11}\right].$

27. Let $P = \begin{bmatrix} .16 & .28 & .56 \\ .43 & .12 & .45 \\ .86 & .05 & .09 \end{bmatrix}$, and let **V** be the probability vector $[v_1 \quad v_2 \quad v_3].$

$$[v_1 \quad v_2 \quad v_3]\begin{bmatrix} .16 & .28 & .56 \\ .43 & .12 & .45 \\ .86 & .05 & .09 \end{bmatrix} = [v_1 \quad v_2 \quad v_3]$$

$.16v_1 + .43v_2 + .86v_3 = v_1$
$.28v_1 + .12v_2 + .05v_3 = v_2$
$.56v_1 + .45v_2 + .09v_3 = v_3$
Simplify these equations by the Gauss-Jordan method to obtain
$v_1 = \frac{7783}{16,799}, v_2 = \frac{2828}{16,799},$ and $v_3 = \frac{6188}{16,799}$
The equilibrium vector is
$\left[\frac{7783}{16,799} \quad \frac{2828}{16,799} \quad \frac{6188}{16,799}\right],$
or $[.4633 \quad .1683 \quad .3684]$ in decimal form.

29. Let $P = \begin{bmatrix} .44 & .31 & .25 \\ .80 & .11 & .09 \\ .26 & .31 & .43 \end{bmatrix}$, and let **V** be the probability vector $\begin{bmatrix} v_1 & v_2 & v_3 \end{bmatrix}$.

$$\begin{bmatrix} v_1 & v_2 & v_3 \end{bmatrix} \begin{bmatrix} .44 & .31 & .25 \\ .80 & .11 & .09 \\ .26 & .31 & .43 \end{bmatrix} = \begin{bmatrix} v_1 & v_2 & v_3 \end{bmatrix}$$

$$\begin{aligned} .44v_1 + .80v_2 + .26v_3 &= v_1 \\ .31v_1 + .11v_2 + .31v_3 &= v_2 \\ .25v_1 + .09v_2 + .43v_3 &= v_3 \end{aligned}$$

Simplify these equations to get the system

$$\begin{aligned} -.56v_1 + .80v_2 + .26v_3 &= 0 \\ .31v_1 - .89v_2 + .31v_3 &= 0 \\ .25v_1 + .09v_2 - .57v_3 &= 0 \end{aligned}$$

Since **V** is the probability vector,

$v_1 + v_2 + v_3 = 1.$

This gives us a system of four equations in three variables.

$$\begin{aligned} v_1 + v_2 + v_3 &= 1 \\ -.56v_1 + .80v_2 + .26v_3 &= 0 \\ .31v_1 - .89v_2 + .31v_3 &= 0 \\ .25v_1 + .09v_2 - .57v_3 &= 0 \end{aligned}$$

This system can be solved by the Gauss-Jordan method. Start with the augmented matrix

$$\left[\begin{array}{ccc|c} 1 & 1 & 1 & 1 \\ -.56 & .80 & .26 & 0 \\ .31 & -.89 & .31 & 0 \\ .25 & .09 & -.57 & 0 \end{array}\right].$$

The solution of this system is

$v_1 = .4872, v_2 = .2583,$ and $v_3 = .2545,$ so the equilibrium vector is $\begin{bmatrix} .4872 & .2583 & .2545 \end{bmatrix}$.

31. The following solution assumes the use of a TI-82 graphing calculator. Similar results can be obtained from other graphing calculators.
Store the given transition matrix as matrix [A].

$$A=\begin{bmatrix}.3 & .2 & .3 & .1 & .1\\ .4 & .2 & .1 & .2 & .1\\ .1 & .3 & .2 & .2 & .2\\ .2 & .1 & .3 & .2 & .2\\ .1 & .1 & .4 & .2 & .2\end{bmatrix};\quad A^2=\begin{bmatrix}.23 & .21 & .24 & .17 & .15\\ .26 & .18 & .26 & .16 & .14\\ .23 & .18 & .24 & .19 & .16\\ .19 & .19 & .27 & .18 & .17\\ .17 & .2 & .26 & .19 & .18\end{bmatrix}$$

$$A^3=\begin{bmatrix}.226 & .192 & .249 & .177 & .156\\ .222 & .196 & .252 & .174 & .156\\ .219 & .189 & .256 & .177 & .159\\ .213 & .192 & .252 & .181 & .162\\ .213 & .189 & .252 & .183 & .163\end{bmatrix}$$

$$A^4=\begin{bmatrix}.2205 & .1916 & .2523 & .1774 & .1582\\ .2206 & .1922 & .2512 & .1778 & .1582\\ .2182 & .1920 & .2525 & .1781 & .1592\\ .2183 & .1909 & .2526 & .1787 & .1595\\ .2176 & .1906 & .2533 & .1787 & .1598\end{bmatrix}$$

$$A^5=\begin{bmatrix}.21932 & .19167 & .25227 & .17795 & .15879\\ .21956 & .19152 & .25226 & .17794 & .15872\\ .21905 & .19152 & .25227 & .17818 & .15898\\ .21880 & .19144 & .25251 & .17817 & .15908\\ .21857 & .19148 & .25253 & .17824 & .15918\end{bmatrix}$$

The entry in row 2, column 4 of A^5 is .17794, which gives the probability that state 2 changes to state 4 after 5 repetitions.

33. **a.** The transition matrix is

$$\begin{array}{l} \\ \text{Approve} \\ \text{Didn't Approve}\end{array}\begin{array}{c}\text{Approve}\quad\text{Didn't Approve}\\ \begin{bmatrix}.9 & \quad\quad .10\\ .3 & \quad\quad .70\end{bmatrix}\end{array}$$

b. $[.35\ \ .65]\begin{bmatrix}.90 & .10\\ .30 & .70\end{bmatrix}=[.51\ \ .49]$

c. $[v_1\ \ v_2]\begin{bmatrix}.9 & .1\\ .3 & .7\end{bmatrix}=[.9v_1+.3v_2\ \ \ .1v_1+.7v_2]$

We obtain the equations

$$.9v_1+.3v_2=v_1$$
$$.1v_1+.7v_2=v_2$$

which simplify to

$$-.1v_1+.3v_2=0$$
$$.1v_1-.3v_2=0.$$

$v_1+v_2=1 \Rightarrow v_1=1-v_2.$

Substituting $1-v_2$ for v_1 in the second equation gives

$.1(1-v_2)-.3v_2=0 \Rightarrow .1-.4v_2=0 \Rightarrow$

$v_2=\frac{.1}{.4}=.25$ and $v_1=1-v_2=.75.$

The long-range trend is that .75 of the voters will favor the initiative and .25 will not favor the initiative. This is represented as $[.75,\ .25]$.

35. From the transition matrix, we obtain $.81v_1 + .77v_2 = v_1$ and $.19v_1 + .23v_2 = v_2$. Collecting like terms, we find that

$-.19v_1 + .77v_2 = 0$ or $v_1 = \frac{77}{19}v_2$.

Substitute this into $v_1 + v_2 = 1$, which yields

$\frac{96}{19}v_2 = 1$.

Therefore, $v_2 = \frac{19}{96} \approx .198$ and $v_1 = \frac{77}{96} \approx .802$.

The line works about .802 of the time.

37. Cross pink with color on left.

Resulting Color

Red Pink White

$$\begin{matrix} \text{Red} \\ \text{Pink} \\ \text{White} \end{matrix} \begin{bmatrix} \frac{1}{2} & \frac{1}{2} & 0 \\ \frac{1}{4} & \frac{1}{2} & \frac{1}{4} \\ 0 & \frac{1}{2} & \frac{1}{2} \end{bmatrix}$$

$$\frac{1}{2}v_1 + \frac{1}{4}v_2 = v_1$$
$$\frac{1}{2}v_1 + \frac{1}{2}v_2 + \frac{1}{2}v_3 = v_2$$
$$\frac{1}{4}v_2 + \frac{1}{2}v_3 = v_3$$

Also, $v_1 + v_2 + v_3 = 1$.

Solving this system, we obtain

$v_1 = \frac{1}{4}, v_2 = \frac{1}{2}, v_3 = \frac{1}{4}$.

The equilibrium vector is $\begin{bmatrix} \frac{1}{4} & \frac{1}{2} & \frac{1}{4} \end{bmatrix}$.

The long range prediction is $\frac{1}{4}$ red, $\frac{1}{2}$ pink, and $\frac{1}{4}$ white snapdragons.

39. a. $$[.6 \quad .395 \quad .005]\begin{bmatrix} .90 & .10 & 0 \\ .09 & .909 & .001 \\ 0 & .34 & .66 \end{bmatrix} \approx [.576 \quad .421 \quad .004]$$

About 57.6% will own a home, 42.1% will rent, and .4% will be homeless.

b. $$[v_1 \quad v_2 \quad v_3]\begin{bmatrix} .9 & .10 & 0 \\ .09 & .909 & .001 \\ 0 & .34 & .66 \end{bmatrix} = [v_1 \quad v_2 \quad v_3]$$

$$\begin{aligned} .90v_1 + .09v_2 + 0v_3 &= v_1 \\ .10v_1 + .909v_2 + .34v_3 &= v_2 \\ 0v_1 + .001v_2 + .66v_3 &= v_3. \end{aligned}$$

Therefore, we have the system

$$\begin{aligned} v_1 + v_2 + v_3 &= 1 \\ .90v_1 + .09v_2 + 0v_3 &= v_1 \\ .10v_1 + .909v_2 + .34v_3 &= v_2 \\ 0v_1 + .001v_2 + .66v_3 &= v_3. \end{aligned}$$

Solve this system by the Gauss-Jordan method to obtain

$v_1 \approx .473$, $v_2 = .526$, $v_3 = .002$. Therefore, 47.3% will be homeowners, 52.6% will be renters, and there will be 0.2% homeless in the long-range.

41. The transition matrix is

$$\begin{bmatrix} .85 & .10 & .05 \\ .15 & .75 & .10 \\ .10 & .30 & .60 \end{bmatrix}.$$

The square of the transition matrix is

$$\begin{bmatrix} .85 & .10 & .05 \\ .15 & .75 & .10 \\ .10 & .30 & .60 \end{bmatrix}\begin{bmatrix} .85 & .10 & .05 \\ .15 & .75 & .10 \\ .10 & .30 & .60 \end{bmatrix} = \begin{bmatrix} .7425 & .175 & .0825 \\ .25 & .6075 & .1425 \\ .19 & .415 & .395 \end{bmatrix}.$$

The cube of the transition matrix is

$$\begin{bmatrix} .85 & .10 & .05 \\ .15 & .75 & .10 \\ .10 & .30 & .60 \end{bmatrix}\begin{bmatrix} .7425 & .175 & .0825 \\ .25 & .6075 & .1425 \\ .19 & .415 & .395 \end{bmatrix} = \begin{bmatrix} .665625 & .23025 & .104125 \\ .317875 & .523375 & .15875 \\ .26325 & .44875 & .288 \end{bmatrix}.$$

a. $$[50{,}000 \quad 0 \quad 0]\begin{bmatrix} .85 & .10 & .05 \\ .15 & .75 & .10 \\ .10 & .30 & .60 \end{bmatrix} = [42{,}500 \quad 5000 \quad 2500]$$

The numbers in the groups after 1 year are 42,500, 5000, and 2500.

b. $$[50,000 \quad 0 \quad 0]\begin{bmatrix} .7425 & .175 & .0825 \\ .25 & .6075 & .1425 \\ .19 & .415 & .395 \end{bmatrix}$$
$$=[37,125 \quad 8750 \quad 4125]$$
The numbers in the groups after 2 years are 37,125, 8750, and 4125.

c. $$[50,000 \quad 0 \quad 0]\begin{bmatrix} .665625 & .23025 & .104125 \\ .317875 & .523375 & .15875 \\ .26325 & .44875 & .288 \end{bmatrix}$$
$$=[33,281 \quad 11,513 \quad 5206]$$
The numbers in the groups after 3 years are 33,281, 11,513, and 5206.

d. The system of equations is
$$\begin{aligned} v_1 + v_2 + v_3 &= 1 \\ .85v_1 + .15v_2 + .10v_3 &= v_1 \\ .10v_1 + .75v_2 + .30v_3 &= v_2 \\ .05v_1 + .10v_2 + .60v_3 &= v_3 \end{aligned}$$
Solve this system by the Gauss-Jordan method to obtain
$$v_1 = \frac{28}{59} \approx .475,\ v_2 = \frac{22}{59} \approx .373, \text{ and}$$
$$v_3 = \frac{9}{59} \approx .152.$$
$$\mathbf{V} = \begin{bmatrix} \frac{28}{59} & \frac{22}{59} & \frac{9}{59} \end{bmatrix} \text{ or } [.475 \quad .373 \quad .152].$$
In the last four years, the probabilities of no accidents, one accident, and more than one accident are .475, .373, and .152, respectively.

43. a.
$$[.137 \quad .102 \quad .761]\begin{bmatrix} .95 & .02 & .03 \\ .04 & .92 & .04 \\ .09 & .07 & .84 \end{bmatrix}$$
$$\approx [.203 \quad .150 \quad .647].$$
The probability that a vehicle was purchased from Toyota in the next year was about .203.

b. From the transition matrix we obtain
$$\begin{aligned} .95v_1 + .04v_2 + .09v_3 &= v_1 \\ .02v_1 + .92v_2 + .07v_3 &= v_2 \\ .03v_1 + .04v_2 + .84v_3 &= v_3 \end{aligned}$$
and $v_1 + v_2 + v_3 = 1$.
Solve using the Gauss-Jordan method to obtain $v_1 = .541, v_2 = .286,$ and $v_3 = .173$.
Thus, the long-term probability that a vehicle is purchased from Honda is .286.

45. a. If there is no one in line, then after 1 minute there will be either 0, 1, or 2 people in line with probabilities $p_{00} = .4, p_{01} = .3,$ and $p_{02} = .3$. If there is one person in line, then that person will be served and either 0, 1 or 2 new people will join the line, with probabilities $p_{10} = .4, p_{11} = .3,$ and $p_{12} = .3$. If there are two people in line, then one of them will be served and either 1 or 2 new people will join the line, with probabilities $p_{21} = .5$ and $p_{22} = .5$; it is impossible for both people in line to be served, so $p_{20} = 0$. Therefore, the transition matrix, A, is
$$\begin{array}{c} \\ 0 \\ 1 \\ 2 \end{array}\begin{array}{c} \begin{array}{ccc} 0 & 1 & 2 \end{array} \\ \begin{bmatrix} .4 & .3 & .3 \\ .4 & .3 & .3 \\ 0 & .5 & .5 \end{bmatrix} \end{array}.$$

b. The transition matrix for a two-minute period is
$$A^2 = \begin{bmatrix} .4 & .3 & .3 \\ .4 & .3 & .3 \\ 0 & .5 & .5 \end{bmatrix}\begin{bmatrix} .4 & .3 & .3 \\ .4 & .3 & .3 \\ 0 & .5 & .5 \end{bmatrix}$$
$$= \begin{bmatrix} .28 & .36 & .36 \\ .28 & .36 & .36 \\ .20 & .40 & .40 \end{bmatrix}.$$

c. The probability that a queue with no one in line has two people in line 2 min later is .36 since that is the entry in row 1, column 3 of A^2.

47. The following solution presupposes the use of a TI-82 graphics calculator. Similar results can be obtained from other graphics calculators. Store the given transition matrix as matrix [*A*] and the probability vector $[.5 \quad .5 \quad 0 \quad 0]$ as matrix [*B*]. Multiply [*B*] by successive powers of [*A*] to obtain $[0 \quad 0 \quad .102273 \quad .897727]$.
This gives the long-range prediction for the percent of employees in each state for the company training program.

Section 9.6 Decision Making

1. a. An optimist should choose the coast; $150,000 is the largest profit.

b. A pessimist should choose the highway; the worst case of $30,000 is better than –$40,000 if the coast is chosen.

c. If the possibility of heavy opposition is .8, the probability of light opposition is .2. Find his expected profit for each strategy.
Highway:
70,000(.2) + 30,000(.8) = $38,000
Coast:
150,000(.2) + –(40,000)(.8) = –$2000
He should choose the highway for an expected profit of $38,000

d. If the probability of heavy opposition is .4, the probability of light opposition is .6. Find his expected profit for each strategy.
Highway:
70,000(.6) + 30,000(.4) = $54,000
Coast:
150,000(.6) + (–40,000)(.4) = $74,000
He should choose the coast.

3. Note that the costs given in the payoff matrix are given in hundreds of dollars.

a. An optimist should make no upgrade; minimum cost is $2800.

b. A pessimist should make the upgrade; a worst case of $13,000 is better than a possible cost of $45,000 if no upgrade is made.

c. Find the expected cost of each strategy.
Make upgrade:
.7(130) + .2(130) + .1(130) = 130, or $13,000
Make no upgrade:
.7(28) + .2(180) + .1(450) = 100.6 or $10,600
He should not upgrade. The expected cost to the company if this strategy is chosen is $10,060.

5. a.

$$\begin{array}{r} \\ \text{Overhaul} \\ \text{Don't Overhaul} \end{array} \begin{array}{c} \text{Fails} \quad \text{Doesn't Fail} \\ \begin{bmatrix} -\$8600 & -\$2600 \\ -\$6000 & \$0 \end{bmatrix} \end{array}$$

b. Find the expected cost under each strategy.
Overhaul:
.1(–8600) + .9(–2600) = –$3200
Don't Overhaul:
.3(–6000) + .7(0) = –$1800
To minimize his expected costs, the business should not overhaul the machine before shipping.

7. a.

$$\begin{array}{r} \\ \text{Tent} \\ \text{No tent} \end{array} \begin{array}{c} \text{No Rain} \quad \text{Rain} \\ \begin{bmatrix} \$2500 & \$1500 \\ \$3000 & \$0 \end{bmatrix} \end{array}$$

b. Find her expected revenue for each scenario.
No rain:
.6(2500) + .4(1500) = $2100
Rain:
.6(1500) + .4(0) = $900
She should rent the tent since the expected revenue is $2100.

9. Find the expected utility under each strategy.
Jobs:
(.35)(40) + (.65)(–10) = 7.5
Environment:
(.35)(–12) + (.65)(30) = 15.3
She should emphasize the environment. The expected utility of this strategy is 15.3.

Chapter 9 Review Exercises

1. $E(x) = 0(.22) + 1(.54) + 2(.16) + 3(.08)$
$= 1.1$

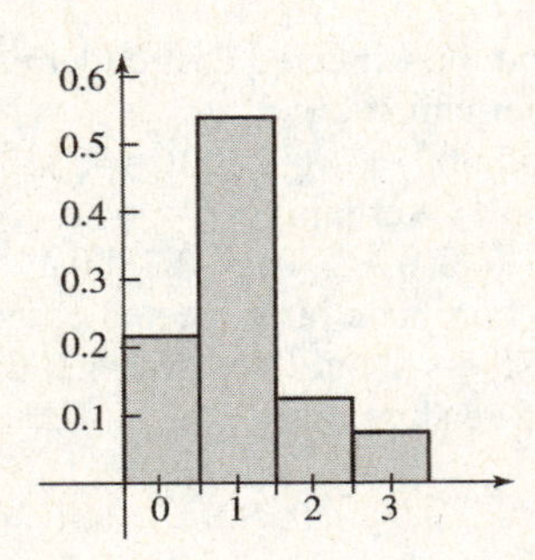

2. $E(x) = -3(.15) - 2(.20) - 1(.25) + 0(.18) + 1(.12) + 2(.06) + 3(.04)$
$= -.74$

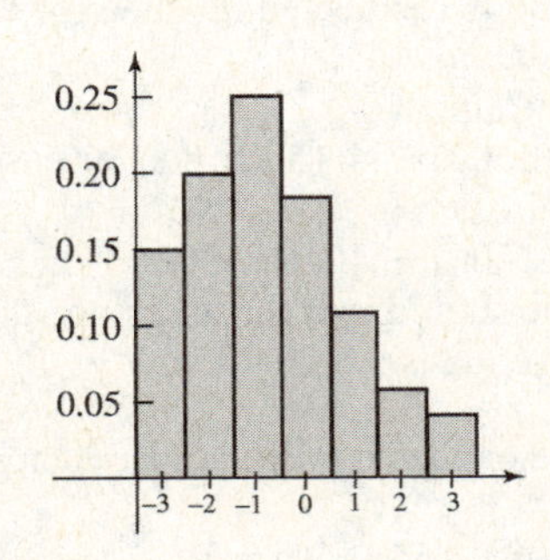

3. $E(x) = -10(.333) + 0(.333) + 10(.333) = 0$

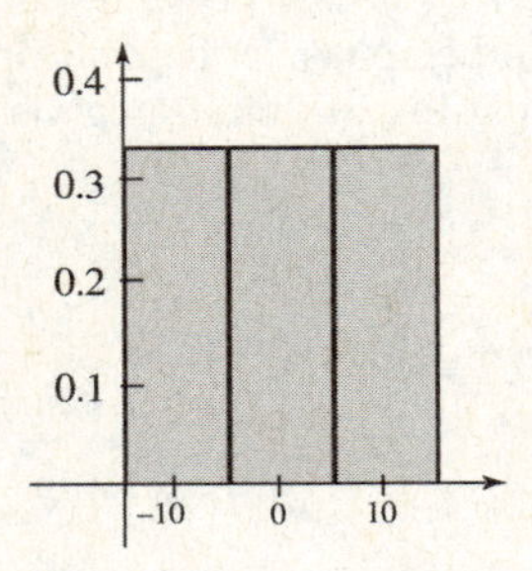

4. $E(x) = 0(.35) + 2(.15) + 4(.2) + 6(.3)$
$= 2.9$

5. $E(x) = 1(.84513) + 2(.15221) + 3(.00251) + 4(.00014)$
$= 1.15764$

6. $E(x) = 0(.0115) + 1(.1324) + 2(.2671) + 3(.3819) + 4(.1655) + 5(.0355) + 6(.0053) + 7(.0006) + 8(.0001)$
$= 2.6886$

7. a. In selecting 2 bouquets from a group of 10, there are ${}_{10}C_2 = 45$ ways.

$$P(0 \text{ roses}) = \frac{{}_3C_0 \cdot {}_7C_2}{{}_{10}C_2} = \frac{21}{45} = \frac{7}{15}$$

$$P(1 \text{ rose}) = \frac{{}_3C_1 \cdot {}_7C_1}{{}_{10}C_2} = \frac{21}{45} = \frac{7}{15}$$

$$P(2 \text{ roses}) = \frac{{}_3C_2 \cdot {}_7C_0}{{}_{10}C_2} = \frac{3}{45} = \frac{1}{15}$$

x	0	1	2
$P(x)$	$\frac{7}{15}$	$\frac{7}{15}$	$\frac{1}{15}$

b. $E(x) = 0\left(\frac{7}{15}\right) + 1\left(\frac{7}{15}\right) + 2\left(\frac{1}{15}\right) = .6$

8. a. Three of the 10 members did not do their homework, and 3 members of the 10 are selected. Three members of the 10 can be selected in ${}_{10}C_3 = 120$ ways. Let $P(n)$ represent the probability of selecting n students who did not do homework.

$$P(0) = \frac{{}_3C_0 \cdot {}_7C_3}{{}_{10}C_3} = \frac{35}{120} = .292$$

$$P(1) = \frac{{}_3C_1 \cdot {}_7C_2}{{}_{10}C_3} = \frac{63}{120} = .525$$

$$P(2) = \frac{{}_3C_2 \cdot {}_7C_1}{{}_{10}C_3} = \frac{21}{120} = .175$$

$$P(3) = \frac{{}_3C_3 \cdot {}_7C_0}{{}_{10}C_3} = \frac{1}{120} = .008$$

x	0	1	2	3
$P(x)$	.292	.525	.175	.008

b. $E(x) = 0(.292) + 1(.525) + 2(.175) + 3(.008)$
$= .899$

9. Probability of getting 3 hearts
$= \frac{_{13}C_3}{_{52}C_3} = \frac{286}{22,100} \approx .0129$
Probability of not getting 3 hearts
$= 1 - .0129 = .9871.$
Let x = the amount to pay for the game.
If you win, you get $100 - x$.
If you lose, you get $-x$. The expected value must be 0.
$(100 - x)(.0129) + (-x)(.9871) = 0 \Rightarrow$
$1.29 - .0129x - .9871x = 0 \Rightarrow 1.29 = x$
You must pay \$1.29.

10. The probability of winning if you bet "under" is $\frac{15}{36}$ and has a net return of \$2. The probability of 7 is $\frac{6}{36}$ and has a net return of \$4. Betting "over" has a probability of $\frac{15}{36}$ with a net return of \$2.

$$P(\text{sum} < 7) = P(2) + P(3) + P(4) + P(5) + P(6) = \frac{1}{36} + \frac{2}{36} + \frac{3}{36} + \frac{4}{36} + \frac{5}{36} = \frac{15}{36}$$
$$P(\text{sum} = 7) = \frac{6}{36}$$
$$P(\text{sum} > 7) = P(8) + P(9) + P(10) + P(11) + P(12) = \frac{5}{36} + \frac{4}{36} + \frac{3}{36} + \frac{2}{36} + \frac{1}{36} = \frac{15}{36}$$

The expected return for each type of bet is as follows:
Under:
$$E(x) = 2\left(\frac{15}{36}\right) - 2\left(\frac{21}{36}\right) = -\$.33$$
Exactly 7:
$$E(x) = 4\left(\frac{6}{36}\right) - 2\left(\frac{30}{36}\right) = -\$1.00$$
Over:
$$E(x) = 2\left(\frac{15}{36}\right) - 2\left(\frac{21}{36}\right) = -\$.33$$

11. $E(x) = 0(.8573) + 1(.1354) + 2(.0071) + 3(.0001) \approx .15$

12. $E(x) = 0(.3277) + 1(.4096) + 2(.2048) + 3(.0512) + 4(.0064) + 5(.0003) = .9999$

13. Since order is important, this is a permutation. Eight taxis can line up in $_8P_8 = 8! = 40,320$ ways.

14. Since order is important, this is a permutation. If there are 8 finalists, there are
$$_8P_3 = \frac{8!}{(8-3)!} = 8 \cdot 7 \cdot 6 = 336 \text{ variations.}$$

15. Since order is not important, this is a combination. Three monitors can be selected from 12 monitors in
$$_{12}C_3 = \frac{12!}{3!9!} = \frac{12 \cdot 11 \cdot 10}{3 \cdot 2 \cdot 1} = 220 \text{ ways.}$$

16. **a.** One of the 4 broken monitors can be selected in $_4C_1 = 4$ ways. The remaining 2 must come from the 8 nonbroken monitors, and can be selected in $_8C_2 = 28$ ways. By the multiplication principle, the selection can be made in $112 = 4 \times 28$ ways.

b. All 3 monitors must come from the non-broken group of 8. This can be accomplished in $_8C_3 = 56$ ways.

c. At least one broken monitor can be accomplished by selecting 1, 2, or 3 defective monitors. If 1 monitor is broken, 2 must be non-broken. If 2 are broken, 1 must be non-broken. If 3 are broken, then 0 must be non-broken. The number of ways to select
1 broken: $_4C_1 \cdot {_8C_2} = 4 \times 28 = 112$
2 broken: $_4C_2 \cdot {_8C_1} = 6 \times 8 = 48$
3 broken: $_4C_3 \cdot {_8C_0} = 4 \times 1 = 4$
is then $112 + 48 + 4 = 164$.

17. Since order is important, this is a permutation. There are 30 choices for the first seat, 29 for the second, 28 for the third, 27 for the fourth, 26 for the fifth, and 25 for the sixth, or $_{30}P_6 = 427,518,000$ ways.

18. The first seat will be occupied by the given student. That leaves 29 choices for the second seat, 28 choices for the third seat, 27 choices for the fourth seat, 26 choices for the fifth seat, and 25 for the sixth seat, or $_{29}P_5 = 14,250,600$ possible ways.

19. a. Since there are 15 students in each major, there are ${}_{15}P_3 = 2730$ ways to arrange the students within each major. There are also ${}_2P_2 = 2$ ways to arrange the two different groups. Thus, the total number of arrangements is $2 \times 2730 \times 2730 = 14{,}905{,}800$.

b. Assume the odd seats are occupied by science majors, then because order is important, there are ${}_{15}P_3 = 2730$ possible arrangements for that group. Then the even seats would be occupied by the business majors with ${}_{15}P_3 = 2730$ possible arrangements By the multiplication principle, there would be $2730 \times 2730 = 7{,}452{,}900$ possible arrangements. However, if the odd seats were occupied by the business majors and the even by the science majors, there would also be $2730 \times 2730 = 7{,}452{,}900$ possible arrangements. So, there must be $7{,}452{,}900 + 7{,}452{,}900 = 14{,}905{,}800$ possibilities.

20. Answers vary.

21. Answers vary.

22. There are 26 black cards in the deck.

$$P(\text{both black}) = \frac{{}_{26}C_2}{{}_{52}C_2} = \frac{325}{1326} \approx .245$$

23. There are 13 hearts in a deck.

$$P(\text{both hearts}) = \frac{{}_{13}C_2}{{}_{52}C_2} = \frac{78}{1326} \approx .059$$

24. To get exactly one face card, you must have one non-face card. There are 12 face cards and 40 non-face cards in a deck.

$$P(\text{exactly one face card}) = \frac{{}_{12}C_1 \cdot {}_{40}C_1}{{}_{52}C_2} = \frac{480}{1326} \approx .362$$

25. There are 4 aces in a deck of cards.

$$P(\text{at most one ace}) = \frac{{}_4C_0 \cdot {}_{48}C_2}{{}_{52}C_2} + \frac{{}_4C_1 \cdot {}_{48}C_1}{{}_{52}C_2} = \frac{1128}{1326} + \frac{192}{1326} = \frac{1320}{1326} \approx .995$$

26. There are 12 possible selections, 6 of them being ice cream, and 6 not ice cream.

$$P(3 \text{ ice cream}) = \frac{{}_6C_3 \cdot {}_6C_0}{{}_{12}C_3} = \frac{20 \cdot 1}{220} \approx .091$$

27. There are 12 possible selections, 4 custard and 8 non-custard.

$$P(4 \text{ custard}) = \frac{{}_4C_3 \cdot {}_8C_0}{{}_{12}C_3} = \frac{4 \cdot 1}{220} \approx .018$$

28. There are 12 possible selections, 2 frozen yogurt and 10 non-yogurt.

$$P(\text{at least 1 yogurt}) = 1 - P(\text{no yogurt}) = 1 - \frac{{}_2C_0 \cdot {}_{10}C_3}{{}_{12}C_3} = 1 - \frac{1 \cdot 120}{220} = \frac{100}{220} \approx .455$$

29. There are 12 possible selections, 4 custard, 6 ice cream, and 2 frozen yogurt.

$$P(1 \text{ custard, 1 ice cream, 1 yogurt}) = \frac{{}_4C_1 \cdot {}_6C_1 \cdot {}_2C_1}{{}_{12}C_3} = \frac{4 \cdot 6 \cdot 2}{220} = \frac{48}{220} \approx .218$$

30. There are 12 possible selections, 6 ice cream, and 6 not ice cream.

$$P(\text{at most 1 ice cream}) = P(0) + P(1) = \frac{{}_6C_0 \cdot {}_6C_3}{{}_{12}C_3} + \frac{{}_6C_1 \cdot {}_6C_2}{{}_{12}C_3} = \frac{20}{220} + \frac{90}{220} = \frac{110}{220} = .5$$

31. a. The number of subsets of size 0 is ${}_nC_0$ or 1. The number of subsets of size 1 is ${}_nC_1$ or n. The number of subsets of size 2 is ${}_nC_2$ or $\frac{n(n-1)}{2}$. The number of subsets of size n is ${}_nC_n$ or 1.

b. The total number of subsets of a set with n elements is ${}_nC_0 + {}_nC_1 + {}_nC_2 + \text{L} + {}_nC_n$.

32. a. Answers vary. Possible Answer: Each element is either in or out of the subset, so there are 2^n ways to make these choices.

b. Let $n = 4$.

$${}_4C_0 + {}_4C_1 + {}_4C_2 + {}_4C_3 + {}_4C_4 = 1+4+6+4+1 = 16$$

Since $2^4 = 16$, the equation from part a holds.

Let $n = 5$.

$${}_5C_0 + {}_5C_1 + {}_5C_2 + {}_5C_3 + {}_5C_4 + {}_5C_5 = 1+5+10+10+5+1 = 32$$

Since $2^5 = 32$, the equation from part c holds.

33. a. $n = 7,\ p = .75,\ 1-p = .25$

x	$P(x)$	
0	.0001	$= {}_7C_0(.75)^0(.25)^7$
1	.0013	$= {}_7C_1(.75)^1(.25)^6$
2	.0115	$= {}_7C_2(.75)^2(.25)^5$
3	.0577	$= {}_7C_3(.75)^3(.25)^4$
4	.1730	$= {}_7C_4(.75)^4(.25)^3$
5	.3115	$= {}_7C_5(.75)^5(.25)^2$
6	.3115	$= {}_7C_6(.75)^6(.25)^1$
7	.1335	$= {}_7C_7(.75)^7(.25)^0$

b. $E(x) = 0(.00001)+1(.0013)+2(.0115)+3(.0577)+4(.1730)+5(.3115)+6(.3115)+7(.1335) \approx 5.25$

34. a. $n = 10,\ p = .75,\ 1-p = .25$

x	$P(x)$	
0	.000001	$= {}_{10}C_0(.75)^0(.25)^{10}$
1	.000029	$= {}_{10}C_1(.75)^1(.25)^9$
2	.000386	$= {}_{10}C_2(.75)^2(.25)^8$
3	.00309	$= {}_{10}C_3(.75)^3(.25)^7$
4	.016222	$= {}_{10}C_4(.75)^4(.25)^6$
5	.058399	$= {}_{10}C_5(.75)^5(.25)^5$
6	.145998	$= {}_{10}C_6(.75)^6(.25)^4$
7	.250282	$= {}_{10}C_7(.75)^7(.25)^3$
8	.281568	$= {}_{10}C_8(.75)^8(.25)^2$
9	.187712	$= {}_{10}C_9(.75)^9(.25)^1$
10	.056314	$= {}_{10}C_{10}(.75)^{10}(.25)^0$

b. $E(x) = 0(.000001)+1(.000029)+2(.000386)+3(.00309)+4(.016222)+5(.058399)+6(.145998)+7(.250282)+8(.281568)+9(.187712)+10(.056314) \approx 7.5$

35. a. $n = 6,\ p = .145,\ 1-p = .855$

$P(0) = {}_6C_0(.145)^0(.855)^6 \approx .3907$

b. $P(\text{at least }2) = P(2)+P(3)+P(4)+P(5)+P(6)$
$= 1 - P(0) - P(1)$
$= 1 - {}_6C_0(.145)^0(.855)^6 - {}_6C_1(.145)^1(.855)^5$
$= 1 - .3907 - .3975$
$\approx .2118$

c. $P(\text{at most }4)$
$= P(0)+P(1)+P(2)+P(3)+P(4)$
$= {}_6C_0(.145)^0(.855)^6 + {}_6C_1(.145)^1(.855)^5 + {}_6C_2(.145)^2(.855)^4 + {}_6C_3(.145)^3(.855)^3 + {}_6C_4(.145)^4(.855)^2$
$= .39065 + .39751 + .16854 + .03811 + .00485$
$\approx .9997$

36. a. $n = 4,\ p = .22,\ 1-p = .78$

$P(4) = {}_4C_4(.22)^4(.78)^0 \approx .002$

b. $P(\text{at least }1) = P(1)+P(2)+P(3)+P(4)$
$= 1 - P(0)$
$= 1 - {}_4C_0(.22)^0(.78)^4$
$\approx 1 - .370 = .630$

c. $P(\text{at most }2)$
$= P(0)+P(1)+P(2)$
$= {}_4C_0(.22)^0(.78)^4 + {}_4C_1(.22)^1(.78)^3 + {}_4C_2(.22)^2(.78)^2$
$\approx .3701 + .4176 + .1767 \approx .964$

37. **a.** $n = 5,\ p = .614,\ 1 - p = .386$

x	$P(x)$	
0	.0086	$= {}_5C_0(.614)^0(.386)^5$
1	.0682	$= {}_5C_1(.614)^1(.386)^4$
2	.2168	$= {}_5C_2(.614)^2(.386)^3$
3	.3449	$= {}_5C_3(.614)^3(.386)^2$
4	.2743	$= {}_5C_4(.614)^4(.386)^1$
5	.0873	$= {}_5C_5(.614)^5(.386)^0$

b. $$E(x) = 0(.0086) + 1(.0682) + 2(.2168) + 3(.3449) + 4(.2743) + 5(.0873) \approx 3.07$$

38. $\begin{bmatrix} 0 & 1 \\ .77 & .23 \end{bmatrix}$

This is a regular transition matrix because
$$\begin{bmatrix} 0 & 1 \\ .77 & .23 \end{bmatrix}\begin{bmatrix} 0 & 1 \\ .77 & .23 \end{bmatrix} = \begin{bmatrix} .77 & .23 \\ .1771 & .8229 \end{bmatrix},$$
in which all entries are positive.

39. $\begin{bmatrix} -.2 & .4 \\ .3 & .7 \end{bmatrix}$

This is not a regular transition matrix because there is a negative entry in the first row and first column. In fact, this makes it not even a transition matrix.

40. $\begin{bmatrix} .21 & .15 & .64 \\ .50 & .12 & .38 \\ 1 & 0 & 0 \end{bmatrix}$

This is a regular transition matrix because
$$\begin{bmatrix} .21 & .15 & .64 \\ .50 & .12 & .38 \\ 1 & 0 & 0 \end{bmatrix}\begin{bmatrix} .21 & .15 & .64 \\ .50 & .12 & .38 \\ 1 & 0 & 0 \end{bmatrix} = \begin{bmatrix} .7591 & .0495 & .1914 \\ .545 & .0894 & .3656 \\ .21 & .15 & .64 \end{bmatrix},$$
in which all entries are positive.

41. $\begin{bmatrix} .22 & 0 & .78 \\ .40 & .33 & .27 \\ 0 & .61 & .39 \end{bmatrix}$

This is a regular transition matrix because
$$\begin{bmatrix} .22 & 0 & .78 \\ .40 & .33 & .27 \\ 0 & .61 & .39 \end{bmatrix}\begin{bmatrix} .22 & 0 & .78 \\ .40 & .33 & .27 \\ 0 & .61 & .39 \end{bmatrix} = \begin{bmatrix} .0484 & .4758 & .4758 \\ .220 & .2736 & .5064 \\ .244 & .4392 & .3168 \end{bmatrix},$$
in which all entries are positive.

42. $P = \begin{bmatrix} .35 & .15 & .50 \\ .30 & .35 & .35 \\ .15 & .30 & .55 \end{bmatrix}$

$I = [.2\ \ .4\ \ .4]$

a. The distribution after one month is
$$[.2\ \ .4\ \ .4]\begin{bmatrix} .35 & .15 & .50 \\ .30 & .35 & .35 \\ .15 & .30 & .55 \end{bmatrix} = [.25\ \ .29\ \ .46].$$

b. $I = [.2\ \ .4\ \ .4]$
$$A = \begin{bmatrix} .35 & .15 & .50 \\ .30 & .35 & .35 \\ .15 & .30 & .55 \end{bmatrix}$$
$$A^2 = \begin{bmatrix} .2425 & .2550 & .5025 \\ .2625 & .2725 & .4650 \\ .2250 & .2925 & .4825 \end{bmatrix}$$
The distribution after 2 months is given by $IA^2 = [.2435\ \ .277\ \ .4795]$.

c. To find the long-range distribution, we use the system

$$\begin{aligned} v_1 + v_2 + v_3 &= 1 \\ .35v_1 + .3v_2 + .15v_3 &= v_1 \\ .15v_1 + .35v_2 + .3v_3 &= v_2 \\ .5v_1 + .35v_2 + .55v_3 &= v_3. \end{aligned}$$

Simplify these equations to obtain the system

$$\begin{aligned} v_1 + v_2 + v_3 &= 1 \\ -.65v_1 + .3v_2 + .15v_3 &= 0 \\ .15v_1 - .65v_2 + .3v_3 &= 0 \\ .5v_1 + .35v_2 - .45v_3 &= 0. \end{aligned}$$

Solve this system by the Gauss-Jordan method to obtain

$v_1 = \frac{75}{313}, v_2 = \frac{87}{313},$ and $v_3 \frac{151}{313}.$

The long-range distribution is

$[.240 \quad .278 \quad .482].$

43. a. $I = [.15 \quad .60 \quad .25];\quad A = \begin{bmatrix} .80 & .14 & .06 \\ .04 & .85 & .11 \\ .03 & .13 & .84 \end{bmatrix}$

The distribution after one month is

$$[.15 \quad .60 \quad .25]\begin{bmatrix} .80 & .14 & .06 \\ .04 & .85 & .11 \\ .03 & .13 & .84 \end{bmatrix} = [.1515 \quad .5635 \quad .2850]$$

b. $$A^2 = \begin{bmatrix} .80 & .14 & .06 \\ .04 & .85 & .11 \\ .03 & .13 & .84 \end{bmatrix}\begin{bmatrix} .80 & .14 & .06 \\ .04 & .85 & .11 \\ .03 & .13 & .84 \end{bmatrix} = \begin{bmatrix} .6474 & .2388 & .1138 \\ .0693 & .7424 & .1883 \\ .0544 & .2239 & .7217 \end{bmatrix}$$

$$A^3 = \begin{bmatrix} .80 & .14 & .06 \\ .04 & .85 & .11 \\ .03 & .13 & .84 \end{bmatrix}\begin{bmatrix} .6474 & .2388 & .1138 \\ .0693 & .7424 & .1883 \\ .0544 & .2239 & .7217 \end{bmatrix} = \begin{bmatrix} .530886 & .308410 & .160704 \\ .090785 & .665221 & .243994 \\ .074127 & .291752 & .634121 \end{bmatrix}$$

The distribution after 3 years is given by

$IA^3 = [.1526 \quad .5183 \quad .3290]$

c. To find the long range distribution, we use the system

$$\begin{aligned} v_1 + v_2 + v_3 &= 1 \\ .80v_1 + .04v_2 + .03v_3 &= v_1 \\ .14v_1 + .85v_2 + .13v_3 &= v_3 \\ .06v_1 + .11v_2 + .84v_3 &= v_3 \end{aligned}$$

This system yields

$$\begin{aligned} v_1 + v_2 + v_3 &= 1 \\ -.20v_1 + .04v_2 + .03v_3 &= 0 \\ .14v_1 - .15v_2 + .13v_3 &= 0 \\ .06v_1 + .11v_2 - .16v_3 &= 0 \end{aligned}$$

Solving this system yields $v_1 = \frac{97}{643}$, $v_2 = \frac{302}{643}$ and $v_3 = \frac{244}{643}$. Thus the long range distribution is

$[.1509 \quad .4697 \quad .3795]$.

44. a. Since the candidate is an optimist, look for the biggest value in the matrix, which is 5000. Hence, she should oppose it.

b. A pessimistic candidate wants to find the best of the worst things that can happen. If she favors, the worst is –4000. If she waffles the worst is –500. If she opposes, then worst is 0. Since the best of these is 0, she should oppose.

c. Since there is a 40% chance the opponent favors the plant and a 35% that he will waffle, the chance he will oppose is
$1 - .4 - .35 = .25$
Expected gain if she favors:
$(0)(.4) + (-1000)(.35) + (-4000)(.25) = -1350$
Expected gain if she waffles:
$(1000)(.4)+0(.35)+(-500)(.25) = 275$
Expected gain if she opposes:
$(5000)(.4) + (2000)(.35) + (0)(.25) = 2700$
She should oppose and get 2700 additional votes.

d. Now the opponent has 0 probability of favoring, .7 of waffling and .3 of opposing.
Expected gain if she favors:
$(0)(0) + (.7)(-1000) + (.3)(-4000) = -1900$
Expected gain if she waffles:
$(0)(1000) + (.7)(0) + (.3)(-500) = -150$
Expected gain if she opposes:
$(0)(5000) + (.7)(2000) + (.3)(0) = 1400$
She should oppose and get 1400 additional votes.

45. a. Since the department chair is an optimist, look for the biggest value in the matrix, which is 100. Hence, she should use active learning.

b. A pessimistic wants to find the best of the worst things that can happen. If she lectures, the worst is –80. If she uses active learning, the worst is –30. Since the best of these is
–30, she should use active learning.

c. Since there is a 75% chance the class will prefer the lecture format, there is a 25% chance the class will support the active learning format.
Expected gain for lecture format:
$50(.75)-80(.25)=17.5$ points.
Expected gain for active learning format:
$-30(.75)+100(.25)=2.5$ points. She should use the lecture format where the expected gain is 17.5.

d. Since there is a 60% chance the class will prefer the active learning format, there is a 40% chance the class will support the lecture format.
Expected gain for lecture format:
$50(.4)-80(.6)=-28$ points.
Expected gain for active learning format:
$-30(.4)+100(.6)=48$ points. She should use the active learning format where the expected gain is 48.

46. P(product is successful) = .5
P(product is unsuccessful) = .5
P(successful product passing quality control) = .8
P(unsuccessful product passing quality control) = .25
P(successful product and passes quality control) = (.5)(.8) = .4
P(unsuccessful product and passes quality control) = (.5)(.25) = .125
P(passes quality control) = .4 + .125 = .525
P(successful product given that passes quality control) $=\frac{.4}{.525}=.7619$
P(unsuccessful product given that passes quality control) $=\frac{.125}{.525}=.2381$

$$E(x)=(40{,}000{,}000)(.7619)+(-15{,}000{,}000)(.2381)\approx 27{,}000{,}000$$

The expected net profit is (e) $27 million.

47. If a box is good (probability .9) and the merchant samples an excellent piece of fruit from that box (probability .80), then he will accept the box and earn a $200 profit on it. If a box is bad (probability .1) and he samples an excellent piece of fruit from the box (probability .30), then he will accept the box and earn a –$1000 profit on it.
If the merchant ever samples a non-excellent piece of fruit, he will not accept the box. In this case he pays nothing and earns nothing, so the profit will be $0.
Let x denote the merchant's earnings.
Note that $.9(.80)=.72$,
$.1(.30)=.03$,
and $1-(.72+.03)=.25$.
The probability distribution is as follows.

x	200	–1000	0
$P(x)$	.72	.03	.25

The expected value when the merchant samples the fruit is

$$E(x)=200(.72)+(-1000)(.03)+0(.25)=144-30+0=114$$

We must also consider the case in which the merchant does not sample the fruit. Let x again denote the merchant's earnings. The probability distribution is as follows.

x	200	–1000
$P(x)$	.9	.1

The expected value when the merchant does not sample the fruit is
$E(x)=200(.9)+(-1000)(.1)=180-100=\$80$.
Combining these two results, the expected value of the right to sample is $114 – $80 = $34, which corresponds to choice (c).

48. Let $I(x)$ represent the airline's net income if x people show up.
$I(0) = 0;\ I(1) = 100$
$I(2) = 2 \cdot 100 = 200$
$I(3) = 3 \cdot 100 = 300$
$I(4) = 3 \cdot 100 - 100 = 200$
$I(5) = 3 \cdot 100 - 2 \cdot 100 = 100$
$I(6) = 3 \cdot 100 - 3 \cdot 100 = 0$
Let $P(x)$ represent the probability that x people will show up. Use the binomial probability formula to find the values of $P(x)$.

$P(0) = \binom{6}{0}(.6)^0(.4)^6 = .004$ $\quad P(3) = \binom{6}{3}(.6)^3(.4)^3 = .276$ $\quad P(6) = \binom{6}{6}(.6)^6(.4)^0 = .047$

$P(1) = \binom{6}{1}(.6)^1(.4)^5 = .037$ $\quad P(4) = \binom{6}{4}(.6)^4(.4)^2 = .311$

$P(2) = \binom{6}{2}(.6)^2(.4)^4 = .138$ $\quad P(5) = \binom{6}{5}(.6)^5(.4)^1 = .187$

On the basis of all calculations, the table given in the exercise is completed as follows.

x	0	1	2	3	4	5	6
Income	0	100	200	300	200	100	0
$P(x)$	.004	.037	.138	.276	.311	.187	.047

a. $E(I) = 0(.004) + 100(.037) + 200(.138) + 300(.276) + 200(.311) + 100(.187) + 0(.047) = \$195$

b. $n = 3$

x	0	1	2	3
Income	0	100	200	300
$P(x)$	$P(0) = \binom{3}{0}(.6)^0(.4)^3 = .064$	$P(0) = \binom{3}{1}(.6)^1(.4)^2 = .288$	$P(0) = \binom{3}{2}(.6)^2(.4)^1 = .432$	$P(0) = \binom{3}{3}(.6)^3(.4)^0 = ..216$

$E(I) = 0(.064) + 100(.288) + 200(.432) + 300(.216) = \$180$

$n = 4$

$P(0) = \binom{4}{0}(.6)^0(.4)^4 = .0256$ $\quad P(1) = \binom{4}{1}(.6)^1(.4)^3 = .1536$ $\quad P(2) = \binom{4}{2}(.6)^2(.4)^2 = .3456$

$P(3) = \binom{4}{3}(.6)^3(.4)^1 = .3456$ $\quad P(4) = \binom{4}{4}(.6)^4(.4)^0 = .1296$

x	0	1	2	3	4
Income	0	100	200	300	200
$P(x)$	.0256	.1536	.3456	.3456	.1296

$E(I) = 0(.0256) + 100(.1536) + 200(.3456) + 300(.3456) + 200(.1296) = \$214.08$

$n = 5$

$P(0) = \binom{5}{0}(.6)^0(.4)^5 = .01024$ $P(1) = \binom{5}{1}(.6)^1(.4)^4 = .0768$ $P(2) = \binom{5}{2}(.6)^2(.4)^3 = .2304$

$P(3) = \binom{5}{3}(.6)^3(.4)^2 = .3456$ $P(4) = \binom{5}{4}(.6)^4(.4)^1 = .2592$ $P(5) = \binom{5}{5}(.6)^5(.4)^0 = .07776$

x	0	1	2	3	4	5
Income	0	100	200	300	200	100
$P(x)$	.01024	.0768	.2304	.3456	.2592	.07776

$E(I) = 0(.01024) + 100(.0768) + 200(.2304) + 300(.3456) + 200(.2592) + 100(.07776) = \$217.06$

Since $E(I)$ is greatest when $n = 5$, the airlines should book 5 reservations to maximize revenue.

Case 9 Quick Draw® from the New York State Lottery

1. Using the probabilities determined in the case study, we have

x	Net winnings	$P(x)$
0	2(0) – 1 = –$1	.16660
1	2(0) – 1 = –$1	.36349
2	2(0) – 1 = –$1	.30832
3	2(1) – 1 = $1	.12982
4	2(6) – 1 = $11	.02854
5	2(55) – 1 = $109	.00310
6	2(1000) – 1 = $1999	.00013

$$E(W) = -1(.16660) - 1(.36349) - 1(.30832) + 1(.12982) + 11(.02854) + 109(.00310) + 1999(.00013) \approx .2031$$

2. It is not in the state's interest to offer such a promotion because the players win about 20 cents on average.

3.

x	$P(x)$
0	$\frac{{}_{20}C_0 \cdot {}_{60}C_4}{{}_{80}C_4} \approx .30832$
1	$\frac{{}_{20}C_1 \cdot {}_{60}C_3}{{}_{80}C_4} \approx .43273$
2	$\frac{{}_{20}C_2 \cdot {}_{60}C_2}{{}_{80}C_4} \approx .21264$
3	$\frac{{}_{20}C_3 \cdot {}_{60}C_1}{{}_{80}C_4} \approx .04325$
4	$\frac{{}_{20}C_4 \cdot {}_{60}C_0}{{}_{80}C_4} \approx .00306$

For exercises 4 and 5, carry all decimal places in order to compute the expected winnings.

4.

x	Net winnings	$P(x)$
0	–$1	.30832
1	–$1	.43273
2	$0	.21264
3	$4	.04325
4	$54	.00306

$$E(W) = -1(.30832) - 1(.43273) + 0(.21264) + 4(.04325) + 54(.00306) \approx -.40281$$

5.

x	Net winnings	$P(x)$
0	$2(0) - 1 = -\$1$	.30832
1	$2(0) - 1 = -\$1$	.43273
2	$2(1) - 1 = \$1$	.21264
3	$2(5) - 1 = \$9$	.04325
4	$2(55) - 1 = \$109$	.00306

$$\begin{aligned} E(W) &= -1(.30832) - 1(.43273) + 1(.21264) \\ &\quad + 9(.04325) + 109(.00306) \\ &\approx .19438 \end{aligned}$$

Chapter 10 Introduction to Statistics

Section 10.1 Frequency Distributions

1. a–b. Since 0–.9 is to be the first interval and there are 10 numbers between 0 and .9 inclusive, we will let all nine intervals be of size 10. The other eight intervals are 1.0–1.9, 2.0–2.9, 3.0–3.9, 4.0–4.9, 5.0–5.9, 6.0–6.9, 7.0–7.9, and 8.0–8.9. Keeping a tally of how many data values lie in each interval leads to the following frequency distribution.

Interval	Frequency
0–.9	11
1.0–1.9	15
2.0–2.9	3
3.0–3.9	4
4.0–4.9	2
5.0–5.9	4
6.0–6.9	0
7.0–7.9	0
8.0–8.9	1

c. Draw the histogram. It consists of 9 bars of equal width, having heights as determined by the frequency of each interval.

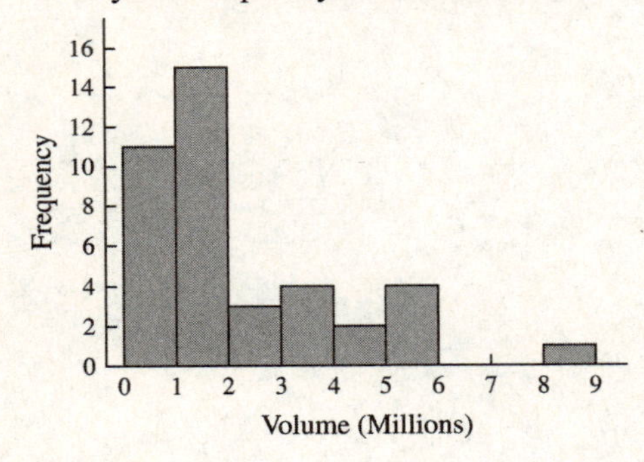

3. a–b. Since – 5.00–(–.01) is to be the first interval, we let all the intervals be of size 5. The largest data value is 23.48, so the last interval that will be needed is 20.00–24.99. The frequency distribution is as follows.

Interval	Frequency
– 5.00–(–.01)	2
0–4.99	30
5.00–9.99	6
10.00–14.99	1
15.00–19.99	0
20.00–24.99	1

c. Draw the histogram. It consists of 6 bars of equal width and having heights as determined by the frequency of each interval.

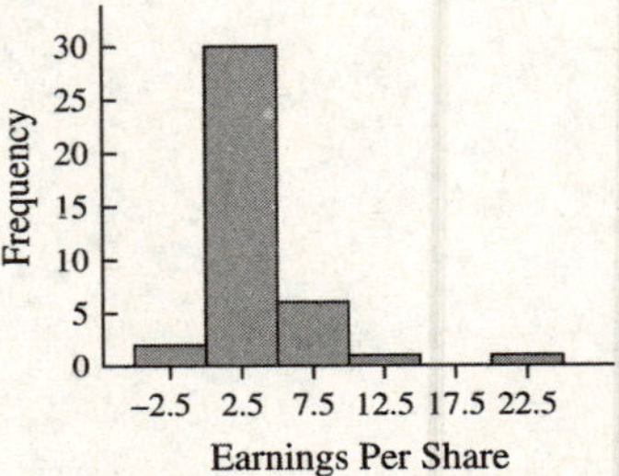

5. The data ranges from a low of 1 to a high of 197, so we will use intervals of size 30 starting with 0–29. The frequency distribution is as follows.

Interval	Frequency
0–29	12
30–59	3
60–89	10
90–119	2
120–149	0
150–179	1
180–209	2

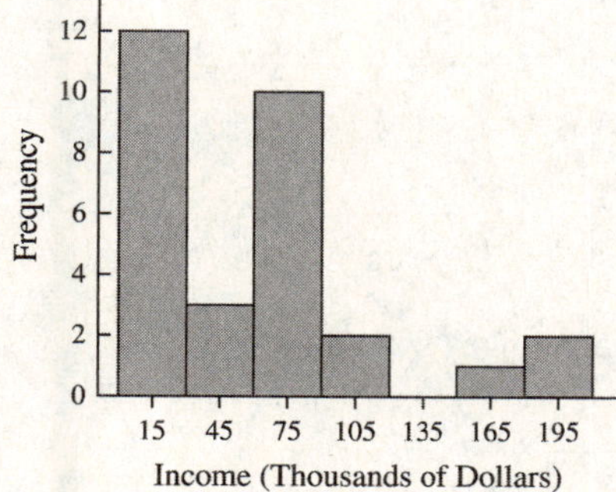

7. The data ranges from a low of 30 to a high of 315, so we will use intervals of size 50 starting with 0–49. The frequency distribution is as follows.

Interval	Frequency
0–49	5
50–99	8
100–149	10
150–199	4
200–249	2
250–299	0
300–349	1

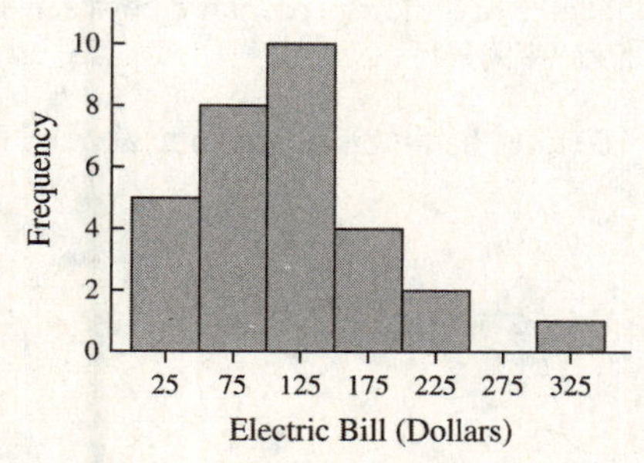

9. The data ranges from a low of 50 to a high of 72, so we will use intervals of size 5 starting with 40–54. The frequency distribution is as follows.

Interval	Frequency
50–54	3
55–59	15
60–64	6
65–69	5
70–74	1

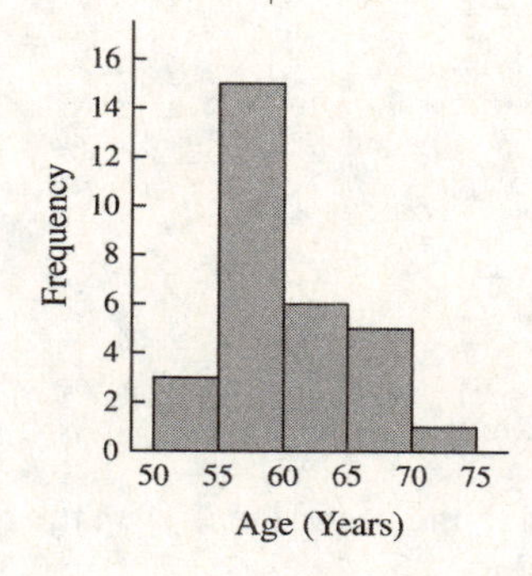

11. The data ranges from .1 to 8.7.

STEM	LEAVES
0	13444556799
1	122333344777889
2	046
3	0157
4	56
5	4666
6	
7	
8	7

Units: 8|7 = 8.7 million

13. Round to the nearest ten. Then the data ranges from 0 to 200.

STEM	LEAVES
0	0011111
0	22223
0	444
0	666777
0	8888
1	01
1	
1	5
1	
1	8
2	0

Units: 2|0 = $200

15. Round the data to the nearest ten. Then the data ranges from 30 to 320.

STEM	LEAVES
0	33444
0	55567789
1	001112334
1	55667
2	22
2	
3	2

Units: 3|2 = $320

17. The data ranges from 50 to 72.

STEM	LEAVES
5	0
5	23
5	55
5	6677
5	888899999
6	0111
6	3
6	455
6	667
6	
7	
7	2

Units: 7|2 = 72 years

19. The data ranges from 81 to 92.

STEM	LEAVES
8	111
8	222333
8	4444555
8	66667777
8	8888899999
9	000000001111
9	2222

Units: 9|2 = 92%

21. uniform

23. left skewed

25. **a.** The distribution is right skewed.

b. 8 states had their percentage between 10% and 14.9%.

c. 10 states had their percentage above 30%.

27. a. The distribution is right skewed.

b. 13 savings institutions had total deposits below $100 million.

c. 5 savings institutions had total deposits above $400 million.

29. a. The distribution is right skewed.

b. 6 states had production below 1 billion pounds.

c. 6 states had production above 2 billion pounds.

Section 10.2 Measures of Center

1. $\Sigma x = 21{,}900 + 22{,}850 + 24{,}930 + 29{,}710 + 28{,}340 + 40{,}000 = 167{,}730$

The mean of the 6 numbers is

$\overline{x} = \frac{167{,}730}{6} = 27{,}955$.

3. $\Sigma x = 3.5 + 4.2 + 5.8 + 6.3 + 7.1 + 2.8 + 3.7 + 4.2 + 4.2 + 5.7 = 47.5$

$\overline{x} = \frac{47.5}{10} = 4.75 \approx 4.8$

5. $\Sigma x = 9.2 + 10.4 + 13.5 + 8.7 + 9.7 = 51.5$

$\overline{x} = \frac{51.5}{5} = 10.3$

7.

Value	Frequency	Value × Frequency	
19	3	$19 \cdot 3$	57
20	5	$20 \cdot 5$	100
21	25	$21 \cdot 25$	525
22	8	$22 \cdot 8$	176
23	2	$23 \cdot 2$	46
24	1	$24 \cdot 1$	24
28	1	$28 \cdot 1$	28
	Total: 45	Total:	956

The mean is $\overline{x} = \frac{956}{45} \approx 21.2$.

9.

x	f	xf
9	5	45
11	10	110
15	12	180
17	9	153
20	6	120
28	1	28
Totals	43	636

$\overline{x} = \frac{636}{43} \approx 14.8$

11. First arrange the numbers in numerical order, from smallest to largest.
$28458, $29679, $33679, $38400, $39720
There are 5 numbers; the median is the middle term, in this case $33,679.

13. First arrange the numbers in numerical order, from smallest to largest.
94.1, 96.8, 97.4, 98.6, 98.4, 98.7, 99.2, 99.9
There are 8 numbers; the median is the mean of the 2 middle numbers, which is
$\frac{98.4 + 98.6}{2} = \frac{197}{2} = 98.5$.

15. 1, 2, 2, 1, 2, 2, 1, 1, 2, 2, 3, 4, 2, 3, 4, 2, 3, 2, 3,
The mode is the number that occurs most often.
The mode is 2.

17. 62, 65, 71, 74, 71, 76, 71, 63, 59, 65, 65, 64, 72, 71, 77, 63, 65
The mode is the number that occurs most often. There are two modes, 65 and 71, since they both appear four times.

19. 3.2, 2.7, 1.9, 3.7, 3.9
There is no one number that occurs more times than any other number; so, there is no mode.

21. Answers vary. Possible answer: The mode has advantages of being easily found and not being influenced by data that are very large or very small compared to the rest of the data. It is often used in samples where the data to be "averaged" are not numerical.

23.

Interval	Midpoint, x	f	xf
0–9.9	4.95	9	44.55
10.0-19.9	14.95	20	299
20.0–29.9	24.95	6	149.7
30.0–39.9	34.95	2	69.9
40.0–49.9	44.95	0	0
50.0–59.9	54.95	1	54.95
60.0–69.9	64.95	0	0
70.0–79.9	74.95	1	74.95
80.0–89.9	84.95	1	84.95
	Totals:	40	778

Mean $\overline{x} = \dfrac{778}{40} = 19.45$

The modal class is 10.0–19.9.

25. a. Mean $\overline{x} = \dfrac{2300+1615+1312+1000+893+811+786+764+718+716}{10} = \dfrac{10{,}915}{10} = \$1091.5$ million.

b. The middle data when the data is in ascending order are 811 and 893. The median value is $\dfrac{811+893}{2} = \$852$ million.

27. a. Mean $\overline{x} = \dfrac{761+659+623+535+475+461+448+441+435+423+423+415}{12}$

$= \dfrac{6099}{12} = \$508.25$ million

b. The two middle receipts are \$448 million and \$461 million. The median revenue is $\dfrac{448+461}{2} = \$454.5$ million

29. a. Mean $\overline{x} = \dfrac{4075.5+5294.3+6369.3+7786.9+9411.5+10{,}383.0+9775+10{,}707+11{,}700+13{,}300}{10}$

$= \dfrac{88{,}802.5}{10} = \$8880.25$ million

The two middle data are 9411.5 and 9775.

Median $= \dfrac{9411.5+9775}{2} = \$9593.25$ million

b. The revenue in 2007 was closest to the mean.

31. The average monthly low temperatures in ascending order: 30, 32, 33, 40, 42, 48, 49, 57, 62, 65, 68, 69

$$\overline{x} = \frac{\sum x}{n} = \frac{595}{12} \approx 49.6°\text{F}$$

$$\text{Median} = \frac{48+49}{2} = 48.5°\text{F}$$

33. The distribution is right skewed, so the median is the better measure of center. There are 30 values, so the median is the average of the 15th and 16th values. These values fall in the 500–749 range, so the median is the midpoint of that range, $624.5.

35. **a.** The distribution is right skewed.

b. There are 50 values, so the median is the average of the 25th and 26th values, so the median value is the average of 22 and 23, which is 22.5%.

Section 10.3 Measures of Variation

Note: Answers may vary slightly throughout this section due to the number of decimal places carried throughout the computations.

1. Answers vary. Possible answer: The standard deviation of a sample of numbers is the square root of the variance of the sample.

3. Range $= 35.1 - 3.6 = 31.5$

$$\overline{x} = \frac{24.9+34.8+12+29.1+3.6+32.8+35.1}{7} = \frac{172.3}{7} = 24.61$$

x	$x-\overline{x}$	$(x-\overline{x})^2$
24.9	.3	.09
34.8	10.2	104.04
12.0	–12.6	158.76
29.1	4.5	20.25
3.6	–21	441
32.8	8.2	67.24
35.1	10.5	110.25
Total		901.63

$$s = \sqrt{\frac{901.63}{7-1}} = \sqrt{150.272} \approx 12.26$$

5. Range $= 5.1 - .1 = 5.0$

$$\overline{x} = \frac{5.1+2.6+4.6+3.5+.1+4+3}{7} = \frac{22.9}{7} = 3.3$$

x	$x-\overline{x}$	$(x-\overline{x})^2$
5.1	1.8	3.24
2.6	–.7	.49
4.6	1.3	1.69
3.5	.2	.04
.1	–3.2	10.24
4.0	.7	.49
3.0	–.3	.09
	Total:	16.28

$$s = \sqrt{\frac{16.28}{7-1}} = \sqrt{2.713} \approx 1.65$$

7. Range = 8.6 − .7 = 7.9

$$\bar{x} = \frac{3.4+8.2+2.9+3.7+.7+4.8+8.6}{7}$$

$$= \frac{32.3}{7} = 4.6$$

x	$x-\bar{x}$	$(x-\bar{x})^2$
3.4	−1.2	1.44
8.2	3.6	12.96
2.9	−1.7	2.89
3.7	−.9	.81
.7	−3.9	15.21
4.8	.2	.04
8.6	4	16
	Total:	49.35

$$s = \sqrt{\frac{49.35}{7-1}} = \sqrt{8.225} \approx 2.87$$

9. Range = 2.3 − .2 = 2.1

$$\bar{x} = \frac{1+2.3+.6+2.3+.2+1.8+.7}{7} = \frac{8.9}{7} = 1.3$$

x	$x-\bar{x}$	$(x-\bar{x})^2$
1.0	−.3	.09
2.3	1.0	1
.6	−.7	.49
2.3	1.0	1
.2	−1.1	1.21
1.8	.5	.25
.7	−.6	.36
	Total:	4.4

$$s = \sqrt{\frac{4.4}{7-1}} = \sqrt{.733} \approx .86$$

11. Expand the table to include columns for the midpoint x of each interval, and for fx, x^2, and fx^2.

Interval	f	x	fx	x^2	fx^2
0–24	4	12	48	144	576
25–49	3	37	111	1369	4107
50–74	6	62	372	3844	23,064
75–99	3	87	261	7569	22,707
100–124	5	112	560	12,544	62,720
125–129	9	137	1233	18,769	168,921
Totals	30		2585		282,095

The mean of the grouped data is

$$\bar{x} = \frac{\Sigma fx}{n} = \frac{2585}{30} \approx 86.2.$$

The standard deviation for the grouped data is

$$s = \sqrt{\frac{\Sigma fx^2 - n\bar{x}^2}{n-1}} = \sqrt{\frac{282{,}095 - 30(86.2)^2}{30-1}}$$

$$\approx \sqrt{2040.8} \approx 45.2$$

13. This exercise should be completed using a computer or calculator. The solution may vary according to the computer program or calculator that is used. Using a TI-83, we have $\bar{x} = 75.08825$ and $s \approx 72.425$.

15. Use Chebyshev's theorem with $k = 2$.

$$1 - \frac{1}{2^2} = 1 - \frac{1}{4} = \frac{3}{4}$$

So, at least $\frac{3}{4}$ of the numbers lie within 2 standard deviations of the mean.

17. Use Chebyshev's theorem with $k = 1.5$.

$1 - \frac{1}{(1.5)^2} = 1 - \frac{4}{9} = \frac{5}{9}$, so at least $\frac{5}{9}$ of the numbers lie within 1.5 standard deviations of the mean.

19. Between 26 and 74, $\bar{x} = 50$ and $s = 6$; we have $26 = 50 - 4 \cdot 6$ and $74 = 50 + 4 \cdot 6$, so $k = 4$.

At least $1 - \frac{1}{k^2} = \frac{15}{16} = 93.75\%$ of the numbers lie between 26 and 74.

21. Less than 32 or more than 68
From exercise 18, 88.9% of the data lie between 32 and 68. So, no more than 100% – 88.9% = 11.1% of the data are less then 32 or more than 68.

23. Aerobic shoe sales

$$\bar{x} = \frac{261+262+280+260+223+249+252}{7} = \frac{1787}{7} \approx 255.2857$$

The mean sales for aerobic shoes is about $255.2857 million.

x	$x-\bar{x}$	$(x-\bar{x})^2$
261	5.7	32.49
262	6.7	44.89
280	24.7	610.09
260	4.7	22.09
223	–32.3	1043.29
249	–6.3	39.69
252	–3.3	10.89
	Total:	1803.43

$$s = \sqrt{\frac{1803.43}{7-1}} \approx 17.3$$

The standard deviation is about $17.3 million.

25. Cross-training shoe sales

$$\bar{x} = \frac{1437+1516+1584+1626+2071+2121+2143}{7} = \frac{12{,}498}{7} \approx 1785.4286$$

The mean sales for cross-training shoes is about $1785.4286 million.

x	$x-\bar{x}$	$(x-\bar{x})^2$
1437	–348.4	121,382.56
1516	–269.4	72,576.36
1584	–201.4	40,561.96
1626	–159.4	25,408.36
2071	285.6	81,567.36
2121	335.6	112,627.36
2143	357.6	127,877.76
	Total:	582,001.72

$$s = \sqrt{\frac{582{,}001.72}{7-1}} \approx 311.4$$

The standard deviation is about $311.4 million.

27. Cross-training shoes have more variation because they have a higher standard deviation.

29. Let x be the midpoint of the interval.

Interval	f	x	fx	x^2	fx^2
0–29	12	14.5	174	210.25	2523
30–59	3	44.5	133.5	1980.25	5940.75
60–89	10	74.5	745	5550.25	55,502.5
90–119	2	104.5	209	10,920.25	21,840.5
120–149	0	134.5	0	18,090.25	0
150–179	1	164.5	164.5	27,060.25	27,060.25
180–209	2	194.5	389	37,830.25	75,660.5
Totals	30		1815		188,527.5

$$\bar{x} = \frac{\Sigma(fx)}{30} = \frac{1815}{30} = 60.5$$

$$s = \sqrt{\frac{188{,}527.5 - 30(60.5)^2}{30-1}} \approx \sqrt{2714.483} \approx 52.1$$

31. a. $\bar{x} = \frac{16+12+11+9+8+7+7+6+5+5+4+4+2}{13}$

$= \frac{96}{13} \approx 7.4$

x	$x-\bar{x}$	$(x-\bar{x})^2$
16	8.6	73.96
12	4.6	21.16
11	3.6	12.96
9	1.6	2.56
8	.6	.36
7	−.4	.16
7	−.4	.16
6	−1.4	1.96
5	−2.4	5.76
5	−2.4	5.76
4	−3.4	11.56
4	−3.4	11.56
2	−5.4	29.16
	Total:	177.08

$s = \sqrt{\frac{177.08}{13-1}} \approx \sqrt{14.8} \approx 3.8$

b. One standard deviation from the mean consists of the values 3.58 to 11.18. Ten animals have blood types within 1 standard deviation of the mean.

33. a, b. 1: $\bar{x} = \frac{2-2+1}{3} = \frac{1}{3}$

$s = \sqrt{\frac{\left(2-\frac{1}{3}\right)^2 + \left(-2-\frac{1}{3}\right)^2 + \left(1-\frac{1}{3}\right)^2}{3-1}}$

$= \sqrt{\frac{26}{6}} \approx 2.1$

2: $\bar{x} = \frac{3-1+4}{3} = 2$

$s = \sqrt{\frac{(3-2)^2 + (-1-2)^2 + (4-2)^2}{3-1}}$

$= \sqrt{\frac{14}{2}} \approx 2.6$

3: $\bar{x} = \frac{-2+0+1}{3} = -\frac{1}{3}$

$s = \sqrt{\frac{\left(-2+\frac{1}{3}\right)^2 + \left(0+\frac{1}{3}\right)^2 + \left(1+\frac{1}{3}\right)^2}{3-1}}$

$= \sqrt{\frac{14}{6}} \approx 1.5$

4: $\bar{x} = \frac{-3+1+2}{3} = 0$

$s = \sqrt{\frac{(-3-0)^2 + (1-0)^2 + (2-0)^2}{3-1}}$

$= \sqrt{\frac{14}{2}} \approx 2.6$

5: $\bar{x} = \frac{-1+2+4}{3} = \frac{5}{3}$

$s = \sqrt{\frac{\left(-1-\frac{5}{3}\right)^2 + \left(2-\frac{5}{3}\right)^2 + \left(4-\frac{5}{3}\right)^2}{3-1}}$

$= \sqrt{\frac{38}{6}} \approx 2.5$

6: $\bar{x} = \frac{3+2+2}{3} = \frac{7}{3}$

$s = \sqrt{\frac{\left(3-\frac{7}{3}\right)^2 + \left(2-\frac{7}{3}\right)^2 + \left(2-\frac{7}{3}\right)^2}{3-1}}$

$= \sqrt{\frac{2}{6}} \approx .6$

(*continued next page*)

$7: \bar{x} = \dfrac{0+1+2}{3} = 1$

$s = \sqrt{\dfrac{(0-1)^2+(1-1)^2+(2-1)^2}{3-1}}$

$= \sqrt{\dfrac{2}{2}} = 1$

$8: \bar{x} = \dfrac{-1+2+3}{3} = \dfrac{4}{3}$

$s = \sqrt{\dfrac{\left(-1-\frac{4}{3}\right)^2+\left(2-\frac{4}{3}\right)^2+\left(3-\frac{4}{3}\right)^2}{3-1}}$

$= \sqrt{\dfrac{26}{6}} \approx 2.1$

$9: \bar{x} = \dfrac{2+3+2}{3} = \dfrac{7}{3}$

$s = \sqrt{\dfrac{\left(2-\frac{7}{3}\right)^2+\left(3-\frac{7}{3}\right)^2+\left(2-\frac{7}{3}\right)^2}{3-1}}$

$= \sqrt{\dfrac{2}{6}} \approx .6$

$10: \bar{x} = \dfrac{0+0+2}{3} = \dfrac{2}{3}$

$s = \sqrt{\dfrac{\left(0-\frac{2}{3}\right)^2+\left(0-\frac{2}{3}\right)^2+\left(2-\frac{2}{3}\right)^2}{3-1}}$

$= \sqrt{\dfrac{8}{6}} \approx 1.2$

c. $\bar{X} = \dfrac{\Sigma \bar{x}}{n} \approx \dfrac{11.3}{10} = 1.13$

d. $\bar{s} = \dfrac{\Sigma s}{n} = \dfrac{16.8}{10} = 1.68$

e. The upper control limit for the sample means is

$\bar{x} + 1.954\bar{s} = 1.13 + (1.954)(1.68) \approx 4.41$

The lower control limit for the sample means is

$\bar{x} - 1.954\bar{s} = 1.13 - (1.954)(1.68) \approx -2.15$

One of the measurements, –3, is outside of these limits, so the process is out of control.

35.

Interval	f	x	xf	x^2	$x^2 f$
50–59	2	54.5	109	2970.25	5940.5
60–69	4	64.5	258	4160.25	16,641
70–79	7	74.5	521.5	5550.25	38,851.75
80–89	9	84.5	760.5	7140.25	64,262.25
90–99	8	94.5	756	8930.25	71,442
Totals:	30		2405		197,137.5

$\bar{x} = \dfrac{2405}{30} \approx 80.17$

$s = \sqrt{\dfrac{197{,}137.5 - 30(80.17)^2}{30-1}} \approx 12.2$

For individualized instruction, the mean is 80.17 and the standard deviation is 12.2.

37. Answers vary. Possible answer: You would expect the mean of the traditional instruction to be smaller, since 13 of the 34 students lie in the first two intervals while only 6 of 30 students lie in the same intervals for individualized instruction. As far as standard deviation is concerned, there is a small difference between the two, 1.1. This gives the impression that the data is pretty much spread out in nearly the same fashion in both types of instruction.

Section 10.4 Normal Distributions and Boxplots

1. The peak in a normal curve occurs directly above the mean.

3. Answers vary. Possible answer:
If a normal distribution has mean μ and standard deviation σ, then the z-score for the number x is

$z = \dfrac{x-\mu}{\sigma}.$

5. By looking up 1.75 in Table 2, we get .4599 = 45.99%. Thus, the percent of the total area between the mean and 1.75 standard deviations from the mean is 45.99%.

7. –.43 indicates that we are below the mean. Looking up .43 in Table 2, we get .1664 = 16.64%. Thus, the percent of the total area between the mean and .43 standard deviations from the mean is 16.64%.

9. The entry corresponding to $z = 1.41$ is .4207, and the entry for $z = 2.83$ is .4977, so the area between $z = 1.41$ and $z = 2.83$ is $.4977 - .4207 = .0770$ or 7.7%.

11. For $z = 2.48$, the entry is .4934, and for $z = .05$, the entry is .0199. The area between $z = -2.48$ and $z = -.05$ is $.4934 - .0199 = .4735$ or 47.35%.

13. To find the area between $z = -3.05$ and $z = 1.36$, add the area between $z = -3.05$ and $z = 0$ to the area between $z = 1.36$ and $z = 0$. The area is $.4989 + .4131 = .9120 = 91.20\%$.

15. 5% of the total area is to the right of z. The mean divides the area in half or .5. The area from the mean to z standard deviations is $.5 - .05 = .45$. Using Table 2 backwards, we get the z-score corresponding to the area .45 as 1.64 or 1.65 (approximately).

17. 15% of the total area is to the left of z. The area from the mean to z standard deviations is $.5 - .15 = .35$. The z-score from the table is 1.04. As the area is to the left of the mean, $z = -1.04$.

19. To find $P(x \le \mu)$ and $P(x \ge \mu)$ consider the normal curve. The curve is symmetric about the vertical line through μ, so $P(x \le \mu)$ represents half the area under the curve. Since the area under the curve is 1, $P(x \le \mu) = .5$. Similarly, $P(x \ge \mu) = .5$.

21. Using Chebyshev's theorem with $k = 3$:

$$1 - \frac{1}{3^2} = 1 - \frac{1}{9} = \frac{8}{9} \approx .889$$

Using the normal distribution, the area between $z = -3$ and $z = 3$ is $.4987 + .4987 = .9974$
The probability a number will lie within 3 standard deviations of the mean is greater as indicated by the normal distribution at .9974 than Chebyshev's theorem shows at .889.

23. Let x represent the number of grams of peanut butter in a jar.
$\mu = 453$, $\sigma = 10.1$
Find the z-score for $x = 450$.

$$z = \frac{x - \mu}{\sigma} = \frac{450 - 453}{10.1} \approx -.30$$

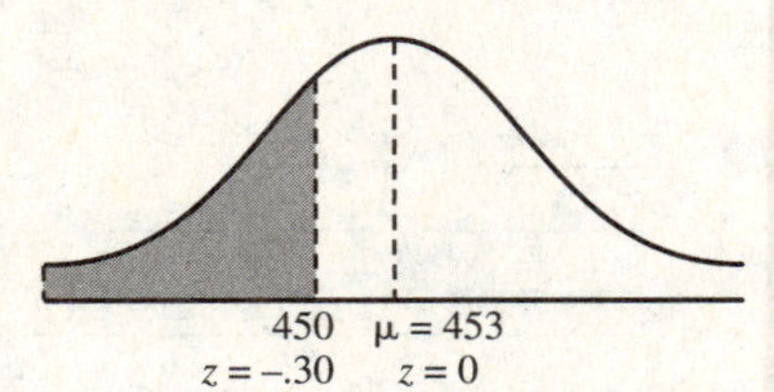

To find the area to the left of $z = -.30$, find the area between $z = 0$ and $z = -.30$ and subtract that answer from .5. The area is $.5 - .1179 = .3821$. Thus, the probability that a jar will contain less than 450 grams is .3821.

25. Let x represent the number of lumens.
$\mu = 1640$, $\sigma = 62$

For $x = 1600$, $z = \frac{x - \mu}{\sigma} = \frac{1600 - 1640}{62} \approx -.65$.

For $x = 1700$, $z = \frac{x - \mu}{\sigma} = \frac{1700 - 1640}{62} \approx .97$.

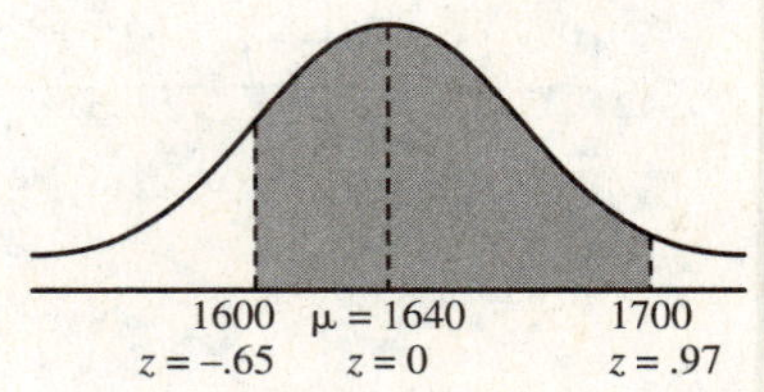

To find the area between $z = -.65$ and $z = .97$, find the area between $z = 0$ and $z = -.65$. Then find the area between $z = 0$ and $z = .97$. Add these two answers together.
The area is $.2422 + .3340 = .5762$. Thus, the probability that a 100-watt bulb will have a brightness between 1600 and 1700 lumens is .5762.

27. Let x represent HDL cholesterol level.
$\mu = 52.6$, $\sigma = 15.5$

For $x = 60$, $z = \dfrac{60 - 52.6}{15.5} \approx .48$

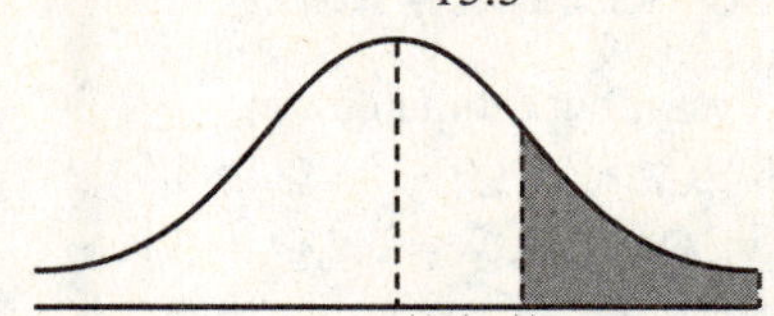

To find the area to the right of $z = .48$, find the area between $z = .48$ and $z = 0$ and subtract that from 0.5. The area is $0.5 - .1844 = .3156$. Thus, the probability that an individual will have an HDL cholesterol level greater than 60 mg/dL is .3156.

29. Let x represent the starting salaries for accounting majors.
$\mu = 53{,}300$, $\sigma = 3200$

For $x = 60{,}000$, $z = \dfrac{60{,}000 - 53{,}300}{3200} \approx 2.09$.

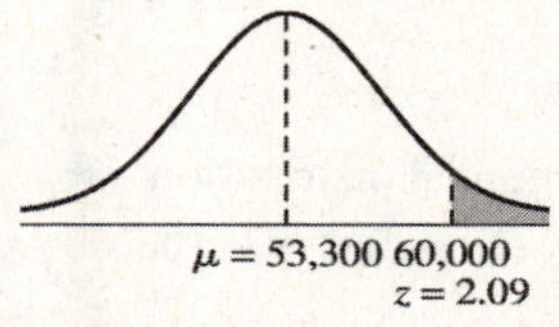

To find the area to the right of $z = 2.09$, find the area between $z = 0$ and $z = 2.09$ and subtract it from .5. The area is $.5 - .4817 = .0183$. Thus, the probability that an individual will have a starting salary above $60,000 is .0183.

For exercises 31–37, $\mu = 540$ and $\sigma = 100$. Refer to Table 2 in the back of the text.

31. For $x = 700$, $z = \dfrac{700 - 540}{100} = 1.6$.
The area between $z = 0$ and $z = 1.6$ is .4452. Thus, the probability that a GMAT test taker earns a score in the range 540–700 is .4452.

33. From exercise 31, we know that $z = 1.6$ for $x = 700$ and $P = .4452$. From exercise 32, we know that $z = -2.4$ for $x = 300$ and $P = .4918$. Therefore, the probability that a GMAT test taker earns a score in the range 300–700 is $.4452 + .4918 = .9370$.

35. For $x = 750$, $z = \dfrac{750 - 540}{100} = 2.1$.
The area between $z = 0$ and $z = 2.1$ is .4821. The area to the right of $z = 0$ is .5, so the area to the right of $z = 2.1$ is $.5 - .4821 = .0179$. Thus, the probability that a GMAT test taker earns a score greater than 750 is .0179.

37. For $x = 300$, $z = \dfrac{300 - 540}{100} = -2.4$.

For $x = 400$, $z = \dfrac{400 - 540}{100} = -1.4$.
The area between $z = 0$ and $z = -2.4$ is .4918, while the area between $z = 0$ and $z = -1.4$ is .4192. Therefore, the area between $z = -2.4$ and $z = -1.4$ is $.4918 - .4192 = .0726$, which is the probability that a GMAT test taker earns a score in the range 300–400.

39. If 85% of the total area is to the left of z, then $z > 0$ and 35% of the area is between z and the mean. Look for a value of z in Table 2 with an area of .35. The closest match is $z = 1.04$.
Let x stand for the speed.

$\dfrac{x - 40}{5} = 1.04 \Rightarrow x - 40 = 5.2 \Rightarrow x = 45.2$

The 85th percentile speed is 45.2 mph.

41.

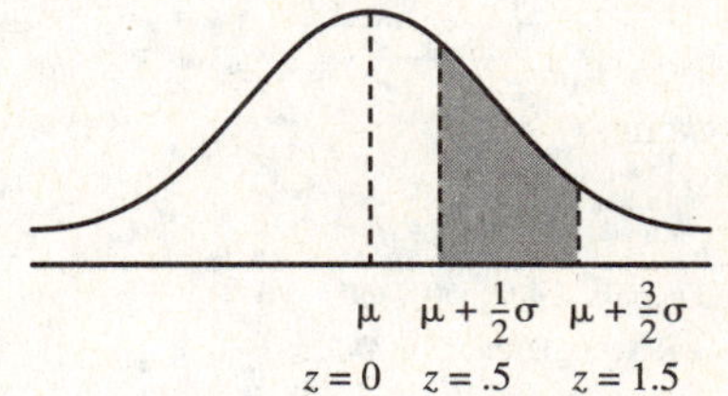

To find the area between $z = .5$ and $z = 1.5$, first find the area between $z = 1.5$ and $z = 0$. Then subtract the area between $z = 0$ and $z = .5$. The area is $.4332 - .1915 = .2417 = 24.17\%$, which means 24.17% of the students receive a B.

43. Answers vary. Possible answer: This system would be more fair in a large freshman class in psychology than in a graduate seminar of five students since the large class is more apt to have grades ranging the entire spectrum. The graduate students might have all grades in the 90's which would mean that the graduate student with the lowest score in the 90's would get an F.

45. $\mu = 550$ units; $\sigma = 45$ units
The recommended daily allowance is
$\mu + 2.5\sigma = 550 + (2.5)(46) = 665$ units

47. $\mu = 155$ units; $\sigma = 14$ units
The recommended daily allowance is
$\mu + 2.5\sigma = 155 + (2.5)(14) = 190$ units.

49. To find the area to the right of $z = 1$, find the area between $z = 0$ and $z = 1$, and subtract that from .5. The area is .5 – .3413 = .1587, which means 15.87% of the students had scores more than 1 standard deviation above the mean.

51. Arrange the data in ascending order:
27.0, 29.3, 32.6, 35.1, 39.5, 43.2, 43.3, 57.8, 65.5, 66.5
Minimum = 27.0; maximum = 66.5

$$\text{Median} = Q_2 = \frac{39.5 + 43.2}{2} = 41.35$$

$n = 10$, so $.25 \cdot 10 = 2.5$, which rounds up to 3. Count to the third data point to obtain $Q_1 = 32.6$.
$.75 \cdot 10 = 7.5$, which rounds up to 8, so count to the eighth data point to obtain $Q_3 = 57.8$.

53.

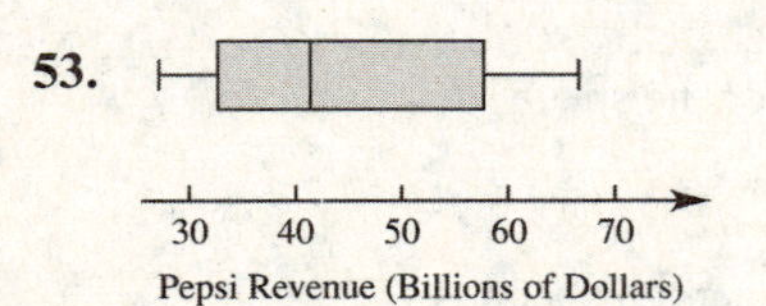

55.

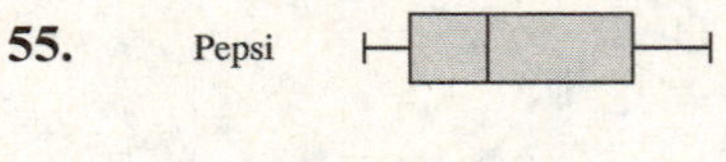

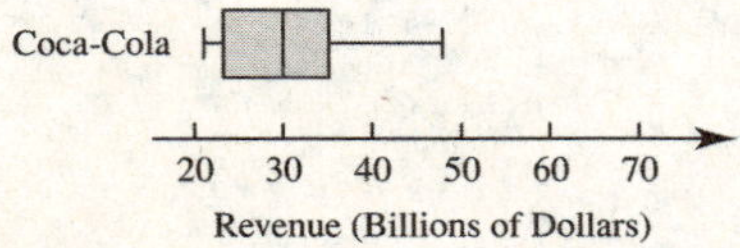

Pepsi has greater variability among its annual revenue amounts.

57. Arrange the expenditures on education data in ascending order:
5.3, 9.3, 13.2, 13.2, 13.4, 14.2, 18.4, 23.7, 25.0, 29.1, 31.8, 32.8
Minimum = 5.3; maximum = 32.8

$$\text{Median} = Q_2 = \frac{14.2 + 18.4}{2} = 16.3.$$

$n = 12$, so $.25 \cdot 12 = 3$. Count to the third data point to obtain $Q_1 = 13.2$. $.75 \cdot 12 = 9$, so count to the ninth data point to obtain $Q_3 = 25.0$.

59.

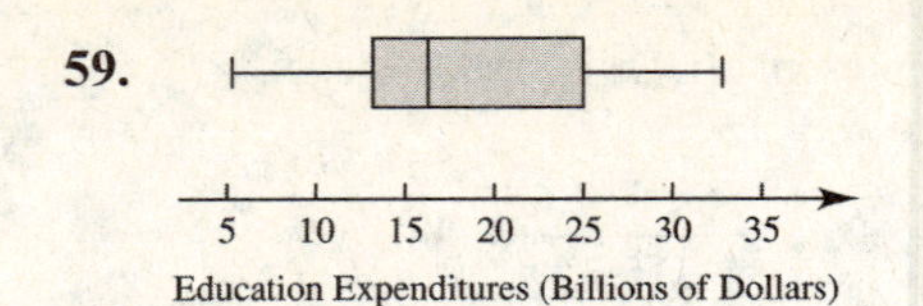

61. Answers vary, but estimates are:
Hospital : $Q_1 \approx 2.2, Q_2 \approx 3.1, Q_3 \approx 4.5$;
Highway: $Q_1 \approx 2.4, Q_2 \approx 3.3, Q_3 \approx 3.9$.

63. Police protection; The distance from minimum to maximum is greater and the distance from Q_1 to Q_3 also is greater.

Section 10.5 Normal Approximation to the Binomial Distribution

1. To find the mean and standard deviation of a binomial distribution, you must know n, the number of independent repeated trials, and p, the probability of a success in a single trial.

3. $n = 16, p = .5, (1 - p) = .5$

a. Using the binomial distribution
$P(x = 8) = {}_{16}C_8(.5)^8(.5)^8 \approx .1964$

b. Since we are using a normal curve approximation.

$$\mu = np = 16\left(\frac{1}{2}\right) = 8 \text{ and}$$

$$\sigma = \sqrt{np(1-p)} = \sqrt{16\left(\frac{1}{2}\right)\left(1 - \frac{1}{2}\right)} = 2.$$

The required probability is the area of the region corresponding to the x-values of 7.5 and 8.5.

$$z = \frac{x - \mu}{\sigma} = \frac{8.5 - 8}{2} = .25$$

From Table 2, the corresponding area is .0987. Therefore, the total area is 2(.0987) = .1974. This is the probability of getting exactly 8 heads.

5. $n = 16, p = .5, (1 - p) = .5$

a. Using the binomial distribution,

$$P(x > 13) = P(x = 14) + P(x = 15) + P(x = 16)$$

$$P(x = 14) = {}_{16}C_{14}(.5)^{14}(.5)^2 \approx .00183$$

$$P(x = 15) = {}_{16}C_{15}(.5)^{15}(.5)^1 \approx .00024$$

$$P(x = 16) = {}_{16}C_{16}(.5)^{16}(.5)^0 \approx .00002$$

$$P(x > 13) \approx .00183 + .00024 + .00002 \approx .00209 \approx .0021$$

b. For more than 13 tails, the desired area is to the right of 13.5.

$\mu = np = 16(.5) = 8$

$\sigma = \sqrt{16(.5)(.5)} = 2$

For $x = 13.5$, $z = \dfrac{13.5 - 8}{2} = 2.75$ and $A = .4970$.

$P(x > 13) = .5 - .4970 = .0030$

7. For 500 heads, we need to find the area of the region corresponding to the x-values of 499.5 and 500.5.

$$z = \frac{500.5 - 500}{15.8} = .03$$

The area is .0120. Since the region is symmetric about the mean $\mu = 500$, the probability (area) is $2(.0120) = .0240$.

9. 475 heads or more

$p = \dfrac{1}{2}$, $\mu = 500$, and $\sigma = 15.8$.

The required area is to the right of the x-value 474.5.

$$z = \frac{474.5 - 500}{15.8} = -1.61$$

The area is .4463. This area is to the left of the mean. So, the required area (probability) is $.4463 + .5 = .9463$.

11. Exactly 20 fives

The probability of a five is $p = \dfrac{1}{6}$, and $n = 120$.

$$\mu = np = 120\left(\frac{1}{6}\right) = 20$$

$$\sigma = \sqrt{120\left(\frac{1}{6}\right)\left(\frac{5}{6}\right)} = 4.08$$

$$z = \frac{20.5 - 20}{4.08} = .12$$

The area is .0478. So, the area between the x-values 19.5 and 20.5 is $2(.0478) = .0956$. This is the required probability.

13. More than 17 threes

$p = \dfrac{1}{6}$, and the required area is to the right of $x = 17.5$.

$$z = \frac{17.5 - 20}{4.08} = -.61$$

The area is .2291. So, for more than 17 threes, the probability is $.2291 + .5 = .7291$.

15. $n = 130,\ p = \dfrac{1}{6},\ 1 - p = \dfrac{5}{6}$

$$\mu = np = 130\left(\frac{1}{6}\right) \approx 21.6667$$

$$\sigma = \sqrt{np(1-p)} = \sqrt{130\left(\frac{1}{6}\right)\left(\frac{5}{6}\right)} \approx 4.2492$$

The required area is to the right of $x = 25.5$.

$$z = \frac{25.5 - 21.6667}{4.2492} \approx .90$$

The area for $z = .90$ is .3159, so the probability is $.5 - .3159 = .1841$.

17. $\mu = np = 120(.6) = 72$

$\sigma = \sqrt{np(1-p)} = \sqrt{120(.6)(.4)} \approx 5.37$

a. For exactly 80 units of food, the area is between 79.5 and 80.5. For $x = 79.5$,

$z = \dfrac{79.5 - 72}{5.37} \approx 1.40$, which gives an area of .4192. For $x = 80.5$,

$z = \dfrac{80.5 - 72}{5.37} \approx 1.58$, which gives an area of .4430.

The probability is $.4430 - .4192 = .0238$.

b. For at least 70 units, the area is to the right of 69.5. For $x = 69.5$, $z = \frac{69.5 - 72}{5.37} = -.47$ which gives an area of .1808.
The probability is $.5 + .1808 = .6808$.

19. $p = .86$, $1 - p = .14$, and $n = 500$
$\mu = np = 500(.86) = 430$
$\sigma = \sqrt{np(1-p)} = \sqrt{430(.14)} \approx 7.759$

For $x = 440$, $z = \frac{439.5 - 430}{7.759} \approx 1.22$.
The area between $z = 0$ and $z = 1.22$ is .3888, so the probability of $z \geq 1.22$ is $.5 - .3888 = .1112$.

21. $p = .24$, $1 - p = .76$, $n = 500$
$\mu = np = 500(.24) = 120$
$\sigma = \sqrt{np(1-p)} = \sqrt{120(.76)} \approx 9.550$
For at most 140 or less, let $x = 140.5$
$z = \frac{140.5 - 120}{9.550} \approx 2.15$
The area is .4842, so the probability is $.5 + .4842 = .9842$.

23. $\mu = np = (134)(.20) = 26.8$
$\sigma = \sqrt{np(1-p)} = \sqrt{(134)(.20)(.80)} = 4.63$

a. $P(x = 12) = P(11.5 < x < 12.5)$
If $x = 11.5$, $z = \frac{11.5 - 26.8}{4.63} \approx -3.31$ and
$A = .4995$.
If $x = 12.5$, $z = \frac{12.5 - 26.8}{4.63} = -3.09$ and
$A = .4990$
$P(11.5 < x < 12.5) = .4995 - .4990 = .0005$

b. $P(\text{no more than } 12) = P(x < 12.5)$
If $x = 12.5$, $z = -3.09$ and $A = .4990$ from (a). $P(x < 12.5) = .5 - .4990 = .001$

c. $P(x = 0) = P(x < .5)$
If $x = .5$, $z = \frac{.5 - 26.8}{4.63} = -5.680$ and $A = .5$.
$P(x \leq .5) = P(z < -5.680) = .5 - .5 = .0000$.

25. $E(x) = np = (1000)(.37) = 370$. We can expect 370 Verizon customers.

27. $p = .37$, $1 - p = .63$, $n = 1000$
$\mu = np = 1000(.37) = 370$
$\sigma = \sqrt{np(1-p)} = \sqrt{370(.63)} \approx 15.268$
For at least 400, let $x = 399.5$
$z = \frac{399.5 - 370}{15.268} \approx 1.93$
The area is .4732, so the probability is $5 - .4732 = .0268$.

29. $p = .37$, $1 - p = .63$, $n = 1000$
$\mu = np = 1000(.37) = 370$
$\sigma = \sqrt{np(1-p)} = \sqrt{370(.63)} \approx 15.268$
For at most 325, let $x = 325.5$
$z = \frac{325.5 - 370}{15.268} \approx -2.91$
The area is .4982, so the probability is $5 - .4982 = .0018$.

31. a. $n = 1000$, $p = .006$, $1 - p = .994$
$\mu = np = (1000)(.006) = 6$
$\sigma = \sqrt{np(1-p)} = \sqrt{1000(.006)(.994)}$
≈ 2.4421
The required area is to the right of $x = 9.5$.
$z = \frac{9.5 - 6}{2.4421} \approx 1.43$
The area for $z = 1.43$ is .4236, so the probability is $.5 - .4236 = .0764$.

b. $n = 1000$, $p = .015$, $1 - p = .985$
$\mu = np = (1000)(.015) = 15$
$\sigma = \sqrt{np(1-p)} = \sqrt{1000(.015)(.985)}$
≈ 3.8438
The area required is between $x = 19.5$ and $x = 40.5$.
If $x = 19.5$, $z = \frac{19.5 - 15}{3.8438} \approx 1.17$ and
$A = .3790$
If $x = 40.5$, $z = \frac{40.5 - 15}{3.8438} \approx 6.63$ and
$A = .5000$.
$P(20 \leq x \leq 40) = .5000 - .3790 = .121$

c. $n = 500, p = .015, 1 - p = .985$

$\mu = np = 7.5$

$\sigma = \sqrt{np(1-p)} = \sqrt{500(.015)(.985)}$
≈ 2.7180

For $x = 14.5$, $z = \dfrac{14.5 - 7.5}{2.7180} \approx 2.58$ and $A = .4951$, so the probability is $.5 - .4951 = .0049$.

Yes, this town does appear to have a higher than normal number of B– donors, since the probability of getting these kinds of results from a normal town is only .0049 or .49%.

Chapter 10 Review Exercises

1. a.

Interval	Frequency
0–9.9	23
10–19.9	11
20–29.9	2
30–39.9	2
40–49.9	0
50–59.9	1
60–69.9	1

b.

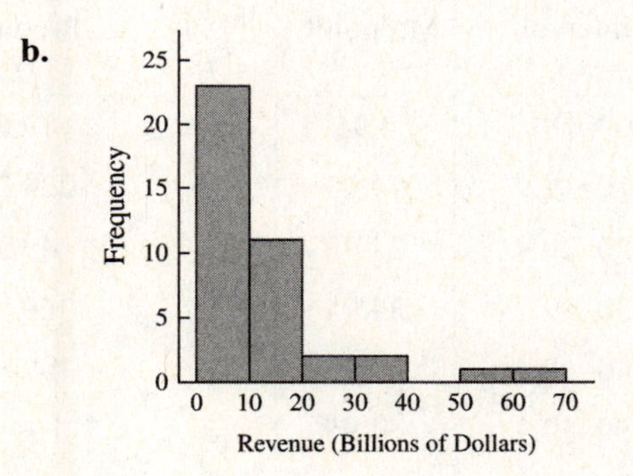

2. a.

Interval	Frequency
0–.99	7
1.00–1.99	7
2.00–2.99	9
3.00–3.99	8
4.00–4.99	2
5.00–5.99	3
6.00–6.99	2
7.00–7.99	1
8.00–8.99	1

b.

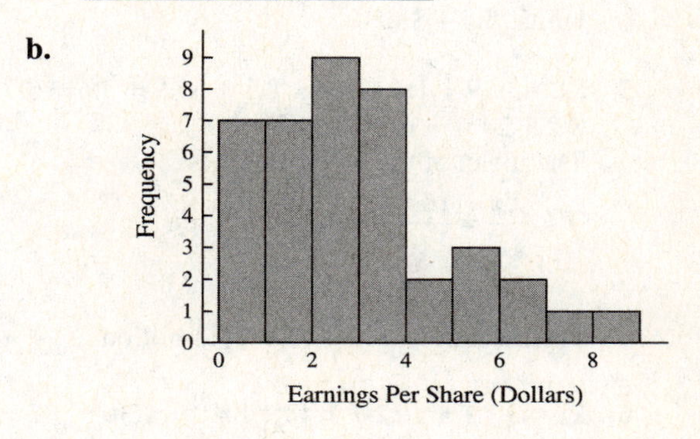

3.

STEM	LEAVES
0	12222334444444557889
1	0000112356789
2	49
3	3
4	0
5	8
6	6

Units: 6|6 = $66 billion

4.

STEM	LEAVES
0	34799
1	001234666
2	556677899
3	02245578
4	6
5	0157
6	16
7	5
8	5

Units: 8|5 = \$8.50

5. $\Sigma x = 39.9 + 1.4 + 5.0 + 8.1 + 65.9 + 3.6 + 6.7 + 8.2 + 24.4 + 1.6 = 164.8$
The mean of the 10 numbers is
$\overline{x} = \frac{\Sigma x}{10} = \frac{164.8}{10} = \$16.48$ billion .

The median is $\frac{6.7+8.1}{2} = \$7.4$ billion

6. $\Sigma x = 3.72 + 1.12 + 1.43 + 8.48 + 1.30 + .87 + 2.86 + 5.47 + 2.67 + 1.22 = 29.14$
$\overline{x} = \frac{\Sigma x}{n} = \frac{29.14}{10} \approx \$2.914$
The median is $\frac{1.43+2.67}{2} = \$2.05$

7.

Interval	Midpoint, x	Frequency, f	Product, xf
0–499	249.5	18	4491
500–999	749.5	11	8244.5
1000–1499	1249.5	2	2499
1500–1999	1749.5	2	3499
2000–2499	2249.5	2	4499
2500–2999	2749.5	2	5499
3000–3499	3249.5	0	0
3500–3999	3749.5	0	0
4000–4499	4249.5	0	0
4500–4999	4749.5	0	0
5000–5499	5249.5	1	5249.5
5500–5999	5749.5	2	11,499
	Total:	40	45,480

The mean of this collection of grouped data is $\overline{x} = \frac{45,480}{40} = \$1137$ million . The median is the average of the 20th and 21st term. Both of these terms are in the 500 – 749 interval, so the median would be \$749.5 million.

8.

Interval	Midpoint, x	Freq, f	Product xf
0–9.9	4.95	16	79.2
10–19.9	14.95	15	224.24
20–29.9	24.95	2	49.9
30–39.9	34.95	2	69.9
40–49.9	44.95	1	44.95
50–59.9	54.95	2	109.9
60–69.9	64.95	0	0
70–79.9	74.95	0	0
80–89.9	84.95	2	169.9
	Total:	40	748

Use the formula for the mean of a grouped frequency distribution.
$\overline{x} = \frac{\Sigma xf}{n} = \frac{748}{40} = \$18.7$ billion The median is the average of the 20th and 21st term. Both of these terms are in the 10 – 19.9 interval, so the median would be \$14.95 billion.

9. Answers vary. Possible answer: Grouped frequency distribution would be used when the data becomes too varied, when the range is a large number.

10. Answers vary. Possible answer: Mean, median, and mode are all types of averages. The mode is the value that occurs the most. The median is the middle value when the data are ranked from highest to lowest. The mean is the sum of the data divided by the number of data items.

11. There are 10 numbers here; the median is the mean of the 2 middle numbers after the data is placed in order, which is $\frac{40+45}{2} = 42.5$ hours. .
The mode = 40 and 50, which both occur 2 times.

12. There are 10 numbers here; the median is the mean of the 2 middle numbers after the data is placed in order, which is $\frac{27+28}{2} = 27.5\%$. .
The mode = 26, 28, and 36, which all occur 2 times.

13. The modal class is the interval with the greatest frequency. For the distribution of Exercise 7, the modal class is 0–499.

14. The modal class for the distribution of Exercise 8 is the interval 0–9.9, since it contains more data values than any of the other intervals.

15. right skewed

16. uniform

17. normal

18. left skewed

19. Answers vary. Possible answer: The range of a distribution is the difference between the largest and smallest data values.

20. Answers vary. Possible answer: The standard deviation is the square root of the variance. The standard deviation measures how spread out the data are from the mean.

21. The range is the difference between the largest and smallest numbers. For this distribution, the range is 70 – 16 = 54.
To find the standard deviation, the first step is to find the mean.

$$\bar{x} = \frac{30+45+16+20+70+40+48+40+50+50}{10}$$
$$= \frac{409}{10} = 40.9$$

Now complete the following chart.

x	x^2
30	900
45	2025
16	256
20	400
70	4900
40	1600
48	2304
40	1600
50	2500
50	2500
Total:	18,985

$$s = \sqrt{\frac{\Sigma x^2 - n\bar{x}^2}{n-1}} = \sqrt{\frac{18{,}985 - 10(40.9)^2}{9}}$$
$$= \sqrt{250.767} \approx 15.8$$

22. The range is 36 – 20 = 16, the difference of the highest and lowest numbers in the distribution.
The mean is $\bar{x} = \frac{\Sigma x}{n} = \frac{279}{10} = 27.9$.
Construct a table with the values of x, $x - \bar{x}$, and $(x - \bar{x})^2$.

(*continued next page*)

X	$x-\overline{x}$	$(x-\overline{x})^2$
22	−5.9	34.81
28	.1	.01
26	−1.9	3.61
20	−7.9	62.41
30	2.1	4.41
36	8.1	65.61
36	8.1	65.61
28	.1	.01
26	−1.9	3.61
27	−.9	.81
Totals: 279		240.9

The standard deviation is

$$s=\sqrt{\frac{240.9}{10-1}}\approx\sqrt{26.767}\approx 5.2.$$

23. Recall that when working with grouped data, x represents the midpoint of each interval. Complete the following table, which extends the table from Exercise 7.

Interval	f	x	xf	x^2	fx^2
0–499	18	249.5	4491	62,250.25	1,120,504.5
500–999	11	749.5	8244.5	561,750.25	61,792,252.75
1000–1499	2	1249.5	2499	1,561,250.25	3,122,500.5
1500–1999	2	1749.5	3499	3,060,750.25	6,121,500.5
2000–2499	2	2249.5	4499	5,060,250.25	10,120,500.5
2500–2999	2	2749.5	5499	7,559,750.25	15,119,500.5
3000–3499	0	3249.5	0	10,559,250.25	0
3500–3999	0	3749.5	0	14,058,750.25	0
4000–4499	0	4249.5	0	18,058,250.25	0
4500–4999	0	4749.5	0	22,557,750.25	0
5000–5499	1	5249.5	5249.5	27,557,250.25	27,557,250.25
5500– 5999	2	5749.5	11,499	33,056,750.25	66,113,500.5
Totals:	40		45,480		135,454,510

Use the formulas for grouped frequency distributions to find the mean and then the standard deviation. (The mean was also calculated in Exercise 7.)

$$\overline{x}=\frac{\Sigma xf}{n}=\frac{45,480}{40}=1137$$

$$s=\sqrt{\frac{\Sigma fx^2-n\overline{x}^2}{n-1}}=\sqrt{\frac{135,454,510-40(1137)^2}{39}}\approx 1465.4$$

24. Start with the frequency distribution that was the answer to Exercise 8, and expand the table to include columns for the midpoint x of each interval, and for xf, x^2, and fx^2.

Interval	f	x	xf	x^2	fx^2
0–9.9	16	4.95	79.2	24.5025	392.04
10–19.9	15	14.95	224.25	223.5025	3352.5375
20–29.9	2	24.95	49.9	622.5025	1245.005
30–39.9	2	34.95	69.9	1221.5025	2443.005
40–49.9	1	44.95	44.95	2020.5025	2020.5025
50–59.9	2	54.95	109.9	3019.5025	6039.005
60–69.9	0	64.95	0	4218.5025	0
70–79.9	0	74.95	0	5617.5025	0
80–89.9	2	84.95	169.9	7216.5025	14,433.005
Totals:	40		748		29,925.1

The mean of the grouped data is

$$\bar{x} = \frac{\Sigma xf}{n} = \frac{748}{40} = 18.7.$$

The standard deviation for the grouped data is

$$s = \sqrt{\frac{\Sigma fx^2 - n\bar{x}^2}{n-1}} = \sqrt{\frac{29{,}925.1 - 40(18.7)^2}{40-1}} \approx \sqrt{408.654} \approx 20.2$$

25. To find the standard deviation, the first step is to find the mean.

$$\bar{x} = \frac{1.74 + 1.36 + 1.40 + 2.15 + 2.23}{5}$$
$$= \frac{8.28}{5} = 1.776$$

Now complete the following chart.

x	x^2
1.74	3.0276
1.36	1.8496
1.40	1.96
2.15	4.6225
2.23	4.9729
Total:	16.4326

$$s = \sqrt{\frac{\Sigma x^2 - n\bar{x}^2}{n-1}} = \sqrt{\frac{16.4326 - 5(1.776)^2}{4}}$$
$$= \sqrt{.16543} \approx .41$$

26. To find the standard deviation, the first step is to find the mean.

$$\bar{x} = \frac{220 + 203 + 192 + 205 + 228}{5}$$
$$= \frac{1048}{5} = 209.6$$

Now complete the following chart.

x	x^2
220	48,400
203	41,209
192	36,864
205	42,025
228	51,984
Total:	220,482

$$s = \sqrt{\frac{\Sigma x^2 - n\bar{x}^2}{n-1}} = \sqrt{\frac{220{,}482 - 5(209.6)^2}{4}}$$
$$= \sqrt{205.3} \approx 14.3$$

27. Answers vary. Possible answer: A normal distribution is a continuous distribution with the following properties:

- The highest frequency is at the mean;
- The graph is symmetric about a vertical line through the mean; and
- The total area under the curve, above the x-axis, is 1.

28. Answers vary. Possible answer: A distribution in which the peak is not at the center, or mean, is called skewed.

29. Between $z = 0$ and $z = 1.35$
By Table 2, the area between $z = 0$ and $z = 1.35$ is .4115.

30. To the left of $z = .38$
The area between $z = 0$ and $z = .38$ is .1480, so the area to the left of $z = .38$ is $.5 + .1480 = .6480$.

31. Between $z = -1.88$ and $z = 2.41$
The area between $z = -1.88$ and $z = 0$ is .4700. The area between $z = 0$ and $z = 2.41$ is .4920. The total area between $z = -1.88$ and $z = 2.41$ is $.4700 + .4920 = .9620$.

32. Between $z = 1.53$ and $z = 2.82$
The area between $z = 2.82$ and $z = 1.53$ is $.4976 - .4370 = .0606$.

33. Since 8% of the area is to the right of the z-score, 42% or .4200 is between $z = 0$ and the appropriate z-score. Use Table 2 to find an appropriate z-score whose area is .4200. The closest approximation is $z = 1.41$.

34. Answers vary. Possible answer: The normal distribution is not a good approximation of a binomial distribution that has a value of p close to 0 or 1 because the histogram of such a binomial distribution is skewed and therefore not close to the shape of a normal distribution.

35. For $x = 70$, $z = \dfrac{70 - 63.2}{2.9} \approx 2.34$.
The area between $z = 0$ and $z = 2.34$ is .4904. The area to the right of $z = 0$ is .5, so the area to the right of $z = 2.34$ is $.5 - .4904 = .0096$. Thus, the probability that a female is taller than 70 inches is .0096.

36. For $x = 62$, $z = \dfrac{62 - 68.7}{3.1} \approx -2.16$.
The area between $z = 0$ and $z = -2.16$ is .4846. The area to the left of $z = 0$ is .5, so the area to the left of $z = -2.16$ is $.5 - .4846 = .0154$. Thus, the probability that a male is shorter than 62 inches is .0154.

37. For $x = 60$, $z = \dfrac{60 - 63.2}{2.9} \approx -1.10$.
The area between $z = 0$ and $z = -1.10$ is .3643. The area to the left of $z = 0$ is .5, so the area to the left of $z = -1.10$ is $.5 - .3643 = .1357$. Thus, the probability that a female is shorter than 60 inches is .1357.

38. For $x = 72$, $z = \dfrac{72 - 68.7}{3.1} \approx 1.06$.
The area between $z = 0$ and $z = 1.06$ is .3554. The area to the right of $z = 0$ is .5, so the area to the right of $z = 1.06$ is $.5 - .3554 = .1446$. Thus, the probability that a male is taller than 72 inches is .1446.

39. a. April 2012

x	x^2
214	45,796
180	32,400
141	19,881
178	31,684
122	14,884
71	5041
62	3844
22	484
48	2304
26	676
24	576
38	1444
1126	159,014

$$\bar{x} = \frac{1126}{12} \approx 93.8$$

$$s = \sqrt{\frac{\Sigma x^2 - n\bar{x}^2}{n-1}} = \sqrt{\frac{159{,}014 - 12(93.8)^2}{12-1}}$$
$$\approx 69.7$$

b. Nissan is closest to the mean sales.

40. **a.** April 2013

x	x^2
238	56,644
212	44,944
157	24,649
176	30,976
131	17,161
88	7744
63	3969
20	400
48	2304
33	1089
25	625
34	1156
1225	191,661

$$\bar{x} = \frac{1225}{12} \approx 102.1$$

$$s = \sqrt{\frac{\Sigma x^2 - n\bar{x}^2}{n-1}} = \sqrt{\frac{191{,}661 - 12(102.1)^2}{12-1}}$$
$$\approx 77.8$$

b. Nissan is closest to the mean sales.

41. April 2012:
Arrange the data in ascending order:
22, 24, 26, 38, 48, 62, 71, 122, 141, 178, 180, 214
Minimum: 22, maximum: 214

$$Q_2 = \frac{62+71}{2} = \frac{133}{2} = 66.5$$

$n = 12$, so $Q_1 =$ the third data entry, 26, and
$Q_3 =$ the ninth data entry, 141.

42. April 2013:
Arrange the data in ascending order:
20, 25, 33, 34, 48, 63, 88, 131, 157, 176, 212, 238
Minimum: 20, maximum: 238

$$Q_2 = \frac{63+88}{2} = \frac{151}{2} = 75.5$$

$n = 12$, so $Q_1 =$ the third data entry, 33, and
$Q_3 =$ the ninth data entry, 157.

43.

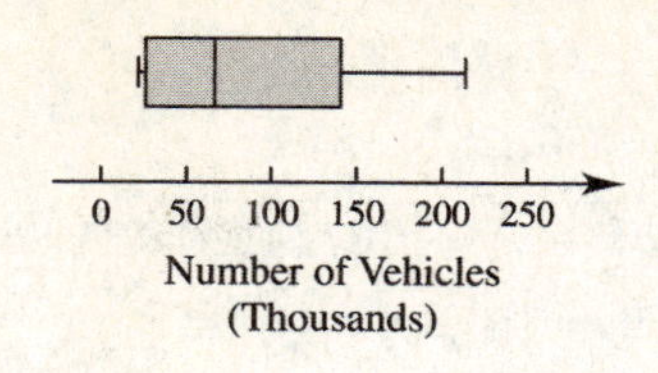

44.

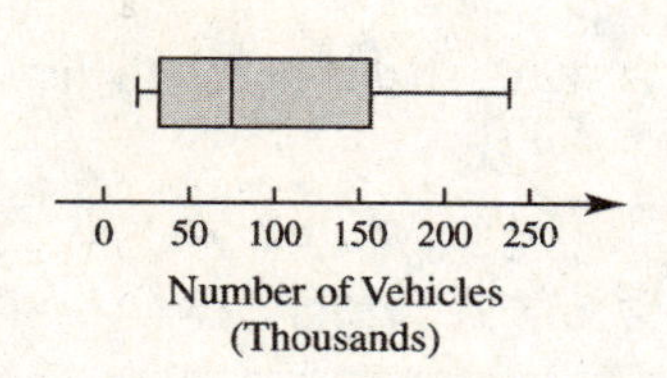

45. Arrange the data in ascending order:
5, 20, 21, 35, 40, 40, 44, 45, 53, 60
Minimum: 5, maximum: 60

$$Q_2 = \frac{40+40}{2} = \frac{80}{2} = 40$$

$n = 10$, so $Q_1 =$ the third data entry, 21, and
$Q_3 =$ the eighth data entry, 45.

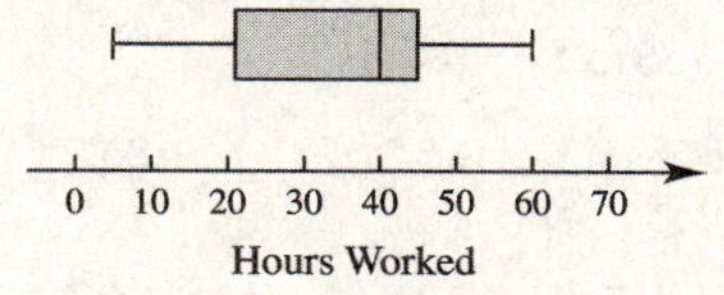

46. Arrange the data in ascending order:
81, 83, 85, 85, 85, 87, 88, 90, 92, 92
Minimum: 81, maximum: 92

$$Q_2 = \frac{85+87}{2} = \frac{172}{2} = 86$$

$n = 10$, so $Q_1 =$ the third data entry, 85, and
$Q_3 =$ the eighth data entry, 90.

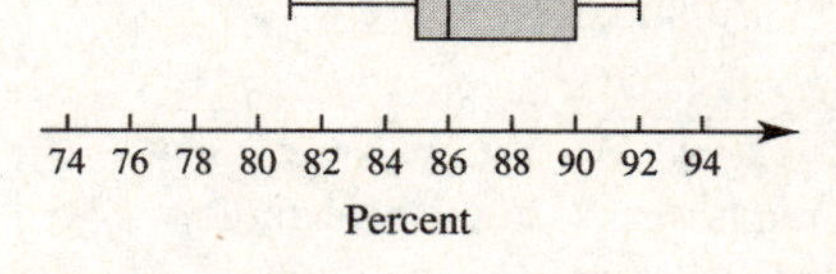

47. $n = 750, p = .25, 1 - p = .75$
$\mu = np = 750(.25) = 187.5$
$\sigma = \sqrt{np(1-p)} = \sqrt{187.5(.75)} \approx 11.86$
For more than 200, let $x = 200.5$

$$z = \frac{200.5 - 187.5}{11.86} \approx 1.10$$

The area is .3643, so the probability is
5 – .3643 = .1357.

48. $n = 750, p = .25, 1 - p = .75$
$\mu = np = 750(.25) = 187.5$
$\sigma = \sqrt{np(1-p)} = \sqrt{187.5(.75)} \approx 11.86$
For less than 150, let $x = 149.5$
$z = \dfrac{149.5 - 187.5}{11.86} \approx -3.20$
The area is .4993, so the probability is $5 - .4993 = .0007$.

49. $n = 750, p = .25, 1 - p = .75$
$\mu = np = 750(.25) = 187.5$
$\sigma = \sqrt{np(1-p)} = \sqrt{187.5(.75)} \approx 11.86$
For less than 190, let $x = 189.5$
$z = \dfrac{189.5 - 187.5}{11.86} \approx .17$
The area is .0675, so the probability is $5 + .0675 = .5675$.

50. $n = 750, p = .25, 1 - p = .75$
$\mu = np = 750(.25) = 187.5$
$\sigma = \sqrt{np(1-p)} = \sqrt{187.5(.75)} \approx 11.86$
For more than 160, let $x = 160.5$
$z = \dfrac{160.5 - 187.5}{11.86} \approx -2.28$
The area is .4887, so the probability is $5 + .4887 = .9887$.

51. $n = 1000, p = .78, 1 - p = .22$
$\mu = np = 1000(.78) = 780$
$\sigma = \sqrt{np(1-p)} = \sqrt{780(.22)} \approx 13.100$
For at least 800, let $x = 799.5$
$z = \dfrac{799.5 - 780}{13.100} \approx 1.49$
The area is .4319, so the probability is $5 - .4319 = .0681$.

52. $n = 1000, p = .78, 1 - p = .22$
$\mu = np = 1000(.78) = 780$
$\sigma = \sqrt{np(1-p)} = \sqrt{780(.22)} \approx 13.100$
For at most 750, let $x = 750.5$
$z = \dfrac{750.5 - 780}{13.100} \approx -2.25$
The area is .4878, so the probability is $5 - .4878 = .0122$.

53. $n = 1000, p = .78, 1 - p = .22$
$\mu = np = 1000(.78) = 780$
$\sigma = \sqrt{np(1-p)} = \sqrt{780(.22)} \approx 13.100$
For at most 795, let $x = 795.5$
$z = \dfrac{795.5 - 780}{13.100} \approx 1.18$
The area is .3810, so the probability is $5 + .3810 = .8810$.

54. $n = 1000, p = .78, 1 - p = .22$
$\mu = np = 1000(.78) = 780$
$\sigma = \sqrt{np(1-p)} = \sqrt{780(.22)} \approx 13.100$
For at least 760, let $x = 759.5$
$z = \dfrac{759.5 - 780}{13.100} \approx -1.56$
The area is .4406, so the probability is $5 + .4406 = .9406$.

Case 10 Standard Deviation as a Measure of Risk

1. The mean monthly return for the Janus fund is $\overline{x} = \dfrac{\Sigma x}{n} = \dfrac{45.71}{36} \approx 1.2697$. The annual return is $12\overline{x} = 12(1.2697) \approx 15.24$.

2. The mean monthly return for the S&P 500 fund is $\overline{x} = \dfrac{\Sigma x}{n} = \dfrac{49.91}{36} \approx 1.3864$. The annual return is $12\overline{x} = 12(1.3864) \approx 16.64$.

3. The mean monthly return for the Large Growth fund is $\overline{x} = \dfrac{\Sigma x}{n} = \dfrac{44.76}{36} \approx 1.2433$. The annual return is $12\overline{x} = 12(1.2433) \approx 14.92$.

4. The S&P 500 fund has the highest annual return.

5. Through the use of a computer or calculator, the standard deviation for the Janus fund will be $s \approx 4.6841$. The annual return is $s\sqrt{12} = (4.6841)\sqrt{12} \approx 16.23$.

6. Through the use of a computer or calculator, the standard deviation for the S&P 500 will be $s \approx 4.0472$. The annual return is $s\sqrt{12} = (4.0472)\sqrt{12} \approx 14.02$.

7. Through the use of a computer or calculator, the standard deviation for the Large Growth fund will be $s \approx 4.5071$. The annual return is $s\sqrt{12} = (4.5071)\sqrt{12} \approx 15.61$.

8. The Janus fund has the highest level of risk.

9. The bounds at the 68% level are found by $\overline{x} \pm s$. Therefore the bounds are:
$(15.24 - 16.23, 15.25 + 16.23)$
$(-.99, 31.47)$

10. The bounds at the 95% level are found by $\overline{x} \pm 2s$. Therefore the bounds are:
$(15.24 - 2 \cdot 16.23, 15.25 + 2 \cdot 16.23)$
$(-17.22, 47.70)$